Modern Mechanics and Robotics

Modern Mechanics and Robotics

Edited by **Jared Kroff**

Willford Press

New York

Published by Willford Press,
118-35 Queens Blvd., Suite 400,
Forest Hills, NY 11375, USA
www.willfordpress.com

Modern Mechanics and Robotics
Edited by Jared Kroff

International Standard Book Number: 978-1-68285-266-8 (Hardback)

Printed in China.

Contents

Preface

This book aims to highlight the current researches and provides a platform to further the scope of innovations in this area. This book is a product of the combined efforts of many researchers and scientists from different parts of the world. The objective of this book is to provide the readers with the latest information in the field.

Robotics has simplified human lives and robots are now being used across a lot of industries. This book contains some path-breaking studies in the field of robotics conducted by internationally renowned experts. The various sub-fields of robotics along with technological progress that has future implications are glanced at in this book through detailed analysis of topics like continuum robotics, compliant mechanisms, torque control, designing techniques, etc. This book is compiled in such a manner, that it will provide in-depth knowledge about the theory and practices of robotics.

I would like to express my sincere thanks to the authors for their dedicated efforts in the completion of this book. I acknowledge the efforts of the publisher for providing constant support. Lastly, I would like to thank my family for their support in all academic endeavors.

Editor

1

A simple design rule for 1st order form-closure of underactuated hands

S. Krut[1] **and V. Bégoc**[1,*]

[1]LIRMM (CNRS & Univ. Montpellier 2), Montpellier, France
[*]now at: B+ Equipment, Gemenos, France

Abstract. The property of form-closure of a grasp, as generally defined in the literature, is based on the assumption that contact points between the hand and the object are fixed in space. However, this assumption is false when considering a grasp exerted by an underactuated hand, since in this case, it is not possible to control the position of each phalanx independently. In spite of researchers' interest in studying form-closure, none of the available published work on this subject takes into consideration the particular kinematics of underactuated hands. Actually, there are few available tools to qualify or quantify the stability of a grasp exerted by an underactuated hand, thus the design of underactuated hands mostly results from an intuitive approach. This paper aims to reduce this gap.

A classification of underactuated hands is proposed, based on the expression of contact forces. This highlights the influence of non-backdrivable mechanisms introduced in the transmission of the closing motion of the hand on the stability of the grasp. The way to extend the original definition of form-closure to underactuated grasps is illustrated. A more general definition is formulated, which checks the stability of the set "object + hand". Using this new definition, a simple rule is proposed for designing a hand capable of achieving 1st order form-closed grasps.

This paper was presented at the IFToMM/ASME International Workshop on Underactuated Grasping (UG2010), 19 August 2010, Montréal, Canada.

1 Introduction

The concept of underactuation makes it possible to create grippers which automatically adapt to the geometry of the grasped object, without requiring a large number of sensors or actuators, nor a complex control strategy. Thus, using a simple binary control, such as the one usually used to drive the closing/opening motion of a parallel-jaw gripper, an underactuated gripper permits to increase the number of contact points, resulting in an enveloping grasp that should be, a-priori, more robust than a two-contact grasp. However, underactuated grippers are rarely used in the industry, first because most of gripping operations can be achieved using simple grippers dedicated to unique objects, but also because, in certain cases, underactuated grippers exhibit aberrant behavior which leads to an unstable grasp.

In Laliberté et al. (2002), a characteristic phenomenon of underactuated hands, the ejection phenomenon, is described. In certain configurations of the finger, the distribution of contact forces degenerates, i.e. some phalanxes must exert a negative force on the object in order to guarantee static equilibrium of the finger. Since a contact action is unidirectional, equilibrium cannot be attained and the ejection of the object from the hand is initiated. The ejection phenomenon highlights the need for a more in-depth investigation of the stability of a grasp exerted by an underactuated hand, considering the particular kinematics of this type of mechanism. Indeed, there has been little work dedicated to studying the stability of a grasp exerted by an underactuated hand. There exist then only few tools for designing such an underactuated hand in order to maximize its ability to stabilize an object. As a result, their design often results from an intuitive approach.

There are two main criteria which can be used to characterize the robustness of a grasp: *form-closure* and *force-closure*. These two properties permit to perform a local and static study of the stability of a grasp. Form-closure describes the capability of a hand to prevent any motion of a grasped object. Force-closure describes the capability of a hand to counterbalance any external disturbances exerted on a grasped object, by applying a combination of contact forces which respect the capabilities of actuators and the condition of friction (Bicchi, 1995). In contrast to force-closure, form-closure is a purely geometric property since it does not depend on actuators capability nor on the eventual presence of frictional forces between the gripper and the object.

These two properties have been the subject of many investigations but, to the best of our knowledge, these have never been extended to the particular case of underactuated grasps[1] (even work in Luo et al. (2004) on analysis of underactuated grasps does not deal with the particular kinematics of underactuated hands). This paper focuses on form-closure and aims to propose a simple design rule for 1st order form-closure of underactuated hands. It is based on some preliminary work introduced in Bégoc et al. (2006) and Krut et al. (2010).

2 Form-closure

2.1 Original definition

Form-closure can be described as the capability of a hand, or more generally of a set of constraining contacts, to prevent motion of a grasped object. The following definition is used for form-closure, a definition which is largely accepted in the literature:

"A grasp is said to be form-close if, and only if, for every motion of the object, at least one contact constraint is violated." (Bicchi, 1995)

The term "contact constraint" relates to the fact that the relative motion of two solid bodies in contact is constrained by the condition of non-interpenetration. This contact constraint is generally approximated to 1st order for sake of simplicity. In the major part of the literature, the configuration of the hand is assumed to be fixed, which permits to avoid the kinematic study of the hand. This assumption implies that the control position of each contact phalanx is infinitely rigid and that motors are oversized in comparison with any disturbances likely to be exerted on the object.

2.2 1st and 2nd order form-closure

A distinction must be made between two definitions: 1st order form-closure and 2nd order form-closure (Rimon and

[1] As an abuse of language, a grasp realized by an underactuated mechanism is referred to as an underactuated grasp.

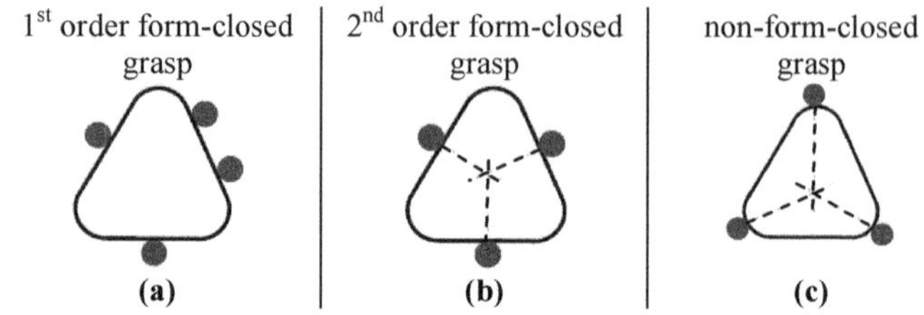

Figure 1. Illustration of grasp types (figures taken from Cheong (2006)): in situation **(a)** the object is totally immobilized, in situation **(b)** it can move with an infinitesimal rotation around the point of concurrency of the contact normals, in situation **(c)** the object is free to rotate around that point. This phenomenon can be modeled considering the local curvatures of the contact surfaces as explained in Van der Stappen et al. (2000).

Burdick, 1996). This is illustrated using the three grasps depicted in Fig. 1, where the object to be grasped remains the same but where the locations of contact points differ. In all three situations, contact points prevent the object from translating. However, if they also prevent it from rotating in case (a), they do not in cases (b) and (c): the object is free to rotate around the point of concurrency of the contact normals. After an infinitesimal rotation, the object is stopped from rotating in case (b), whereas it is free to escape in case (c). This phenomenon can be modeled considering the local curvatures of the contact surfaces as explained in Carrozza et al. (2005). Grasp (a) is 1st order form-closed, because a 1st order modeling of the grasp is sufficient to demonstrate form-closure, whereas grasp (b) is 2nd order form-closed, because 2nd order modeling is required for demonstrating form-closure.

3 Underactuation

3.1 Definition

"A mechanism is said to be underactuated, when it has fewer actuators than degrees of freedom."
(Birglen et al., 2008)

An actuator is a sub-assembly which causes relative motion between the parts to which it is attached in response to a signal (IFToMM, 2010). The number of degrees of freedom of a system is defined by the number of independent generalized coordinates required to define completely the configuration of a system at any instant of time (IFToMM, 2010). Hence, determining the number of degrees of freedom (dof) of a hand requires considering the coupling that is often introduced between the rotations of interphalanx joints. This is the case for several bionic hands, such as the DLR hands (Butterfass et al., 2001), the Robonaut's hand (Ambrose et al., 2000), the NAIST hand (Ueda et al., 2005) and the hand of humanoid robot HRP-3 (Kaneko et al., 2007).

The Grübler-Kutzbach equation cannot be applied starightforwadly to determine the degree of freedom of an underactuated hand (Krut, 2005). In fact, this method involves determining the order of a system of kinematic linear equations and it is therefore not applicable to underactuated hands which:

- use unidirectional mechanisms (non-backdrivable mechanisms or anti-return mechanisms) in the transmission of the opening/closing motion of the fingers, because these are modeled using inequalities. This is the case for many underactuated hands such as the Barrett Hand (Townsend, 2000) and the SARAH Hand (Lopez, 2001),
- use compliant bodies to transfer the motor's torque to the fingers as is the case of the RTR II hand (Massa et al., 2002), because the classical system of kinematic equations does not hold anymore.

3.2 Classification of underactuated hands

There is a large diversity of mechanical devices which enable a hand to adapt to the geometry of an object. In this paper, we propose to classify underactuated hands based on the expression of contact forces as a function of the torques exerted by the actuators on the mechanism. Hence, an underactuated hand can be:

1. differential,
2. compliant,
3. self-locking,
4. or have a combination of these three properties.

3.2.1 Differential underactuation

The first category can be characterized by the formulation given in Krut (2005). A mechanism is said to be differential when its behavior, from a kinetostatic point of view, can be described by the following equations:

$$F_{\mathrm{a}} = F_1^{\mathrm{a}}\, r_1 = \cdots = F_n^{\mathrm{a}}\, r_n \tag{1}$$

$$\dot{\theta}_a + \sum_{i=1}^{n} \frac{\dot{\theta}_i^{\mathrm{a}}}{r_i^{\mathrm{a}}} = 0 \tag{2}$$

where F_{a} is the force, or the actuation torque, exerted on the input of the differential mechanism and F_i^{a} is the force or torque transmitted to the output i. $\left(\theta_{\mathrm{a}}, \theta_1^{\mathrm{a}}, \cdots, \theta_n^{\mathrm{a}}\right)$ describes the configuration of the differential mechanism, $\left(\dot{\theta}_{\mathrm{a}}, \dot{\theta}_1^{\mathrm{a}}, \cdots, \dot{\theta}_n^{\mathrm{a}}\right)$ the joint velocities and $\left(r_1^{\mathrm{a}}, \cdots, r_n^{\mathrm{a}}\right)$ the transmission ratios. A review of the state of the art of differential mechanisms used in underactuated grippers can be found in Birglen and Gosselin (2006).

3.2.2 Compliant underactuation

A mechanism is said to be compliant when the forces exerted on its output can be written as a function of the configuration of the mechanism and the stiffness of the mechanical elements introduced into the transmission of the closing motion of the hand:

$$\left[\begin{array}{cccc} F_{\mathrm{a}} & F_1^{\mathrm{a}} & \cdots & F_n^{\mathrm{a}} \end{array}\right]^T = \boldsymbol{R} \left[\begin{array}{cccc} \theta_{\mathrm{a}} & \theta_1^{\mathrm{a}} & \cdots & \theta_n^{\mathrm{a}} \end{array}\right]^T \tag{3}$$

where $\mathbf{R}$ is the stiffness matrix of the mechanism.

3.2.3 Self-locking underactuation

The category of self-locking underactuated mechanisms is introduced in order to characterize hands which have unidirectional mechanisms in the transmission of the fingers' closing motion. Indeed, this has a strong influence on the expression of contact forces and, as a consequence, on the characteristics of the grasp since this type of mechanisms prevents all return motion of the fingers or phalanxes whatever the disturbances exerted on the grasped object. Theoretically, the contact forces could then be infinite. We define a mechanism to be self-lockable, when the configuration of the mechanism is constrained by a set of inequalities of the type:

$$\delta_i \dot{\theta}_i^{\mathrm{a}} \geq 0,\ \forall i = \{1, \cdots, n\} \tag{4}$$

$\delta_i = \pm 1$ depending on the case.

3.3 Classification of unidirectional mechanisms

The unidirectional mechanisms, on which the self-locking hands are based, provide unidirectional transmission of motion, which can be modeled by:

$$\dot{\theta}_{\mathrm{out}} \geq 0 \tag{5}$$

where $\dot{\theta}_{\mathrm{out}}$ represents the output velocity of the mechanisms. Unidirectional mechanisms can be spited in two main categories: (1) anti-return mechanisms and (2) non-backdrivable mechanisms.

3.3.1 Anti-return mechanisms

Among the available anti-return mechanisms are:

- blocking mechanisms by obstacle, such as the "pawl and ratchet" mechanism or the sliding gear,
- locking mechanisms by bracing or wedging such as a freewheel (with ramp or cam), wrap spring couplings or sliding/bracing mechanisms, such as those used in the underactuated hand designed by Lopez (2001),
- a non-return valve when fluid energy is used.

In the case of the free-wheel mechanism and wrap spring coupling, the unilateral condition is described differently:

$$\theta_{\text{out}} \geq \theta_{\text{in}} \tag{6}$$

where θ_{out} and θ_{in} are the output and input velocities of the mechanism respectively. This condition is identical to condition (5) when the input velocity is zero.

3.3.2 Non-backdrivable mechanisms

Non-backdrivable mechanisms provide also the same unilateral condition (5). A mechanism is said to be non-backdrivable when it is incapable of transmitting motion and power from the output to the input (IFToMM, 2010), as for instance:

- the "triangular wedge" mechanism,
- The "wheel and worm drive" mechanism,
- The "rack and worm drive" mechanism,
- The "lead screw and nut" mechanism.

A non-backdrivable mechanism operates differently from an anti-return mechanism. Nevertheless, it is possible to impose a unidirectional transmission of motion $\theta_{\text{out}} \geq 0$ by maintaining a positive or null force on the input of the non-backdrivable mechanism.

In practice, it is noticeable that designers of underactuated hands prefer to introduce non-backdrivable mechanisms. The advantage of these mechanisms is that it is sufficient to invert the direction of rotation at the input in order to invert the sense of rotation at the output, and hence to enable opening of the gripper. In the case of a unidirectional mechanism, a mechanism is required to deactivate the anti-return function, as for instance the solution proposed in Lopez (2001) to deactivate a sliding/bracing mechanism.

4 Extension of form-closure to underactuated hands

The definition of form-closure, as it is usually encountered in the literature, is based on the assumption that contact points are fixed in space. When considering a robotic hand, this assumption implies to consider that the position control of each phalanx is infinitely rigid, so that the configuration of the hand remains unchanged whatever the amplitude of disturbances exerted on the object. This greatly simplifies the study of the grasp, by avoiding the kinematic study of the hand. This assumption is justified for the majority of robotic hands, for which the position of each phalanx can be independently controlled. In contrast, it is not possible to independently control the position of each phalanx when using an underactuated hand. The assumption of fixed contact points is then false.

In Krut et al. (2010), the definition of form-closure was reformulated so that it applies to the case of underactuated grasps. Since immobility of contact points is not guaranteed, not only the motions of the object have to be studied but also the motions of the grasp, i.e. the motions of the overall system including the object and the hand. Thus, checking form-closure of an underactuated grasp now requires verifying that any variation in the configuration of the grasp is prevented by a set of unilateral kinematic constraints, i.e. the contact constraints and the constraints imposed by the non-backdrivable mechanisms. The definition of form-closure was reformulated as follows:

"A grasp is said to be form-closed if, and only if, for any variation of the configuration of the grasp at least one of the unilateral kinematic constraints is violated."

To illustrate this definition, two simple cases of two-finger grasps are considered (see Fig. 2). These grasps differ in terms of type of transmission used to produce the closing motion of fingers. In order to simplify this analysis, in a way that will make it intuitive, an operational space of dimension one only is considered, i.e. only the translation of the object along x is studied.

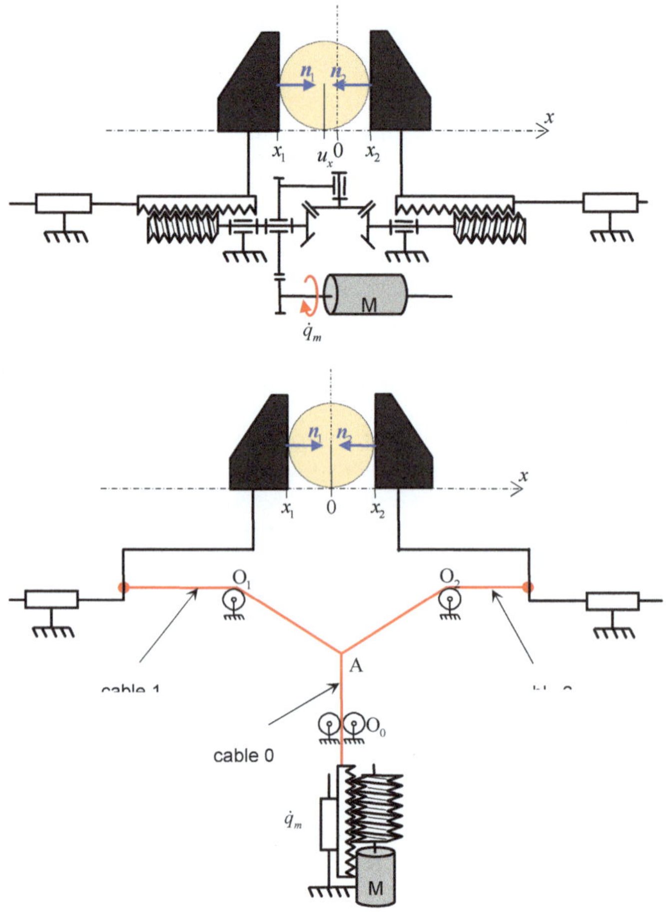

Figure 2. Diagram representing a two contact grasp.

4.1 Mechanism A

The first mechanism is underactuated by a geared differential mechanism (Fig. 2a). The closing motion of each jaw is transmitted via a "wheel and worm drive" non-backdrivable mechanism. Once the two jaws are in contact, they cannot move back, meaning that the object is completely immobilized. The grasp is form-closed.

4.2 Mechanism B

The second mechanism is underactuated by a "pulley/tendon" differential mechanism. A single actuator drives the closing motion of the gripper by tightening cable 0. This tension is transmitted to cables 1 and 2 which are connected to cable 0 at point A (Fig. 2b). This device tends to move the grasped object in the centered position during the closing motion when cable 0 is tightened. Once centered, the object is immobilized. However, contrarily to the grasp exerted by mechanism (a), infinitesimal motions can still occur in this centered configuration. The grasp is 2nd order form-closed. To make it 1st order form close, extra non-backdrivable mechanisms should be added in the transmission of the closing motion of the hand. This aspect will be emphasize in next section while introducing simple design rules to achieve 1st order form-closure.

5 Design rules for 1st order form-closure of underactuated hands

5.1 The minimum number of unilateral constraints

As illustrated by these two examples, the introduction of non-backdrivable mechanisms into the transmission of the closing motion of an underactuated gripper has a direct influence on the stability of the grasp.

In this paper, mathematical formulations that could justify the above assertions are not given (see Krut et al. (2010) to get them). These assertions are based on intuition, and should help us to understand that there is an analogy between the underactuated grasps illustrated in Fig. 2 and the classical grasps illustrated in Fig. 1:

- grasps shown on Fig. 1a and Fig. 2a are both 1st order form-closed, whereas
- grasps shown on Fig. 1b and Fig. 2b are both 2nd order form closed.

This analogy allows understanding how it is possible to extend an important result from classical grasps (exerted by fully actuated grippers) to underactuated grasps: that of the minimum number of contact points necessary for 1st order form-closure. These works were initiated by Reulaux (1976) and Somov (1900), and later demonstrated by Lakshminarayana (1978), Markenscoff et al. (1990) and Reulaux (1976). They demonstrate that $d+1$ contacts are necessary to achieve 1st order form-closure of classical grasps:

$$n_c \geq d+1 \tag{7}$$

where n_c is the number of unilateral contact constraints and d is the dimension of the configuration space of the object (3 for the planar examples depicted in Fig. 1, 6 in the general case).

By analogy, the following inequality can be written for underactuated grasps:

$$n_k \geq g+1 \tag{8}$$

where n_k is the number of unilateral constraints of the problem and g is the dimension of the grasp configuration space.

This can be stated the following way:

"At least g+1 unilateral constraints are necessary for 1st order form-closure."

The unilateral constraints of the problem are composed (see Krut et al. (2010) for more details) of the unilateral contact constraints and the unilateral constraints imposed by the unidirectional mechanisms (such as worm gears or free wheels). Hence:

$$n_k = n_u + n_c \tag{9}$$

with n_u the number of unilateral constraints imposed by the unidirectional mechanisms, and n_c is the number of unilateral contact constraints.

Additionally, the grasp configuration space is made of the configuration space of the object and the hand considered separately. Hence:

$$g = n_p + d \tag{10}$$

with n_p the dimension of the hand configuration space, that is, the number of degrees of freedom of the hand, and d is the dimension of the configuration space of the object.

5.2 The minimum number of unidirectional mechanisms

Relation (8) can be rewritten as:

$$n_u + n_c \geq n_p + d + 1 \tag{11}$$

Considering a hand with as many DOF as phalanxes, and a grasp where each phalanx exerts one contact point with the object including the palm, leads to:

$$n_c = n_p + 1 \tag{12}$$

Equations (11) and (12) enable a simple rule for designing underactuated hands, capable of performing 1st order form-closed grasps:

Table 1. Characteristics of underactuated robotic hands. The symbol – indicates that there is no underactuation or no non-backdriveable mechanisms depending on the column.

Name	No. of DOF	No. of act. for closure	No. of act. for config.	Type of underact. between fingers	Type of underact. between phalanxes	Non-backdriveable mechanisms	Reference
Soft Gripper	18	2	0	differential pulleys/cables	differential pulleys/cables	–	Hirose and Umetani (1978)
Hand of the LMS	9	1	0	differential with cables	differential pulleys/cables	–	Champeau (1991)
TU Delft Hand	5	1	0	differential with cables	differential pulleys/cables	–	De Visser and Herder (2000)
Barrett Hand	7	3	1	–	self-locking	6 wheels and worm drives	Townsend (2000)
TBM Prosthesis	15	1	0	compliant	differential four-bar	–	Dechev et al. (2001)
Southampton Hand Prosthesis	15	4	2	–	differential four-bar	6 wheels and worm drives	Light and Chappell (2000)
Lopez Hand	9	1	0	self-locking	differential pulleys/cable	9 bracing devices	Lopez (2001)
MARS	12	3	3	–	differential four-bar	–	Laliberté et al. (2002)
SARAH	10	1	1	self-locking	differential four-bar	3 wheels and worm drives	Laliberté et al. (2002)
Laval Univ. Pneumatic Hand	10	3	1	–	differential four-bar	3 motorised anti-return valves	Laliberté and Gosselin (2003)
RTR II	9	1	1	compliant	differential pulleys/cables	1 wheel and worm drive	Massa et al. (2002)
SPRING Hand	8	1	0	compliant	differential pulleys/cables	1 wheel and worm drive	Carrozza et al. (2004)
RTR IV	14	1	0	compliant	differential with cables	–	Carrozza et al. (2005)
Nasser Prosthesis	15	1	0	self-locking	differential four-bar	4 wheels and worm drives	Nasser et al. (2006)
Cyberhand	16	5	1	–	differential pulleys/cables	–	Cipriani et al. (2006)
Laval Univ. Prosthesis	17	1	0	differential pulleys/cables	differential pulleys/cables	–	Birglen and Gosselin (2006)
SDM Hand	8	1	0	differential pulleys/cables	differential with cables	–	Dollar and Howe (2007)
iLimb	11	5	1	–	differential with tendons	5 wheels and worm drives	Gow (2007)
Southampton End Effector Hand	8	1	0	differential four-bar	differential four-bar	–	Crowder (1991)

"When considering that each phalanx exerts one contact point with the object including the palm,at least d unidirectional mechanisms are required for an underactuated hand to be capable of performing 1st order form-closed grasps."

The TWIX hand, proposed by the authors, was designed according to this simple design rule (Bégoc et al., 2007). Note that for the grasps illustrated in Fig. 2, as there is no palm and as the dimension of the configuration space is $d = 1$, at least 2 unidirectional mechanisms are required for an underactuated hand to perform 1st order form-closed grasps. This witnesses why the grasp illustrated in Fig. 2a., performed by an underactuated hand having 2 unidirectional mechanisms, is 1st order form closed, while the one of Fig. 2b., performed by an underactuated hand having only one unidirectional mechanism (the actuator), is only 2nd order form-closed.

5.3 Discussion and analysis of some underactuated hands

Table 1 lists the main characteristics of various existing underactuated hands. For each hand, the table lists the number of dof, the number of actuators used to drive the closing/opening motion of the hand and the number of actuators used to modify the configuration of the hand. The type of underactuation used for the finger inputs and the phalanx inputs is listed separately, often being of different types. Finally, the number and type of the non-backdrivable mechanisms introduced in the mechanical transmission of the closing motion of the hand are indicated.

One of the major limitations of underactuated hands lies in the ejection phenomenon described in Laliberté et al. (2002). As already mentioned, in certain configurations of the finger, the force exerted by certain phalanxes on the object must be negative to achieve static equilibrium of the finger. Because the contact forces are unilateral, this equilibrium cannot be attained, which causes a backward motion of the corresponding phalanxes. Introducing a unidirectional mechanism in the transmission of the closing motion of each phalanx prevents any backward motion of the phalanx and consequently eliminates the ejection phenomenon. This explains why a large number of underactuated hands described in Table 1 use non-backdrivable mechanisms without, however, justifying it.

On the basis of the design rule introduced in the previous section, let us review the existing underactuated hands listed in Table 1 It can be noted that, out of the 19 hands, only 3 integrate at least six unidirectional mechanisms: the Lopez hand, the Southampton hand, and the Barrett hand. The other hands in the table cannot achieve 1st order form-closed grasps. Of course they are capable of performing stable grasps, but exhibiting other types of stability such as higher order form-closure or force-closure.

Following this simple rule, when designing underactuated hands, may improve the quality of the grasp in some circumstances. For us, the best proof is that the Barrett hand, that satisfies this simple rule, is the only one that encountered success in industry. However the systematic adoption of that design rule can lead to an increased complexity of the overall mechanical structure of the hand, thus compromising the main appeal of underactuation, that is, simplicity, low cost and so on.

6 Conclusions

In this paper, the inadequation of the definition of form-closure, as largely accepted in the literature, with the case of underactuated grasps has been pointed out. The definition is based on the assumption that contact points are fixed in space; this hypothesis is no longer true when the grasp is exerted by an underactuated hand. Therefore, a more general definition has been introduced, which consists in verifying that the system formed by the hand and the object is immobilized by a set of kinematic constraints. Among these constraints are the contact constraints to which the constraints imposed by unidirectional mechanisms have been added. Indeed, it has been shown, using several simple examples, that unilateral mechanisms have a relevant influence on the stability of the grasp. Hence, a new classification of underactuated hands has been proposed, based on the study of the expression of contact forces. Numerous existing underactuated hands use unilateral mechanisms, without justifying it.

Finally, this study allowed us to state a simple design rule on the minimum number of unidirectional mechanisms to be introduced into the transmission of the closing motion of the fingers of an underactuated hand, so that it is capable of producing 1st order form-closed grasps.

Edited by: J. L. Herder
Reviewed by: two anonymous referees

References

Ambrose, R. O., Aldridge, H., Askew, R. S., Burridge, R. R., Bluethmann, W., Diftler, M., Lovchik, C., Magruder, D., and Rehnmark, F.: Robonaut: NASA's space humanoid, IEEE Intell. Syst., 15(4), 57–63, 2000.

Bégoc, V., Durand, C., Krut, S., Dombre, E., and Pierrot, F.: On the Form-Closure Capability of Robotic Underactuated Hands, Proc. of ICARCV 2006: 9th International Conference on Control, Automation, Robotics and Vision, Grand Hyatt, Singapore, 5–8 December 2006, 2011–2018, 2006.

Bégoc, V., Krut, S., Dombre, E., Durand, C., and Pierrot, F.: Mechanical design of a new pneumatically driven underactuated hand, Proc. of IEEE Int. Conf. on Robotics and Automation (ICRA'07), Roma, Italy, 927–933, 2007.

Bicchi, A.: On the closure properties of robotic grasping, Int. J. Robot. Res., 14(4), 319–334, 1995.

Birglen, L. and Gosselin, C.: Force analysis of connected differential mechanisms: application to grasping, Int. J. Robot. Res., 25(10), 1033–1046, 2006.

Birglen, L., Laliberté, T., and Gosselin, C.: Underactuated Robotic Hands, Springer-Verlag, ISBN 978-3-540-77458-7, 2008.

Butterfass, J., Grebenstein, M., Liu, H., and Hirzinger, G.: DLR-hand II next generation of a dextrous robot hand, Proc. of IEEE Int. Conf. on Robotics and Automation (ICRA'01), Seoul, Korea, 21–26 May 2001, 109–120, 2001.

Carrozza, M., Suppo, C., Sebastiani, F., Massa, B., Vecchi, F., Lazzarini, R., Cutkosky, M., and Dario, P.: The SPRING hand: development of a self-adaptive prosthesis for restoring natural grasping, Autonomous Robots, 16(2), 125–141, 2004.

Carrozza, M., Cappiello, G., Stellin, G., Zaccone, F., Vecchi, F., Micera, S., and Dario, P.: A cosmetic prosthetic hand with tendon driven under-actuated mechanism and compliant joints: ongoing research and preliminary results, Proc. of IEEE Int. Conf. on Robotics and Automation (ICRA'05), Barcelona, Spain, 18–22 April 2005, 2672–2677, 2005.

Champeau, H.: Conception et réalisation d'une main mécanique anthropomorphe en vue de son implantation sur le robot manus, M.S. report, Poitiers University of Sciences, France, 1991.

Cheong, J.: Immobilizing grasps for two- and three-dimensional objects, Ph.D. Thesis, Utrecht University, The Netherlands, 2006.

Cipriani, C., Zaccone, F., Stellin, G., Beccai, L., Cappiello, G., Carrozza, M. C., and Dario, P.: Closed-loop controller for a bio-inspired multi-fingered underactuated prosthesis, Proc. of IEEE Int. Conf. on Robotics and Automation (ICRA'06), Orlando, Florida, USA, 15–19 May 2006, 2111–2116, 2006.

Crowder, R. M.: An anthropomorphic robotic end effector, Robot. Auton. Syst., 7(4), 253–268, 1991.

De Visser, H. and Herder, J.: Force directed design of a voluntary closing hand prosthesis, J. Rehabil. Res. Dev., 37(3), 261–271, 2000.

Dechev, N., Cleghorn, W. L., and Naumann, S.: Multiple finger, passive adaptive grasp prosthetic hand, Mech. Mach. Theory, 36(10), 1157–1173, 2001.

Dollar, A. M. and Howe, R. D.: Simple, robust autonomous grasping in unstructured environments, Proc. of IEEE Int. Conf. on Robotics and Automation (ICRA'07), Roma, Italy, 10–14 April 2007, 4693–4700, 2007.

Gow, D. J.: Prostheses with Mechanically Operable Digit Members, Touch Emas LTD., Int. patent WO2007063266 (A1), 7 June 2007.

Hirose, S. and Umetani, Y.: The development of soft gripper for the versatile robot hand, Mech. Mach. Theory, 351–359, 1978.

IFToMM website: http://www.iftomm.3me.tudelft.nl/, version 2.3, April 2010.

Kaneko, K., Harada, K., and Kanehiro, F.: Development of multi-fingered hand for life-size humanoid robots, Proc. of IEEE Int. Conf. on Robotics and Automation (ICRA'07), Roma, Italy, 10–14 April 2007, 913–920, 2007.

Krut, S.: A force-isotropic underactuated finger, Proc. of IEEE Int. Conf. on Robotics and Automation (ICRA'05), Barcelona, Spain, 18–22 April 2005, 2314–2319, 2005.

Krut, S., Bégoc, V., Dombre, E., and Pierrot, F.: Extension of the Form-Closure Property to Underactuated Hands, IEEE T-RO: Transaction on Robotics, 26(5), 853–866, October 2010.

Lakshminarayana, K.: Mechanics of form closure, ASME Technical paper 78-DET-32, 2–8, 1978.

Laliberté, T. and Gosselin, C.: Actuation system for highly underactuated gripping mechanism, US patent no. US6.505.870, 2003.

Laliberté, T., Birglen, L., and Gosselin, C.: Underactuation in robotic grasping hands, Japanese Journal of Machine Intelligence and Robotic Control, Special Issue on Underactuated Robots, 4(3), 77–87, 2002.

Light, C. and Chappell, P.: Development of a lightweight and adaptable multiple-axis hand prosthesis, Med. Eng. Phys., 22(10), 679–684, 2000.

Lopez, G.: Préhenseur anthropomorphe à serrage automatique, French patent no. FR2.822.404, 2001.

Luo, M., Mei, T., Wang, X., and Yu, Y.: Grasp Characteristics of an Underactuated Robot hand, Proc. of IEEE Int. Conf. on Robotics and Automation (ICRA'04), New Orleans, LA, USA, 26 April–1 May 2004, 2236–2241, 2004.

Markenscoff, X., Ni, L., and Papadimitriou, C. H.: The geometry of grasping, Int. J. Robot. Res., 9(1), 61–74, 1990.

Massa, B., Roccella, S., Carrozza, M. C., and Dario, P.: Design and development of an underactuated prosthetic hand, Proc. of IEEE Int. Conf. on Robotics and Automation (ICRA'02), Washington, USA, 11–15 May 2002, 3374–3379, 2002.

Nasser, S., Rincon, D., and Rodriguez, M.: Design of an anthropomorphic underactuated hand prosthesis with passive-adaptive grasping capabilities, Florida Conf. on Recent Advances in Robotics and Robot Showcase, Miami, Florida, USA, 25–26 May 2006.

Reulaux, F.: The kinematics of machinery, MacMillan, NY, 1976.

Rimon, E. and Burdick, J. W.: On force and form closure for multiple finger grasps, Proc. of IEEE Int. Conf. on Robotics and Automation (ICRA'96), Minneapolis, MN, USA, 22–28 April 1996, 1795–1800, 1996.

Somov, P.: Über Gebiete von Schraubengeschwindigkeiten eines starren Korpers bei verschiedener Zahl von Stutzflachen, Zeitschrift für Mathematik und Physik, 45, 245–306, 1900.

Townsend, W.: The Barrett Hand grasper – programmably flexible part handling and assembly, Ind. Robot, 27(3), 181–188, 2000.

Ueda, J., Ishida, Y., Kondo, M., and Ogasawara, T.: Development of the NAIST-hand with vision-based tactile fingertip sensor, Proc. of IEEE Int. Conf. on Robotics and Automation (ICRA'05), Barcelona, Spain, 18–22 April 2005, 2332–2337, 2005.

Van der Stappen, A., Wentink, C., and Overmars, M.: Computing immobilizing grasps of polygonal parts, Int. J. Robot. Res., 19(5), 467–479, 2000.

2

Reaction Null Space of a multibody system with applications in robotics

D. N. Nenchev

Graduate School of Engineering, Tokyo City University, Tamazutsumi 1-28-1, Setagaya-ku, Tokyo 158-8557, Japan

Correspondence to: D. N. Nenchev (nenchev@tcu.ac.jp)

Abstract. This paper provides an overview of implementation examples based on the Reaction Null Space formalism, developed initially to tackle the problem of satellite-base disturbance of a free-floating space robot, when the robot arm is activated. The method has been applied throughout the years to other unfixed-base systems, e.g. flexible-base and macro/mini robot systems, as well as to the balance control problem of humanoid robots. The paper also includes most recent results about complete dynamical decoupling of the end-link of a fixed-base robot, wherein the end-link is regarded as the unfixed-base. This interpretation is shown to be useful with regard to motion/force control scenarios. Respective implementation results are provided.

1 Introduction

In articulated multibody systems (MBS), such as robots or smart structures, the force imposed on a specific link via reactions from the motion of other links may need to be controlled in an appropriate way to ensure desired performance. We focus here on the field of robotics exclusively, with representative examples such as free-floating space robots, manipulator(s) fixed to a mobile or flexible base, and humanoid robots. Consider for example a free-floating space robot consisting of one or more manipulators mounted on a satellite base. In order to avoid loss of communication of the system with a distantly located control station, it is highly desirable to maintain the orientation of the satellite base during manipulator operation. As another example, consider the case when the manipulators are mounted on a flexible base. Then, the goal would be avoiding base deflection and/or base vibrations during manipulator operation; otherwise manipulator task performance may deteriorate significantly. Yet as another example, consider a biped humanoid robot – a MBS that also lacks a fixed base. The limbs of such a biped robot need to be controlled in a way that balance is maintained during such operations as: walking, "reflexive" motion in response to an unexpected external force, motion/force control tasks whereby the end-links contact the environment etc., in order to prevent the robot from falling down.

The common way for achieving such seemingly different control objectives, for different robotic systems, is proper path planning and control of the manipulator/limb movements, such that the respective spatial reaction forces (wrenches) imposed on the unfixed base would be minimized or controlled in some other specific way. In the case of a free-floating space robot, satellite-base orientation can be maintained when the reaction moments at the base are minimized. In the case of a flexible-base robot, on the other hand, the spatial reaction forces at the flexible base should be minimized to avoid inducing vibrations. Finally, in the case of a humanoid robot, the ground reaction force should be controlled appropriately to maintain the balance and/or to generate appropriate propulsion and other desirable forces. Significant research has been carried out to address the above control objectives. Representative works in the field of free-floating space robots are Masutani et al. (1989); Papadopoulos and Dubowsky (1991); Torres and Dubowsky (1992). *Inertial damping* has been exploited to deal with vibration suppression of flexible-base robots (Lee and Book, 1990; Hanson and Tolson, 1995; Torres et al., 1996) and with so-called macro-micro manipulator systems (Book and Lee, 1989; Yoshikawa et al., 1993; Sharf, 1996; Parsa et al., 2005). In the field of biped humanoid robots, dynamic postural control in unknown environments has been discussed, see e.g. Gorce

(1999); Stephens (2007); Hyon et al. (2009) and also, the problem of compliant control of multicontact was addressed (Sentis and Khatib, 2010).

The aim of this work is to present the formalism called "Reaction Null Space", or RNS for short, summarizing thereby research results from the past 25 yr and demonstrating the usefulness of the method when dealing with a wide range of problems like those outlined above. The RNS was initially introduced in Nenchev et al. (1988) to tackle the problem of free-floating space robot control (see also Nenchev et al., 1992). Later, it was applied to reactionless motion generation and vibration control with flexible-base robots (Nenchev et al., 1996, 1999; Yoshida et al., 1996) and long-reach manipulators (Gouo et al., 1998). The potential of the method has also been explored by others, e.g. for modeling and control of the biped gait for the design of interactive orthesis in rehabilitation tasks (Finat and Gonzalez-Sprinberg, 2002), reactionless satellite capture (Dimitrov and Yoshida, 2004; Xu et al., 2007; Piersigilli et al., 2010; Cong and Sun, 2010; Nguyen-Huynh and Sharf, 2011), design of reactionless manipulators (Fattah and Agrawal, 2005), design and control of a dual-stage feed drive (Elfizy et al., 2005), wire-suspended manipulator control (Osumi and Saitoh, 2006; Lampariello et al., 2006), end-point control (Cheng, 2005) and compliance control of flexible-base robots (Ott et al., 2006), optimal motion planning and control (Shui et al., 2009), adaptive reaction control (Abiko and Yoshida, 2010), and reaction torque control of redundant space robots (Cocuzza et al., 2010), impacts with a humanoid robot (Konno et al., 2011), optimal motion planning for space robots with base disturbance (Kaigom et al., 2011), shaking force minimization of high-speed robots (Briot et al., 2012), and so on.

We explain how to implement RNS based methods within an existing robotic MBS, e.g. the first free-floating space robot ETS-VII (Oda, 2000), the Japanese Experiment Module Remote Manipulator System (JEMRMS) on the International Space Station (Sato and Wakabayashi, 2001; Morimoto et al., 2001), the DLR robot Rollin' Justin with a humanoid upper body mounted on a mobile base with flexible suspension (Borst et al., 2009). In addition, we will introduce our recent result on how the RNS can be applied to a fixed-base manipulator for ensuring full dynamic decoupling of the end-link and the usefulness of this property in motion/force control tasks. As an example, we present an implementation with a small humanoid robot performing a surface cleaning task.

The paper is organized as follows. The following section introduces briefly some of the notation. Details of the Reaction Null Space derivation are given in Sect. 3, using a serial-link chain as a representative of a free-floating space robot. Section 4 discusses the implementation in flexible-base and macro/mini manipulation systems. Section 5 shows the usefulness of the RNS formalism within an operational-space motion/force control framework. Section 6 discusses the application to humanoid robots, both for balance control in response to unknown external disturbances and for motion/force control tasks. Finally, Sect. 7 gives our conclusions.

2 Notation

In this work, we deal with a robotic MBS that comprises one or more manipulators/limbs attached to an unfixed base and arranged in a tree-like structure. We assume that the manipulators/limbs are made of rigid-body links and include n single-DOF joints. The respective joint coordinates will be denoted by $\boldsymbol{\theta} \in \Re^n$. Hence, the system can be described with $6+n$ generalized coordinates $\boldsymbol{q} = (\mathcal{X}, \boldsymbol{\theta})$, where $\mathcal{X} \in SE(3)$ denotes the position/orientation of the unfixed base w.r.t. the inertial frame. Note that lower-case bold characters denote vectors, upper-case bold characters are used for matrices, and spatial quantities like the spatial velocity of and the spatial force acting at a rigid body, will be denoted by calligraphic symbols, e.g. $\mathcal{V}_O, \mathcal{F}_O \in \Re^6$, respectively. The convention for spatial vectors composed of 3-D quantities is: linear part followed by angular, e.g. $\mathcal{V}_O = \begin{bmatrix} \boldsymbol{v}_O^T & \boldsymbol{\omega}^T \end{bmatrix}^T$ and $\mathcal{F}_O = \begin{bmatrix} \boldsymbol{f}^T & \boldsymbol{n}_O^T \end{bmatrix}^T$, where $\boldsymbol{v}, \boldsymbol{\omega}, \boldsymbol{f}, \boldsymbol{n}$ denote 3-D vectors of body velocity, angular velocity, force and moment, respectively. Note also that spatial velocities are transformed via the operator:

$$ {}^{\mathrm{k}}\mathbf{T}_{\mathrm{l}} = \begin{bmatrix} {}^{\mathrm{k}}\mathbf{R}_{\mathrm{l}} & -{}^{\mathrm{k}}\mathbf{R}_{\mathrm{l}}{}^{\mathrm{k}}\boldsymbol{r}_{\mathrm{l}}^{\times} \\ \mathbf{0} & {}^{\mathrm{k}}\mathbf{R}_{\mathrm{l}} \end{bmatrix} \in \Re^{6\times 6}, \qquad (1) $$

${}^{\mathrm{k}}\mathbf{R}_{\mathrm{l}} \in \Re^{3\times 3}$ denoting the orientation of coordinate frame {l} w.r.t. {k}, ${}^{\mathrm{k}}\boldsymbol{r}_{\mathrm{l}}^{\times} \in \Re^{3\times 3}$ standing for the skew-symmetric operator associated with vector ${}^{\mathrm{k}}\boldsymbol{r}_{\mathrm{l}} \in \Re^3$ expressing the position of {l} w.r.t. {k}. The transpose of operator (1) transforms spatial forces.

3 The Reaction Null Space of a free-floating serial-link robot in zero gravity

We will introduce the RNS formalism with a simple example: a free-floating serial-link chain in zero gravity. This model will be used to derive the basic notations. The most relevant application would be a free-floating space robot comprising a rigid-link manipulator arm mounted on a rigid-body satellite – the floating base of the system (see Fig. 1).

The manipulator joints are assumed actuated while the base is not. First, we will derive the RNS formalism at velocity level, based on the momentum conservation condition. Then, the full dynamics will be taken into consideration.

3.1 Momentum-based derivation

Consider first the spatial momentum of the free-floating robot represented as an *composite rigid body* (CRB)

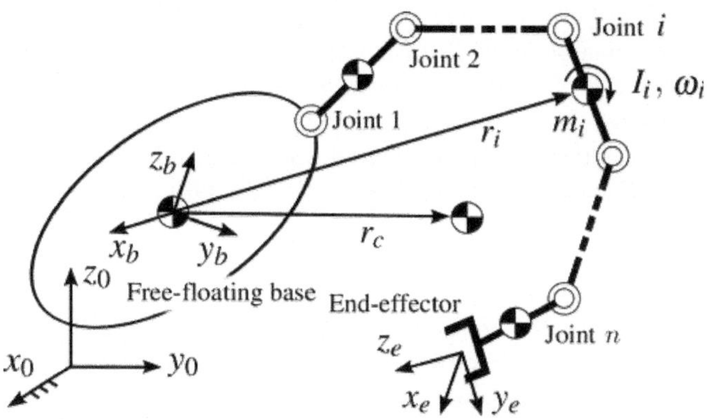

Figure 1. Model of a free-floating base serial-link chain.

(Featherstone, 2008), wherein the angular momentum part is written w.r.t. the CoM of the CRB:

$$\mathcal{L}_c \equiv \begin{bmatrix} \boldsymbol{p} \\ \boldsymbol{l}_c \end{bmatrix} = \mathbf{M}_c \mathcal{V}_c. \tag{2}$$

Subscript $(\circ)_c$ denotes quantities defined w.r.t. coordinate frame {c} fixed at the CRB CoM, which is an inertial (non-accelerating) frame in the absence of external forces, and thus under momentum conservation. The linear part of momentum is $\boldsymbol{p} = \sum_{j=0}^{n} m_j \dot{\boldsymbol{r}}_j = m_t \dot{\boldsymbol{r}}_c$ while the angular part is $\boldsymbol{l}_c = \sum_{j=0}^{n} \left(\mathbf{I}_j \omega_j + m_j \boldsymbol{r}_j \times \dot{\boldsymbol{r}}_j \right)$, where m_j, $\mathbf{I}_j$, $\boldsymbol{r}_j$, ω_j, stand for link j mass, inertia matrix, CoM position and angular velocity, respectively, the latter three given in inertial coordinates. m_t denotes the total mass of the CRB system, $\boldsymbol{r}_c$ and $\mathcal{V}_c$ stand for its CoM position and spatial velocity, respectively. Matrix $\mathbf{M}_c(\boldsymbol{q})$ is a 6×6 block-diagonal matrix having $\mathbf{M}_v \equiv m_t \mathbf{U}$ and ${}^c\mathbf{M}_\omega \equiv \sum_{j=0}^{n} \left(\mathbf{I}_j + m_j {}^j\boldsymbol{r}_c^{\times c} \boldsymbol{r}_j^{\times} \right)$ as upper and lower blocks, respectively, where $\mathbf{U}$ is the 3×3 unit matrix[1]. Henceforth, constants will be denoted by an over-bar. Since the above momentum is a conserved quantity, we then denote it as $\bar{\mathcal{L}}_c$.

With the above notations, the momentum conservation equation has the simplest possible form $\mathbf{M}_c \mathcal{V}_c = \bar{\mathcal{L}}_c$. Nevertheless, it is desirable to employ inertial properties familiar from fixed-base manipulator descriptions. For this purpose, it is necessary to redefine spatial momentum w.r.t. the base frame {b}:

$$\mathcal{L}_b = \begin{bmatrix} \boldsymbol{p} \\ {}^b\boldsymbol{r}_c \times \boldsymbol{p} + \boldsymbol{l}_c \end{bmatrix}, \tag{3}$$

The motion of the robot can then be represented at the velocity level as:

$$\mathbf{M}_b \mathcal{V}_b + \mathbf{M}_{bm} \dot{\boldsymbol{\theta}} = \mathcal{L}_b, \tag{4}$$

where $\mathcal{V}_b$ denotes the spatial velocity of the base. Matrix

$$\mathbf{M}_b(\boldsymbol{q}) = \begin{bmatrix} \mathbf{M}_v & \mathbf{M}_{v\omega} \\ \mathbf{M}_{v\omega}^T & \mathbf{M}_\omega \end{bmatrix} \in \Re^{6 \times 6}$$

[1]Note that ${}^j\boldsymbol{r}_c^{\times} = ({}^c\boldsymbol{r}_j^{\times})^T = -{}^c\boldsymbol{r}_j^{\times}$.

is the inertia matrix of the system regarded as an CRB, matrix

$$\mathbf{M}_{bm}(\boldsymbol{q}) = \begin{bmatrix} \mathbf{M}_{vm}^T & \mathbf{M}_{\omega m}^T \end{bmatrix}^T \in \Re^{6 \times n}$$

is a block submatrix of the system inertia matrix, called henceforth *coupling inertia matrix*. The block submatrices in the above terms are:

$$\mathbf{M}_{v\omega} \equiv -m_t {}^b\boldsymbol{r}_c^{\times} \tag{5}$$

$$\mathbf{M}_{vm} \equiv m_t {}^b\mathbf{J}_c \tag{6}$$

$$\mathbf{M}_\omega \equiv {}^b\mathbf{I}_0 + \sum_{j=1}^{n} \left({}^b\mathbf{I}_j + m_j {}^j\boldsymbol{r}_b^{\times b} \boldsymbol{r}_j^{\times} \right) \tag{7}$$

$$\mathbf{M}_{\omega m} \equiv \sum_{j=1}^{n} \left({}^b\mathbf{I}_j {}^b\mathbf{J}_{\omega j} + m_j {}^b\boldsymbol{r}_j^{\times b} \mathbf{J}_{vj} \right) \tag{8}$$

where it was assumed that link 0 is the base. $\mathbf{J}_c, \mathbf{J}_{vj}$ denote the CRB CoM and link-j CoM velocity Jacobians, respectively, while $\mathbf{J}_{\omega j}$ is the link-j angular velocity Jacobian. As can be inferred from the leading superscript, all these quantities are defined w.r.t. the base frame {b} (see also Masutani et al., 1989). We should note that the coupling inertia submatrix $\mathbf{M}_{vm}$, contributing to translational motion of the CRB, is identical to the system CoM Jacobian, up to a multiplicative constant (the total mass). For a free-floating system, translational motion is considered less important than rotational. But for the other MBS discussed below, e.g. flexible-base and humanoid robots, the case is just the opposite. We should also note that usually, it is assumed that initial spatial momentum is null.

The momentum component $\mathbf{M}_b \mathcal{V}_b$, appearing on the l.h.s. of Eq. (4), can be interpreted twofold. First, assuming a nonzero initial momentum and immobilized manipulator joints, the component has the meaning of conserved CRB momentum. This is a trivial case. Usually, and henceforth, zero initial momentum will be assumed, such that the condition $\bar{\mathcal{L}}_c = \mathbf{0} = \mathcal{L}_b$ holds. Then, the above component has the meaning of CRB momentum occurring in reaction to the manipulator motion (s.t. $\mathbf{M}_b \mathcal{V}_b = -\mathbf{M}_{bm} \dot{\boldsymbol{\theta}}, \dot{\boldsymbol{\theta}} \neq \mathbf{0}$). Therefore, we will henceforth refer to $\mathbf{M}_b \mathcal{V}_b$ as *reaction momentum*. The other momentum component $\mathbf{M}_{bm} \dot{\boldsymbol{\theta}}$, on the other hand, has the meaning of momentum *imposed* upon the CRB (i.e. including the base) via manipulator motion. We will refer to it as the *coupling momentum* (Nenchev et al., 1996) and denote it as:

$$\mathcal{L}_{bm} \equiv \mathbf{M}_{bm} \dot{\boldsymbol{\theta}}. \tag{9}$$

Equation (4) can be solved for the manipulator joint velocities, needed as input variables for velocity-based system motion control. Since the equation is linear in the velocities, its solution type depends upon the number of manipulator joints n. In the case of a six-DOF manipulator ($n = 6$), for example, the unique solution is:

$$\dot{\boldsymbol{\theta}} = -\mathbf{M}_{bm}^{-1} \mathbf{M}_b \mathcal{V}_b. \tag{10}$$

More interesting is the case of a kinematically redundant manipulator ($n > 6$). We have then an underdetermined system, with the general solution (Nenchev, 1989):

$$\dot{\theta} = -\mathbf{M}^{\#}_{\mathrm{bm}}\mathbf{M}_{\mathrm{b}}\mathcal{V}_{\mathrm{b}} + \mathbf{P}_{\mathrm{bm}}\dot{\theta}_{\mathrm{a}}, \tag{11}$$

where $(\circ)^{\#}$ denotes a generalized inverse, $\mathbf{P}_{(\circ)}$ stands for a null-space projector and $\dot{\theta}_{\mathrm{a}}$ is an arbitrary vector dimensioned as joint velocity.

Apparently, the coupling inertia matrix comprises a nontrivial kernel. The infinite set of joint velocities from the kernel can be derived via the second term on the r.h.s. $\{\dot{\theta}_{\mathrm{rl}} : \mathbf{P}_{\mathrm{bm}}\dot{\theta}_{\mathrm{a}}, \forall\dot{\theta}_{\mathrm{a}}\}$. This is the set of *reactionless joint velocities*; these velocities do not impose any momenta on the base, and thus, manipulator motion will be *completely dynamically decoupled* from the motion of the base. Note also that the reactionless joint velocities constitute the set of solutions of homogeneous equation

$$\mathbf{M}_{\mathrm{bm}}\dot{\theta} = \mathbf{0}, \tag{12}$$

which stands for coupling momentum conservation. Hence, we can conclude that reactionless motion (and hence, complete dynamical decoupling) can be achieved if and only if the coupling momentum is conserved (Nenchev et al., 1996). Further on, note that the set $\{\dot{\theta}_{\mathrm{rl}}\}$ has the structure of a manifold in joint space, e.g. similarly to the selfmotion manifold known from studies on kinematically redundant manipulators (Burdick, 1989). We will call it the *reactionless motion manifold*. The manifold depends upon the rank of the RNS projector: $\mathrm{rank}\mathbf{P}_{\mathrm{bm}} = n-6$. With a seven-DOF articulated manipulator, for example, the manifold will be just one-dimensional. Hence, reactionless motions can be derived via the differential equation:

$$\dot{\theta} = \mathrm{b}\boldsymbol{n}_{\mathrm{bm}}, \tag{13}$$

where b is an arbitrary scalar with dimension of angular speed. $\boldsymbol{n}_{\mathrm{bm}}(\boldsymbol{q}) \in \ker\mathbf{M}_{\mathrm{bm}}$ will be called *reactionless vector field*. The integral curves, projected onto workspace via the direct kinematics, will be referred to as *reactionless paths*.

In general, it is desirable to have a larger set of reactionless paths. One possibility to achieve this is increasing the number of manipulator joints. Another possibility is to redefine the coupling inertia matrix (and thus its kernel) w.r.t. a suitable subset of base coordinates. For a free-floating space robot, most important is the orientation of the satellite base. Hence, we may redefine the above equations to ignore base translation variables. Then, the rank of the null-space projector will increase to $n-3$. Because of its fundamental role, the kernel has been named as the *Reaction Null Space* (Nenchev et al., 1999).

Let us focus now on the other joint velocity component in Eq. (11), i.e. reaction momentum mapped via a generalized inverse of the coupling inertia matrix. Recall first that velocity-based redundancy resolution schemes, similar to that in Eq. (11), are known from studies of kinematically redundant manipulators (Konstantinov et al., 1981) – the so-called "task-of-priority" type schemes (Nakamura et al., 1987). Such schemes give the possibility to address dual-task control scenarios: typically end-link motion control via the generalized-inverse component, plus an additional control task (e.g. optimization of a suitable measure, obstacle avoidance etc), via the null-space component. Note also that quite often the Moore-Penrose generalized inverse (the pseudoinverse) is used in such schemes, since it yields a locally optimal solution for the joint-velocity norm (Nenchev, 1989). Also, in this case, the two components of the general solution (11) become orthogonal, yielding joint-space decomposition into two orthogonal complementary subspaces. In our case, when the pseudoinverse is used in Eq. (11), the respective joint velocity component yields optimal inertial coupling in terms of minimizing that part of the total kinetic energy, that is due to the dynamical coupling between the base and the rest of the links. We will refer to this energy as the *coupling kinetic energy*: $\mathcal{V}^T_{\mathrm{b}}\mathbf{M}_{\mathrm{bm}}\dot{\theta}$. Such energy minimization is a highly desirable feature. The reason is that typical motion control scenarios are dual-task ones: e.g. reactionless motion control via the null space component (a feedforward control component), plus error compensation control for small base attitude errors, via the pseudoinverse component (a feedback control component)[2]. Coupling energy optimization will thereby yield better performance with regard to error dynamics.

In conclusion, via the null space and the pseudoinverse, we obtain a *decomposition formalism* that can be quite useful for motion analysis, planning and control of various unfixed-base systems, as will be shown henceforth with a few more examples. This decomposition is the essence of the RNS method.

3.2 Dynamics-based derivation

To account for the presence of external forces, we consider the full dynamics of the free-floating robot:

$$\begin{bmatrix} \mathbf{M}_{\mathrm{b}} & \mathbf{M}_{\mathrm{bm}} \\ \mathbf{M}^T_{\mathrm{bm}} & \mathbf{M}_{\mathrm{m}} \end{bmatrix} \begin{bmatrix} \dot{\mathcal{V}}_{\mathrm{b}} \\ \ddot{\theta} \end{bmatrix} + \begin{bmatrix} \mathcal{C}_{\mathrm{b}} \\ \boldsymbol{c}_{\mathrm{m}} \end{bmatrix} = \begin{bmatrix} \mathcal{F}_{\mathrm{b}} \\ \boldsymbol{\tau} \end{bmatrix} + \begin{bmatrix} {}^{\mathrm{b}}\mathbf{T}^T_{\mathrm{e}} \\ \mathbf{J}^T \end{bmatrix} \mathcal{F}_{\mathrm{e}}, \tag{14}$$

where quantities, not yet introduced, are:

$\mathbf{M}_{\mathrm{m}} \in \Re^{n\times n}$: manipulator inertia matrix
$\mathbf{J} \in \Re^{6\times n}$: manipulator Jacobian matrix
$\boldsymbol{c}_{\mathrm{m}} \in \Re^{n}$: manipulator nonlinear force
$\mathcal{C}_{\mathrm{b}} \in \Re^{6}$: CRB nonlinear force
$\boldsymbol{\tau} \in \Re^{n}$: joint torque vector
$\mathcal{F}_{\mathrm{b}} \in \Re^{6}$: external force at the base
$\mathcal{F}_{\mathrm{e}} \in \Re^{6}$: external force at the end-link

[2] See e.g. the discussion on possible motion control tasks in Nenchev et al. (1992).

Hereby, we assumed that external forces may act upon the base ($\mathcal{F}_b$) and/or the end-link ($\mathcal{F}_e$). In fact, the base force $\mathcal{F}_b$ term could be assigned a broader role to include base constraint and/or actuator forces. This will allow us, in what follows, to model other types of systems with the same equation, e.g. a free-flying space robot with attitude controlled base (i.e. using reaction/momentum wheels as actuators) and/or flexible appendages, a flexible-base manipulator, a humanoid robot, and others.

Let us focus now on the upper part of the equation of motion. It can be rewritten as:

$$\mathbf{M}_b\dot{\mathcal{V}}_b + \mathbf{M}_{bm}\ddot{\boldsymbol{\theta}} + \mathcal{C}_b = \mathcal{F}_{ext}, \tag{15}$$

where $\mathcal{F}_{ext} = \mathcal{F}_b + {}^b\mathbf{T}_e^T\mathcal{F}_e$ denotes the *external forces*. This equation represents the dynamics of the CRB since only external forces are present. The dynamic equilibrium can then be expressed as $\mathcal{F}_d - \mathcal{F}_{ext} = \mathbf{0}$, where $\mathcal{F}_d$ is the *dynamical force* obtained as time derivative of CRB momentum:

$$\mathcal{F}_d \equiv \frac{d}{dt}\mathcal{L}_b = \mathbf{M}_b\dot{\mathcal{V}}_b + \mathbf{M}_{bm}\ddot{\boldsymbol{\theta}} + \dot{\mathbf{M}}_b\mathcal{V}_b + \dot{\mathbf{M}}_{bm}\dot{\boldsymbol{\theta}}. \tag{16}$$

The last two terms on the r.h.s. denote the CRB nonlinear force $\mathcal{C}_b \equiv \dot{\mathbf{M}}_b\mathcal{V}_b + \dot{\mathbf{M}}_{bm}\dot{\boldsymbol{\theta}}$. The two manipulator motion components, on the other hand, represent the spatial force:

$$\mathcal{F}_{bm} \equiv \frac{d}{dt}\mathcal{L}_{bm} = \mathbf{M}_{bm}\ddot{\boldsymbol{\theta}} + \dot{\mathbf{M}}_{bm}\dot{\boldsymbol{\theta}}. \tag{17}$$

We will refer to $\mathcal{F}_{bm}$ as the *imposed force*, in the sense that the force is imposed upon the CRB via manipulator motion. It should be apparent then that any motion along a reactionless path (i.e. reactionless motion), conserves coupling momentum (cf. Eq. 12), and implies hence a null imposed force $\mathcal{F}_{bm} = \mathbf{0}$.

From (15), it is straightforward to derive manipulator joint acceleration, needed in dynamical control schemes, e.g. resolved acceleration control (Luh et al., 1980) or computed torque control (Craig, 2004). We will skip the trivial nonredundant case and focus on the more interesting kinematically redundant manipulator case:

$$\ddot{\boldsymbol{\theta}} = \mathbf{M}_{bm}^{+}(\mathcal{F}_{ext} - \mathbf{M}_b\dot{\mathcal{V}}_b - \mathcal{C}_b) + \mathbf{P}_{bm}\ddot{\boldsymbol{\theta}}_a, \tag{18}$$

where $\ddot{\boldsymbol{\theta}}_a$ denotes an arbitrary n-vector with dimension of joint acceleration. With the help of this vector, reactionless manipulator motion can be generated in a feedforward manner, since the respective joint acceleration component $\mathbf{P}_{bm}\ddot{\boldsymbol{\theta}}_a$ yields coupling momentum conservation. In addition, the pseudoinverse acceleration component will be useful in dual-task scenarios, e.g. for feedback compensation of small attitude errors, as already explained. Thereby, full dynamic decoupling between the two subtasks will be ensured, as already discussed in the last subsection.

In computed torque controllers, the joint torque is used as control input. It can be derived by inserting the above joint acceleration into the lower part of Eq. (14):

$$\begin{aligned}\boldsymbol{\tau} &= \mathbf{M}_{bm}^T\dot{\mathcal{V}}_b + \mathbf{M}_m\ddot{\boldsymbol{\theta}} - \mathbf{J}^T\mathcal{F}_e + \boldsymbol{c}_m \\ &= \left(\mathbf{M}_{bm}^T - \mathbf{M}_m\mathbf{M}_{bm}^{+}\mathbf{M}_b\right)\dot{\mathcal{V}}_b \\ &\quad + \left(\mathbf{M}_m\mathbf{M}_{bm}^{+}\,{}^b\mathbf{T}_e^T - \mathbf{J}^T\right)\mathcal{F}_e \\ &\quad + \boldsymbol{c}_m + \mathbf{M}_m\mathbf{M}_{bm}^{+}(\mathcal{F}_b - \mathcal{C}_b) + \mathbf{M}_m\mathbf{P}_{bm}\ddot{\boldsymbol{\theta}}_a.\end{aligned} \tag{19}$$

Figure 2. Artistic rendering of the ETS-VII space robot mission. A six-DOF manipulator arm is mounted on the larger satellite thus constituting a free-floating space robot with serial arm structure. (Courtesy of JAXA – the Japan Aerospace Exploration Agency, Oda, 2000).

3.3 Implementation issues

The RNS method has been experimentally verified via both simulations and on-orbit experiments with the ETS-VII space robot system (see Fig. 2). The goal was to confirm the usefulness of reactionless manipulator motion on-orbit. Since the space robot was freely floating in micro-gravity environment, the presence of external forces (e.g. solar pressure, air drag, gravity gradient) has been ignored during the relatively short time interval of the experiment (about 20 min). It was possible then to employ the momentum conservation condition to obtain a suitable velocity-based reactionless motion generator, via the null-space solution (13) in a feedforward manner. The trajectories were generated in advance off-line and then transferred to the robot arm on-orbit for execution. The experimental results can be found e.g. in (Yoshida et al., 2000).

4 Application to flexible-base and macro/mini manipulator systems

A variety of flexible-base manipulator systems exist. Consider, for example, a serial-link manipulator mounted at the distant end of a long flexible beam (Lew et al., 1995). Such a system is useful as a "long-reach manipulator", e.g. to gain

Figure 3. An example of a flexible-base robot: Rollin' Justin – a robot with a humanoid upper body mounted on a wheeled mobile base with flexible suspension (Courtesy of DLR – the German Aerospace Center, Borst et al., 2009).

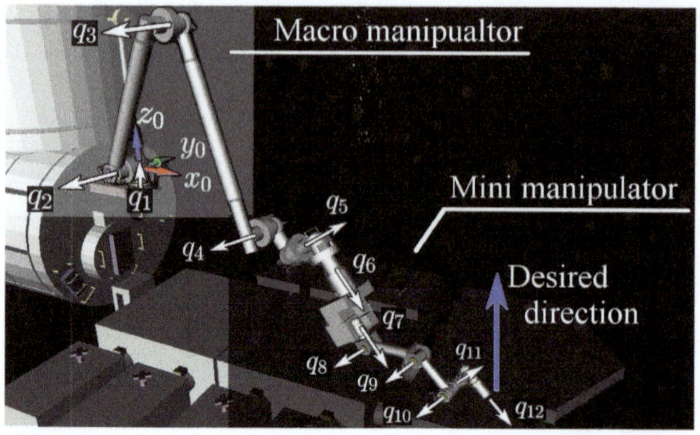

Figure 4. An example of a macro/mini manipulator system: model of the Japanese Experiment Module (JEM) "Kibo" on the International Space Station with the large Remote Manipulator System (JEMRMS) and the Small Fine Arm (SFA) attached.

access to a dangerous site. Another example is a robot comprising a humanoid upper body mounted on a mobile base via flexible suspension, e.g. the robot Rollin' Justin designed at DLR (Borst et al., 2009) (see Fig. 3). There are also so-called macro/mini manipulator systems, consisting of a dexterous manipulator (the mini part) attached to the end-link of a large manipulator (the macro part). The latter ensures positioning capability of the mini part within a large workspace. Due to its structure, the large arm usually has inherent flexibilities in the links and/or joints. Hence, the end-link of the large arm can be thought of as a flexible base for the mini part, whereby, the flexible base can be characterized as a composite rigid body. Two such systems exist on the International Space Station: the large Canadarm 2 with the Special Purpose Dexterous Manipulator "Dextre" (Coleshill et al., 2009) and the large Japanese Experiment Module Remote Manipulator System (JEMRMS) with the Small Fine Arm (SFA) (see Fig. 4) (Sato and Wakabayashi, 2001).

Such flexible-base robots require a special motion generation technique and respective control in order to minimize the reactions at the flexible base. Otherwise, essential high-precision positioning and/or path tracking capabilities of the dexterous manipulator(s) may degrade significantly. The Reaction Null Space method with its joint-space decomposition formalism can be employed in a straightforward manner to ensure reactionless motion, via the RNS component, in combination with inertial damping control of flexible-base vibrations, via the pseudoinverse component.

4.1 Single-body flexible base

The simplest possible case is a serial rigid-link manipulator attached to a single-body flexible base (see Fig. 5). First, we will assume that the spatial elastic forces, constraining the motion of the flexible base, are expressed via the following $\mathcal{F}_b$ appearing in Eq. (14):

$$\mathcal{F}_b = -\mathbf{D}_b \mathcal{V}_b - \mathbf{K}_b \Delta \mathcal{X}_b. \tag{20}$$

$\mathbf{D}_b$, $\mathbf{K}_b \in \Re^{6\times 6}$ denote base spatial viscous damping and stiffness, respectively, and $\Delta \mathcal{X}_b$ stands for base spatial displacement from the equilibrium. With this notation, the CRB dynamics (Eq. 15) become:

$$\begin{aligned}\mathbf{M}_b \dot{\mathcal{V}}_b + \mathbf{D}_b \mathcal{V}_b + \mathbf{K}_b \Delta \mathcal{X}_b &= -\mathbf{M}_{bm} \ddot{\theta} - \dot{\mathbf{M}}_{bm} \dot{\theta} \\ &= -\mathcal{F}_{bm}\end{aligned} \tag{21}$$

where we assume that no external force acts at the end-link and that the nonlinear term $\dot{\mathbf{M}}_b \mathcal{V}_b$ is ignorable. From this relation, it becomes apparent that by designing a suitable imposed force $\mathcal{F}_{bm}$, additional damping can be injected into the system, e.g.:

$$\mathcal{F}_{bm}^{ref} = \mathbf{G}_b \mathcal{V}_b, \tag{22}$$

$\mathbf{G}_b$ denoting the additional spatial damping gain. In the case of a redundant manipulator, the control input is the joint acceleration:

$$\ddot{\theta} = \mathbf{M}_{bm}^{+}(\mathbf{G}_b \mathcal{V}_b - \dot{\mathbf{M}}_{bm} \dot{\theta}) + \mathbf{P}_{bm} \ddot{\theta}_a. \tag{23}$$

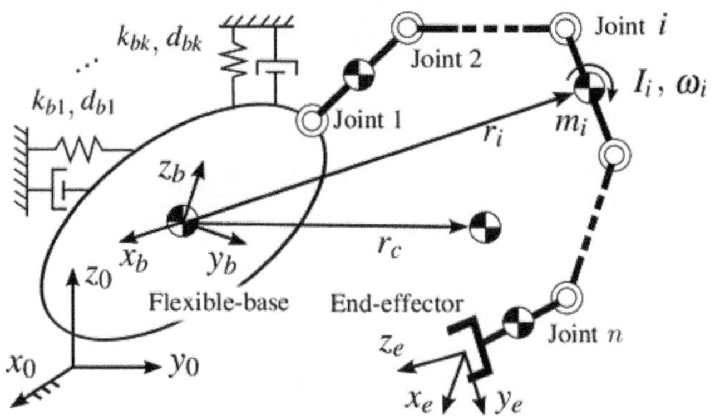

Figure 5. Model of a single-body flexible base manipulator system.

This equation has the same structure as Eq. (18). Hence, a dual-task control scenario can be achieved wherein the two subtasks will be completely dynamically decoupled: the RNS component $\mathbf{P}_{\mathrm{bm}}\dot{\boldsymbol{\theta}}_{\mathrm{a}}$ ensures reactionless motion, while the pseudoinverse component is useful for inertial damping control of any existing vibrations, whereby the coupling energy will be minimized. In addition, the equation is suitable for both velocity control and torque control. In the former case, velocities are obtain via integration; in the latter case, the joint torques are obtained in a similar way as Eq. (19), i.e. by substitution of the joint acceleration into the lower part of the equation of motion.

Further on, it is easy to confirm that the closed-loop dynamics are:

$$\mathbf{M}_{\mathrm{b}}\dot{\mathcal{V}}_{\mathrm{b}} + (\mathbf{D}_{\mathrm{b}} + \mathbf{G}_{\mathrm{b}})\mathcal{V}_{\mathrm{b}} + \mathbf{K}_{\mathrm{b}}\Delta\mathcal{X}_{\mathrm{b}} = \mathbf{0}, \tag{24}$$

i.e. they appear in the form of unforced dynamics of a spatial mass-damper-spring system. Then, proper damping can be achieved with a suitably chosen damping gain $\mathbf{G}_{\mathrm{b}}$ (Hara et al., 2010).

4.2 Composite rigid-body flexible base (macro/mini system)

The macro part of a macro/mini manipulator system can be thought of as an composite rigid-body flexible base, under the assumption that the joints and/or links have inherent flexibilities. Having in mind the JEMRMS/SFA system, we will further assume that both macro and mini system parts comprise a serial-link structure. The generalized coordinates will be denoted as $\boldsymbol{q} = \begin{bmatrix}\boldsymbol{q}_{\mathrm{M}}^T & \boldsymbol{q}_{\mathrm{m}}^T\end{bmatrix}^T$, $\boldsymbol{q}_{\mathrm{M}} \in \Re^k$ and $\boldsymbol{q}_{\mathrm{m}} \in \Re^n$ standing for the joint variables of the macro and mini part, respectively. The structure of the equation of motion resembles that of a single-body base, whereby subscript $(\circ)_{\mathrm{M}}$ replaces the base subscript $(\circ)_{\mathrm{b}}$ to denote quantities associated with the macro part:

$$\begin{bmatrix}\mathbf{M}_{\mathrm{M}} & \mathbf{M}_{\mathrm{Mm}} \\ \mathbf{M}_{\mathrm{Mm}}^T & \mathbf{M}_{\mathrm{m}}\end{bmatrix}\begin{bmatrix}\ddot{\boldsymbol{q}}_{\mathrm{M}} \\ \ddot{\boldsymbol{q}}_{\mathrm{m}}\end{bmatrix} + \begin{bmatrix}\boldsymbol{c}_{\mathrm{M}} \\ \boldsymbol{c}_{\mathrm{m}}\end{bmatrix} = \begin{bmatrix}-\mathbf{D}_{\mathrm{M}}\dot{\boldsymbol{q}}_{\mathrm{M}} - \mathbf{K}_{\mathrm{M}}\Delta\boldsymbol{q}_{\mathrm{M}} \\ \boldsymbol{\tau}\end{bmatrix}. \tag{25}$$

Note that, usually, it is assumed that the macro part joints are passive (Morimoto et al., 2001). Hence, their joint torque does not appear in the equation. Joint damping and stiffness, however are present; they are expressed via diagonal matrices $\mathbf{D}_{\mathrm{M}}, \mathbf{K}_{\mathrm{M}} \in \Re^{k\times k}$, respectively. Other quantities associated with the macro part include joint space inertia $\mathbf{M}_{\mathrm{M}} \in \Re^{k\times k}$ and nonlinear force $\boldsymbol{c}_{\mathrm{M}} \in \Re^k$.

From the upper part of the equation, the macro/mini dynamics are expressed as:

$$\mathbf{M}_{\mathrm{M}}\ddot{\boldsymbol{q}}_{\mathrm{M}} + \mathbf{D}_{\mathrm{M}}\dot{\boldsymbol{q}}_{\mathrm{M}} + \mathbf{K}_{\mathrm{M}}\Delta\boldsymbol{q}_{\mathrm{M}} = \boldsymbol{\tau}_{\mathrm{M}}, \tag{26}$$

where $\boldsymbol{\tau}_{\mathrm{M}} \equiv -\mathbf{M}_{\mathrm{Mm}}\ddot{\boldsymbol{q}}_{\mathrm{m}} - \boldsymbol{c}_{\mathrm{M}}$ plays the role of a torque imposed upon the joints of the macro part via the motion of the mini part. We focus again on inertia coupling, represented via matrix $\mathbf{M}_{\mathrm{Mm}} \in \Re^{k\times n}$. If we assume that the mini part has more joints than the macro[3], then the kernel of this matrix is nontrivial. Reactionless motion, and hence, complete dynamical macro/mini decoupling, can then be achieved via joint accelerations derived from the kernel – the Reaction Null Space of the macro/mini system. Using the RNS joint space decomposition property and following the derivations introduced with the previous example, we can devise a control law for the mini part, in resemblance to Eq. (23), to ensure a dual-task control scenario involving reactionless motion in combination with inertial damping of the macro vibration:

$$\ddot{\boldsymbol{q}}_{\mathrm{m}} = -\mathbf{M}_{\mathrm{Mm}}^{+}(\mathbf{G}_{\mathrm{M}}\dot{\boldsymbol{q}}_{\mathrm{M}} + \dot{\mathbf{M}}_{\mathrm{Mm}}\dot{\boldsymbol{q}}_{\mathrm{m}}) + \mathbf{P}_{\mathrm{Mm}}\ddot{\boldsymbol{q}}_{\mathrm{ma}}, \tag{27}$$

where $\mathbf{G}_{\mathrm{M}} \in \Re^{k\times k}$ is a diagonal gain matrix for additional damping injection. The second term on the r.h.s. represents the reactionless joint acceleration component, $\mathbf{P}_{\mathrm{Mm}}$ denoting the RNS projector and $\ddot{\boldsymbol{q}}_{\mathrm{ma}}$ standing for an arbitrary vector dimensioned as mini-part joint acceleration.

4.3 Humanoid upper body on flexible base

This example is interesting because the system has a tree-like structure comprising two arms of seven DOFs each, attached to the upper end of a torso of three DOF. The other end of the torso is attached to the flexible base that is represented as a single body with two elastic DOFs (pitch and roll), see Fig. 7. There is an abundant 15 degree of redundancy w.r.t. reactionless motion. From a practical viewpoint, additional constraints are to be imposed. A simple example, as discussed in (Wimbock et al., 2009), is specific motion task assignment for one of the arms, let's say the right arm, while using the other arm, or the other arm and the torso, to compensate for the disturbance imposed on the base. For these two scenarios, the degree of redundancy reduces to five or eight, respectively. The equation of motion can be written

[3] In the JEMRMS/SFA model we used (Hara et al., 2010) $k = 3$ and $n = 9$ (in Fig. 4, the respective joint sets are $\{q_1, q_2, q_3\}$ and $\{q_4, q_5, \ldots, q_{12}\}$).

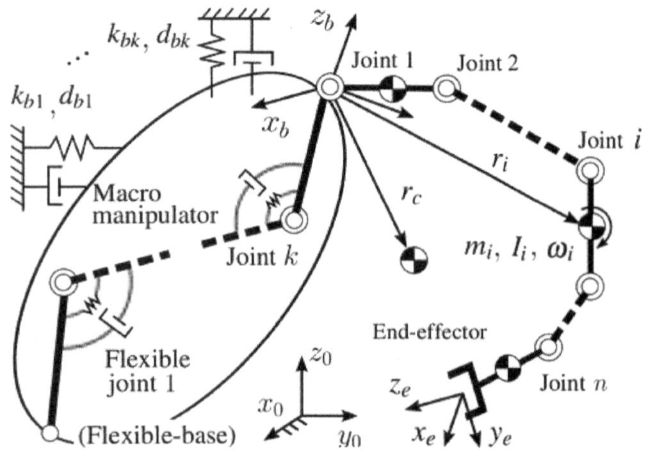

Figure 6. Model of an composite rigid-body flexible base system (a macro/mini manipulator).

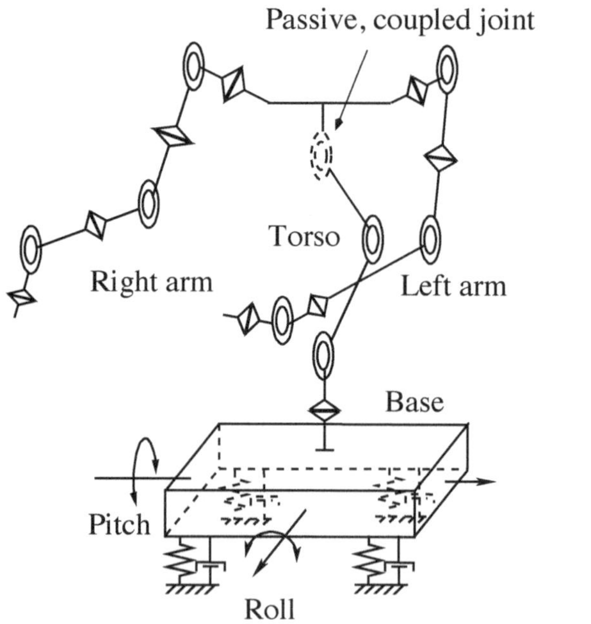

Figure 7. Model of Rollin' Justin – humanoid upper body on flexible base (Wimbock et al., 2009).

as:

$$\begin{bmatrix} \mathbf{M}_b & \mathbf{M}_{br} & \mathbf{M}_{bc} \\ \mathbf{M}_{br}^T & \mathbf{M}_r & \mathbf{M}_{rc} \\ \mathbf{M}_{bc}^T & \mathbf{M}_{rc}^T & \mathbf{M}_c \end{bmatrix} \begin{bmatrix} \dot{\mathcal{V}}_b \\ \ddot{q}_r \\ \ddot{q}_c \end{bmatrix} + \begin{bmatrix} \mathcal{C}_b \\ c_r \\ c_c \end{bmatrix} + \begin{bmatrix} \mathcal{G}_b \\ g_r \\ g_c \end{bmatrix} = \begin{bmatrix} -\mathbf{D}_b\mathcal{V}_b - \mathbf{K}_b\Delta\mathcal{X}_b \\ \tau_r \\ \tau_c \end{bmatrix}, \tag{28}$$

where newly appearing subscripts $(\circ)_r$ and $(\circ)_c$ stand for "right arm" and "compensating subsystem" (i.e. left arm or left arm plus torso), respectively. The "g"-terms denote gravity forces. The CRB dynamics are selected from the upper part:

$$\mathbf{M}_b\dot{\mathcal{V}}_b + \mathbf{D}_b\mathcal{V}_b + \mathbf{K}_b\Delta\mathcal{X}_b = -\mathbf{M}_{br}\ddot{q}_r - \mathbf{M}_{bc}\ddot{q}_c - \mathcal{N}, \tag{29}$$

where $\mathcal{N}$ collects all nonlinear and gravity terms. Under the assumption that the right arm acceleration $\ddot{q}_r$ is known from the task assignment, the control acceleration for the compensating subsystem can be selected as:

$$\ddot{q}_c = \mathbf{M}_{bc}^{+}(\mathbf{G}_c\dot{q}_c - \mathbf{M}_{br}\ddot{q}_r - \mathcal{N}) + \mathbf{P}_{bc}\ddot{q}_{ca}, \tag{30}$$

where $\ddot{q}_{ca}$ is an arbitrary vector dimensioned as compensating subsystem joint acceleration. The form of this equation is the same as in the previous examples, Eqs. (23) and (27). Additional damping can be injected via gain $\mathbf{G}_c$, and also, reactionless motion can be achieved via the RNS term $\mathbf{P}_{bc}\ddot{q}_{ca}$. Experimental data can be found in Wimbock et al. (2009).

5 Application as an Operational Space Method

So far, we have confirmed that via the RNS decomposition, complete dynamical decoupling between the unfixed base and the rest of the links can be achieved. An interesting question is whether the same approach can be applied to the manipulator's end-effector e.g. of a serial-link chain, instead to the base. The answer is trivial; the implication is an alternative to the Operational Space formulation (OSF) (Khatib, 1987). This sections gives the details of the derivation and a qualitative comparison between the two formulations.

5.1 Brief overview of the Operational Space formulation

The importance of the OSF, e.g. in motion/force control scenarios for fixed-base manipulators, is quite well known. More recently, the formulation is being also applied to more sophisticated MBS like humanoid robots (Sentis and Khatib, 2010). A brief overview is included here for completeness. The underlying equation is:

$$\mathbf{M}_e(\boldsymbol{\theta})\dot{\mathcal{V}} + \mathcal{C}_e(\boldsymbol{\theta}, \dot{\boldsymbol{\theta}}) + \mathcal{G}_e(\boldsymbol{\theta}) = \mathcal{F}, \tag{31}$$

where $\mathcal{V}$ denotes end-link spatial velocity,

$$\mathbf{M}_e(\boldsymbol{\theta}) = \left((\mathbf{J}(\boldsymbol{\theta})\mathbf{M}_m^{-1}(\boldsymbol{\theta})\mathbf{J}^T(\boldsymbol{\theta})\right)^{-1} \in \Re^{6\times 6}, \tag{32}$$

is the *operational space inertia* (Khatib, 1987), $\mathbf{M}_m(\boldsymbol{\theta})$ and $\mathbf{J}(\boldsymbol{\theta})$ standing for the manipulator inertia matrix and the manipulator Jacobian, respectively. $\mathcal{C}_e$, $\mathcal{G}_e$ and $\mathcal{F}$ denote the Coriolis and centrifugal force, the gravity force and the force imposed on the end-link, respectively. These forces are expressed in end-link coordinates and are obtained from the manipulator joint-space dynamics via the transpose of the inertia-weighted generalized inverse[4] of the Jacobian:

$$\mathbf{J}_M^{\#}(\boldsymbol{\theta}) = \mathbf{M}_m^{-1}(\boldsymbol{\theta})\mathbf{J}^T(\boldsymbol{\theta})\mathbf{M}_e(\boldsymbol{\theta})$$

[4]Referred to also as the "dynamically consistent" inverse (Khatib, 1995).

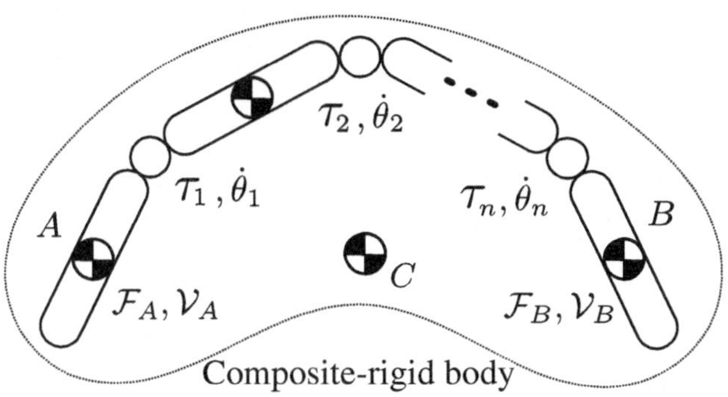

Figure 8. Model of a unfixed-base serial-link chain. The system constitutes a single composite rigid body, C denoting its CoM.

as the underlying transform. It should be apparent that the end-link dynamics formulation (31) will fail when the manipulator Jacobian becomes rank-deficient, i.e. at a *kinematic singularity*.

5.2 RNS-based end-link dynamics formulation

We will use the free-floating dynamics notation from Sect. 3.2. To avoid confusion, we rename the two end-links of the free-floating chain as A and B (see Fig. 8). Without loss of generality, we will pick up end-link A as the link of reference. The equation of motion is then:

$$\begin{bmatrix} \mathbf{M}_{\mathrm{A}} & \mathbf{M}_{\mathrm{Am}} \\ \mathbf{M}_{\mathrm{Am}}^T & \mathbf{M}_{\mathrm{m}} \end{bmatrix} \begin{bmatrix} \dot{\mathcal{V}}_{\mathrm{A}} \\ \ddot{\boldsymbol{\theta}} \end{bmatrix} + \begin{bmatrix} \mathcal{C}_{\mathrm{A}} \\ \boldsymbol{c}_{\mathrm{m}} \end{bmatrix} + \begin{bmatrix} \mathcal{G}_{\mathrm{A}} \\ \boldsymbol{g}_{\mathrm{m}} \end{bmatrix} = \begin{bmatrix} \mathcal{F}_{\mathrm{A}} \\ \boldsymbol{\tau} \end{bmatrix} + \begin{bmatrix} {}^{\mathrm{A}}\mathbf{T}_{\mathrm{B}}^T \\ \mathbf{J}^T \end{bmatrix} \mathcal{F}_{\mathrm{b}}, \quad (33)$$

gravity terms inclusively. The upper part is the CRB dynamics; the coordinates are those of end-link A but the inertial properties are those of the entire system. The lower part, on the other hand, contains generalized force components of a fixed-base manipulator, link A being the "fixed base". The manipulator is composed of all bodies except link A; because end-link A coordinates are used, quantities $\mathbf{M}_{\mathrm{m}}$, $\boldsymbol{c}_{\mathrm{m}}$ and $\mathbf{J}$ are those of the fixed-base manipulator, link B being its end-link.

Henceforth we switch the roles of the two end-links: link A becomes the "real" end-link, while link B is the (unfixed) base. The latter will be later on constrained to obtain a fixed-base system. In this way, results comparable to the fixed-base OSF dynamics can be obtained. With this preparation, it is apparent that the CRB dynamics

$$\mathbf{M}_{\mathrm{A}}\dot{\mathcal{V}}_{\mathrm{A}} + \mathcal{C}_{\mathrm{A}} + \mathcal{G}_{A} = \mathcal{F}_{\mathrm{A}} + {}^{\mathrm{A}}\mathbf{T}_{\mathrm{B}}^T\mathcal{F}_{\mathrm{B}} - \mathbf{M}_{\mathrm{Am}}\ddot{\boldsymbol{\theta}} \quad (34)$$

represent system dynamics in terms of end-link coordinates, i.e. similar to Eq. (31) in OSF. Several remarks are due. First, note that in the above equation, end-link acceleration $\dot{\mathcal{V}}_{\mathrm{A}}$ and force $\mathcal{F}_{\mathrm{A}}$ are explicitly present. Therefore, there was no need to invoke a transformation from joint-space dynamics, i.e. via the inertia-weighted generalized inverse, as it was the case with the OSF. Second, differently from the OSF end-link dynamics (Eq. 31), the above equation can be applied even at kinematic singularities. Third, the inverse dynamics problem for the joint accelerations can be solved directly from the above dynamical relation in end-link coordinates. This is due to the term $-\mathbf{M}_{\mathrm{Am}}\ddot{\boldsymbol{\theta}}$ appearing on the r.h.s. of the equation. This term stands for inertial coupling between the end-link and the rest of the links; it makes joint motion explicitly visible as an end-link dynamical force. The respective joint torque can then be derived from the lower part of Eq. (33).

These properties of the RNS end-link dynamics can be considered as advantageous, e.g. when compared to the OSF end-link dynamics. There are also other differences: under the RNS formulation, end-link force $\mathcal{F}_{\mathrm{A}}$ is a "real" external force; it is not merely a mapping of the joint torque vector as in the OSF. Similarly, nonlinear force $\mathcal{C}_{\mathrm{A}}$ and gravity force $\mathcal{G}_A$ are "true" CRB forces, and not merely mappings of the respective joint-space terms. Also, the presence of force $\mathcal{F}_{\mathrm{B}}$ may be of some advantage, e.g. in situations when reaction force control will be needed (such as in the case of a humanoid robot).

5.3 Inverse dynamics and controller design

The inverse dynamics problem plays an important role in model-based control design, i.e. in computed-torque control methods. The OSF provides the possibility to design controllers that ensure complete dynamical decoupling of the end-effector. A motion/force controller, proposed in (Khatib, 1987), calculates the end-link reference force as:

$$\begin{aligned} \mathcal{F}^{\mathrm{ref}} &= \mathcal{F}_{\mathrm{m}}^{\mathrm{ref}} + \mathcal{F}_{\kappa}^{\mathrm{ref}}, \\ \mathcal{F}_{\mathrm{m}}^{\mathrm{ref}} &= \mathbf{M}_{\mathrm{e}}(\boldsymbol{\theta})\mathbf{S}\dot{\mathcal{V}}^{\mathrm{ref}} + \mathcal{C}_{\mathrm{e}}(\boldsymbol{\theta},\dot{\boldsymbol{\theta}}) + \mathcal{G}_{\mathrm{e}}(\boldsymbol{\theta}) + \mathbf{M}_{\mathrm{e}}(\boldsymbol{\theta})\dot{\mathbf{S}}\mathcal{V}, \\ \mathcal{F}_{\kappa}^{\mathrm{ref}} &= \mathbf{S}^{\perp}\mathcal{F}_{\mathrm{c}}^{\mathrm{ref}}, \end{aligned} \quad (35)$$

where $\mathcal{F}_{\mathrm{m}}^{\mathrm{ref}}$ and $\mathcal{F}_{\kappa}^{\mathrm{ref}}$ are two components referring to end-link motion and contact force, respectively. $\mathbf{S}$ is a selection matrix suitably defined to specify the unconstrained (motion) directions, while $\mathbf{S}^{\perp}$ is its complement, specifying the constrained directions from the reference contact force $\mathcal{F}_{\mathrm{c}}$. In the case of a redundant manipulator, the respective computed-torque controller can be written as:

$$\boldsymbol{\tau} = \mathbf{J}^T\mathcal{F}^{\mathrm{ref}} + \left(\mathbf{U} - \mathbf{J}^T\mathbf{J}_{\mathrm{M}}^{\#T}\right)\boldsymbol{\tau}_{\mathrm{a}}^{\mathrm{ref}}, \quad (36)$$

where $\boldsymbol{\tau}_{\mathrm{a}}^{\mathrm{ref}}$ denotes an arbitrary vector dimensioned as joint torque. Torque component $\mathbf{J}^T\mathcal{F}^{\mathrm{ref}}$ is the nominal joint torque obtained from the static force/torque relation. The other component is a null-space torque that does not affect the imposed end-link force since it is constructed from a generalized inverse of the transposed Jacobian. As pointed out in Khatib (1995); Featherstone and Khatib (1997), there is an infinite number of such inverses, however, only the inertia-weighted generalized inverse of the Jacobian yields a *dynamically consistent force/torque relation* – i.e. complete dynamical decoupling of the two components.

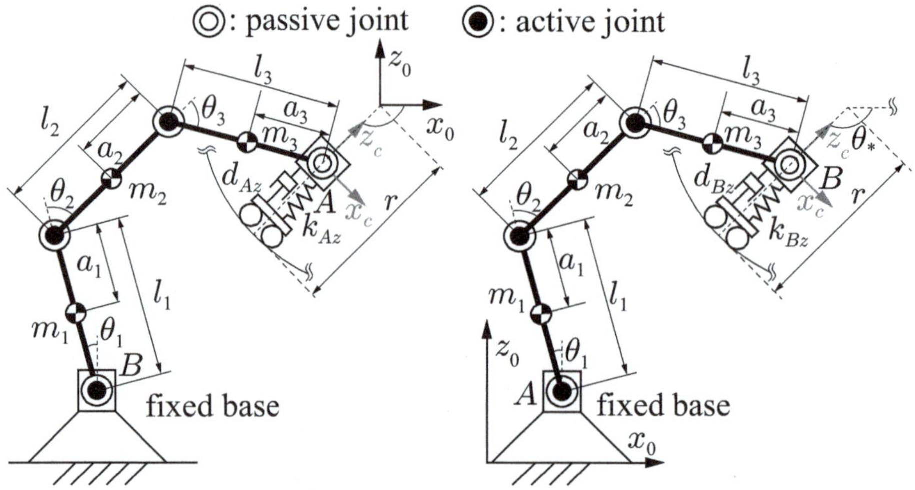

Figure 9. Equivalent planar five-link four-joint manipulator models used in: **(a)** RNS and **(b)** Operational Space formulations. In **(a)**, the "root" link is end-link *A* (the end-effector); In **(b)**, the "root" link is end-link *A* (the base). The passive joint at the end-link appears since the end-link orientation is ignored.

Below we will show how similar dynamically-consistent relations can be derived under the RNS formulation. The joint acceleration can be obtained from Eq. (34) as:

$$\ddot{\boldsymbol{\theta}} = \mathbf{M}^{\#}_{\mathrm{Am}}\left(\mathcal{F}_{\mathrm{A}} - \mathbf{M}_{\mathrm{A}}\dot{\mathcal{V}}_{\mathrm{A}} + {}^{\mathrm{A}}\mathbf{T}^{T}_{\mathrm{B}}\mathcal{F}_{\mathrm{b}} - \mathcal{C}_{\mathrm{A}} - \mathcal{G}_{A}\right) + \mathbf{P}_{\mathrm{Am}}\ddot{\boldsymbol{\theta}}_{\mathrm{a}}, \quad (37)$$

where $(\circ)^{\#}$ denotes a generalized inverse of the coupling inertia matrix. The second term on the r.h.s. is a vector from the kernel of this matrix, $\ddot{\boldsymbol{\theta}}_{\mathrm{a}}$ denoting an arbitrary vector with dimension of joint acceleration. Hence, there is an infinite set of joint accelerations $\{\ddot{\boldsymbol{\theta}}_{\mathrm{rl}} : \ddot{\boldsymbol{\theta}}_{\mathrm{rl}} = \mathbf{P}_{\mathrm{Am}}\ddot{\boldsymbol{\theta}}_{\mathrm{a}}, \forall\ddot{\boldsymbol{\theta}}_{\mathrm{a}}\}$ that would not disturb the state of the end-link. Any acceleration from the above set can therefore be characterized as *reactionless joint acceleration* w.r.t. the state of the end link. This implies complete dynamical decoupling of the end link from the rest of the links. Henceforth, we refer to the kernel as the *Reaction Null Space w.r.t. the end-link*.

It follows that there is also an infinite set of joint torque vectors $\{\boldsymbol{\tau}_{\mathrm{rl}} = \mathbf{M}_{\mathrm{m}}\ddot{\boldsymbol{\theta}}_{\mathrm{rl}}, \forall\ddot{\boldsymbol{\theta}}_{\mathrm{a}}\}$ that do not affect the imposed end-link force and hence, maintain the state of the end-link. The joint torque is obtained by inserting joint acceleration (37) into the lower part of (33):

$$\boldsymbol{\tau} = \boldsymbol{\tau}_{\mathrm{n}} + \boldsymbol{\tau}_{\mathrm{rl}}, \quad (38)$$

where

$$\begin{aligned}\boldsymbol{\tau}_{\mathrm{n}} = {} & \left(\mathbf{M}^{T}_{\mathrm{Am}} - \mathbf{M}_{\mathrm{m}}\mathbf{M}^{\#}_{\mathrm{Am}}\mathbf{M}_{A}\right)\dot{\mathcal{V}}_{\mathrm{A}} \\ & + \mathbf{M}_{\mathrm{m}}\mathbf{M}^{\#}_{\mathrm{Am}}\mathcal{F}_{\mathrm{A}} + \left(\mathbf{M}_{\mathrm{m}}\mathbf{M}^{\#}_{\mathrm{Am}}\,{}^{\mathrm{A}}\mathbf{T}^{T}_{\mathrm{B}} - \mathbf{J}^{T}\right)\mathcal{F}_{\mathrm{b}} \\ & + \boldsymbol{c}_{\mathrm{m}} + \boldsymbol{g}_{\mathrm{m}} - \mathbf{M}_{\mathrm{m}}\mathbf{M}^{\#}_{\mathrm{Am}}(\mathcal{C}_{\mathrm{A}} + \mathcal{G}_{A})\end{aligned} \quad (39)$$

is the nominal component of the solution. It includes end-link *A*'s acceleration and force that can be used as reference inputs in a motion/force controller similar to Eq. (36).

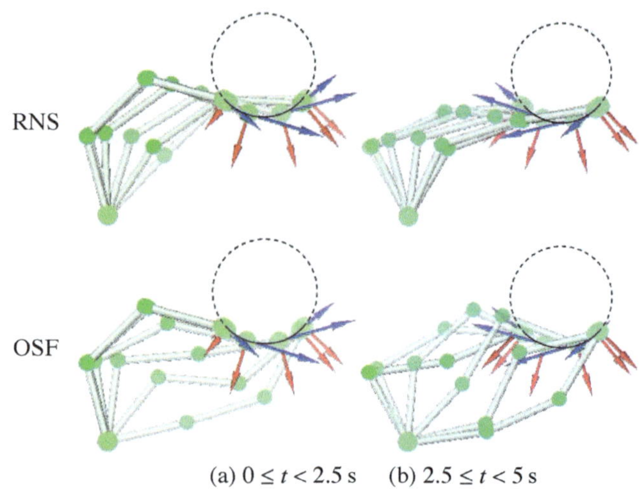

Figure 10. Snapshots from the simulation with two equivalent planar 3R redundant manipulators tracking a semi-circular path and applying a desired force. The blue/red arrows denote the desired motion/force vectors, respectively. No null-space motion is involved. In the RNS simulation (upper graphs) no significant arm reconfiguration is observed; in the OSF simulation (lower graphs), on the other hand, significant arm reconfiguration (spurious link motion) is observed (Hara et al., 2012).

Since the other joint-toque component $\boldsymbol{\tau}_{\mathrm{rl}}$ is reactionless w.r.t. the end-effector, it should be apparent that we obtained a dynamically-consistent relationship in the sense of Khatib.

Note, however, that in our formulation the nominal component $\boldsymbol{\tau}_{\mathrm{n}}$ represents a dynamical torque; it does *not* stem from a static relation, as in Eq. (36). The consequence is that even when the manipulator/limb performs self motion, e.g. due to a nonzero $\boldsymbol{\tau}_{\mathrm{rl}}$, the complete dynamical decoupling

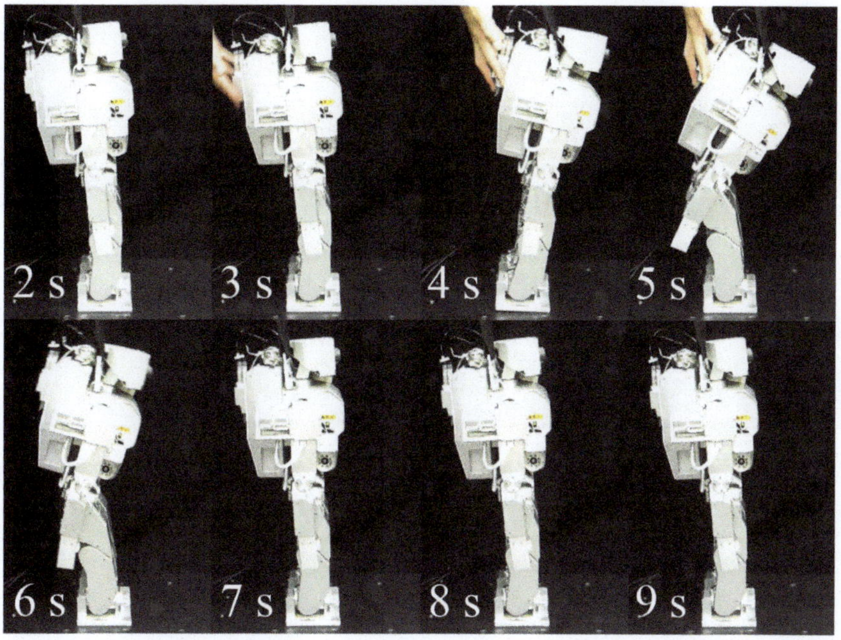

Figure 11. Snapshots from the experiment with a HOAP-2 robot responding to an unknown continuous-force disturbance on the back (sagittal plane). The so-called Hip strategy has been realized under velocity-based reactionless motion with a three-link two-joint model in the sagittal plane (Nenchev and Nishio, 2008; Kanamiya et al., 2010).

property will not be lost. Actually, $\boldsymbol{\tau}_{\rm n}$ constitutes an infinite number of dynamically-consistent relationships because there is an infinite set of generalized inverses for $\mathbf{M}_{\rm Am}$, namely $\{\mathbf{M}^{\#}_{\rm Am} : \mathbf{M}_{\rm Am}\mathbf{M}^{\#}_{\rm Am}\mathbf{M}_{\rm Am}\}$. Each specific generalized inverse will provide a specific dynamically-consistent scheme. Such schemes would be usually constructed to support task-dependent redundancy resolution, quite in a similar fashion as known from past studies on kinematically redundant manipulators (Nenchev, 1989).

As an example, let us pick up the Moore-Penrose generalized inverse (pseudoinverse). Local optimality can then be achieved, the minimized quantity being that part of total kinetic energy that is due to the dynamical coupling between end-link A and the rest of the links: $\mathcal{V}_{\rm A}^T\mathbf{M}_{\rm Am}\dot{\boldsymbol{\theta}}$. Note that under the Operational Space formulation, the minimized quantity is the total kinetic energy (Khatib, 1987): $\frac{1}{2}\mathcal{V}^T\mathbf{M}_{\rm e}\mathcal{V}$. As apparent from Eq. (32), this is a highly nonlinear function due to the inverse of a quadratic form of the Jacobian matrix. This means that in the (not necessarily small) vicinity of kinematic singularities, excessive fluctuation in the joint velocity can be expected. In contrast, the coupling kinetic energy, minimized under the RNS formulation, is quite well behaved, even within a relatively small vicinity of ill-defined inertial coupling, i.e. where the coupling inertia matrix becomes rank deficient. We can summarize then: with the RNS formulation we can expect better performance in terms of joint motion than with the OSF, and equal performance in terms of end-link motion/force control. This has been examined experimentally, with two simple models (see Fig. 9) performing the same motion/force control task, realized with the respective nominal component (no null-space motion) (Hara et al., 2012). The significant difference in terms of joint-space motion can be confirmed from the animation snapshots in Fig. 10. With the RNS motion/force control formulation, link motion does not deviate much from the initial configuration. This indicates the lack of large peaks in joint velocity. With the OSF, on the other hand, spurious link motion is observable, which is due to the highly nonlinear dynamic transform (the inertia-weighted generalized inverse) used in the formulation and the resulting peak joint velocities in the vicinity of kinematic singularities.

6 Application to humanoid robots

A humanoid is an underactuated system and its balance control can be achieved only via the imposed/reaction forces, when the feet are in contact with the ground. The prevailing research approach is to make use of the Zero Moment Point[5] (Vukobratović and Borovac, 2004) that can provide information about roll/pitch momenta on the feet. These momenta are sufficient for balance control on flat ground, when frictional forces are ignored. A more interesting situation, however, is balancing/walking on uneven ground, and also, when considering the presence of friction and other unknown disturbances. To deal with such problems, full spatial force control

[5]In the static case, the ZMP is the projection of the total CoM on ground. When the robot is in motion and the CoM accelerates, the ZMP accelerates in accordance with the respective inertia force.

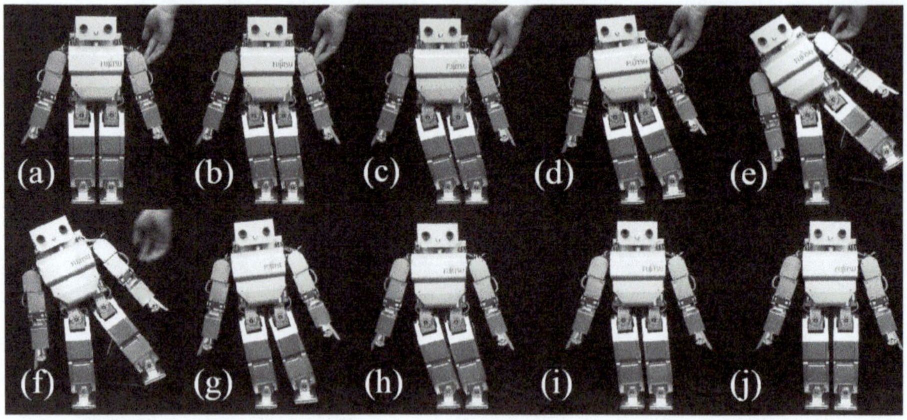

Figure 12. Snapshots from the experiment with a HOAP-2 robot responding to an unknown continuous-force disturbance on the shoulder (frontal plane). (**a–c, i** and **j**) ankle strategy; (**d–g**) lift-leg strategy; (**h**) transition between the two strategies. Two different models in the frontal plane are used; Ankle strategy: three-link two-joint model; Lift-leg strategy: four-link three-joint model (Yoshida et al., 2011).

at the feet via the imposed/reaction force relation is necessary. This can be achieved with the RNS formulation in a straightforward manner, as should be apparent from the applications discussed so far.

6.1 Planar humanoid models and balance strategies

The RNS method has been applied to the balance problem in terms of both joint velocity, i.e. using momentum balance (Nenchev and Nishio, 2008; Kanamiya et al., 2010; Yoshida et al., 2011), and joint torque (Tamegaya et al., 2008). In the former case, balance strategies in response to an unknown external disturbance (continuous or impact type) have been devised, based on the analysis of human behavior under similar circumstances. For example, when the disturbance is applied on the back while standing upright, the so-called hip strategy may be invoked (Shumway-Cook and Horak, 1989), i.e. bending in the hips and motion in the ankles in the opposite direction. This strategy can be readily realized with a simple three-link (foot, leg, upper body), two-joint (ankle, hip) planar model in the sagittal plane. The unfixed-base motion dynamics are then represented as:

$$\begin{bmatrix} \mathbf{M}_{\mathrm{f}} & \mathbf{M}_{\mathrm{fm}} \\ \mathbf{M}_{\mathrm{fm}}^T & \mathbf{M}_{\mathrm{m}} \end{bmatrix} \begin{bmatrix} \dot{\mathcal{V}}_{\mathrm{f}} \\ \ddot{\boldsymbol{\theta}} \end{bmatrix} + \begin{bmatrix} \mathcal{C}_{\mathrm{f}} \\ \boldsymbol{c}_{\mathrm{m}} \end{bmatrix} + \begin{bmatrix} \mathcal{G}_{\mathrm{f}} \\ \boldsymbol{g}_{\mathrm{m}} \end{bmatrix} = \begin{bmatrix} \mathcal{F}_{\mathrm{f}} \\ \boldsymbol{\tau} \end{bmatrix} + \begin{bmatrix} {}^{\mathrm{f}}\mathbf{T}_{\mathrm{e}}^T \\ \mathbf{J}_{\mathrm{e}}^T \end{bmatrix} \mathcal{F}_{\mathrm{e}}, \quad (40)$$

where subscripts "f" and "e" stand for "foot" and "external", respectively. With this notation, the foot is considered the unfixed base, $\mathcal{F}_{\mathrm{f}}$ denoting the force at the foot resulting from the specific ground contact conditions and including three components: vertical ground reaction force, horizontal frictional force and foot moment, the latter being the most important for balance.

First, we consider the following simplifying assumptions: the foot is always in contact with the ground (the acceleration of the CoM vertically upward is restricted) and also,

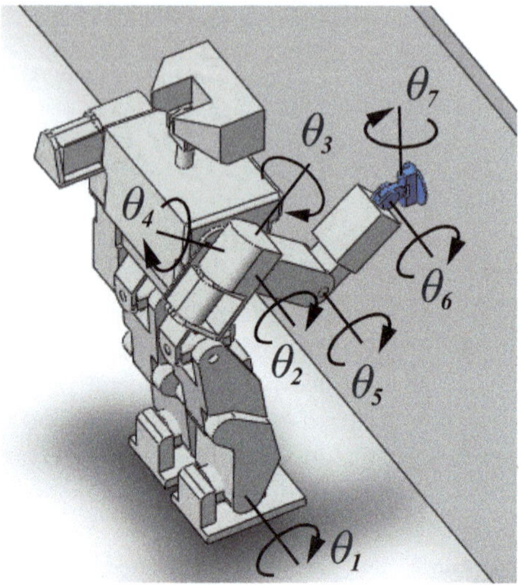

Figure 13. A small humanoid robot cleaning a vertical surface: spatial seven-DOF model, with active (θ_1 through θ_5) and passive (θ_6 and θ_7) joint coordinates.

the horizontal frictional force is sufficiently large. The conditions are then similar to those when using ZMP-based control. We can then ignore the two force components at the foot and rewrite the dynamics to include just the foot moment. Then, the coupling inertia matrix $\mathbf{M}_{\mathrm{fm}} \in \Re^{1\times 2}$ will comprise a nontrivial kernel. The related null-space projector will be denoted as $\mathbf{P}_{\mathrm{fm}}$. Further on, if we assume as initial conditions static equilibrium and null foot moment, i.e. $\mathcal{V}_{\mathrm{f}}, \dot{\mathcal{V}}_{\mathrm{f}}$ and $\mathcal{F}_{\mathrm{f}}$ all zero, then this state (and thus balance) can be maintained with reactionless motion. In terms of joint accelerations, reactionless motions are derived from the CRB dynamics (i.e. the upper part of the equation of motion):

$$\ddot{\boldsymbol{\theta}} = -\mathbf{M}_{\mathrm{fm}}^{+}\dot{\mathbf{M}}_{\mathrm{fm}}\dot{\boldsymbol{\theta}} + \mathbf{P}_{\mathrm{fm}}\ddot{\boldsymbol{\theta}}_{\mathrm{a}}. \quad (41)$$

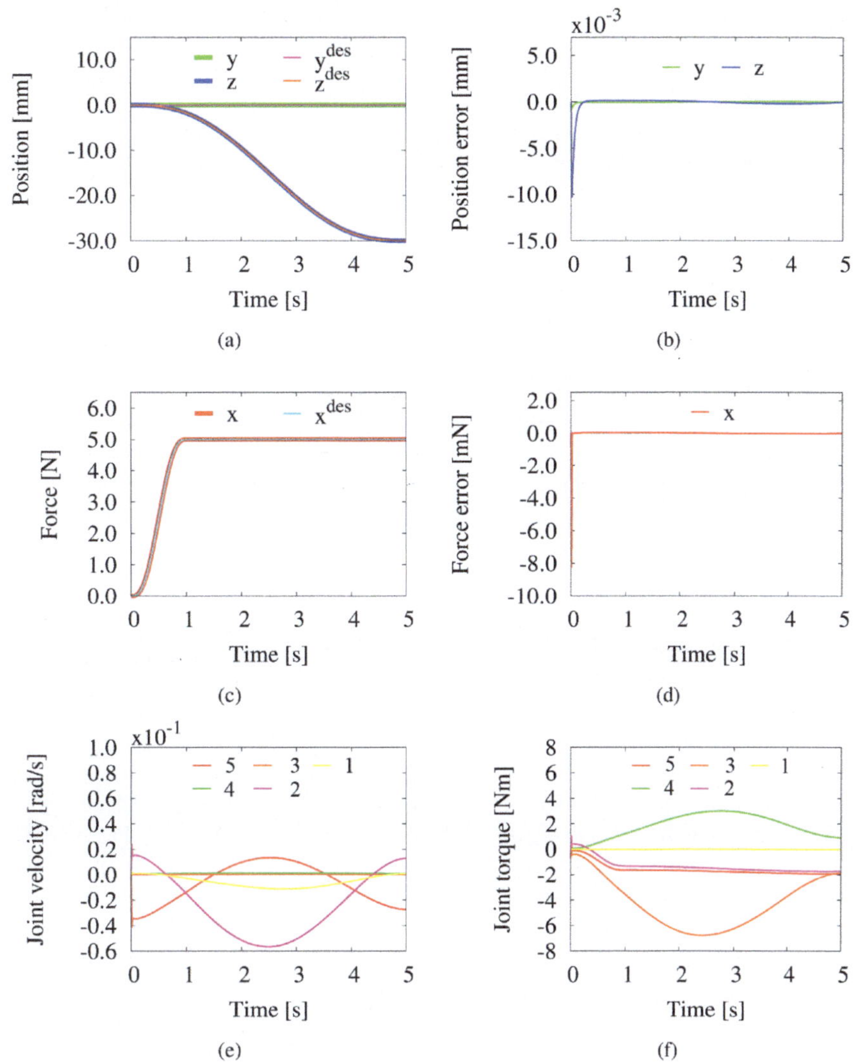

Figure 14. Simulation results under RNS-based dynamical motion/force feedback control: the desired motion trajectory in hand coordinates is a 30 mm straight line downwards (along -z); the hand force (x component) is regulated thereby to 5 Nm.

This is the reactionless joint acceleration set that was used to generate a compliant response to the disturbance by bending in the hips, followed up by standing upright configuration recovery, after the disturbance disappeared (see Fig. 11). Further on, since reactionless motion implies coupling momentum conservation, as explained in Sect. 3.1, reactionless motion generation in terms of joint velocity is also possible. This property was used to realize the Hip strategy under velocity (and thus position) control.

The same approach was adopted with regard to disturbances at the shoulder, within the frontal plane (Yoshida et al., 2011). Snapshots from respective experiments are shown in Fig. 12. Initially, again, a three-link model was used to ensure compliant upper-body response, with parallel motion of the legs (considered as a single-link motion – the Ankle strategy: (a)–(c), (i) and (j)). When the disturbance persisted, the robot responded by shifting the CoM further over the right foot and lifting the left leg (Lift-leg strategy: (d)–(g)). Thereby, the model was extended by one more link and a joint.

It should be noted that in practice, dynamical models and hence reactionless motions, cannot be perfect. Therefore, a dual-task control scenario should be envisioned, similar to those mentioned in the previous examples. The full CRB dynamics should then be used to account for foot rotational acceleration due to small errors, in a feedback control loop. We have designed a computed torque feedback controller and confirmed its satisfactory performance w.r.t. to a larger variety of external disturbances. Results will be reported elsewhere.

6.2 Spatial humanoid models and motion/force control

As mentioned in Sect. 5, the RNS formalism suits especially operational-space type motion/force control task scenarios with humanoid robots. As an example, consider a spatial humanoid model with seven DOFs: six for the arm and one for the ankle joint (see Fig. 13a). The robot's task is to clean a vertical surface. Three hand coordinates are involved to complete the task: two tangential coordinates (y and z) for trajectory tracking within the vertical plane, and one normal coordinate (x) for force tracking (Sato et al., 2011). Note that the wrist comprises a passive U-joint; its joint coordinates (θ_6 and θ_7) are available via the loop-closure equation. The remaining five joint coordinates (θ_1 through θ_5) are actively controlled. Thus, the system has two redundant DOFs. The equations from Sect. 5 can be applied in a straightforward manner, whereby end-links A and B denote the robot hand and the foot, respectively. Below we present simulation data, wherein the joint acceleration is computed via Eq. (37). Thereby, the pseudoinverse is used as a generalized inverse, the feet are assumed stationary and fixed (i.e. the six-dimensional spatial force $\mathcal{F}_B$ is determined via the Lagrange multiplier method). No use of the null space acceleration is made ($\ddot{\boldsymbol{\theta}}_a = \mathbf{0}$). Thus, feedback control only is applied to track the desired y-z hand (unfixed-base) motion trajectory (a 30 mm straight line downwards), regulating thereby the desired force to 5 Nm. The graphs are shown in Fig. 14. It is seen that the motion/force control task could be performed in a stable manner, and without excessive joint velocity and torque. The null-space component, though not used currently, is available e.g. for balance control (roll/pitch control) of the feet. This will be confirmed with upcoming experiments.

7 Conclusions

This study summarizes results based on the application of the Reaction Null Space approach. The RNS, defined as the kernel of the coupling inertia – a submatrix of the system inertia matrix – provides a joint space decomposition formalism that can be quite useful for motion analysis, motion generation and motion control of various unfixed-base systems, such as free-floating space robots, flexible-base and macro/mini robot systems, as well as humanoid robots. Via the null space and the pseudoinverse mapping of the coupling inertia, the joint space can be decomposed into two complementary orthogonal subspaces. Thus, two orthogonal joint space components in terms of joint velocity, joint acceleration and joint torque can be derived, with the properties of complete dynamical decoupling and locally optimal dynamical coupling, respectively. The former component induces a reactionless vector field and the respective reactionless-motion manifold in joint space. The latter component, on the other hand, imposes a spatial-force constraint via manipulator motion, that can be used for locally optimized control of the unfixed base, e.g. base orientation, base vibration suppression or foot reaction control of a free-floating space robot, a flexible-base robot and a humanoid robot, respectively.

Acknowledgements. The contribution of Naoyuki Hara and Ryohei Okawa with figures and HOAP-2 model simulation data is acknowledged.

Edited by: A. Müller
Reviewed by: I. Sharf and one anonymous referee

References

Abiko, S. and Yoshida, K.: Adaptive Reaction Control for Space Robotic Applications with Dynamic Model Uncertainty, Adv. Robotics, 24, 1099–1126, doi:10.1163/016918610X501264, 2010.

Book, W. and Lee, S. H.: Vibration Control of a Large Flexible Manipulator by a Small Robotic Arm, in: American Control Conference, 1377–1380, 1989.

Borst, C., Wimbock, T., Schmidt, F., Fuchs, M., Brunner, B., Zacharias, F., Giordano, P. R., Konietschke, R., Sepp, W., Fuchs, S., Rink, C., Albu-Schaffer, A., and Hirzinger, G.: Rollin' Justin – Mobile platform with variable base, in: 2009 IEEE International Conference on Robotics and Automation, 1597–1598, doi:10.1109/ROBOT.2009.5152586, 2009.

Briot, S., Arakelian, V., and Le Baron, J.-P.: Shaking force minimization of high-speed robots via centre of mass acceleration control, Mech. Mach. Theory, 57, 1–12, doi:10.1016/j.mechmachtheory.2012.06.006, 2012.

Burdick, J.: On the inverse kinematics of redundant manipulators: characterization of the self-motion manifolds, in: 1989 IEEE International Conference on Robotics and Automation, 264–270, doi:10.1109/ROBOT.1989.99999, 1989.

Cheng, D.: End-point Control of a Flexible Structure Mounted Manipulator Based on Wavelet Basis Function Networks, in: 2005 IEEE/RSJ International Conference on Intelligent Robots and Systems, Workshop on Robot Vision for Space Applications, 29–34, 2005.

Cocuzza, S., Pretto, I., and Debei, S.: Reaction torque control of redundant space robotic systems for orbital maintenance and simulated microgravity tests, Acta Astronaut., 67, 285–295, doi:10.1016/j.actaastro.2009.05.007, 2010.

Coleshill, E., Oshinowo, L., Rembala, R., Bina, B., Rey, D., and Sindelar, S.: Dextre: Improving maintenance operations on the International Space Station, Acta Astronaut., 64, 869–874, doi:10.1016/j.actaastro.2008.11.011, 2009.

Cong, P. C. and Sun, Z. W.: Preimpact Configuration Analysis of Dual-Arm Space Manipulator Capturing Object, J. Aerospace Eng., 23, 117–123, doi:10.1061/(ASCE)AS.1943-5525.0000016, 2010.

Craig, J.: Introduction To Robotics: Mechanics & Control, 3rd Edn., Prentice Hall, 2004.

Dimitrov, D. and Yoshida, K.: Utilization of the bias momentum approach for capturing a tumbling satellite, in: 2004 IEEE/RSJ International Conference on Intelligent Robots and Systems (IROS) (IEEE Cat. No.04CH37566), 4, 3333–3338, doi:10.1109/IROS.2004.1389931, 2004.

Elfizy, A., Bone, G., and Elbestawi, M.: Design and control of a dual-stage feed drive, Int. J. Mach. Tool. Manu., 45, 153–165, doi:10.1016/j.ijmachtools.2004.07.008, 2005.

Fattah, A. and Agrawal, S. K.: Design and simulation of a class of spatial reactionless manipulators, Robotica, 23, 75–81, doi:10.1017/S0263574704000670, 2005.

Featherstone, R.: Rigid Body Dynamics Algorithms, Springer US, Boston, MA, doi:10.1007/978-0-387-74315-8, 2008.

Featherstone, R. and Khatib, O.: Load Independence of the Dynamically Consistent Inverse of the Jacobian Matrix, Int. J. Robot. Res., 16, 168–170, doi:10.1177/027836499701600203, 1997.

Finat, J. and Gonzalez-Sprinberg, G.: A semianalytic approach to the modelling and control of the biped gait for the design of interactive orthoses in rehabilitation tasks, in: 11th IEEE International Workshop on Robot and Human Interactive Communication, 350–355, doi:10.1109/ROMAN.2002.1045647, 2002.

Gorce, P.: Dynamic postural control method for biped in unknown environment, IEEE T. Syst. Man Cyb. A, 29, 616–626, doi:10.1109/3468.798065, 1999.

Gouo, A., Nenchev, D., Yoshida, K., and Uchiyama, M.: Motion control of dual-arm long-reach manipulators, Adv. Robotics, 13, 617–631, doi:10.1163/156855399X01846, 1998.

Hanson, M. L. and Tolson, R. H.: Reducing flexible base vibrations through local redundancy resolution, J. Robotic Syst., 12, 767–779, doi:10.1002/rob.4620121106, 1995.

Hara, N., Kanamiya, Y., and Sato, D.: Stable path tracking with JEMRMS through vibration suppression algorithmic singularities using momentum conservation, in: 10th International Symposium on Artificial Intelligence, Robotics and Automation in Space, 214–221, 2010.

Hara, N., Handa, Y., and Nenchev, D.: End-link dynamics of redundant robotic limbs: The Reaction Null Space approach, in: 2012 IEEE International Conference on Robotics and Automation, 299–304, doi:10.1109/ICRA.2012.6224627, 2012.

Hyon, S.-H., Osu, R., and Otaka, Y.: Integration of multi-level postural balancing on humanoid robots, in: 2009 IEEE International Conference on Robotics and Automation, 1549–1556, doi:10.1109/ROBOT.2009.5152434, 2009.

Kaigom, E. G., Jung, T. J., and Rossmann, J.: Optimal Motion Planning of a Space Robot with Base Disturbance Minimization, in: 11th Symposium on Advanced Space Technologies in Robotics and Automation 2011 (Astra2011), 1–6, http://www.mmi.rwth-aachen.de/?public100, http://robotics.estec.esa.int/ASTRA/Astra2011/Papers/04B/FCXNL-11A06-2139450-1-2139450GuiffoKaigom.pdf, 2011.

Kanamiya, Y., Ota, S., and Sato, D.: Ankle and hip balance control strategies with transitions, in: 2010 IEEE International Conference on Robotics and Automation, 3446–3451, doi:10.1109/ROBOT.2010.5509785, 2010.

Khatib, O.: A unified approach for motion and force control of robot manipulators: The operational space formulation, IEEE Journal on Robotics and Automation, 3, 43–53, doi:10.1109/JRA.1987.1087068, 1987.

Khatib, O.: Inertial Properties in Robotic Manipulation: An Object-Level Framework, Int. J. Robot. Res., 14, 19–36, doi:10.1177/027836499501400103, 1995.

Konno, A., Myojin, T., Matsumoto, T., Tsujita, T., and Uchiyama, M.: An impact dynamics model and sequential optimization to generate impact motions for a humanoid robot, Int. J. Robot. Res., 30, 1596–1608, doi:10.1177/0278364911405870, 2011.

Konstantinov, M. S., Markov, M. D., and Nenchev, D. N.: Kinematic Control of Redundant Manipulators, in: 11th International Symposium on Industrial Robots, 561–568, Tokyo, 1981.

Lampariello, R., Heindl, J., Koeppe, R., and Hirzinger, G.: Reactionless Control for two Manipulators Mounted on a Cable-Suspended Platform, in: 2006 IEEE/RSJ International Conference on Intelligent Robots and Systems, 91–97, doi:10.1109/IROS.2006.281777, 2006.

Lee, S. H. and Book, W.: Robot Vibration Control Using Inertial Damping Forces, in: VIII CISM-IFToMM Symposium on the Theory and Practice of Robots and Manipulators (Ro. Man. Sy. '90), 252–259, Cracow, Poland, http://hdl.handle.net/1853/39161, 1990.

Lew, J., Trudnowski, D., Evans, M., and Bennett, D.: Micro manipulator motion control to suppress macro manipulator structural vibrations, in: 1995 IEEE International Conference on Robotics and Automation, 3, 3116–3120, doi:10.1109/ROBOT.1995.525728, 1995.

Luh, J., Walker, M., and Paul, R.: Resolved-acceleration control of mechanical manipulators, IEEE T. Automat. Contr., 25, 468–474, doi:10.1109/TAC.1980.1102367, 1980.

Masutani, Y., Miyazaki, F., and Arimoto, S.: Modeling and sensory feedback control for space manipulators, in: NASA Conf. Space Telerobotics, 1989.

Morimoto, H., Kasama, Y., Doi, S., and Wakabayashi, Y.: Motion Control of the Small Fine Arm for the Japanese Experiment of the International Space Station, in: Int. Conf. Advanced Robotics, 187–195, 2001.

Murray, R. M., Li, Z., and Sastry, S. S.: A Mathematical Introduction to Robotic Manipulation, CRC Press, 1994.

Nakamura, Y., Hanafusa, H., and Yoshikawa, T.: Task-Priority Based Redundancy Control of Robot Manipulators, Int. J. Robot. Res., 6, 3–15, doi:10.1177/027836498700600201, 1987.

Nenchev, D., Yoshida, K., and Umetani, Y.: Introduction of redundant arms for manipulation in space, in: IEEE International Workshop on Intelligent Robots, 88, 679–684, doi:10.1109/IROS.1988.593682, 1988.

Nenchev, D., Umetani, Y., and Yoshida, K.: Analysis of a redundant free-flying spacecraft/manipulator system, IEEE T. Robotic. Autom., 8, 1–6, doi:10.1109/70.127234, 1992.

Nenchev, D., Yoshida, K., and Uchiyama, M.: Reaction null-space based control of flexible structure mounted manipulator systems, in: Proceedings of 35th IEEE Conference on Decision and Control, 4, 4118–4123, doi:10.1109/CDC.1996.577417, 1996.

Nenchev, D., Yoshida, K., Vichitkulsawat, P., and Uchiyama, M.: Reaction null-space control of flexible structure mounted manipulator systems, IEEE T. Robotic. Autom., 15, 1011–1023, doi:10.1109/70.817666, 1999.

Nenchev, D. N.: Redundancy resolution through local optimization: A review, J. Robotic Syst., 6, 769–798, doi:10.1002/rob.4620060607, 1989.

Nenchev, D. N. and Nishio, A.: Ankle and hip strategies for balance recovery of a biped subjected to an impact, Robotica, 26, 643–653, doi:10.1017/S0263574708004268, 2008.

Nguyen-Huynh, T.-C. and Sharf, I.: Adaptive reactionless motion for space manipulator when capturing an unknown tumbling target, in: 2011 IEEE International Conference on Robotics and Automation, 4202–4207, doi:10.1109/ICRA.2011.5980398, 2011.

Oda, M.: Experiences and lessons learned from the ETS-VII robot satellite, in: 2000 IEEE International Conference on Robotics and Automation. (Cat. No. 00CH37065), 1, 914–919, doi:10.1109/ROBOT.2000.844165, 2000.

Osumi, H. and Saitoh, M.: Control of a Redundant Manipulator Mounted on a Base Plate Suspended by Six Wires, in: 2006 IEEE/RSJ International Conference on Intelligent Robots and Systems, 73–78, doi:10.1109/IROS.2006.281669, 2006.

Ott, C., Albu-Schaffer, A., and Hirzinger, G.: A Cartesian Compliance Controller for a Manipulator Mounted on a Flexible Structure, in: 2006 IEEE/RSJ International Conference on Intelligent Robots and Systems, 4502–4508, doi:10.1109/IROS.2006.282539, 2006.

Papadopoulos, E. and Dubowsky, S.: Coordinated manipulator/spacecraft motion control for space robotic systems, in: Proceedings. 1991 IEEE International Conference on Robotics and Automation, April, 1696–1701, IEEE Comput. Soc. Press, doi:10.1109/ROBOT.1991.131864, 1991.

Parsa, K., Angeles, J., and Misra, A. K.: Control of Macro-Micro Manipulators Revisited, J. Dyn. Syst.-T. ASME, 127, 688, doi:10.1115/1.1870039, 2005.

Piersigilli, P., Sharf, I., and Misra, A.: Reactionless capture of a satellite by a two degree-of-freedom manipulator, Acta Astronaut., 66, 183–192, doi:10.1016/j.actaastro.2009.05.015, 2010.

Sato, F., Nishii, T., Takahashi, J., Yoshida, Y., Mitsuhashi, M., and Nenchev, D.: Experimental evaluation of a trajectory/force tracking controller for a humanoid robot cleaning a vertical surface, in: 2011 IEEE/RSJ International Conference on Intelligent Robots and Systems, 3179–3184, doi:10.1109/IROS.2011.6094615, 2011.

Sato, N. and Wakabayashi, Y.: JEMRMS design features and topics from testing, in: 6th International Symposium on Artificial Intelligence and Robotics & Automation in Space, 1–7, 2001.

Sentis, L. and Khatib, O.: Compliant Control of Multicontact and Center-of-Mass Behaviors in Humanoid Robots, IEEE Transactions on Robotics, 26, 483–501, doi:10.1109/TRO.2010.2043757, 2010.

Sharf, I.: Active Damping of a Large Flexible Manipulator With a Short-Reach Robot, J. Dyn. Syst.-T. ASME, 118, 704, doi:10.1115/1.2802346, 1996.

Shui, H., Wang, J., and Ma, H.: Optimal Motion Planning for Free-Floating Space Robots Based on Null Space Approach, 2009 International Conference on Measuring Technology and Mechatronics Automation, 845–848, doi:10.1109/ICMTMA.2009.556, 2009.

Shumway-Cook, A. and Horak, F. B.: Vestibular Rehabilitation: an exercise approach to managing symptoms of vestibular dysfunction, Seminars in Hearing, 10, 196–209, 1989.

Stephens, B.: Humanoid push recovery, in: Humanoids 2007 – 7th IEEE-RAS International Conference on Humanoid Robots, 589–595, doi:10.1109/ICHR.2007.4813931, 2007.

Tamegaya, K., Kanamiya, Y., Nagao, M., and Sato, D.: Inertia-coupling based balance control of a humanoid robot on unstable ground, in: Humanoids 2008 – 8th IEEE-RAS International Conference on Humanoid Robots, 151–156, doi:10.1109/ICHR.2008.4755960, 2008.

Torres, M. and Dubowsky, S.: Minimizing spacecraft attitude disturbances in space manipulator systems, J. Guid. Control Dynam., 15, 1010–1017, doi:10.2514/3.20936, 1992.

Torres, M., Dubowsky, S., and Pisoni, A.: Vibration control of deployment structures' long-reach space manipulators: The P-PED method, in: Proceedings of IEEE International Conference on Robotics and Automation, 3, 2498–2504, doi:10.1109/ROBOT.1996.506538, 1996.

Vukobratović, M. and Borovac, B.: Zero-Moment Point – Thirty Five Years of Its Life, Int. J. Hum. Robot., 01, 157–173, doi:10.1142/S0219843604000083, 2004.

Wimbock, T., Nenchev, D., Albu-Schaffer, A., and Hirzinger, G.: Experimental study on dynamic reactionless motions with DLR's humanoid robot Justin, in: 2009 IEEE/RSJ International Conference on Intelligent Robots and Systems, 5481–5486, doi:10.1109/IROS.2009.5354528, 2009.

Xu, W., Liu, Y., Liang, B., Xu, Y., and Qiang, W.: Autonomous Path Planning and Experiment Study of Free-floating Space Robot for Target Capturing, Journal of Intelligent and Robotic Systems, 51, 303–331, doi:10.1007/s10846-007-9192-3, 2007.

Yoshida, K., Nenchev, D., Vichitkulsawat, P., Kobayashi, H., and Uchiyama, M.: Experiments on the point-to-point operations of a flexible structure mounted manipulator system, Adv. Robotics, 11, 397–411, doi:10.1163/156855397X00399, 1996.

Yoshida, K., Hashizume, K., Nenchev, D., Inaba, N., and Oda, M.: Control of a space manipulator for autonomous target capture - ETS-VII flight experiments and analysis-, in: AIAA Guidance, Navigation and Control Conference and Exhibit, 1297–1306, 2000.

Yoshida, Y., Takeuchi, K., Sato, D., and Nenchev, D.: Postural balance strategies for humanoid robots in response to disturbances in the frontal plane, in: 2011 IEEE International Conference on Robotics and Biomimetics, 1825–1830, doi:10.1109/ROBIO.2011.6181555, 2011.

Yoshikawa, T., Hosoda, K., Doi, T., and Murakami, H.: Quasi-static trajectory tracking control of flexible manipulator by macro-micro manipulator system, in: 1993 IEEE International Conference on Robotics and Automation, 210–215, IEEE Comput. Soc. Press, doi:10.1109/ROBOT.1993.291848, 1993.

3

Continuum robots and underactuated grasping

N. Giri and I. Walker

Department of Electrical and Computer Engineering, Clemson University, Clemson, SC, USA

Abstract. We discuss the capabilities of continuum (continuous backbone) robot structures in the performance of under-actuated grasping. Continuum robots offer the potential of robust grasps over a wide variety of object classes, due to their ability to adapt their shape to interact with the environment via non-local continuum contact conditions. Furthermore, this capability can be achieved with simple, low degree of freedom hardware. However, there are practical issues which currently limit the application of continuum robots to grasping. We discuss these issues and illustrate via an experimental continuum grasping case study.

This paper was presented at the IFToMM/ASME International Workshop on Underactuated Grasping (UG2010), 19 August 2010, Montréal, Canada.

1 Introduction

Grasping has been a core topic in robotics since the inception of the field. The classic "pick and place" strategy for robot manipulators is core to many industrial applications. Typically, in this situation grasping is achieved using a parallel jaw gripper at the end of the manipulator. This is a simple and reliable strategy, but limits the set of graspable objects to those that fit into the shape and scale of the gripper.

Researchers have long sought to improve the versatility and dexterity of robot end effectors. Typically the strategy has been influenced by the readily available case study of the human hand. A great many multi-fingered hand designs have been proposed, constructed, and analyzed through the years. An extensive survey conducted over 30 years ago is reported by Kato and Sadamoto (1977). A more recent survey of the field is given by Bicchi (2000).

However, despite steady improvement in the performance of multi-fingered robot hands, the industrial standard today remains the parallel jaw gripper. This is in part due to the inherent complexity of the human hand that multi-fingered robot hands seek to emulate. It is difficult to produce a dexterous yet reliable hand at the scale desired for most applications. This difficulty is amplified by the need to mount the hands at the end of robot manipulators, which in turn imposes significant restrictions on hand weight and packaging.

In response to the above challenges, researchers have been considering innovative solutions, both in hardware and grasping strategies, based on simpler, under-actuated strategies.

The underlying concept is to "do more with fewer inputs". For example, the three-fingered Barrett Hand (www.barrett.com) features the (quite non-anthropomorphic) design of rotating one finger around its "palm", so it can serve at times as an opposable thumb, and at other times as a conventional finger. Numerous other approaches to hardware complexity reduction, such as coupling of joints to reduce the number of controlled degrees of freedom, have been adopted (Kato and Sadamoto, 1977). On the theoretical side, a significant body of work has considered the underlying nature of machine manipulation, concentrating on situations where analytical understanding (for example of non-holonomic conditions in rolling contact) can be applied to simplify the input space. An overview of this work is given by Bicchi (2000).

In this paper, we consider a different approach to under-actuated grasping. Instead of using the human hand for motivation, we adopt concepts from biological "tongues, trunks, and tentacles". This results in robot grasping based on "invertebrate" continuum robots as opposed to "vertebrate" fingers in conventional robot hands. This leads to simpler, lower degree of freedom designs. Rather than analyzing special classes of grasping, we seek to widen the range of graspable objects via the judicious inclusion of inherent compliance in the hardware.

Figure 1. Octarm Continuum Robot.

The following sub-sections briefly review the state of the art in continuum robots and related literature on under-actuated manipulation with continuum contacts. The potential and reality of continuum robot grasping are explored in the next two sections, followed by discussion and conclusions.

2 Continuum robots – potential

The idea of creating "trunk and tentacle" robots, (in recent years termed continuum robots (Robinson and Davies, 1999)), is not new (Anderson and Horn, 1967). Inspired by the bodies of animals such as snakes (Hirose, 1993), the arms of octopus (Walker et al., 2005), and the trunks of elephants (Cieslak and Morecki, 1999; Hannan and Walker, 2003), researchers have been building prototypes for many years (Fig. 1). A key motivation in this research has been to reproduce in robots some of the special qualities of the biological counterparts. This includes the ability to "slither" into tight and congested spaces, and (of particular interest in this work) the ability to grasp and manipulate a wide range of objects, via the use of "whole arm manipulation" (Salisbury et al., 1988), i.e. wrapping their bodies around objects, conforming to their shape profiles.

Much of the effort thus far has concentrated on medium-scale (length roughly 1 m) continuum manipulators, and on simple grasping with single continuum bodies on that scale. However, some initial work in combining multiple continuum bodies into "multi-fingered" versions has been demonstrated (Suzumori et al., 1991; Lane et al., 1999). In the following sections, we will discuss the fundamental advantages and disadvantages of continuum contact in grasping, covering both cases.

One notable feature of continuum structures is that, while kinematically redundant versions have been developed (Hannan et al., 2003; Chrikjian, 1992), many prototypes have been designed to be under-actuated (in terms of numbers of independent actuators) with respect to their anticipated tasks. This is partly due to the desire to keep the body structures (which, unlike in conventional rigid-link manipulators or fingers, are required to directly contact the environment) "clean and soft", but also to exploit the extra control authority available due to the continuum contact conditions with a minimum number of actuators (Trivedi et al., 2008). It is in this sense that we use the term "under-actuated grasping" in this paper in the context of continuum robots.

3 Continuum contact and under-actuated manipulation

The ability to exploit continuum contact – particularly line contact – to restrain and manipulate objects is well-established. Consider for example the example of spinning tops by pulling on strings initially wrapped around their bodies (www.anirudh.net/courses/emch520/html/), or the dynamic manipulation of objects using whips (Bernstein et al., 1958). The physics of these activities, and related activities such as flycasting (Robson, 1990), are well established. The situation of manipulation of objects using ropes has also been the subject of interest in the robotics community (Donald et al., 2000).

The above real-world examples demonstrate situations in which low (often single) degree of freedom inputs, when well planned and executed, are sufficient to control higher-dimensional behavior of the manipulated environment. It is clear that the ability to "wrap around" an object presents a significant mechanical advantage, which if exploited carefully, can afford a complex behavior from a simple, low degree of freedom actuation strategy. A key issue for developers of continuum robots is to what extent this concept can be used in practice to perform useful grasping with under-actuated systems. This is the main issue explored in the following sections.

4 Continuum grasping – concepts

In grasping, as discussed above, the key attribute of continuum robots is their capability, via their inherent ability to bend at any point along their structure, to adapt their shape to conform to the perimeter of objects to be grasped. In theory, this ability could be exploited, if the continuum robot were sufficiently long and powerful, to "wrap around an object in all directions and completely constrain it". This suggests an alternative to the traditional way of thinking about grasp analysis as the net effect of a finite number of (local) contact locations. Conceptually, continuum (line) contact can be viewed as placing an infinite number of fingers in a tunable line around the surface of an object to be grasped. The inference is that the object can be "surrounded" by contacts over a sufficiently wide range of its surface to achieve, for

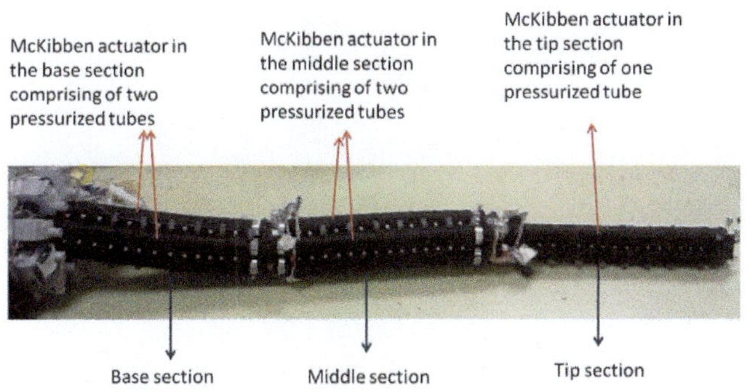

Figure 2. Picture of Octarm VI with sections and actuators marked.

example, full force- or form-closure (Bicchi, 2000), and thus stable grasping.

The engineered situation however can never be quite as above. While continuum contact can be maintained around the complete perimeter of an object, it is not feasible to apply arbitrary forces at given points on the perimeter, as if there were "infinite fingers". To do this would require in general an infinite number of actuators, corresponding to the infinite number of degrees of freedom theoretically available in the robot structure. In practice, although there are numerous different design strategies (Trivedi et al., 2008), continuum robots possess a small finite number of actuators, with the remaining (infinite) degrees of backbone freedom determined at each instant implicitly, via a combination of backbone materials properties, actuation forces, and external loading (Trivedi et al., 2008).

Despite this, continuum robot grasps do indeed tend to be quite robust to external disturbances (Trivedi et al., 2008; McMahan et al., 2006). The passive compliance inherent in almost all continuum manipulators causes them to "squeeze" around the perimeter of the continuum contact, evenly distributing the force resulting from even a single degree of freedom of actuation.

5 The Octarm VI continuum manipulator – kinematics and operation

The Octarm is a biologically inspired continuum manipulator that resembles an elephant's trunk (McMahan et al., 2006). Octarm VI has three sections each comprising of three independently actuated pneumatic actuators also known as McKibben actuators positioned at an angle of 120 degrees with respect to each other (see Fig. 2). These actuators comprise of latex tubes (two tubes per actuator in the base and middle sections and one tube per actuator in the tip section) covered with a plastic mesh sheet that is wound in a double helical manner.

When all the three actuators of a section are pressurized with equal amount of pressure, the section extends along the direction of length of the actuator tubes. When the air pres-

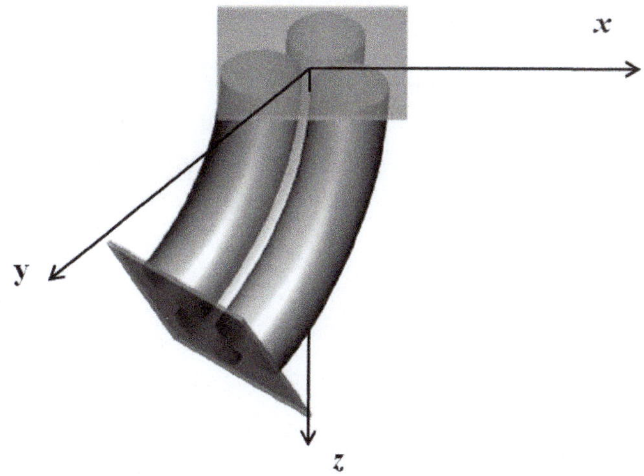

Figure 3. Bending of a section due to difference in actuator pressure levels (Jones and Walker, 2006a).

sure in one of the actuators is reduced, the section bends with constant curvature (illustrated in Fig. 3) . Thus by varying air pressures in the three actuators in suitable proportions determined by an inverse kinematics mapping procedure (Jones and Walker, 2006a), the section can be made to bend in different directions. Thus each section can bend about two axes (x and y) and can extend along a third axis (z) resulting in three degrees of freedom. This gives a total of nine degrees of freedom for the whole manipulator (three per each section).

Each continuum section of the Octarm has three internal parameters – curvature κ, direction of curvature φ and section length s (shown in Fig. 4). The forward kinematics of a section of the manipulator has been developed by approximating each section of the continuum arm to a conventional rigid link robotic arm, noticing that the net transformations are the same and then by expressing the D-H table parameters as a function of the internal parameters of the manipulator (Jones and Walker, 2006b). The homogeneous transformation matrix (Jones and Walker, 2006b) expressed in terms of the internal parameters of the continuum arm is as follows,

$$\mathbf{A}=\begin{pmatrix} \cos^2\varphi(\cos\kappa s-1)+1 & \sin\varphi\cos\varphi(\cos\kappa s-1) & -\cos\varphi\sin\kappa s & \frac{\cos\varphi\cos(\kappa s-1)}{\kappa} \\ \sin\varphi\cos\varphi(\cos\kappa s-1) & \cos^2\varphi(1-\cos\kappa s)+\cos\kappa s & -\sin\varphi\sin\kappa s & \frac{\sin\varphi\cos(\kappa s-1)}{\kappa} \\ \cos\varphi\sin\kappa s & \sin\varphi\sin\kappa s & \cos\kappa s & \frac{\sin\kappa s}{\kappa} \\ 0 & 0 & 0 & 1 \end{pmatrix}$$

From the co-ordinates of the end point of the section, the internal parameters can be calculated using geometry of the section (Csencsits et al., 2005). In such calculations, singularities occur in two configurations – when the curvature is zero and when the tip of the section is at origin. An inverse kinematics mapping procedure for converting the internal parameters (s, κ and φ) of a section to actuator lengths and thereby to air pressures in the three actuators of a section was developed by Jones and Walker (2006b).

The fundamental operations of a continuum robot are achieved by the combination of one or more actuator inputs

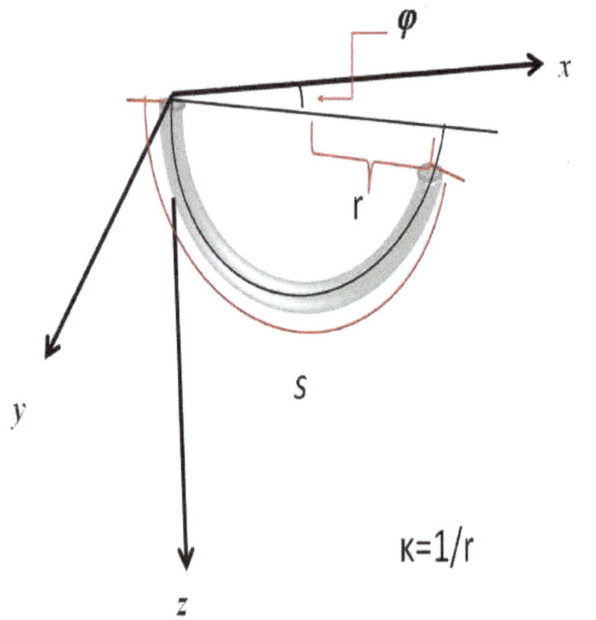

Figure 4. Kinematics of a single section of a continuum arm.

in contrast to conventional rigid link robots wherein each degree of freedom is controlled by an actuator. With a motive to make the control of operations intuitive to the operator, the user-interface for Octarm VI has been designed in a way that only the high-level control operation is obtained as input from a joystick (Csencsits, 2007). A detailed description of various joystick mapping schemes employed to obtain s, κ and φ values of the selected sections from different joystick positions is given by Csencsits et al. (2005). Inverse kinematics procedure (Jones and Walker, 2006b) is implemented to determine the actuator pressure levels required for the desired position of the arm.

6 Continuum grasping – practise

In this section, we use the results of a series of experiments to demonstrate aspects of continuum grasping (that were discussed in previous sections) which are easily achievable, and some of which are less accessible at the present time. The experiments discussed below were conducted in the robotics laboratories at Clemson University, using Octarm VI.

For experiments conducted on continuum grasping and reported on in this paper, the Octarm was placed horizontally on the floor thereby restricting, for each section, one of its degrees of freedom (to bend upwards/downwards). Thus, in such a planar arrangement, each section can bend sideways and extend along its length; thereby the manipulator as a whole has six controllable degrees of freedom. This arrangement is convenient to analyze grasping of stationary as well as moving objects. Also, maximum curvature for each section is achieved when the Octarm is laid on the floor as bending of the air actuators is not opposed by gravitational effects. Due to mechanical constraints (inherent compliance), curvature limitation and sagging are unavoidable when the Octarm operates in the spatial world.

Figure 5. Octarm grasping a spherical object – continuum grasp.

The inputs to the pressure regulators that control the pressure levels in the air actuators of the Octarm are given through a control PC having a Data Acquisition I/O board. This is a Pentium III EBX form-factor Single Board Computer (SBC) running QNX Neutrino real-time Operating System. This is connected to another PC on the operator side that is interfaced to a joystick (Logitech Wingman 3-D). The values of s, κ and φ are obtained as inputs from the joystick position. The mapping from these parameters to air pressures in each of the actuators in a section is implemented in a MATLAB/Simulink code. String encoders that are mounted on the base of the arm provide actuator lengths as feedback.

Throughout these experiments on continuum grasping, one or more number of sections were used and the curvatures were controlled using the joystick. The grasping ability of the Octarm was initially analyzed with a set of stationary objects of different shapes, sizes, textures and orientations. Depending on the size of the object, one or more sections of the Octarm were used for grasping. A firm, continuum contact was observed in grasping spherical and cylindrical objects that aligned with the curvature of the arm. A picture of the Octarm grasping a ball is shown in Fig. 5.

While two sections were required to encircle objects having larger diameter, objects were still firmly held by a single

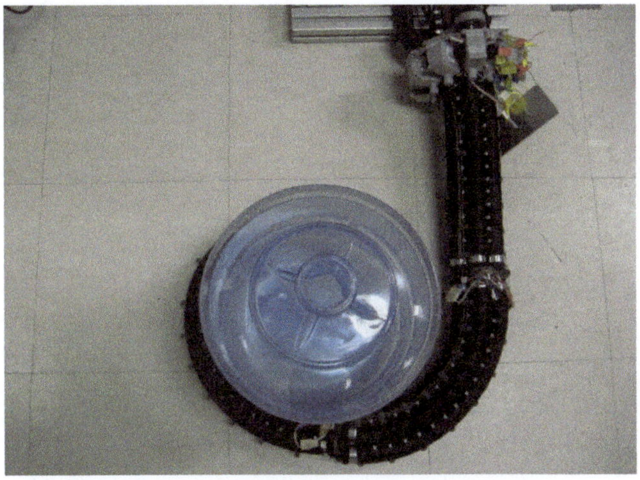

Figure 6a. Two sections of the Octarm used for grasping.

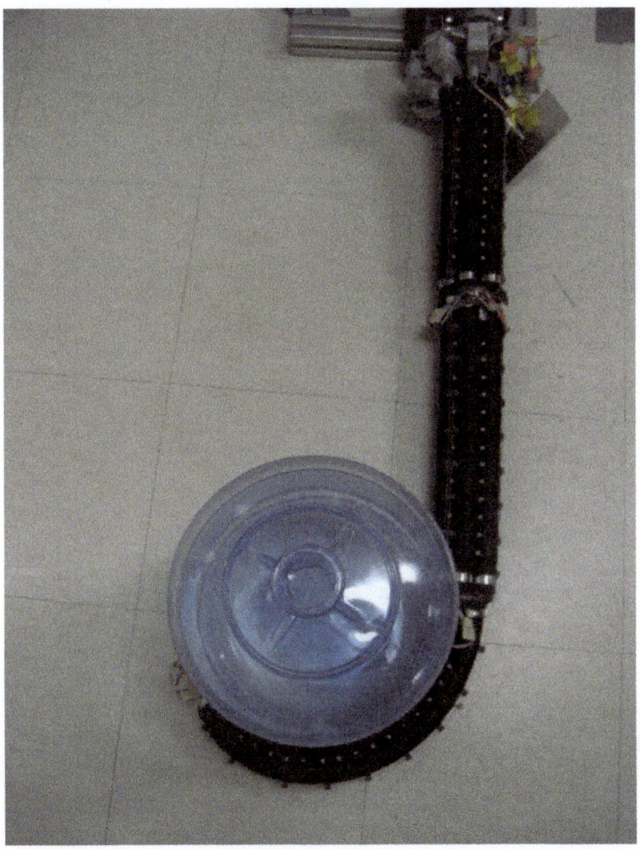

Figure 6b. The same object grasped using the tip section of the Octarm.

section by having a tight grasp, thereby realizing force closure. This is shown in Fig. 6a and b. When multiple sections were used to grasp an object having a circular boundary however, a perfect continuum contact was often not observed as all the three sections did not uniformly bend with the same curvature. This was attributed to the Octarm manipulator's construction which caused all three sections to individually bend with uniform curvatures but not as a whole. Continuum grasping was also seen in the case of soft objects which deformed their shape to confirm to the boundary of the manipulator sections.

Figure 7. Object held by point contacts.

Interesting results were obtained when objects with sharp edges were grasped. While grasping objects of the shapes of cube, cuboid, etc., contacts were made at the edges or at the faces of the object. The number and location of contact points that determined the stability of the grasp were dependent on the initial orientation of the object. Although continuum contact was not possible in this case, the manipulator was able to hold objects through distributed point contacts leading to force closure in the plane (Fig. 7). An increase in tightness of the grasp on a rigid object flattened the curvature of the manipulator thereby increasing the contact surface. Thus, based on the dimensions, orientation and rigidity of the target object, grasping in practice was realized partially by continuum contacts and partially by (locally) point contacts (Fig. 8).

Continuum grasping using the Octarm is particularly attractive in grasping fragile objects (like an egg tray, glass jug) where a soft but firm hold is required (Fig. 9). Potential applications include rescue operations and safe manipulation of delicate objects. Apart from the objects mentioned earlier in this paper, there are also numerous other cases in which the parallel jaw gripper is not plausible. Multifingered robots can provide a more dexterous solution for grasping but with more complicated mechanisms. Continuum grasping, on the other hand derives inspiration from biological structures and redefines grasping by providing a novel and less complicated approach. Its versatility in handling a plethora of objects makes sense intuitively and is also proven from the above experiments.

Continuum manipulators also have an edge over their competitors by their ability to robustly grasp moving objects

Figure 8. Continuum and point contacts.

Figure 9. Grasping fragile objects – using a parallel jaw gripper (unsafe grasp) and the Octarm.

without the requirement of precise knowledge of the relative velocity of the moving object. The inherent structure of a continuum manipulator and its nature of grasp are advantageous in grasping moving objects. The Octarm was able to restrict the motion of and grasp passive but moving objects like a spinning ball, rolling ball and a sliding object as long as the relative motion between object and robot was towards the inner surface of the loop formed by the manipulator (Fig. 10). In the case of passive rolling objects, grasping was successful for objects that were able to drift and not sufficiently heavy to oppose the movement of the manipulator. Although a few practical problems were faced due to slow response of the Octarm relative to fast – moving objects, there exists a huge potential for grasping non-stationary objects with continuum manipulators to be explored in the future.

Another topic of interest in this context that is in line with grasping is "acquisition". Various bio-inspired strategies were developed to use continuum manipulators to grasp an object and bring it towards the base of the manipulator. This is similar to the behavior seen in animals possessing trunks and tentacles in grabbing food objects. Octopus-inspired strategies were developed for the Octarm by McMahan and Walker (2009).

Manipulation of an object grasped is achieved using the base section of the continuum manipulator. Since the Octarm was operated in a plane, manipulation of the base section was restricted to one degree of freedom only.

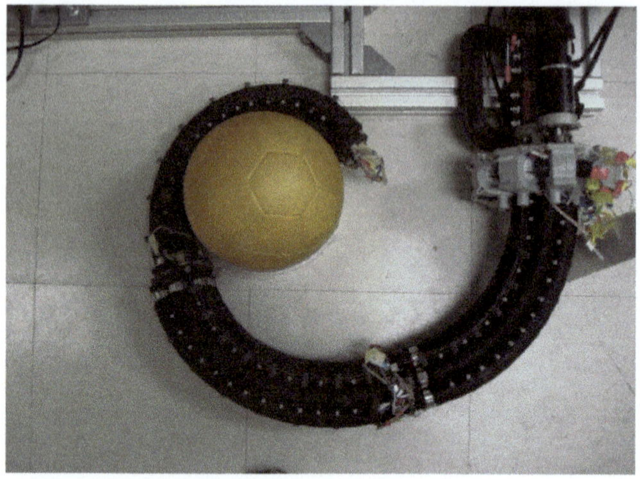

Figure 10. The Octarm - grasping and acquiring a moving object.

Table 1. Compliance of all three sections (measured in a distance perpendicular to z-axis indicated in Fig. 3).

	Compliance (in 10^{-3} m kg^{-1})		
Pressure Levels in the Actuators (in $\times 10^5$ Pa)	Base Section	Middle Section	Tip Section
0.8273	0.5698	0.5835	1.1454
2.1373	0.3655	0.3699	0.8201
4.5505	0.2776	0.2876	0.5705
5.5158	0.2276	0.2302	0.4262

7 Discussion

In this section, we analyze the potential of Octarm to be used for continuum grasping by discussing various aspects of its capabilities as well as a few inherent problems that accompany grasping. The analysis is based on quantified physical data and observations made from the experiments in the previous section.

One of the major advantages of any bio-inspired soft robot is its intrinsic compliance which enables grasping of a wide variety of objects whose shape, size and orientation are not accurately known. While holding an object, the continuum arm no longer maintains its constant curvature bending configuration as it deforms to confine to the shape of the object. This passive adaptation ability of the Octarm relies on its compliance as well as the rigidity of the object that is being grasped. Table 1 gives approximate measures of compliances of the three sections of Octarm VI at different actuator pressure levels.

Table 2. Approximate curvatures of the three sections.

Section	Maximum Curvatures (in m^{-1})	
	planar (2-D) operation	spatial (3-D) operation
Base section	3.98465	2.28228
Middle section	4.16693	3.79449
Tip section	8.33072	8.07598

Table 3. Payload of Octarm.

Weight of Octarm VI (Approx)	6.93996 kg
Payload (Lifting against gravity)	0.90718 kg

The compliances of the base and middle sections are almost the same. The tip section having a thinner structure than the other two sections is more compliant. With an increase in actuator pressure, the stiffness of each section increases thereby making it less compliant.

There are a few potential problems faced with the operation of the Octarm in full 3-D environments which degrade its performance in grasping to certain extent. A comparison of the approximate curvatures of the three sections of the Octarm VI measured in planar (2-D with the effects of gravity eliminated) and in 3-D configurations is shown in Table 2.

The maximum curvatures of the three sections in 3-D configuration are lesser than the ones achievable in 2-D planar configuration. The maximum curvatures that can be reached also give an idea of the dimensions of the object that can be effectively grasped using one or more sections of the Octarm.

In 3-D operation, the manipulator's position deviates slightly from its kinematics which is attributed to sagging effects due to gravity and weight of the arm. Also, when the manipulator carries and lifts objects, its ability to grasp is limited by the payload capacity of the arm. The maximum load that the Octarm VI can sustain without deforming its grasping configuration is given in Table 3.

The time taken for each section to expand to its maximum possible length i.e. for the pressure in the actuators to increase from 0 Pa to a maximum value (5.5158×10^5 Pa) is tabulated in Table 4. Currently, there is no explicit mechanism to regulate the speed of operation of the Octarm. But, faster movement of the Octarm for the same curvature can be observed by varying at maximum section length (s).

Although 3-D operation of the manipulator permits an infinite number of directions of bending for grasping, the inherent design of the manipulator permits maximum curvature to be achieved only in three directions ($\varphi = 30°$, $150°$, $270°$) , when the section bends away from any one of its three actuators. Also, there is a second local maximum curvature

Table 4. Actuator parameters.

Section	Time taken for the pressure levels to increase from 0 Pa to 5.5158×10^5 Pa in all the actuators (in s)	maximum percentage of extension
Base section	5.51	27.18
Middle section	5.66	42.47
Tip section	5.78	38.5

that can be achieved in three directions ($\varphi = 90°, 210°, 330°$) when the section bends away from any two of its three actuators. This imposes a restriction on effective grasping operation of the arm in spatial configuration. For the experiments conducted in planar 2-D space, the Octarm is bent in the direction of maximum curvatures. The curvature values of each section reported in this paper also correspond to the directions of maximum curvatures.

In this paper, a novel idea to use a continuum manipulator for under-actuated continuum grasping has been proposed and demonstrated using the Octarm as a prototype model. Continuum manipulators can be approximated by the operations of a multi-linked under-actuated chain, but the kinematics, actuation and control strategies employed are different from the former. The Octarm lacks links and joints, but, having the air pressurized actuators arranged in a triangular pattern enables the arm to bend in an infinite number of directions. The grasp realized using a continuum arm is more qualified by its flexibility (compliance) and this compensates for the lack of accuracy when compared to rigid link robots in positioning the arm to grasp objects. The arm was manually operated and its potential to grasp was analyzed and quantified using empirical data. We are currently programming different trajectory motions for the Octarm to find a movement of the arm that increases the impact force at the contact point. Having this as the groundwork, in the short term, we intend to develop autonomous grasping algorithms to realize impulsive manipulation with continuum arms.

8 Conclusions

Continuum robot structures are designed to create continuum contact with a grasped object, over a relatively large range of the object surface. This capability arises due to their ability to adapt their shape to that of the object, and can be achieved with relatively few controlled inputs. This combination of increased contact area with fewer degrees of freedom, compared with conventional multi-fingered robot hands, offers the potential for robust and adaptive under-actuated grasping with continuum robot elements. In this paper, we discussed and demonstrated this potential via a series of experimental case study examples. However, as illustrated in the experiments, numerous challenges need to be addressed before under-actuated continuum grasping is a practical option.

Acknowledgements. This research was supported by the US National Science Foundation under grants IIS-0844954 and IIS-0904116.

Edited by: J. L. Herder
Reviewed by: two anonymous referees

References

Anderson, V. C. and Horn, R. C.: Tensor Arm Manipulator Design, Transactions of the ASME, 67-DE-57, 1–12, 1967.

Bernstein, B., Hall, D. A., and Trent, H. M.: On the Dynamics of a Bull Whip, J. Acoust. Soc. Am., 30(12), 1112–1115, 1958.

Bicchi, A.: Hands For Dexterous Manipulation and Robust Grasping: A Difficult Road Towards Simplicity, IEEE T. Robotic. Autom., 16(6), 652–662, 2000.

Chrikjian, G. S.: Theory and Applications of Hyperredundant Robotic Mechanisms, Ph.D. Thesis, Department of Applied Mechanics, California Institute of Technology, 1992.

Cieslak, R. and Morecki, A.: Elephant Trunk Type Elastic Manipulator – a Tool for Bulk and Liquid Type Materials Transportation, Robotica, 17, 11–16, 1999.

Csencsits, M.: Operational Strategies for Continuum Manipulators, M.S. Thesis, Department of Electrical and Computer Engineering, Clemson University, 2007.

Csencsits, M., Jones, B. A., McMahan, W., Iyengar, V., and Walker, I. D.: User Interfaces for Continuum Robot Arms, Proc. IEEE/RSJ International Conference on Intelligent Robots and Systems (IROS), Edmonton, Canada, 3011–3018, August 2005.

Donald, B., Gariepy, L., and Rus, D.: Distributed Manipulation of Multiple Objects using Ropes, Proceedings IEEE International Conference on Robotics and Automation, 450–457, 2000.

Hannan, M. W. and Walker, I. D.: Kinematics and the Implementation of an Elephant's Trunk Manipulator and other Continuum Style Robots, J. Robotic Syst., 20(2), 45–63, 2003.

Hirose, S.: Biologically Inspired Robots, Oxford University Press, 1993.

Jones, B. and Walker, I. D.: Practical Kinematics for Real-Time Implementation of Continuum Robots, IEEE T. Robot., 22, 6, 1087–1099, 2006a.

Jones, B. and Walker, I. D.: Kinematics for Multi-Section Continuum Robots, IEEE T. Robot., 22, 1, 43–55, 2006b.

Kato, I. and Sadamoto, K..: Mechanical Hands Illustrated, Survey Japan, 1977.

Lane, D. M., Davies, J. B. C., Robinson, G., O'Brien, D. J., Sneddon, J., Seaton, E., and Elfstrom, E.: The AMADEUS Dextrous Subsea Hand: Design, Modeling, and Sensor Processing, IEEE J. Oceanic Eng., 24, 1, 96–111, 1999.

McMahan W., Pritts, M., Chitrakaran, V., Dienno, D., Grissom, M., Jones, B., Csencsits, M., Rahn, C. D., Dawson, D., and Walker, I. D.: Field Trials and Testing of "OCTARM" Continuum Robots, Proc. IEEE International Conference on Robotics and Automation, 2336–2341, 2006.

McMahan, W. and Walker, I. D.: Octopus-Inspired Grasp Synergies for Continuum Manipulators, Proc. IEEE International Conference on Robotics and Biomimetics, 945–950, 2009.

Robinson, G. and Davies, J. B. C.: Continuum Robots - A State of the Art, Proc. IEEE International Conference on Robotics and Automation, 2849–2854, 1999.

Robson, J. M.: The Physics of Flycasting, Am. J. Phys., 58(3), 234–240, 1990.

Salisbury, K., Townsend, W., Ebrman, B., and DiPietro, D.: Preliminary Design of a Whole-Arm Manipulation System (WAMS), Proc. IEEE International. Conference on Robotics and Automation, 254–260, 1988.

Suzumori, K., Iikura, S., and Tanaka, H.: Development of Flexible Microactuator and it's Applications to Robotic Mechanisms, Proc. IEEE International Conference on Robotics and Automation, Sacramento, California, 1622–1627, 1991.

Trivedi, D., Rahn, C. D., Kier, W. M., and Walker, I. D.: Soft Robotics: Biological Inspiration, State of the Art, and Future Research, Applied Bionics and Biomechanics, 5(2), 99–117, 2008.

Walker, I. D., Dawson, D., Flash, T., Grasso, F., Hanlon, R., Hochner, B., Kier, W. M., Pagano, C., Rahn, C. D., and Zhang, Q.: Continuum Robot Arms Inspired by Cephalopods, Proc. SPIE Conference on Unmanned Ground Vehicle Technology VII, Orlando, FL, 303–314, 2005.

www.anirudh.net/courses/emch520/html/, last access: 19 January 2011.

www.barrett.com, last access: 19 January 2011.

4

Towards the design of a prosthetic underactuated hand

T. Laliberté, M. Baril, F. Guay, and C. Gosselin
Département de Génie Mécanique, Université Laval, Québec, QC, G1V 0A6, Canada

Abstract. This paper presents recent advances in the design of an underactuated hand for applications in prosthetics. First, the design of the fingers is addressed. Based on previous experiments with prototypes developed in the past, new tendon routings are proposed that lead to a more effective transmission of the forces. A novel elastic tendon routing is also proposed for the passive opening of the hand. A simplified static analysis of the fingers is proposed to support the results. Then, a new kinematic design of the thumb is presented. The thumb is designed to perform out-of-the-plane motions in order to broaden the variety of possible grasps. A mechanism for the implementation of underactuation between the fingers is proposed that alleviates the friction problems encountered in earlier hand designs. Finally, a prototype of the hand is briefly described and typical grasps are shown.

This paper was presented at the IFToMM/ASME International Workshop on Underactuated Grasping (UG2010), 19 August 2010, Montréal, Canada.

1 Introduction

Underactuation will probably be an important feature in future prosthetic hands. Indeed, reducing the number of actuators without significantly compromising performance has many advantages such as reducing weight, cost and complexity of control.

Although research is pursued in order to allow users of prostheses to control many degrees of actuation (Alstrom et al., 1981; Chu et al., 2007; Castellini et al., 2008), the number of degrees of actuation is still limited. In this context, underactuation is a very attractive feature. Underactuation can be introduced in the fingers using link mechanisms (Shimojima et al., 1987; Laliberté and Gosselin, 1998; Dubey and Crowder, 2002). Although large forces can be obtained with this approach, the resulting fingers are significantly thicker than human fingers. In order to obtain more compact fingers, cable-driven mechanisms can be used (Hirose and Umetani, 1978; Crismal et al., 1996). Also, many efforts have been made in order to include underactuation between the fingers of a hand (Rakic, 1989; Fukaya et al., 2000; Birglen and Gosselin, 2006). Finally, the use of underaction in prosthetic devices is increasingly popular (Herder and Visser, 2000; Kyberd et al., 2001; Massa et al., 2002; Dollar and Howe, 2007; Kamikawa and Maeno, 2008). For a review of underactuated hands, see Birglen et al. (2008).

In this paper, the ongoing design of a prosthetic underactuated hand is presented. The design is based on an anthropomorphic underactuated hand previously developed in the Robotics Laboratory at Laval University. The new design includes several improvements that mainly cover the underactuation within fingers, the passive opening of the fingers, the thumb implementation and the actuation transmission between the fingers.

2 Previous work

In Gosselin et al. (2008), an underactuated anthropomorphic robotic hand was presented that includes 15 degrees of freedom and a single actuator. The hand automatically adapts to the shape of objects and the fingers are designed to optimize grasping criteria, mainly the resulting force on the object and force distribution among the phalanges. At rest, the hand is completely open and when it is operated, the four fingers, then the thumb, close together until they make contact with an object.

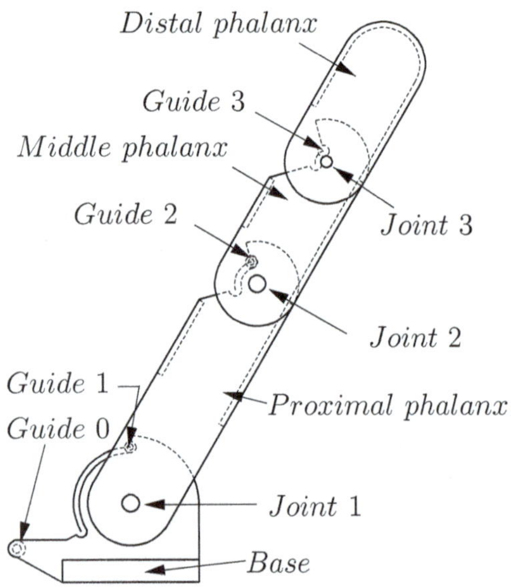

Figure 1. Design of the fingers proposed in Gosselin et al. (2008), which includes guiding rods.

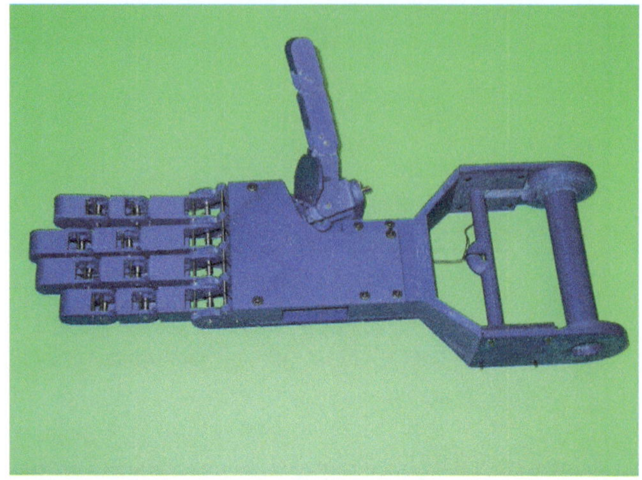

Figure 2. Prototype of underactuated hand proposed in Gosselin et al. (2008).

Like the tendons of a human hand, finger actuation (power transmission) is performed through cables. The torque ratio between the joints is mainly dictated by the position of guiding points on each phalanx. In the prototype presented in Gosselin et al. (2008), the fingers are normally maintained open by torsional springs inserted at each of the joints. The order of closure of each phalanx is dictated by the actuation torque available at each joint and the opening torque from the torsional springs. The geometric design of the fingers is presented in Fig. 1 (see Gosselin et al., 2008, for more details).

It is worth mentioning that two different mechanisms have been studied using the prototype presented in Gosselin et al. (2008) to achieve underactuation and coupling between the fingers. In the first mechanism – which did not work very well mainly because of friction – the thumb and the fingers are all linked by an underactuated mechanism. In the second mechanism – which worked better – the thumb is coupled to the other fingers, which are underactuated. These mechanisms are composed of cables, pulleys, guides and slides, and are described in Gosselin et al. (2008).

The thumb is placed such that it is in opposition with the index finger and middle finger and such that it is oriented to point slightly towards the centre of the hand. A plastic prototype of the hand was built using rapid prototyping. The joint pins were made of metal. Kite cables were used for the tendons, which provided high stiffness, flexibility and relatively low friction. In order to provide manual actuation of the hand, the prototype was equipped with a handle, which is convenient for the testing and demonstration of the hand in a general context. The prototype proposed in Gosselin et al. (2008) is shown in Fig. 2.

Experiments demonstrated that the hand is performing well for enveloping grasps, which involve contact with all the phalanges. The grasps are generally firm and stable and have a human-like appearance.

However, the experiments also revealed several weaknesses. As opposed to what was observed for enveloping grasps, the hand often performed poorly whith pinch grasps of small objects, which involve contact with only the tip of the fingers. This is due to a lack of synchronization of the closing motion of the thumb and fingers as well as to the slanted orientation of the thumb, which is not ideal for pinch grasps. Friction is also an important issue. For instance, the efficiency of the transmission system was estimated to only 35%. This is mainly explained by the use of small rods as guiding points that add friction as the cables rub directly on them. This is also explained by the sliding joints located in the palm. The sliders can even jam the mechanism which makes it difficult to predict the trajectory of the fingers. With the use of guiding pins, the torques and forces in the fingers are not constant during the closing of the fingers since the lever arms are not constant. One of the consequences is that the force needed to start the motion of the fingers is too large. Finally, the use of torsional springs to open the fingers makes the assembly of the fingers quite tedious.

Consequently, a new prototype has been developed to address the above issues. Changes have been made to the fingers, thumb and underactuated mechanisms. Also, the new hand is attached directly to a lower arm prosthesis to verify experimentally the advantages and disadvantages of the hand as a prosthesis.

3 Improved finger design

Several improvements have been made to the fingers of the new prototype. Firstly, the guiding pins of the previous version have been replaced by pulleys. This reduces the friction and provides constant ratios at each joint. This modification adds significant flexibility to the design, since the ratios can be easily controlled by changing the diameter of the pulleys. Optimal combinations are obtained from numerical simulations. Another important change is that the opening of the fingers is now produced by an opening cable rather than torsional springs. The force in the opening cable can be obtained from a spring attached in series with the cable. However, it is more compact to replace the cable and spring by a *bungee cable*, which integrates the cable and spring in the same component. Similarly to the closing mechanism, the ratios of the pulleys in the routing of the opening cable can be easily controlled by changing the diameter of the pulleys. Therefore, there are now two independent tendons per finger, an active tendon for closing and a passive tendon for opening. This significantly simplifies assembly and provides better compactness. This new type of finger is consequently simpler, more efficient and more accurate.

As mentioned above, the fingers are composed of two independent tendons directed through small pulleys. For the actuation tendon, two slightly different routing options have been developed.

3.1 Actuation routing with idler on the middle phalanx

For the actuation tendon, the topology of the routing is similar to that presented in Crismal et al. (1996). However, in order to obtain the desired range of motion – 0° to 90° at each joint – the pulleys are located differently. The tendon is attached to the inside of the distal phalanx, routed down the finger and fixed on the mechanism providing the underactuation between the fingers, located in the palm of the hand. As shown in Fig. 3, the cable is maintained in contact with the pulleys by idlers for any configuration of the finger. This provides a constant torque at all joints for all possible configurations. The distal phalanx does not require sheave since the cable does not move in it. Indeed, the closure of this phalanx simply affects the winding on the distal pulley and this movement generates very little friction.

3.2 Actuation routing without idler on the middle phalanx

In the second routing, illustrated in Fig. 4, an idler is added to the base of the finger in order to displace the input tendon in the plane of the underactuation mechanism located in the palm. Because of the idler in the base, the idler in the proximal phalanx can be displaced close to the middle joint. This allows to keep the cable in contact with the pulleys in all configurations without the use of an idler in the middle phalanx. The removal of the middle idler minimizes friction. This routing has been implemeted in the prototype. It is noted that as long as the cable is maintained in contact with the pulleys, the two mechanisms are kinematically equivalent.

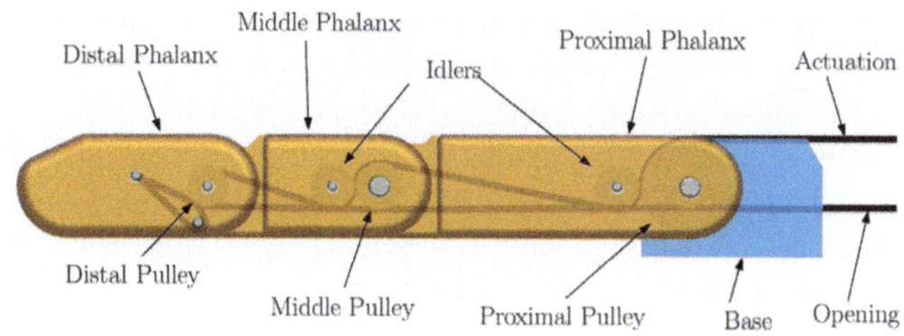

Figure 3. CAD model of the proposed finger with the actuation routing using an idler on the middle phalanx.

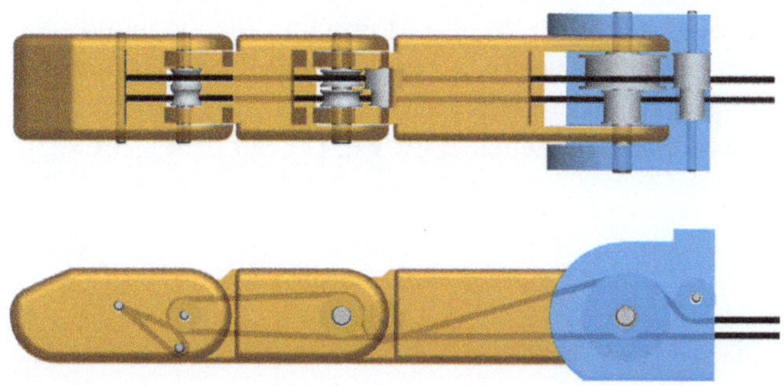

Figure 4. CAD model of the proposed finger with the actuation routing that does not require an idler on the middle phalanx.

3.3 Routing of the opening cable

As illustrated in Figs. 3 and 4, the tendon used to open the finger is also attached to the distal phalanx and routed to the palm where it is fixed. Compared to the actuation tendon, it passes on the opposite side of the joints of each phalanx to produce the opening of the finger. This tendon is also always in contact with the pulleys in the joints without the need of idlers, which minimizes friction. Similarly to the actuation torques, the opening torques are independent from the configuration of the finger and only depend on the tension in the bungee cable.

3.4 Static modelling of the fingers

The static model of the fingers was developed using the procedure given in Birglen et al. (2008). According to the latter reference, the force transmission between the actuator – here the force on the tendon at the base of the finger – and the contact forces on the phalanges can be written as:

$$f = \mathbf{J}^{-T}\mathbf{T}^{-T}t \tag{1}$$

where $f = [f_1, f_2, f_3]^T$ is the vector of phalanx contact forces with the object, matrix $\mathbf{J}$ is a lower triangular matrix characteristic of the contact locations – and friction, if modelled – referred to as the Jacobian matrix of the finger, matrix $\mathbf{T}$ is the transmission matrix, characterizing the transmission used and its geometry, and vector $\boldsymbol{t}$ is the vector of actuating torques at the joints. The analytical expression of the matrices defined above can be found in Birglen et al. (2008) for linkage-driven fingers and tendon-driven fingers.

3.4.1 Closing sequence

A knowledge of the torques acting internally at each joint of a finger was needed in order to determine the closing sequence of the finger. The model developed provides the magnitude of the response torque vector, $\boldsymbol{\tau}$, based on an actuation torque vector, $\boldsymbol{\tau}_a$, an opening torque vector, $\boldsymbol{\tau}_r$, and a torque vector due to gravity, $\boldsymbol{\tau}_g$. Mathematically, Eq. (1) can be reformulated as

$$f = \mathbf{J}^{-T}\boldsymbol{\tau} \tag{2}$$

with

$$\boldsymbol{\tau} = \boldsymbol{\tau}_a + \boldsymbol{\tau}_r + \boldsymbol{\tau}_g \tag{3}$$

Then, the actuation torque vector is written as

$$\boldsymbol{\tau}_a = \mathbf{T}_a^{-T}\boldsymbol{t}_a \tag{4}$$

where $\mathbf{T}_a$ is the transmission matrix which relates the actuation joint velocity to the time derivatives of the joint coordinates. For this design, one has

$$\mathbf{T}_a = \begin{bmatrix} 1 & \frac{-r_{3a}}{r_{1a}} & \frac{-r_{5a}}{r_{1a}} \\ 0 & 1 & 0 \\ 0 & 0 & 1 \end{bmatrix} \tag{5}$$

where r_{1a}, r_{3a} and r_{5a} are the radii of the proximal, middle and distal actuation pulleys respectively and $\boldsymbol{t}_a$ is the input torque vector defined as $\boldsymbol{t}_a = [T_a, 0, 0]^T$. Similarly, the opening torque vector is written as

$$\boldsymbol{\tau}_r = \mathbf{T}_r^{-T}\boldsymbol{t}_r \tag{6}$$

where $\mathbf{T}_r$ is the transmission matrix which, in this design, is exactly the same matrix as for the actuation apart from the pulley radii which are different for the opening routing, namely

$$\mathbf{T}_r = \begin{bmatrix} 1 & \frac{-r_{3r}}{r_{1r}} & \frac{-r_{5r}}{r_{1r}} \\ 0 & 1 & 0 \\ 0 & 0 & 1 \end{bmatrix} \tag{7}$$

where r_{1r}, r_{3r} and r_{5r} are the radii of the proximal, middle and distal opening pulleys respectively and $\boldsymbol{t}_r$ is the opening torque vector defined as $\boldsymbol{t}_r = [T_r, 0, 0]^T$.

Finally, the torque vector due to gravity is written as $\boldsymbol{\tau}_g = [\tau_{g1}, \tau_{g2}, \tau_{g3}]^T$ where the three entries describe the effect of gravity on each joint, namely:

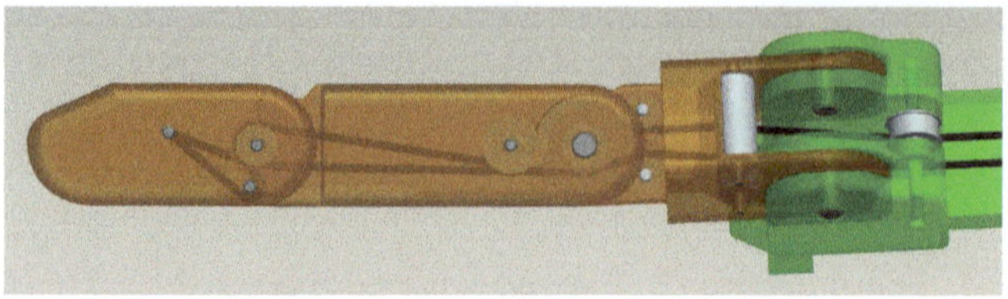

Figure 5. CAD model of the thumb.

$$\begin{aligned} \tau_{g1} &= \cos(\varphi)([m_1 g c_{g1}\cos(\gamma_0+\gamma_1)] \\ &+ [m_2 g(c_{g2}\cos(\gamma_0+\gamma_1+\gamma_2)+l_1\cos(\gamma_0+\gamma_1))] \\ &+ [m_3 g(c_{g3}\cos(\gamma_0+\gamma_1+\gamma_2+\gamma_3)+l_2\cos(\gamma_0+\gamma_1+\gamma_2) \\ &+ l_1\cos(\gamma_0+\gamma_1))]) \end{aligned} \tag{8}$$

$$\begin{aligned} \tau_{g2} &= \cos(\varphi)([m_2 g c_{g2}\cos(\gamma_0+\gamma_1+\gamma_2)] \\ &+ [m_3 g(c_{g3}\cos(\gamma_0+\gamma_1+\gamma_2+\gamma_3) \\ &+ l_2\cos(\gamma_0+\gamma_1+\gamma_2))]) \end{aligned} \tag{9}$$

$$\tau_{g3} = \cos(\varphi)(m_3 g c_{g3}\cos(\gamma_0+\gamma_1+\gamma_2+\gamma_3)) \tag{10}$$

where g is the gravitational acceleration, m_i is the mass of the i-th phalanx, c_{gi} is the distance from the i-th joint to the centre of mass of the i-th phalanx and l_i is the length of the i-th phalanx. Furthermore, φ and γ_0 are the first two angles of the Tait-Bryan's convention which is used here to describe the orientation of the hand (pitch and roll). Finally, angle γ_i is the angle describing the rotation of the i-th joint of the finger.

4 Improvement of the thumb design

Several changes were also made to the thumb in the new prototype. Like the new fingers, the thumb is now composed of two independent sets of cables and pulleys, one for closing and another one for the opening. However, a twist rotation is introduced between the proximal phalanx and the middle joint in order to explore the possibilites of motion out of a single plane and to broaden the variety of possible grasps (see Fig. 5). More precisely, the proximal joint of the thumb is perpendicular to the axes of the fingers, allowing a lateral motion of the thumb. The middle joint is rotated 25 degrees from the axes of the fingers (65 degrees from the proximal axis of the thumb) and points towards the centre of the hand. This rotation allows a more anthropomorphic behaviour and experiments show that it improves the quality of the grasps. In order to cope with the rotation between the proximal and middle joints, four idlers are included to manage the routing of cables. The middle and distal phalanges are similar to those of the other fingers, but they include the first actuation routing and the actuation pulley of the middle joint is slightly larger in order to obtain more force on the thumb. The thumb is illustrated in Fig. 5.

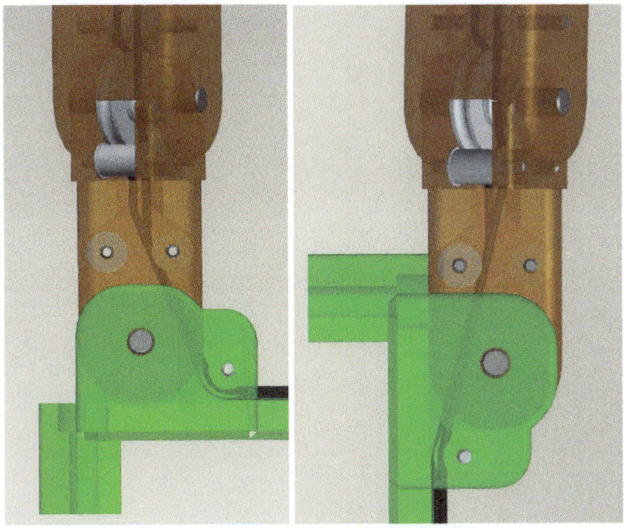

Figure 6. CAD model of the routing of the tendons of the thumb around the proximal joint: opposing configuration (left) and lateral configuration (right).

In this specific implementation, it is desired to obtain two possible configurations of the thumb: one opposing the other fingers and one located lateraly with respect to the other fingers. It is expected that this new feature will allow to pick up some objects on flat surfaces more easily while still allowing the performance of enveloping grasps. The configuration with the thumb facing the other fingers is mainly used for enveloping grasps. It can also be used for pinch grasps with the tip of the fingers, but with limited stability. The configuration with the thumb located lateraly allows to perform pinch grasps using the side of the closed index. In this configuration, the large surface of the side of the closed index allows more stable pinch grasps.

In order to select which configuration to use, a first option would be to include an additional actuator, which significantly complexifies the hand considering that all that is needed is a simple switch between two configurations. The suggested option is to ensure that the thumb freely stays in one of the two configurations and to select the other configuration by pushing laterally on the thumb with the other hand or against a fixed object. In order to obtain this feature, the actuation cables (closing and opening) can pass through the first axis when the thumb is moved laterally (see Fig. 6). Then, for each configuration, the passive opening cable tends to keep the thumb in place. It is noted that the active closing cable is slightly loose when not actuated, which eases the change of configuration. Then, when the thumb is closed, the closing cable further helps to keep the thumb in place.

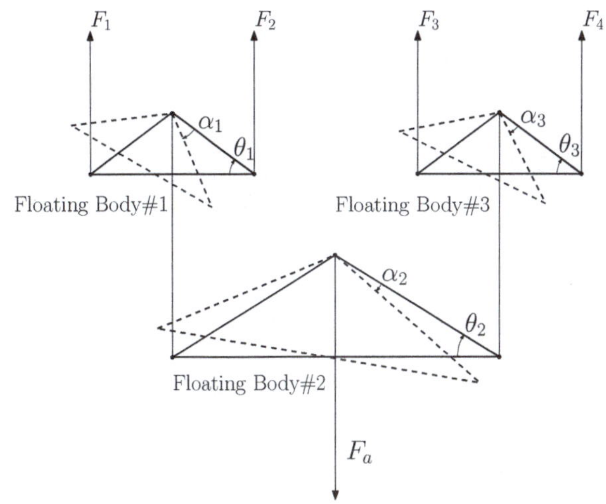

Figure 7. Schematic illustration of the mechanism allowing underactuation between the fingers.

5 Mechanism for the underactuation between the fingers

5.1 Description of the mechanism

The prototype presented here can be useful to test various mechanisms of transmission between the fingers. Indeed, a relatively large space is devoted to include different transmission mechanisms in the palm. Some mechanisms providing underactuation between the fingers have been developed (Baril et al., 2010). One of the most promising mechanisms is implemented and discussed here. As shown in Fig. 7, this mechanism is composed of three small seesaws connected together with cables. The geometric arrangement generates an equal force in each finger. When the hand is activated but does not make contact with an object, it simply closes and the entire mechanism moves downward, without any rotation of the rods. However, if there is a restriction on a finger, the smaller floating bodies will rotate to allow other fingers to continue their trajectory. The same phenomenon occurs if two consecutive fingers are subject to a restriction, except that the rotation will take place at the larger floating body. It is noted that the range of underactuation possible with this mechanism depends on the distance between the outputs. Indeed, the maximum relative motion of two fingers linked by a seesaw is given by the length of the seesaw. As a result, the closing motion of a finger can be slightly limited if its neighbour is kept fully open. A major distinction with the mechanisms developed in Gosselin et al. (2008) is that the mechanisms presented here are fully floating, i.e., they are not guided by sliders. This has the effect of decreasing friction and increasing precision. Also, the mechanism is much simpler since it does not need to be set anywhere in the palm. Finally, the thumb is coupled with the four fingers by attaching its tendon directly to the actuation tendon. As discussed

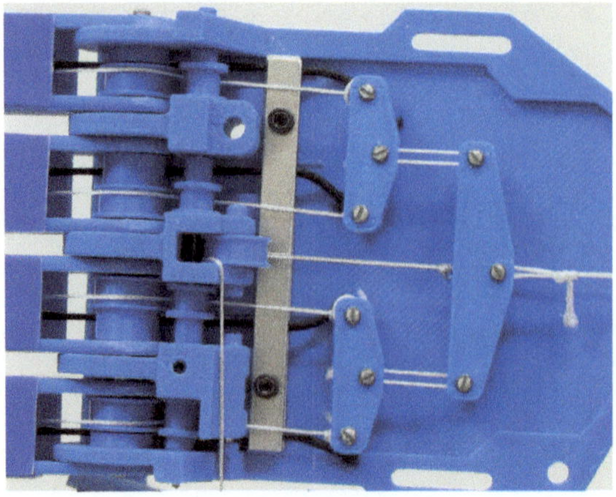

Figure 8. Mechanism allowing underactuation between the fingers and coupling with the thumb. The thumb has been removed for clarity. The thumb tendon is attached at the centre of the large seesaw.

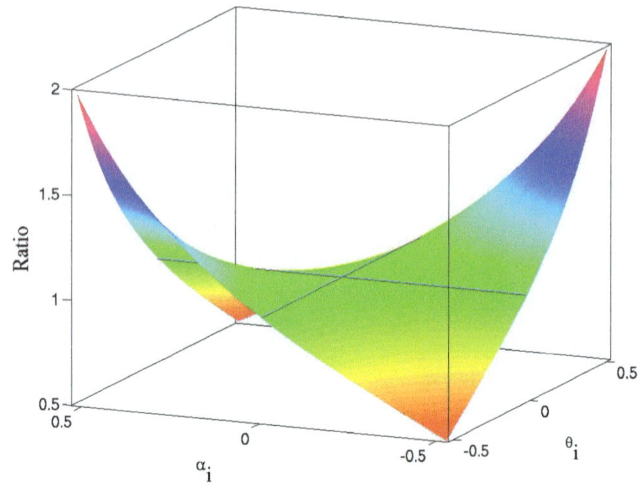

Figure 9. Ratio of the output forces of a triangular seesaw as a function of the floating body geometry (θ_i) and orientation (α_i).

in Gosselin et al. (2008) this configuration gives good practical results, mainly because the thumb works in opposition with the other fingers. The mechanism implemented is illustrated in Fig. 8.

5.2 Static modelling of the underactuation between the fingers

In Fig. 7, it should be noted that if the angle θ_i describing the geometry of a floating body is negative, then the triangle will be pointing towards the bottom instead of the top. This affects the kinematics of the mechanism for the cases of underactuation between the different outputs, as described in the following equations. To simplify the analysis, lateral movements between the triangles are neglected, i.e., it is assumed that the cables remain parallel. This assumption is valid as long as the lateral displacements are significantly smaller than the length of the cables. From the static equilibrium, one has

$$F_1 = \frac{c_{p1}c_{p2}F_a}{c_{p1}c_{p2}+c_{p1}c_{m2}+c_{m1}c_{p2}+c_{m1}c_{m2}} \tag{11}$$

$$F_2 = \frac{c_{m1}c_{p2}F_a}{c_{p1}c_{p2}+c_{p1}c_{m2}+c_{m1}c_{p2}+c_{m1}c_{m2}} \tag{12}$$

$$F_3 = \frac{c_{p3}c_{m2}F_a}{c_{p3}c_{p2}+c_{p3}c_{m2}+c_{m3}c_{p2}+c_{m3}c_{m2}} \tag{13}$$

$$F_4 = \frac{c_{m3}c_{m2}F_a}{c_{p3}c_{p2}+c_{p3}c_{m2}+c_{m3}c_{p2}+c_{m3}c_{m2}} \tag{14}$$

where

$$c_{pi} = \cos(\alpha_i+\theta_i) \quad i=1,2,3 \tag{15}$$

$$c_{mi} = \cos(\alpha_i-\theta_i) \quad i=1,2,3 \tag{16}$$

When the triangular seesaw rotates, its two outputs will not be equal because the lever arms are changing. Figure 9 summarizes the effect of the angles θ_i and α_i on the ratio of the two output forces of a triangular seesaw. As a remark, if $\theta_i = 0$ or if $\alpha_i = 0$, then the two output forces are equal.

It is important to mention that it is preferable to let the triangles point downward ($\theta_i < 0$), since the mechanism is more likely to stay in an equilibrium configuration if a resistive force (for example friction) is slightly larger at one of the outputs. Then, an external perturbation will only slightly desynchronize the output positions. This is due to the fact that the seesaw applies more force on the output offering more resistance. On the contrary, if the triangles point upward ($\theta_i > 0$), the seesaw will flip on its side as soon as the resisting force is slight larger at one of the outputs. Then, a small external perturbation will strongly desynchronize the output positions. This is due to the fact that the seesaw applies a smaller force on the output with more resisting force. In practice, a small negative θ_i is suggested. This generally allows to maintain the synchronization of the mechanism when small perturbations, such as friction, are applied, but leads to relatively equal forces on four outputs locked at different positions.

6 Hand integration

The prototype of prosthetic hand is illustrated in Figs. 10, 11, 12, 13 and 14. It is made of plastic and built using rapid prototyping. The connection between the hand and the prosthesis forearm is through a steel shaft with threads at both ends. This shaft is inserted tightly in a part made of rapid prototyping and the tip pointing inside the hand is fixed with a nut. For the other end, a small adapter is screwed on so the hand is compatible with the prosthesis forearm. For the actuation of the hand, the main cable actuating the mechanism in the

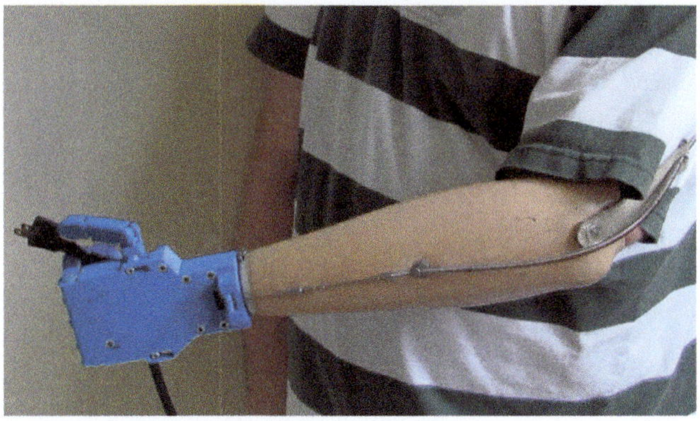

Figure 10. The prosthetic hand and forearm used by an amputee, grasping a plug.

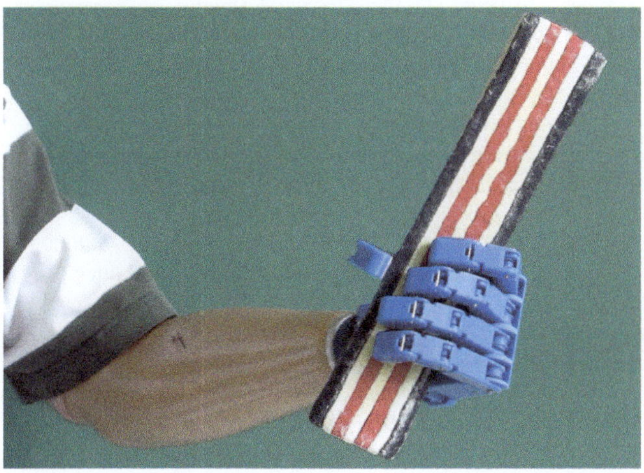

Figure 11. The prosthetic hand and forearm used by an amputee, grasping an eraser.

palm is simply attached to the steel cable of the prosthesis. The hand connected to the prosthesis forearm is illustrated in Figs. 10 and 11. The prosthesis is actuated through a harness by a motion of the shoulders of the amputee. It is noted that a handle can be attached to the hand in order to actuate it manually. Many objects have been grasped with enveloping grasps (ball, wine glass, bottle, etc.), pinch grasps (nut, small box, etc.) and lateral pinch grasps (spring, fork, paper, coin, etc.). As expected, the lateral pinch grasp is more stable than the opposing pinch grasp. Examples are illustrated in Figs. 12, 13 and 14.

In order to estimate the improvement of the efficiency of the prosthetic hand over the previous design, preliminary measurements of the forces on a grasped object have been performed with a setup similar to the one used in Gosselin et al. (2008). For instance, the actuation is applied by suspending weights (the forearm and harness are removed) and the force on the object is measured using a dynamometer.

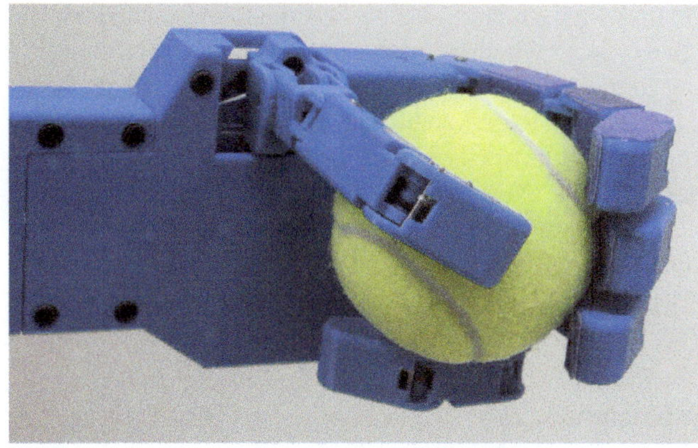

Figure 12. The prosthetic hand grasping a tennis ball with an enveloping grasp.

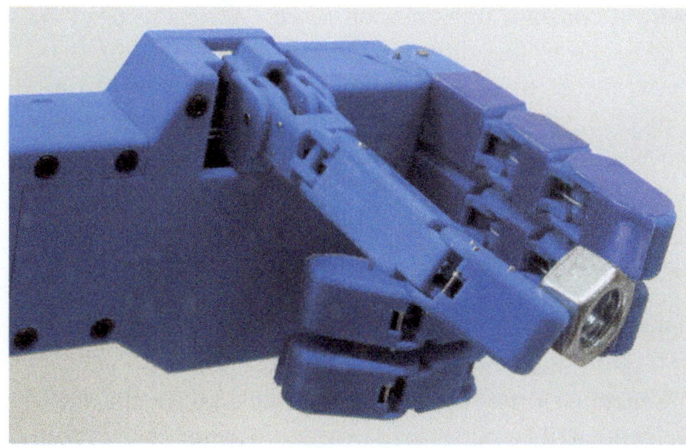

Figure 13. The prosthetic hand grasping a nut with an opposing pinch grasp.

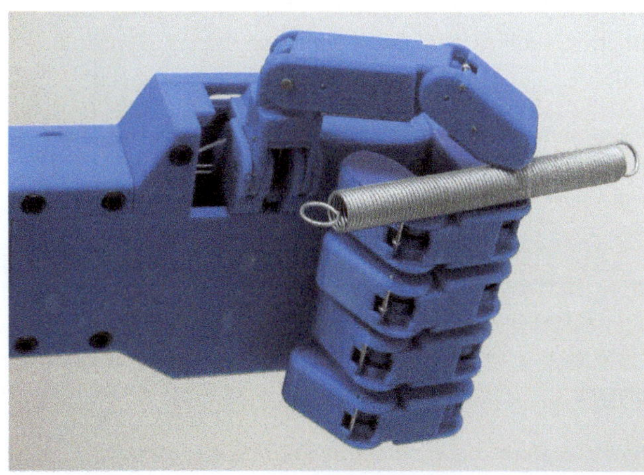

Figure 14. The prosthetic hand grasping a spring with a lateral pinch grasp.

Two situations are studied. First, the fingers are closed towards the object, which reproduces the conditions corresponding to the grasping of an object. Second, the object is pushed towards the fingers, which corresponds to an external force applied on a grasped object. The efficiency can then be estimated from the ratio between the forces measured. The estimated efficiency of the prosthetic hand is 0.70, which is a significant improvement over the efficiency of 0.35 of the previous prototype. This improved efficiency increases the capability to apply grasping forces.

7 Conclusions

In this paper, the ongoing design of a prosthetic hand is presented. The prosthetic hand is based on an anthropomorphic robotic hand previously developed. The actuation mechanism of the fingers is improved in order to reduce friction and obtain a constant transmission force. Also, the prototype is designed to test different mechanisms of underactuation between the fingers. A promising version, based on seesaw mechanisms, is included. It leads to reduced friction and improved behaviour. Finally a new thumb, which allows motion outside of a single plane is introduced. This thumb allows enveloping and pinch grasps with other fingers and lateral pinch grasps between the thumb and the side of the index finger. Preliminary experiments conducted with the prototype demonstrate the effectiveness of each type of grasp. Although preliminary force measurements confirmed a signifiant reduction of the friction in the hand, additional force measurements will be performed in order to better characterize the prosthetic hand. Also, the qualitative performance of the grasp on a variety of objects will be evaluated. Finally, the work presented here is mainly devoted to the underactuated transmission mechanisms. Therefore, significant work is still required in order to address other issues pertaining to prosthetics, such as easthetics, actuation means, covering, and others.

Acknowledgement. The authors gratefully acknowledge the financial support of the Natural Sciences and Engineering Research Council of Canada (NSERC) as well as the Canada Research Chair program.

Edited by: J. L. Herder
Reviewed by: two anonymous referees

References

Almström, C., Herberts, P., and Korner, L.: Experience with Swedish Multifunctional Prosthetic Hands Controlled by Pattern Recognition of Multiple Myoelectric Signals, Int. Orthop., 5, 15–21, 1981.

Baril, M., Laliberté, T., Guay, F., and Gosselin, C.: Static Analysis of Single-Input/Multiple-Output Tendon-Driven Underactuated Mechanisms for Robotic Hands, in: Proceedings of the ASME IDETC/CIE 2010, Montreal, Canada, 2010.

Birglen, L. and Gosselin, C.: Force Analysis of Connected Differential Mechanisms: Application to Grasping, Int. J. Rob. Res., 25, 1033–1046, 2006.

Birglen, L., Laliberté, T., and Gosselin, C.M.: Underactuated Robotic Hands, Springer Tracts in Advanced Robotics, 40, 244 pp., 2008.

Castellini, C., van der Smagt, P., Sandini, G., and Hirzinger, G.: Surface EMG for Force Control of Mechanical Hands, in: Proceedings of the 2008 IEEE International Conference on Robotics and Automation, Pasadena, CA, 2008.

Chu, J. U., Moon, I., Lee, Y. J., Kim, S. K., and Mun, M. S.: A Supervised Feature-Projection-Based Real-Time EMG Pattern Recognition for Multifunction Myoelectric Hand Control, IEEE/ASME Trans. Mechatron., 12, 282–290, 2007.

Crisman, J. D., Kanojia, C., and Zeid, I.: Graspar: A Flexible, Easily Controllable Robotic Hand, IEEE Rob. Autom. Mag., June, 32–38, 1996.

Dollar, A. M. and Howe, R. D.: The SDM Hand as a Prosthetic Terminal Device: A Feasibility Study, in: Proceedings of the 2007 IEEE International Conference on Rehabilitation Robotics (ICORR), Noordwijk, Netherlands, 2007.

Dubey, V. N. and Crowder, R. M.: A Finger Mechanism for Adaptive End Effectors, in: Proceedings of 2002 ASME Design Engineering Technical Conferences, Montreal, Canada, 2002.

Fukaya, N., Toyama, S., Asfour, T., and Dillmann, R.: Design of the Tuat/Karlsruhe Humanoid Hand, in: Proceedings of 2000 IEEE/RSJ International Conference on Advanced Robotics, 3, Takamatsu, Japan, 1754–1759, 2000.

Gosselin, C., Pelletier, F., and Laliberté, T.: An Anthropomorphic Underactuated Robotic Hand with 15 Dofs and a Single Actuator, in: Proceedings of the 2008 IEEE International Conference on Robotics and Automation, Pasadena, CA, 2008.

Herder J. L., Visser H. de: Force Directed Design of a Voluntary Closing Hand Prosthesis, in: Proceedings ASME DETC 26th Biennial Mechanisms and Robotics Conference, Baltimore, MD, 2000.

Hirose, S. and Umetani, Y.: The Development of Soft Gripper for the Versatile Robot Hand, Mech. Mach. Theory, 13, 351–358, 1978.

Kamikawa, Y. and Maeno, T.: Underactuated Five-Finger Prosthetic Hand Inspired by Grasping Force Distribution of Humans, in: Proceedings of the IEEE/RSJ International Conference on Intelligent Robots and Systems, Nice, France, 2008.

Kyberd, P. J., Light, C., Chappell, P. H., Nightingale, J. M., Whatley, D., and Evans, M.: The Design of Anthropomorphic Prosthetic Hands: A Study of the Southampton Hand, Robotica, 19, 593–600, 2001.

Laliberté, T. and Gosselin, C.: Simulation and Design of Underactuated Mechanical Hands, Mech. Mach. Theory, 33, 39–57, 1998.

Massa, B., Roccella, S., Carrozza, M. C., and Dario, P.: Design and Development of an Underactuated Prosthetic Hand, in: Proceedings of the 2002 IEEE International Conference on Robotics and Automation, Washington, DC, 3374–3379, 2002.

Rakic, M.: Multifingered Robot Hand with Selfadaptability, Rob. Comput. Integr. Manuf., 3, 269–276, 1989.

Shimojima, H., Yamamoto, K., and Kawakita, K.: A Study of Grippers with Multiple Degrees of Mobility, JSME Int. J., 30, 515–522, 1987.

5

Analysis of servo-constraint problems for underactuated multibody systems

R. Seifried[1] **and W. Blajer**[2]

[1]Institute of Engineering and Computational Mechanics, University of Stuttgart, Pfaffenwaldring 9, 70569 Stuttgart, Germany

[2]Faculty of Mechanical Engineering, Technical University of Radom, ul. Krasickiego 54, 26-600 Radom, Poland

Correspondence to: R. Seifried (robert.seifried@itm.uni-stuttgart.de)

Abstract. Underactuated multibody systems have fewer control inputs than degrees of freedom. In trajectory tracking control of such systems an accurate and efficient feedforward control is often necessary. For multibody systems feedforward control by model inversion can be designed using servo-constraints. So far servo-constraints have been mostly applied to differentially flat underactuated mechanical systems. Differentially flat systems can be inverted purely by algebraic manipulations and using a finite number of differentiations of the desired output trajectory. However, such algebraic solutions are often hard to find and therefore the servo-constraint approach provides an efficient and practical solution method. Recently first results on servo-constraint problems of non-flat underactuated multibody systems have been reported. Hereby additional dynamics arise, so-called internal dynamics, yielding a dynamical system as inverse model. In this paper the servo-constraint problem is analyzed for both, differentially flat and non-flat systems. Different arising important phenomena are demonstrated using two illustrative examples. Also strategies for the numerical solution of servo-constraint problems are discussed.

1 Introduction

Multibody systems with fewer control inputs than degrees of freedom are called underactuated. Typical examples are multibody systems with passive joints, body flexibility, joint elasticities, aircrafts and cranes. A possible performance task of such systems is output trajectory tracking, e.g. tracking of the end-effector point of flexible manipulators. In order to obtain a good performance in trajectory tracking an accurate and efficient feedforward control is often necessary, which then can be combined with a feedback controller. A feedforward control is an inverse model of the multibody system, providing the necessary control inputs for exact output reproduction. Depending on the system's properties the inverse model might be purely algebraic or might contain a dynamical part. While there is a large amount of various linear and nonlinear feedback control strategies available, there exist much less concepts for feedforward control design of nonlinear systems.

A very appealing and efficient feedforward control design approach for multibody systems is the use of so-called servo-constraints, which are also called programm constraints or control constraints, see Blajer (1992); Campbell (1995); Kirgetov (1967); Rosen (1999); Bajodah et al. (2005). Thereby the equations of motion of the underactuated multibody system are supplemented by a servo-constraint, which enforces the exact reproduction of the desired output trajectory. This yields as set of differential-algebraic equations (DAEs), whose solution provides the searched control input. Due to some similarities to classical constraints, servo-constraint problems have recently been attracted increasing attention in the multibody system dynamics context.

The servo-constraint approach is often applied to so-called differentially flat underactuated mechanical systems, such as cranes, aircrafts and mass-spring chains, as investigated in Betsch et al. (2008); Blajer and Kolodziejczyk (2004, 2007, 2011); Fumagalli et al. (2011). The theory of differential flatness is a differential algebraic approach, which is due to the fundamental work of Fliess et al. (1995). Differential flatness is a structural property, which is determined by the system and the imposed system output. Roughly speaking, in a differentially flat system a system output can be found, from which all states and inputs can be determined without integration. However, a finite number of derivatives of the output might have to be taken. Differentially flat nonlinear systems can be seen as a generalization of linear controllable systems, as discussed by Rothfuss et al. (1997). These systems have the favorable property that they can be inverted purely by algebraic manipulations and using a finite number of differentiations of the system output. However, such algebraic solutions are often hard to find and therefore the servo-constraint approach provides an efficient and practical solution method for model inversion. An extension of this servo-constraint approach is its use in feedback linearization, where the model is formulated in redundant coordinates, see Frye and Fabien (2011).

More recently also first results on the application of servo-constraints to non-flat systems, such as, e.g. flexible manipulators or systems with passive joints, have been reported, see Seifried (2012a); Moberg and Hanssen (2007); Kovács et al. (2011); Masarati et al. (2011). In the inverse model of non-flat systems additional dynamics arise, so-called internal dynamics. Thus the inverse model is a dynamical model. This internal dynamics of the inverse model might be stable or unstable. Therefore it must be analyzed carefully in order to obtain a meaningful solution and is treated extensively in differential-geometric nonlinear control theory, see Isidori (1995); Sastry (1999).

In this paper the servo-constraint problem is analyzed for both, flat and non-flat systems. Two approaches for analyzing and solving the servo-constraint problem are taken. These are a projection approach and a coordinate transformation approach. The projection approach is due to Blajer and Kolodziejczyk (2004, 2007). It allows a straightforward formulation of the servo-constraint problem and simplifies significantly the numerical solution of the arising DAEs. This method has also been applied to differentially flat multibody systems with mixed geometric and servo-constraints as reported by Betsch et al. (2008) and Blajer and Kolodziejczyk (2011). In the coordinate transformation approach the servo-constraint problem is reformulated in new coordinates containing the output. In this way a DAE formulation might be avoided, which significantly simplifies the analysis of the servo-constraint system dynamics. The equivalence of both approaches is discussed in Seifried (2012a). In this paper a slightly different formulation of the coordinate transformation approach is used, see also Blajer and Seifried (2012); Blajer et al. (2013). Based on both formulations of the servo-constraint problem, the various possible situations which can occur in servo-constraint problems are demonstrated. Therefore two illustrative examples are used. These are a mass-spring-damper system on a car and a rotational manipulator arm with passive joint. Finally some remarks on the numerical solution of servo-constraint problems are given. The numerical solution methods depend strongly on the previously analyzed system properties.

2 Trajectory tracking of underactuated multibody systems

Multibody systems with f degrees of freedom and m control inputs are considered. For underactuated multibody systems it is imperative $m < f$. The kinematics of multibody systems is described using generalized coordinates $\boldsymbol{q} \in \mathbb{R}^f$. The control inputs $\boldsymbol{u} \in \mathbb{R}^m$ are assumed to be control forces and torques. Based on d'Alembert's principle the equations of motion in minimal form can be derived using the Newton-Euler-Formalism, see e.g. Schiehlen et al. (2006). The nonlinear equations of motion are given by,

$$\boldsymbol{M}(\boldsymbol{q})\ddot{\boldsymbol{q}} + \boldsymbol{f}(\boldsymbol{q},\dot{\boldsymbol{q}}) = \boldsymbol{B}(\boldsymbol{q})\boldsymbol{u}, \tag{1}$$

where $\boldsymbol{M} \in \mathbb{R}^{f\times f}$ is the symmetrical and positive definite mass matrix and $\boldsymbol{f} \in \mathbb{R}^f$ summarizes all generalized forces. These generalized forces are given by $\boldsymbol{f} = \boldsymbol{k} - \boldsymbol{g}$, whereby $\boldsymbol{k}$ is the vector of generalized gyroscopic, centrifugal and Coriolis forces and $\boldsymbol{g}$ are all applied forces such as gravity. The system inputs $\boldsymbol{u} \in \mathbb{R}^m$ are distributed by the input matrix $\boldsymbol{B} \in \mathbb{R}^{f\times m}$ on the directions of the generalized coordinates. Generally it is assumed that there is no redundant actuation and thus the rank of $\boldsymbol{B}$ equals m. It is often useful to partition the equations of motion of underactuated multibody systems in actuated and unactuated parts,

$$\begin{bmatrix} \boldsymbol{M}_{\mathrm{aa}}(\boldsymbol{q}) & \boldsymbol{M}_{\mathrm{au}}(\boldsymbol{q}) \\ \boldsymbol{M}_{\mathrm{au}}^T(\boldsymbol{q}) & \boldsymbol{M}_{\mathrm{uu}}(\boldsymbol{q}) \end{bmatrix} \begin{bmatrix} \ddot{\boldsymbol{q}}_{\mathrm{a}} \\ \ddot{\boldsymbol{q}}_{\mathrm{u}} \end{bmatrix} + \begin{bmatrix} \boldsymbol{f}_{\mathrm{a}}(\boldsymbol{q},\dot{\boldsymbol{q}}) \\ \boldsymbol{f}_{\mathrm{u}}(\boldsymbol{q},\dot{\boldsymbol{q}}) \end{bmatrix} = \begin{bmatrix} \boldsymbol{B}_{\mathrm{a}}(\boldsymbol{q}) \\ \boldsymbol{B}_{\mathrm{u}}(\boldsymbol{q}) \end{bmatrix} \boldsymbol{u}. \tag{2}$$

Thereby $\boldsymbol{q}_{\mathrm{a}} \in \mathbb{R}^m$ are the actuated generalized coordinates and $\boldsymbol{q}_{\mathrm{u}} \in \mathbb{R}^{f-m}$ are the unactuated coordinates. This partition is based on the requirement that the rank of the submatrix $\boldsymbol{B}_{\mathrm{a}} \in \mathbb{R}^{m\times m}$ equals m. In many instances, e.g. passive joint manipulators or flexible multibody systems, the input submatrices might reduce to $\boldsymbol{B}_{\mathrm{a}} = \boldsymbol{I}$ and $\boldsymbol{B}_{\mathrm{u}} = \boldsymbol{0}$, where $\boldsymbol{I}$ is the identity matrix. In this case each generalized coordinate of $\boldsymbol{q}_{\mathrm{a}}$ is directly collocated with one component of input $\boldsymbol{u}$.

2.1 Output trajectory tracking control

The control task which is considered in this paper is output trajectory tracking. Thereby, a system output $\boldsymbol{y} \in \mathbb{R}^m$ of the multibody system is given by

$$\boldsymbol{y} = \boldsymbol{h}(\boldsymbol{q}). \tag{3}$$

This system output can depend linearly or nonlinearly on all generalized coordinates. A typical example for such a system output is an end-effector position of a manipulator. The dimension of input and output coincide, an assumption often

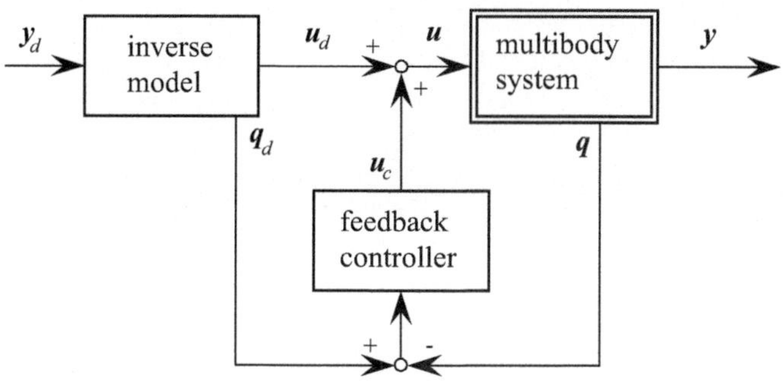

Figure 1. Control structure with feedforward and feedback controller.

required in nonlinear control theory. The velocity and acceleration of the system output follow as

$$\dot{y} = H(q)\dot{q}, \quad (4)$$

$$\ddot{y} = H(q)\ddot{q} + \overline{h}(q, \dot{q}). \quad (5)$$

Thereby $H \in \mathbb{R}^{m \times f}$ is the Jacobian matrix of the system output and $\overline{h} = \dot{H}\dot{q} \in \mathbb{R}^m$. In trajectory tracking the system output (Eq. 3) should track exactly a time-variant output trajectory $y_d(t)$, which is defined in space and time. Thus also velocity $\dot{y}_d(t)$ and acceleration $\ddot{y}_d(t)$ of the system output are specified.

Multibody systems in output trajectory tracking perform often large working motion. Thus, the equations of motion (Eq. 1) are highly nonlinear and in many instances linear control theory cannot be applied. An efficient approach of output trajectory tracking of nonlinear systems is a so-called two design degree-of-freedom control structure, consisting of a feedforward control and an additional feedback control, see Fig. 1. Thereby the feedforward control is an inverse model of the multibody system. It provides for a given output trajectory $y_d(t)$ the associated control inputs u_d and the trajectories q_d of all generalized coordinates. In the absence of any uncertainties and disturbances the control input u_d can be applied to the multibody system and reproduces the desired output trajectory exactly. Since in a real hardware implementation always some parameter uncertainties and disturbances arise, additional feedback control is necessary and provides additional control input u_c. For feedback control design the computed trajectories q_d of the generalized coordinates can be used as reference signal. In trajectory tracking the most control action is provided by the feedforward part and the feedback part has to compensate only small derivations following from uncertainties and disturbances. Therefore, often simple linear control strategies such as PID control might be applicative for the feedback part.

For fully actuated multibody systems, such as fully actuated manipulators, it is $f = m$. Then, the inverse model can be derived easily by pure algebraic manipulations, see e.g. Spong et al. (2006). In this case the inverse model can be split into inverse kinematics and inverse dynamics. In inverse kinematics of a fully actuated system the nonlinear output equation (Eq. 3) can be solved, providing for given y_d the trajectories q_d of the generalized coordinates. This can be achieved by algebraic manipulations, numerical solution or differential kinematics, respectively. For details it is pointed to Siciliano et al. (2010). By using the determined q_d and its derivatives $\dot{q}_d$, $\ddot{q}_d$ in the equations of motion (Eq. 1) the control inputs u_d can be computed algebraically.

For underactuated multibody systems the inverse kinematics following from Eq. (3) is under-determined. Also the inverse dynamics problem cannot be solved since the input matrix B is not invertible. Thus, for underactuated multibody systems the splitting of the model inversion into inverse kinematics and inverse dynamics is in general not possible and both parts must be be solved concurrently. For differentially flat underactuated multibody systems a purely algebraic inverse model can be derived, using a finite number of derivatives of the system output y. In contrast, for non-flat systems the inverse model is a dynamical system. Flatness is a system property determined by the system dynamics and the chosen output, but is independent of the used coordinates to describe the multibody system.

2.2 Servo-constraints in multibody systems

An efficient and straightforward approach for model inversion is the use of servo-constraints. The basic idea of servo-constraints is the enforcement of output trajectory tracking by introducing constraint equations. These servo-constraints can be seen as an extension of classical geometric constraints, which makes this approach so appealing in multibody system dynamics. In order to introduce the concept of servo-constraints, classical constraints are briefly reviewed. For example, consider a multibody system with a kinematic loop, see Fig. 2. In order to obtain its equation of motion the kinematic loop is cut at a suitable joint, removing n constraints. Then, the corresponding equations of motion of the open chain system are derived in minimal form and the kinematic loop is enforced by introducing algebraic loop closing constraints $c_c(q, t) = 0 \in \mathbb{R}^n$. Restricting to a multibody system without control action, the equations of motion of the open chain system yields together with the constraint equations,

$$M(q)\ddot{q} + f(q, \dot{q}) = C^T \lambda, \quad (6)$$

$$c_c(q) = 0. \quad (7)$$

Thereby $C = \partial c_c / \partial q \in \mathbb{R}^{f \times n}$ is the Jacobian matrix of the constraint equation and $\lambda \in \mathbb{R}^n$ are the Lagrangian multipliers. These ensure that the loop closing constraints are met. Equations (6)–(7) form a set of differential algebraic equations (DAE). The numerical solution of the DAE provides the forward dynamics of the closed loop multibody system with trajectories of the states q, $\dot{q}$ and the Lagrangian multipliers λ.

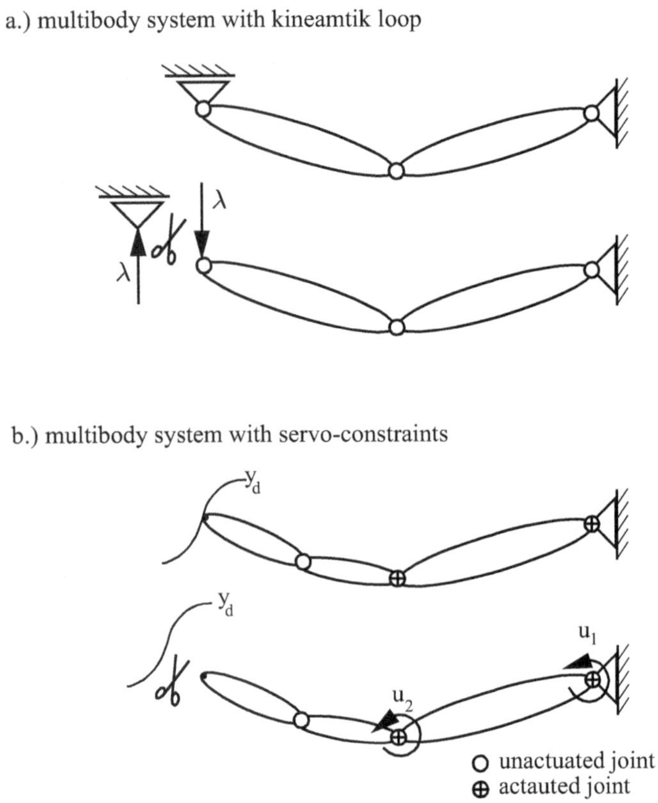

Figure 2. Multibody systems with constraints.

In order to derive an inverse model for underactuated multibody systems a similar approach can be used. The problem of tracking a desired trajectory $\boldsymbol{y}_d$ is induced by introducing m algebraic servo-constraints. The inverse model of an underactuated multibody systems is then given by the equations motion (Eq. 1) and the servo-constraints,

$$\boldsymbol{M}(\boldsymbol{q})\ddot{\boldsymbol{q}} + \boldsymbol{f}(\boldsymbol{q},\dot{\boldsymbol{q}}) = \boldsymbol{B}(\boldsymbol{q})\boldsymbol{u}, \tag{8}$$

$$\boldsymbol{c}(\boldsymbol{q},t) = \boldsymbol{h}(\boldsymbol{q}) - \boldsymbol{y}_d(t) = \boldsymbol{0}, \tag{9}$$

where Eq. (9) represents the servo-constraint. As noticed by Blajer (1997a); Blajer and Kolodziejczyk (2004, 2007) the servo-constraint problem Eqs. (8)–(9) is mathematically equivalent to Eqs. (6)–(7). Thereby the desired trajectory $\boldsymbol{y}_d(t)$ can be interpreted as a drift in time of constraint manifold $\boldsymbol{c}(\boldsymbol{q}) = \boldsymbol{0}$ in the system configuration space, see Blajer (2001). The generalized actuating forces $\boldsymbol{Bu}$ can then be viewed as a generalized reaction forces of the servo constraints. Thus, structurally the generalized actuation forces $\boldsymbol{Bu}$ corresponds to the generalized reaction forces $\boldsymbol{C}^T\boldsymbol{\lambda}$. Therefore, in the servo-constraint approach the control inputs $\boldsymbol{u}$ ensure that the servo-constraints are met. The similarities between both cases are illustrated in Fig. 2.

At first, multibody systems with servo- and physical constraints show many similarities. However, servo-constraint problems can posses more complex properties, which have to be understood to obtain a meaningful solution. For a multibody system with classical constraints the matrix $\boldsymbol{C}^T$ projects the Lagrange multipliers $\boldsymbol{\lambda}$ on the directions orthogonal to the constraint manifold, which is defined by the constraint equation (Eq. 7). Thus, the generalized reaction forces $\boldsymbol{C}^T\boldsymbol{\lambda}$ are orthogonal to the constraint manifold, see Fig. 3. Therefore such a system is called an ideal orthogonal realization.

In contrast, the generalized actuation forces $\boldsymbol{Bu}$ are not necessarily ideal orthogonal to the constraint manifold which is defined by the servo-constraint (Eq. 9). The actuation forces $\boldsymbol{Bu}$ might be non-ideal orthogonal or in the extreme case even tangential to the constraint manifold, see Fig. 3. These cases are called non-ideal orthogonal realization and tangential realization, respectively. Thereby, ideal and non-ideal orthogonal realization have many similarities, and are in the following only distinguished, if different phenomena occur. In the case of both types of orthogonal realization control inputs are explicitly available in all directions orthogonal to the servo-constraints and can directly actuate the servo-constraint condition. However, in the case of non-ideal orthogonal realization the projection with $\boldsymbol{B}$ yields also components in direction tangential to the constraint manifold. In a tangential realization the control inputs are projected in tangential direction. Then, the control inputs $\boldsymbol{u}$ cannot actuate directly the constraint condition, but output tracking of $\boldsymbol{y}_d$ might be still possible due to coupling with other forces of the system, see Blajer and Kolodziejczyk (2004). This tangential projection is often connected to underactuated differentially flat systems. However, as will be shown in section 3 tangential realization can also arise in non-flat systems. Further analysis from a geometric point of view are also found in Blajer (1992), Blajer and Kolodziejczyk (2004) and Blajer and Kolodziejczyk (2007).

For systems with multiple inputs and outputs it might occur that both, orthogonal (ideal or non-ideal) and tangential realization exist. Thus, in so-called mixed orthogonal-tangential realizations only some outputs can be directly influenced by the inputs, while others can only be influenced indirectly. A measure of the control singularity is the deficiency in rank p of the matrix

$$\boldsymbol{P} = \boldsymbol{H}\boldsymbol{M}^{-1}\boldsymbol{B}. \tag{10}$$

The case $p = m$ indicates that all components of the system output $\boldsymbol{y}$ can directly be actuated by the inputs $\boldsymbol{u}$. The case $0 < p < m$ shows that only p components of the output can be regulated in the orthogonal way, while realization of the other $m - p$ output components are without direct involvement of the actuating forces $\boldsymbol{Bu}$. Finally, $p = 0$ refers to a pure tangential realization of servo-constraints, as the system inputs do not directly influence the outputs.

If the solution of the servo constraint problem (Eqs. 8–9) exists, the numerical solution of this DAE provides the trajectories of all states $\boldsymbol{q}_d$, $\dot{\boldsymbol{q}}_d$ as well as the corresponding control inputs $\boldsymbol{u}_d$. For characterizing DAEs often the differentiation index is used. Following Hairer and Wanner (2010) the differentiation index is the number of derivatives of the DAE,

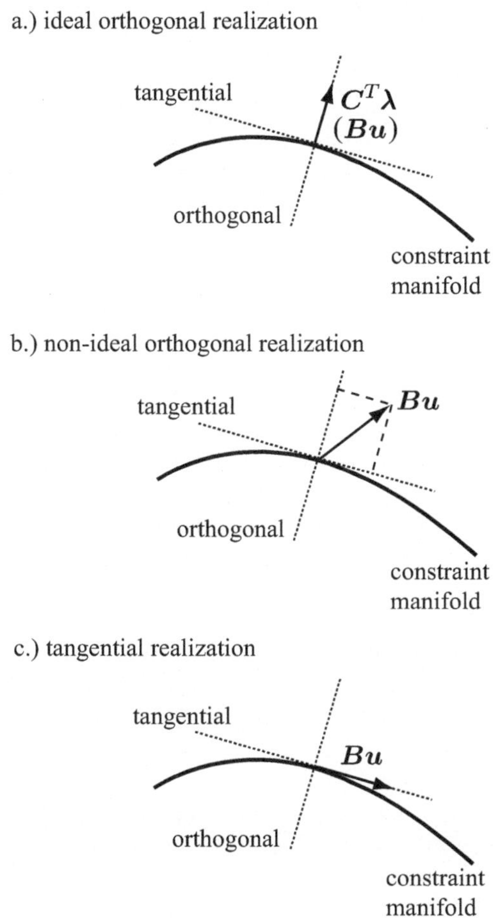

Figure 3. Possible realizations in servo-constraint problems.

or parts of it, until an ordinary differential equation (ODE) for all unknowns are obtained. For multibody systems with classical constraints it is well know that they have differentiation index 3. Here the constraint equation (Eq. 7) must be differentiated three times in order to derive an ODE for the unknown λ. This is provided by the fact that for classical constraints the matrix (Eq. 10) becomes $\boldsymbol{P} = \boldsymbol{C}\boldsymbol{M}^{-1}\boldsymbol{C}^T$, which has full rank if the constraints are independent, see Hairer and Wanner (2010).

In the case of servo-constraints this is not any more necessarily true. In the case of orthogonal realizations $\boldsymbol{P}$ has full rank and index 3 arises. However, if the servo-constraint problem includes a tangential realization $\boldsymbol{P}$ is singular and higher differentiation index arise. For various mechanical systems with servo-constraints the differentiation index is analyzed in Campbell (1995).

The differentiation index is closely related to the relative degree used in differential geometric nonlinear control theory. An extensive treatment of this nonlinear control theory is given in Isidori (1995); Sastry (1999). Restricting to systems with n states and one input and one output, the relative degree r is the number of Lie derivatives of the system output until the first time the control input occurs. If the relative degree is $r = n$, then the system is differentially flat and a purely algebraic inverse model can be extracted. In the case $r < n$ so-called internal dynamics remain and the inverse model will contain a dynamical part. For extension to systems with multiple inputs and outputs it is pointed to the aforementioned nonlinear control literature. In Campbell (1995) it is pointed out that the differentiation index is one higher than the relative degree, if the internal dynamics are not affected by a constraint.

2.3 Projection approach

In Blajer (1997a) it is shown that the equation of motion with additional constraints can be projected into two complementary subspaces in velocity space. These are the constrained and unconstrained subspace. The unconstrained subspace is tangential to the constraint manifold, while the constrained subspace is orthogonal to it, see Fig. 3. The constrained subspace describes in the servo-constraint context the output subspace and follows from projection with the Jacobian matrix $\boldsymbol{H} \in \mathbb{R}^{m\times f}$ of the output, which has rank m. For the second subspace an orthogonal complement $\boldsymbol{D} \in \mathbb{R}^{f\times f-m}$ with rank $f - m$ must be derived, such that

$$\boldsymbol{H}\boldsymbol{D} = \boldsymbol{0} \quad \text{and} \quad \boldsymbol{D}^T\boldsymbol{H}^T = \boldsymbol{0} \tag{11}$$

is satisfied. Using these two matrices the equations of motion are projected into the two subspaces,

$$\begin{bmatrix} \boldsymbol{H}\boldsymbol{M}^{-1} \\ \boldsymbol{D}^T \end{bmatrix}\left(\boldsymbol{M}\ddot{\boldsymbol{q}} + \boldsymbol{f} - \boldsymbol{B}\boldsymbol{u}\right) = \boldsymbol{0}, \tag{12}$$

which yields,

$$\boldsymbol{H}\ddot{\boldsymbol{q}} + \boldsymbol{H}\boldsymbol{M}^{-1}\boldsymbol{f} = \boldsymbol{H}\boldsymbol{M}^{-1}\boldsymbol{B}\boldsymbol{u}, \tag{13}$$

$$\boldsymbol{D}^T\boldsymbol{M}\ddot{\boldsymbol{q}} + \boldsymbol{D}^T\boldsymbol{f} = \boldsymbol{D}^T\boldsymbol{B}\boldsymbol{u}. \tag{14}$$

With the output equation (Eq. 5) at acceleration level the corresponding servo-constraint provides $\boldsymbol{H}\ddot{\boldsymbol{q}} = \ddot{\boldsymbol{y}}_d - \overline{\boldsymbol{h}}$. This relationship can be used in Eq. (13). Introducing the new state $\boldsymbol{v} = \dot{\boldsymbol{q}}$ and adding the servo-constraints at position level, after reordering the projected servo-constraint formulation, provides

$$\dot{\boldsymbol{q}} = \boldsymbol{v} \tag{15}$$

$$\boldsymbol{D}^T\boldsymbol{M}\dot{\boldsymbol{v}} = -\boldsymbol{D}^T\boldsymbol{f} + \boldsymbol{D}^T\boldsymbol{B}\boldsymbol{u} \tag{16}$$

$$\boldsymbol{0} = \boldsymbol{P}\boldsymbol{u} - \boldsymbol{H}\boldsymbol{M}^{-1}\boldsymbol{f} + \overline{\boldsymbol{h}} - \ddot{\boldsymbol{y}}_d \tag{17}$$

$$\boldsymbol{0} = \boldsymbol{h}(\boldsymbol{q}) - \boldsymbol{y}_d. \tag{18}$$

This forms a set of $2f + m$ differential-algebraic equations for the $2f + m$ unknowns $\boldsymbol{q}, \boldsymbol{v}, \boldsymbol{u}$. Equation (17) has dimension m and describes an algebraic equation in $\boldsymbol{q}, \boldsymbol{v}, \boldsymbol{u}$. Together with the m servo-constraints (Eq. 18) there are $2m$ algebraic equations in these DAEs. By this projection and incorporating the servo-constraint at acceleration level, an index reduction by two is achieved, which in general simplifies the numerical solution. For example, for the crane considered in Blajer and Kolodziejczyk (2004, 2007), the differentiation index is reduced from 5 to 3 using this projection approach.

2.4 Coordinate transformation

The numerical solution of Eqs. (15)–(18) provides the model inversion. However, for analysis purpose it might be of advantage to write first the equations of motion in a new set of coordinates containing the system output. In addition, this might also be used to simplify the projections in Eqs. (15)–(18). Also for systems with orthogonal realization an inverse model as ODE can be derived in a straightforward way. For example, as new set of coordinates

$$\overline{\boldsymbol{q}} = \begin{bmatrix} \boldsymbol{y} \\ \boldsymbol{q}_u \end{bmatrix} = \begin{bmatrix} \boldsymbol{h}(\boldsymbol{q}_a, \boldsymbol{q}_u) \\ \boldsymbol{q}_u \end{bmatrix} = \boldsymbol{\phi}(\boldsymbol{q}) \tag{19}$$

might be used, which yields for the velocities

$$\dot{\overline{\boldsymbol{q}}} = \begin{bmatrix} \dot{\boldsymbol{y}} \\ \dot{\boldsymbol{q}}_u \end{bmatrix} = \begin{bmatrix} \boldsymbol{H}_a & \boldsymbol{H}_u \\ \boldsymbol{0} & \boldsymbol{I} \end{bmatrix} \begin{bmatrix} \dot{\boldsymbol{q}}_a \\ \dot{\boldsymbol{q}}_u \end{bmatrix} = \frac{\partial \boldsymbol{\phi}(\boldsymbol{q})}{\partial \boldsymbol{q}} \dot{\boldsymbol{q}}. \tag{20}$$

It is noted that the first row of Eq. (20) is identical to Eq. (4), i.e. $\boldsymbol{H} = [\boldsymbol{H}_a \ \boldsymbol{H}_u]$. In order to be an admissible coordinate transformation, it must be a diffeomorphism, i.e. smooth and invertible. Relationship (Eq. 19) is at least a local diffeomorphism if the Jacobian matrix in Eq. (19) is non-singular. Inspecting Eq. (19) shows, that this is true if the submatrix $\boldsymbol{H}_a$ is nonsingular. This requires that the output equation (Eq. 3) depends on all m actuated coordinates $\boldsymbol{q}_a$. This is for example the case for manipulators with flexible links or passive joints in end-effector tracking. A counterexample is a manipulator with flexible joints in end-effector tracking, see e.g. De Luca (1998).

With the results of Eq. (5) the coordinate transformation (Eq. 19) at acceleration level is,

$$\ddot{\overline{\boldsymbol{q}}} = \begin{bmatrix} \ddot{\boldsymbol{y}} \\ \ddot{\boldsymbol{q}}_u \end{bmatrix} = \begin{bmatrix} \boldsymbol{H}_a & \boldsymbol{H}_u \\ \boldsymbol{0} & \boldsymbol{I} \end{bmatrix} \begin{bmatrix} \ddot{\boldsymbol{q}}_a \\ \ddot{\boldsymbol{q}}_u \end{bmatrix} + \begin{bmatrix} \overline{\boldsymbol{h}} \\ \boldsymbol{0} \end{bmatrix}. \tag{21}$$

From the equations of motion (Eq. 1) follows $\ddot{\boldsymbol{q}} = \boldsymbol{M}^{-1}(\boldsymbol{B}\boldsymbol{u} - \boldsymbol{f})$ which can be inserted in Eq. (21) and yields

$$\ddot{\boldsymbol{y}} = \boldsymbol{P}\boldsymbol{u} - \boldsymbol{H}\boldsymbol{M}^{-1}\boldsymbol{f} + \overline{\boldsymbol{h}}, \tag{22}$$

$$\ddot{\boldsymbol{q}}_u = [\boldsymbol{0} \ \vdots \ \boldsymbol{I}]\boldsymbol{M}^{-1}(\boldsymbol{B}\boldsymbol{u} - \boldsymbol{f}). \tag{23}$$

In all entries of these two equations the original states $\boldsymbol{q}_a$, $\dot{\boldsymbol{q}}_a$ must be replaced by the new states $\boldsymbol{y}$, $\dot{\boldsymbol{y}}$. Therefore the upper part of Eq. (19) must be solved for $\boldsymbol{q}_a$, which is in general nonlinear. Afterwards the velocities $\dot{\boldsymbol{q}}_a$ can be computed from the linear equations provided by Eq. (20).

The two second order differential Eqs. (22)–(23) represent the equations of motion of the multibody system expressed in the new coordinates $\overline{\boldsymbol{q}}$, which include the system output $\boldsymbol{y}$. The equations of motion (Eqs. 22–23) can be helpful in analyzing both, orthogonal and tangential realization, as will be seen in the next subsection.

The coordinate transformation approach is inspired by differential-geometric control theory, which is the basis of feedback linearization and can also be used for feedforward control design, see Isidori (1995); Sastry (1999). Thereby, nonlinear systems are transformed by diffeomorphic coordinate transformations into the so-called nonlinear input-output normal form using new states, containing the output and a finite number of its time derivatives. The application of this nonlinear control theory to underactuated multibody systems in orthogonal realization is given in Seifried (2012a,b).

The following short discussion highlights some correspondence of the servo-constraint approach with aforementioned nonlinear control theory for underactuated multibody systems in orthogonal realization. In this case, the equations of motion (Eqs. 22–23) in new coordinates $\overline{\boldsymbol{q}}$ are identical to the nonlinear input-output normal form. The matrix $\boldsymbol{P} = \boldsymbol{H}\boldsymbol{M}^{-1}\boldsymbol{B}$ is called decoupling matrix in nonlinear control theory. Equation (22) links the input $\boldsymbol{u}$ to the second derivative of the output $\ddot{\boldsymbol{y}}$, describing the input-output relationship. Equation (23) is called in nonlinear control theory the internal dynamics. This is the remaining system dynamics of the inverse model. From this input-output normal form feedback linearization and feedforward control design are easily possible. For a desired output trajectory $\boldsymbol{y}_d$ the necessary control input follows from Eq. (22) as

$$\boldsymbol{u}_d = \boldsymbol{P}^{-1}(\boldsymbol{H}\boldsymbol{M}^{-1}\boldsymbol{f} - \overline{\boldsymbol{h}} + \ddot{\boldsymbol{y}}_d). \tag{24}$$

It is noticed that this is structurally identical to Eq. (17), however expressed in terms of the new coordinates. Equation (24) is an algebraic expression for the input, depending solely on the known values $\boldsymbol{y}_d$, $\dot{\boldsymbol{y}}_d$ and the unknowns $\boldsymbol{q}_u$, $\dot{\boldsymbol{q}}_u$. The later ones follow from solving the internal dynamics by applying Eq. (24) to the ODE Eq. (23), resulting in

$$\begin{aligned} \ddot{\boldsymbol{q}}_u &= [\boldsymbol{0} \ \vdots \ \boldsymbol{I}]\boldsymbol{M}^{-1}[\boldsymbol{B}\boldsymbol{P}^{-1}(\boldsymbol{H}\boldsymbol{M}^{-1}\boldsymbol{f} - \overline{\boldsymbol{h}} + \ddot{\boldsymbol{y}}_d) - \boldsymbol{f}], \\ &= \overline{\boldsymbol{a}}(\boldsymbol{q}_u, \dot{\boldsymbol{q}}_u, \boldsymbol{y}_d, \dot{\boldsymbol{y}}_d, \ddot{\boldsymbol{y}}_d). \end{aligned} \tag{25}$$

These are $f - m$ second order differential equations for $\boldsymbol{q}_u$, driven by the desired output $\boldsymbol{y}_d$ and its derivatives $\dot{\boldsymbol{y}}_d$, $\ddot{\boldsymbol{y}}_d$.

3 Illustrative example 1: mass on car

The different phenomena which might arise in servo-constraint problems of underactuated multibody systems are demonstrated using a spring-mass system mounted on a car, which is shown in Fig. 4. The car with mass m_1 moves along the horizontal e_1 axis and is actuated by the force $u = F$. On the car a mass m_2 is mounted, which moves along an axis which is inclined by the angle α. The system is described by the two generalized coordinates $\boldsymbol{q} = [x, s]$, whereby x is the horizontal car position and s the relative position of the

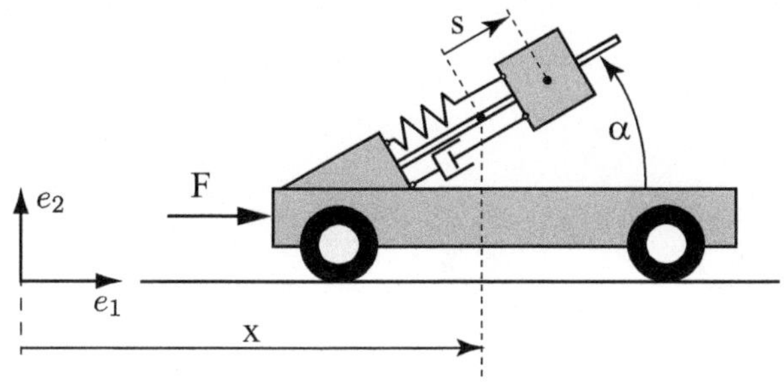

Figure 4. Mass on car system.

mass along the inclined axis. Thereby, $q_a = x$ is the actuated coordinate and $q_u = s$ is the unactuated coordinate. The mass is supported by a parallel spring-damper combination, with spring and damping coefficients k, d, respectively. In the equilibrium position it is $s = 0$. This yields the equations of motion

$$\begin{bmatrix} m_1+m_2 & m_2\cos\alpha \\ m_2\cos\alpha & m_2 \end{bmatrix}\begin{bmatrix} \ddot{x} \\ \ddot{s} \end{bmatrix}+\begin{bmatrix} 0 \\ ks+d\dot{s} \end{bmatrix}=\begin{bmatrix} F \\ 0 \end{bmatrix}. \quad (26)$$

The system output is the horizontal position of the mass,

$$y = x + s\cos\alpha, \quad (27)$$

which should follow a predefined trajectory $y_d(t)$. This yields the servo constraint

$$c(\boldsymbol{q},t) = x + s\cos\alpha - y_d(t). \quad (28)$$

Equations (26) and (28) form the servo-constrained problem. From Eq. (28) follow the two projection matrices,

$$\boldsymbol{H} = \begin{bmatrix} 1 & \cos\alpha \end{bmatrix}, \quad \boldsymbol{D} = \begin{bmatrix} \cos\alpha \\ -1 \end{bmatrix}. \quad (29)$$

With these matrices the projected Eqs. (15)–(18) of the servo-constraint problem can be computed. Thereby it follows from Eq. (18)

$$P = \boldsymbol{H}\boldsymbol{M}^{-1}\boldsymbol{B} = \frac{\sin^2\alpha}{m_1 + m_2\sin^2\alpha}. \quad (30)$$

It becomes apparent that this matrix is nonsingular for $\alpha \neq 0$, i.e. it poses an orthogonal realization. However for $\alpha = 0$ a tangential realization occurs.

In order to analyze this servo-constraint problem in more detail, the coordinate transformation approach presented in Sect. 2.4 is applied. The new set of coordinates containing the output are chosen as $\overline{\boldsymbol{q}} = [y,\, s]$. The system dynamics in new coordinates follows from evaluating Eqs. (22)–(23),

$$\ddot{y} = -\frac{m_1\cos\alpha[ks+d\dot{s}]}{m_1m_2+m_2^2\sin^2\alpha} + \frac{\sin^2\alpha}{m_1+m_2\sin^2\alpha}F, \quad (31)$$

$$\ddot{s} = -\frac{(m_1+m_2)[ks+d\dot{s}]}{m_1m_2+m_2^2\sin^2\alpha} - \frac{\cos\alpha}{m_1+m_2\sin^2\alpha}F. \quad (32)$$

Based on this description of the system dynamics different occurring phenomena are discussed, whereby four different cases are distinct.

Case 1: the relative motion of the mass occurs in vertical direction, i.e. $\alpha = 90°$, see Fig 5a. In this case the system output is identical to the car position $y = x$ and (Eq. 30) reduces to $P = (m_1 + m_2)^{-1} \neq 0$. Thus, the equations of motion Eq. (26) and Eqs. (31)–(32) coincide and provide

$$(m_1+m_2)\ddot{y} = F \quad (33)$$

$$m_2\ddot{s} = -ks - d\dot{s}. \quad (34)$$

Both equations are here fully decoupled. The force F is orthogonal to the constraint manifold, which is described by the servo-constraint (Eq. 28), see Fig. 5a. This is identical to classical geometric constraints and the control force regulates directly the output. The control action which is necessary to reproduce the output trajectory follows from Eq. (33) as $F_d = (m_1 + m_2)\ddot{y}_d$. The dynamics of the mass in vertical direction is described by Eq. (34) and is not influenced by the control force and vice versa. This dynamics of the mass cannot be observed by the system output y and thus in reference to nonlinear control theory this dynamics (Eq. 34) is called internal dynamics, see Isidori (1995) and Sastry (1999).

Case 2: the mass moves along a tilted slope with $0° < \alpha < 90°$, see Fig. 5b. The system dynamics in new coordinates are given by Eqs. (31)–(32) and from Eq. (30) follows $P \neq 0$. This indicates a non-ideal orthogonal realization since the control force has an orthogonal component to the constraint manifold. The control force F can still regulate directly the constraint condition, however it also has a component in tangential direction influencing the relative motion of the mass, see Fig. 5b.

For a given output trajectory y_d Eq. (31) can be solved algebraically for the desired control input,

$$F_d = \frac{m_1+m_2\sin^2\alpha}{\sin^2\alpha}\ddot{y}_d + \frac{m_1\cos\alpha[ks+d\dot{s}]}{m_2\sin^2\alpha}. \quad (35)$$

The control input depends on the second derivative of the system output $\ddot{y}_d$ and the unknown states $s, \dot{s}$ which must be computed from Eq. (32). Replacing in Eq. (32) the control input by Eq. (35) yields after reordering

$$m_2\sin^2\alpha\ddot{s} + d\dot{s} + ks = -\ddot{y}_d m_2\cos\alpha. \quad (36)$$

This is the dynamics of the mass m_2 on the tilted slope under the applied trajectory tracking control force F. In accordance to nonlinear control theory it forms again the internal dynamics. The control force F, and thus the system output y, influence the internal dynamics, while internal dynamics again influence the control force F. In this case

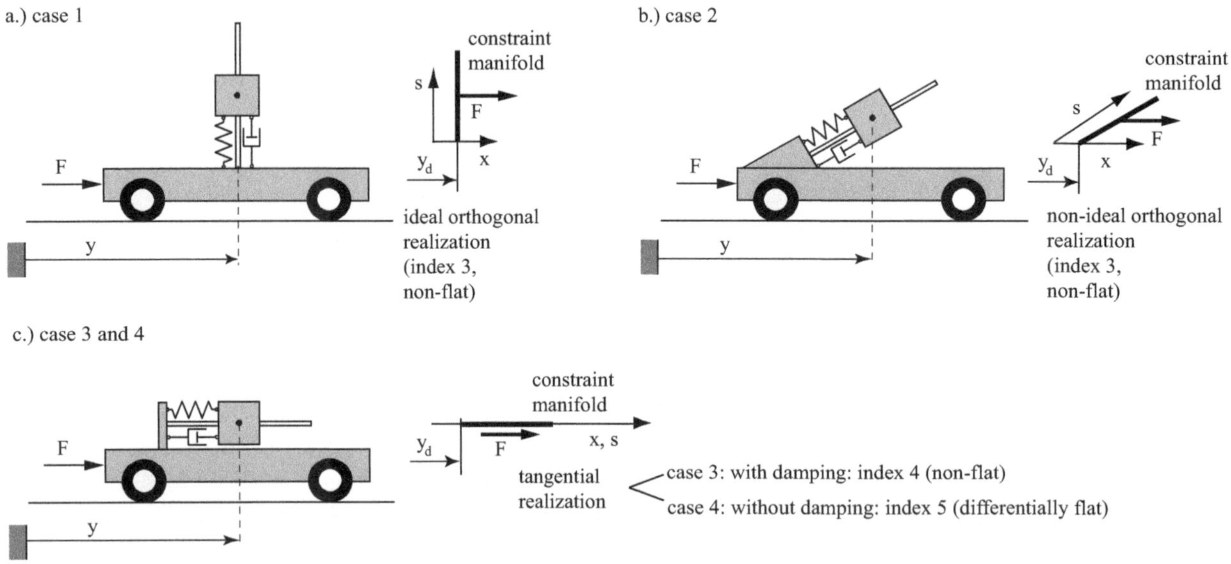

Figure 5. Possible situations arising in servo-constraint problems.

the internal dynamics is a second order differential equation, which is driven by the second derivative of the desired system output trajectory y_d.

Case 3: the mass moves in horizontal direction, i.e. $\alpha = 0°$, see Fig. 5c. For this case the equations of motion in new coordinate follow from Eqs. (31)–(32) after reordering as

$$m_2\ddot{y} = -ks - d\dot{s}, \tag{37}$$

$$m_1 m_2 \ddot{s} = -(m_1 + m_2)[ks + d\dot{s}] - m_2 F. \tag{38}$$

In addition, it follows from Eq. (30) that $P = 0$. This indicates a tangential realization, where the control force F is tangential to the constraint manifold, see Fig. 5c. Thus F cannot directly regulate the servo-constraint, i.e. the system output y. This is also seen from Eq. (37), which contains $\ddot{y}$, but not any more the control input F. However, output tracking of y is still possible due to coupling with other forces of the system, here the spring force and damper force.

For tracking of the desired output trajectory y_d, the necessary control input can be computed form Eq. (38) as

$$F_d = -\frac{(m_1 + m_2)}{m_2}[ks + d\dot{s}] - m_1 \ddot{s}. \tag{39}$$

Firstly, the values of s, $\dot{s}$, $\ddot{s}$ are unknown. For given y_d these can be computed from Eq. (37) as,

$$ks + d\dot{s} = -m_2 \ddot{y}_d. \tag{40}$$

This is a differential equation for s and poses in this case the internal dynamics. In contrast to the previous two cases, the internal dynamics is here a first order differential equation. For given $\ddot{y}_d$ the solution of the internal dynamics Eq. (40) provides the corresponding values s_d. Then, the values $\dot{s}_d$ follow directly form the algebraic solution of Eq. (40) as

$$\dot{s}_d = -\frac{ks_d}{d} - \frac{m_2}{d}\ddot{y}_d. \tag{41}$$

Taking one time-derivative of Eq. (41) yields an algebraic expression for $\ddot{s}_d$,

$$\ddot{s}_d = -\frac{k}{d}\dot{s}_d - \frac{m_2}{d} y_d^{(\mathrm{III})}. \tag{42}$$

Thus, all quantities for evaluating the control force F using Eq. (39) are available. The last equation shows, that in contrast to the previous two cases the third derivative of the desired output trajectory y_d must be available.

Case 4: the mass moves in horizontal direction and no damping is present, i.e. $\alpha = 0°$ and $d = 0$. Similar to case 3 a tangential realization exists since $P = 0$, and the same interpretations apply. The equations of motion in new coordinates simplifies to

$$m_2 \ddot{y} = -ks, \tag{43}$$

$$m_1 m_2 \ddot{s} = -(m_1 + m_2)ks - m_2 F. \tag{44}$$

Similarly to case 3, the control force F_d can be computed algebraically form Eq. (44), where s, $\ddot{s}$ are unknowns. In contrast to case 3 Eq. (43) is now an algebraic expression for computing s_d for given $\ddot{y}_d$,

$$s_d = -\frac{m_2}{k}\ddot{y}_d. \tag{45}$$

Taking two time-derivatives of Eq. (45) yields

$$\ddot{s}_d = -\frac{m_2}{k} y_d^{(\mathrm{IV})}. \tag{46}$$

The last equation shows, that in this case the forth derivative of the output trajectory is necessary. Combing Eqs. (44)–(46) yields the control force

$$F_d = (m_1 + m_2)\ddot{y}_d + \frac{m_1 m_2}{k} y_d^{(\mathrm{IV})}. \tag{47}$$

Thus, in this case the control input $\boldsymbol{u}_d$ and all states $y_d, \dot{y}_d, s_d, \dot{s}_d$ of the system can be computed by purely algebraic manipulations, without the need of solving any differential equations. Thus, since all these quantities are specified by the system output y and its four time-derivatives, this case poses a differentially flat system.

Summary and comparison of cases: the servo-constraint problem for this illustrative example has been analyzed using the coordinate transformation approach. Of course, also the DAEs (Eqs. 8–9) or the projected DAEs (Eqs. 15–18) can be established. Thereby, the previous analysis can be used to analyze the differentiation index. In accordance with the discussion at the end of Sect. 2.2 it can be obtained that the differentiation index of the original DAEs (Eqs. 8–9) is one higher than the highest derivative of the system output y which is necessary to compute the control force F.

The cases 1 and 2 are orthogonal realizations and therefore provide DAEs with differentiation index 3, similar to systems with geometric constraints. This is irrespectively of the existence of damping in the system. The dynamics along the constraint manifold is not specified by the output, which forms the internal dynamics. Thus, these are two differentially non-flat mechanical systems. These cases 3 and 4, with tangential realizations, yield higher differentiation index, which is dependent on the existence of damping. In case 3, where damping is present, the system has index 4. Internal dynamics remain, which in this case is a dynamical system of first order. Thus this example with damping poses a tangential realization for a differentially non-flat system. In case 4 no damping exists and differentiation index 5 arises. Then, the complete motion is specified by the trajectory of the output y and its time-derivatives. No internal dynamics remains and this tangential realization represents the case of a differentially flat underactuated mechanical system. It should be mentioned, that differentially flat systems with higher index exist. Such an example is the n-mass-spring chain as analyzed in Blajer (1997b).

This example is representative for the different possible phenomena in servo-constraint problems of underactuated multibody systems. An orthogonal realization yields index 3 and internal dynamics remains, which are described by $f-m$ differential equations of second order. For tangential realization higher index arise. Thereby an increasing differentiation index indicates a reduced size of the internal dynamics and the need for higher derivatives of the output trajectory. In the limit cases no internal dynamics remains, and the system can be inverted purely algebraically, i.e. the system is differentially flat.

4 Stability of the internal dynamics

The previous discussion highlights that the inverse model might be a dynamical system, namely containing internal dynamics. This can occur in both cases, the orthogonal realization and the tangential realization. In the computation of the inverse model these internal dynamics must be solved. Thereby, the stability of the internal dynamics, i.e. the system dynamics of the servo-constraint problem, must be investigated carefully. For an ideal orthogonal realization, e.g. classical constraints, the stability properties and analysis of multibody systems with classical constraints applies. This ideal orthogonal realization also occurs in underactuated multibody systems with collocated inputs and outputs, i.e. there is a control input at each system output. This occurs for example in case 1 of the mass on car example which is presented in Sect. 3.

For non-ideal orthogonal and tangential realization, the internal dynamics might be more complex, and stability might not be ensured. This is due to the combination of the multibody system with a control, whereby inputs and outputs are not collocated. In the case of unstable internal dynamics, forward time integration of the internal dynamics might yield unbounded states and control inputs, which provides an unfeasible inverse model. Therefore, careful stability analysis of the internal dynamics is necessary. Using the coordinate transformation approach the internal dynamics of the inverse model can be extracted explicitly, which is helpful in system analysis. For example, for the orthogonal realization the internal dynamics is given by Eq. (25).

In general, the internal dynamics is nonlinear and driven by the desired output trajectory $\boldsymbol{y}_d(t)$, posing a nonlinear time-variant system. Since stability analysis of such systems is quite complex, one uses often the concept of zero dynamics, see Isidori (1995); Sastry (1999). The zero dynamics is the internal dynamics under a constant system output, e.g. $\boldsymbol{y}_d = \boldsymbol{0}, \forall t$. This reduces the internal dynamics to a time-invariant nonlinear system. For the orthogonal realization follows from Eq. (25) the zero dynamics,

$$\ddot{\boldsymbol{q}}_\mathrm{u} = \overline{\boldsymbol{a}}(\boldsymbol{q}_\mathrm{u}, \dot{\boldsymbol{q}}_u). \tag{48}$$

Local stability of the zero dynamics can then be checked, e.g. using Lyapunov's indirect method. Local asymptotic stability requires that the linearized zero dynamics has only eigenvalues with negative real part. If at least one eigenvalue has a positive real part the system is unstable. If there are both, eigenvalues with negative real part and purely imaginary eigenvalues, then Lyapunov's indirect method is inconclusive. In this case stability can be checked e.g. using Lyapunov's direct method, see e.g. Khalil (2002), or the center manifold method, see e.g. Isidori (1995).

For a linear system it can be shown that the eigenvalues of the zero dynamics are precisely the zeros of the transfer function, see Isidori (1995). Further details about the close connection between the concept of zero dynamics of

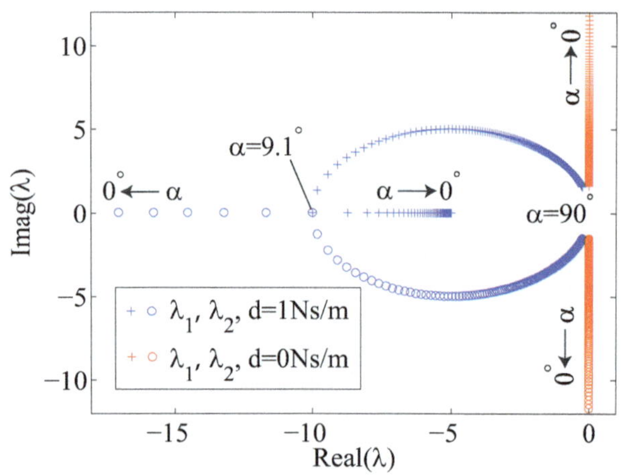

Figure 6. Eigenvalues of zero dynamics for mass on car system.

Table 1. Properties for mass on car system.

$m_1 = 1$ kg	$m_2 = 2$ kg	$k = 5\,\mathrm{N\,m^{-1}}$	$d = 0\,\mathrm{Ns\,m^{-1}}$ and $d = 1\,\mathrm{Ns\,m^{-1}}$

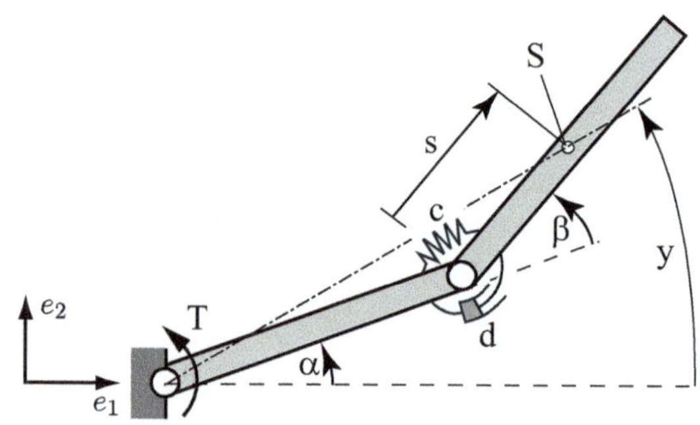

Figure 7. Rotational arm with passive joint.

nonlinear systems and zeros of linear systems are discussed in Isidori and Moog (1988). A linear system is called minimum phase, if all its zeros are in the open left half plane, i.e. its zero dynamics is exponentially stable. This definition is also extended to nonlinear systems. In nonlinear control theory systems with asymptotical stable zero dynamics are called asymptotically minimum phase, otherwise non-minimum phase, see Isidori (1995). It is important to notice, that local asymptotic stability of the zero dynamics is a necessary but not sufficient condition for stability of the driven internal dynamics. This last step is often very complex and from a practical point one restricts often to verify local asymptotic stability of the zero dynamics.

4.1 Illustrative example 1: stability of mass on car

For the mass on car example in Sect. 3 internal dynamics remains for cases 1–3. Figure 6 shows the location of the eigenvalues of the zero dynamics for $0° \leq \alpha \leq 90°$, i.e. case 1 and 2. The system properties are summarized in Table 1. The eigenvalues of the zero dynamics follow from the internal dynamics (Eq. 36) with $\ddot{y}_d = 0$. The internal dynamics is in these cases similar to a spring-mass system and is therefore asymptotically stable for the damped case and stable for the undamped case. For $\alpha = 90°$ it is identical to a free damped mass-spring system and mass m_2 vibrates freely in vertical direction. Hereby, in the undamped case the eigenfrequency of the free vibration is $\omega = 0.25\,\mathrm{Hz}$. However, with the inclination angle α the dynamic behavior of the internal dynamics changes. Thus for example, for $\alpha = 5°$ and $d = 0\,\mathrm{Ns\,m^{-1}}$ the eigenfrequency of the zero dynamics increases to $2.88\,\mathrm{Hz}$. Also in the damped case the behavior changes dramatically, and for $\alpha < 9.1°$ and $d = 1\,\mathrm{Ns\,m^{-1}}$ an over-damped behavior occurs. Thus, due to the servo-constraint, the dynamical behavior of the internal dynamics can be quite different from the dynamics of the uncontrolled underactuated multibody system.

4.2 Illustrative example 2: rotational manipulator arm

As simple example for an underactuated multibody system with unstable internal dynamics a single rotational manipulator arm with a passive joint is considered, see Fig. 7. The rotational arm consists of two identical homogenous links with length l, center of mass $s = l/2$, mass m and inertia $I = ml^2/12$. The first link is actuated by the control torque $u = T$. The second link is connected by a passive joint to the first link, which is supported by a linear spring-damper combination with spring constant c and damper coefficient d. Thus, the passive joint manipulator has its elasticity parallel to the joint and shows very many similar properties as manipulators with flexible links, see Seifried et al. (2013). In contrast, such a passive joint system is quite different to a so-called flexible joint manipulator with drive train elasticities, where the flexibility is located between a link of the manipulator and its motor, see De Luca (1998).

The rotational manipulator arm is described by the generalized coordinates $\boldsymbol{q} = [\alpha, \beta]$, whereby $q_\mathrm{u} = \beta$ denotes the unactuated coordinate. The arm moves perpendicular to the direction of gravity and the equations of motion are given by

$$l^2 m \begin{bmatrix} \frac{5}{3} + \cos\beta & \frac{1}{3} + \frac{1}{2}\cos\beta \\ \frac{1}{3} + \frac{1}{2}\cos\beta & \frac{1}{3} \end{bmatrix} \begin{bmatrix} \ddot{\alpha} \\ \ddot{\beta} \end{bmatrix} + \begin{bmatrix} -0,5 l^2 m\dot{\beta}(2\dot{\alpha} + \dot{\beta})\sin\beta \\ c\beta + d\dot{\beta} + 0,5 l^2 m\dot{\alpha}^2 \sin\beta \end{bmatrix} = \begin{bmatrix} T \\ 0 \end{bmatrix}. \quad (49)$$

The position of a point S on the second link is described in the body fixed coordinate system by $0 < s < l$, see Fig. 7. The control goal is the tracking of the position of point S. For small β the position can be described approximately by the system output

$$y = \alpha + \frac{s}{s + l}\beta. \quad (50)$$

This system output is a linear combination of the two generalized coordinates and can be seen as an auxiliary angle y, see Fig. 7. For example, with this auxiliary angle the position in e_2 direction is

$$r_{s2} = l\sin\alpha + s\sin(\alpha+\beta) \approx (l+s)\sin y. \tag{51}$$

This approximation holds for small β, which can be verified by computing the Jacobian linearization around $\beta = 0$, see Seifried (2012b). It is noted, that the following computations and analysis for the exact position of S follow the same steps and yield the same results. However, the use of the linearly combined system output (Eq. 50) simplifies the expressions significantly and allows an easier discussion of the results. Similar linearly combined outputs are also often used in end-effector tracking of flexible manipulators, see e.g. De Luca (1998); Seifried et al. (2011).

For this rotational arm the servo-constraint is given by

$$c(\boldsymbol{q},t) = \alpha + \frac{s}{s+l}\beta - y_d(t) \tag{52}$$

and its Jacobian matrix is

$$\boldsymbol{H} = \left[\begin{array}{cc} 1 & \frac{s}{l+s} \end{array}\right]. \tag{53}$$

With this matrix follows from Eq. (18)

$$P = \boldsymbol{H}\boldsymbol{M}^{-1}\boldsymbol{B} = \frac{2l(3s\cos\beta - 2l)}{ml^2(l+s)(9\cos(2\beta)-23)}. \tag{54}$$

For $s^* = 2l/(3\cos\beta)$ it is $P = 0$, i.e. only for this cases a tangential realization occurs. For small angles β the position s^* approximates the center of percussion. In Fliess et al. (1995) it is shown, that for a manipulator with one passive joint the center of percussion might yield a differentially flat system output. In contrast, in the following the case $s \neq s^*$ is considered, i.e. the case of an orthogonal realization. If $s = 0$ then the input and output are collocated since $y = \alpha$. In this case an ideal orthogonal realization occurs, otherwise a non-ideal orthogonal realization exists. From the previous discussions it is obvious that internal dynamics remain.

To analyze the internal dynamics and the servo-constraint problem in more detail, the coordinate transformation approach presented in Sect. 2.4 is applied. The new set of coordinates containing the output are chosen as $\overline{\boldsymbol{q}} = [y, \beta]$. The system dynamics in new coordinates follows from evaluating Eqs. (22)–(23). For a given output trajectory y_d the required control input u_d follows from Eq. (24). Then, the system dynamics of the inverse model, i.e. the internal dynamics, are described by Eq. (25). For its analysis the zero dynamics is derived, which follows from the internal dynamics with $y = 0, \forall t$. For this example the zero dynamics turns out to be

$$\begin{aligned} &ml^2(s+l)(2l-3s\cos\beta)\ddot{\beta} \\ &\quad = -3ms^2l^2\sin\beta\dot{\beta}^2 - 6(s+l)^2(c\beta + d\dot{\beta}). \end{aligned} \tag{55}$$

For analysis the zero dynamics (Eq. 55) is linearized around its equilibrium point $\beta = 0$ yielding

$$\underbrace{ml^2(2l-3s)}_{a_2}\ddot{\tilde{\beta}} + \underbrace{6d(s+l)}_{a_1}\dot{\tilde{\beta}} + \underbrace{6c(s+l)}_{a_0}\tilde{\beta} = 0. \tag{56}$$

Table 2. Properties for rotational manipulator arm.

$l = 1$ m	$m = 3.438$ kg	$k = 50$ Nm rad^{-1}	$d = 0.25$ Nms rad^{-1}

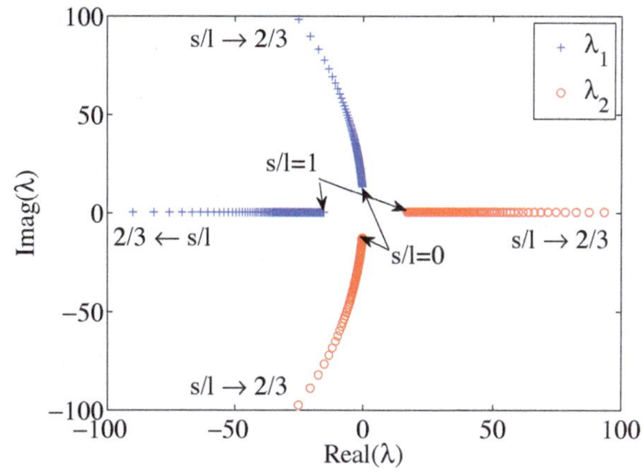

Figure 8. Eigenvalues of zero dynamics for rotational arm.

Thereby a_2, a_1, a_0 correspond to the coefficients of the characteristic polynomial. Using Stodola's criterion the linearized zero dynamics (Eq. 56) of the rotational arm is only asymptotically stable, if all coefficients a_2, a_1, a_0 have the same sign and are non-zero. The constants c, d of the spring-damper combination and the dimensions l, s are by nature positive, yielding positive constants a_0, a_1. Thus also the coefficient a_2 must be positive to obtain stable internal dynamics. From this follows that for $s/l < 2/3$ the internal dynamics is locally asymptotically stable, while for $s/l > 2/3$ the internal dynamics is unstable. The location of the eigenvalues for different $0 \le s \le 1$ is shown in Fig. 8. The used system parameters are summarized in Table 2. This shows clearly that the dynamics of a servo-constraint problem, namely its internal dynamics, might be fundamentally different from the dynamics of the uncontrolled mechanical system. End-effector tracking, i.e. $s = l$, is in robotic manipulator applications the most interesting case. The presented analysis shows that in this example unstable internal dynamics occur in end-effector trajectory tracking.

The analysis of the rotational arm shows that the stability of the internal dynamics can depend on the choice of the system output location. In addition, if a non-homogenous design for the links is admitted, the stability of the internal dynamics also depends on the mass distribution of the links, see the analysis in Seifried (2012b). Then, the mass distribution can also be designed in such a way that for end-effector trajectory tracking stable internal dynamics remain.

The presented analysis and the discussed structural properties for this small servo-constraint system are representative for many multibody systems with passive joints and flexible manipulators in end-effector trajectory tracking. Analysis of such systems are given e.g. in Seifried (2012b) and Seifried et al. (2011), respectively.

5 Numerical solution

In most cases the inverse model requires a numerical solution, whereby the solution method depends on the previously analyzed system properties. In the following some basic numerical solution methods are summarized for differentially flat systems, systems with stable internal dynamics and systems with unstable internal dynamics. For demonstration purpose these methods are applied to the two presented illustrative examples. The following presentation highlights some solution issues and demonstrates also the effect of the different system properties. However, it is not meant as a full in depth investigation of numerical time integrators for servo-constraint problems.

5.1 Differentially flat systems

For differentially flat systems, a purely algebraic inverse model can be derived. However, this requires often a large number of symbolic time-derivations and manipulations of the output equation and the equations of motion. This might be possible for small systems, such as in case 4 of the mass on car system, but for larger systems these symbolic computations might become very burdensome. Therefore, also for differentially flat systems a numerical solution based on the servo-constraint approach might be useful.

Due to the tangential realization in underactuated differentially flat systems the original servo-constraint formulation (Eqs. 8–9) has differentiation index greater 3. For the numerical solution the use of the projection approach presented in Sect. 2.3 might be advantageous, yielding a index reduction. Then, the set of differential-algebraic Eqs. (15)–(18) must be solved numerically. For readability these equations are summarized as

$$\dot{\boldsymbol{q}} = \boldsymbol{v} \tag{57}$$

$$\boldsymbol{A}\dot{\boldsymbol{v}} = \boldsymbol{a}(\boldsymbol{v},\boldsymbol{q},\boldsymbol{u}) \tag{58}$$

$$\boldsymbol{0} = \boldsymbol{b}(\boldsymbol{v},\boldsymbol{q},\boldsymbol{u},\ddot{\boldsymbol{y}}_d) \tag{59}$$

$$\boldsymbol{0} = \boldsymbol{c}(\boldsymbol{q},\boldsymbol{y}_d). \tag{60}$$

The solution of this set of $2f+m$ equations are variations in time of the m control inputs $\boldsymbol{u}(t)$ which are required for the exact reproduction of the desired output trajectory $\boldsymbol{y}_d(t)$, and variations of the $2f$ states $\boldsymbol{q}(t)$, $\boldsymbol{v}(t)=\dot{\boldsymbol{q}}(t)$ in the specified motion.

Since differentially flat systems can completely be inverted algebraically, the output specifies completely the entire motion of the system and Eqs. (57)–(60) do not contain any internal dynamics. This allows the efficient use of rather simple solution formulas. A simple numerical solution schema for solving the DAEs can be based on the Euler backward differentiation scheme, as proposed by Blajer and Kolodziejczyk

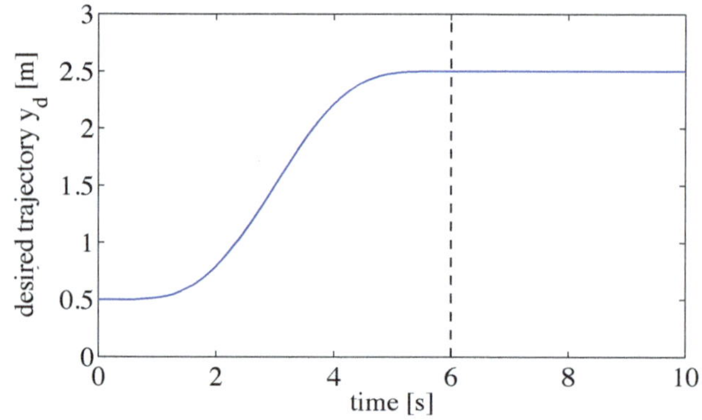

Figure 9. Desired output trajectory y_d for mass m_2.

(2004). Thereby the time derivatives $\dot{\boldsymbol{q}}$ and $\dot{\boldsymbol{v}}$ are approximated with their backward differences $(\boldsymbol{q}_{n+1}-\boldsymbol{q}_n)/\Delta t$ and $(\boldsymbol{v}_{n+1}-\boldsymbol{v}_n)/\Delta t$, respectively. Thereby Δt is the constant integration time step, such that $t_{n+1}=t_n+\Delta t$. With the known values $\boldsymbol{q}_n$, $\boldsymbol{v}_n$ at time t_n, the solution $\boldsymbol{q}_{n+1}$, $\boldsymbol{v}_{n+1}$, $\boldsymbol{u}_{n+1}$ at time t_{n+1} can be obtained from the solution to the set of nonlinear algebraic equations

$$\boldsymbol{q}_{n+1}-\boldsymbol{q}_n-\Delta t\boldsymbol{v}_{n+1} = \boldsymbol{0} \tag{61}$$

$$\boldsymbol{A}(\boldsymbol{q}_{n+1})[\boldsymbol{v}_{n+1}-\boldsymbol{v}_n]-\Delta t\boldsymbol{a}(\boldsymbol{v}_{n+1},\boldsymbol{q}_{n+1},\boldsymbol{u}_{n+1}) = \boldsymbol{0} \tag{62}$$

$$\boldsymbol{b}(\boldsymbol{v}_{n+1},\boldsymbol{q}_{n+1},\boldsymbol{u}_{n+1},\ddot{\boldsymbol{y}}_{d,n+1}) = \boldsymbol{0} \tag{63}$$

$$\boldsymbol{c}(\boldsymbol{q}_{n+1},\boldsymbol{y}_{d,n+1}) = \boldsymbol{0}. \tag{64}$$

This schema can also be used for systems with mixed geometric and servo-constraints as presented in Blajer and Kolodziejczyk (2011).

This very basic solution schema is applied to the mass on car system. Mass 2 should follow the output trajectory

$$y_d = y_0 + \Big[126\Big(\frac{t}{t_f-t_0}\Big)^5 - 420\Big(\frac{t}{t_f-t_0}\Big)^6 + 540\Big(\frac{t}{t_f-t_0}\Big)^7 - 315\Big(\frac{t}{t_f-t_0}\Big)^8 + 70\Big(\frac{t}{t_f-t_0}\Big)^9\Big](y_f-y_0), \tag{65}$$

whereby starting point $y_0=0.5\,\mathrm{m}$ and finial point $y_f=2.5\,\mathrm{m}$ are chosen. After reaching the final point at time $t_f=6\,\mathrm{s}$ the output is at rest. The complete simulation time is 10 s. This trajectory is designed in such a way, that also its higher derivatives are sufficiently smooth. The trajectory is shown in Fig. 9.

It is noticed that the original servo-constraint problem has differentiation index 5, while the projection approach yields index 3. Figure 10 shows the control forces u_d computed with the projection approach and compared to the analytical solution from Eq. (47). This shows, that this rough computational scheme is of acceptable accuracy for appropriately small values of Δt. Here, the Euler backward schema yields nearly identical control inputs as the analytical solution for time sizes 1 ms and 10 ms. Only after increasing the time step size to 100 ms significant errors in form of a time delay are obvious. For differentially flat systems, the inverse model is

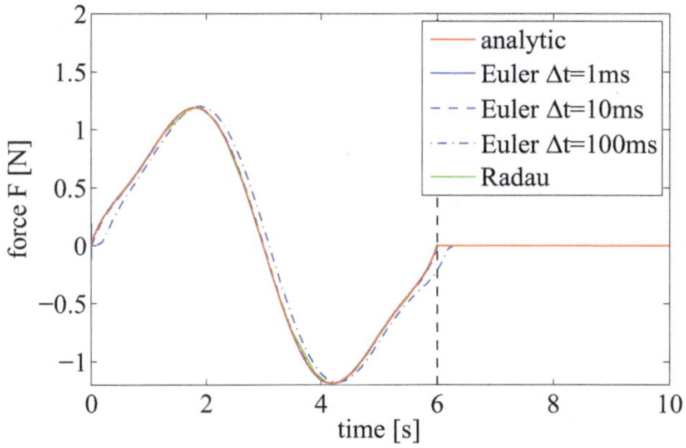

Figure 10. Mass on car – case 4, flat system: computed control input $\boldsymbol{u}_d$.

completly specified by the output. Thus, it can be argued that in this case the issue of numerical damping, which for such a simple discretization schema normally appears, might not pose a significant problem here.

More accurate time discretization schema can be used for the solution of the servo-constraint problem of differentially flat underactuated multibody systems. Therefore, in Fig. 10 also the solution obtained by a *Radau IIa* schema is added. This method is capable of handling index 3 DAEs, see Hairer and Wanner (2010). Hereby an accurate solution is obtained by using only 56 time points. This compares to 1000–10 000 points using the Euler backward schema. In literature further solution methods have been proposed for differentially flat servo-constraint problems. Betsch et al. (2007, 2008) use a energy conserving schema for solving the projected Eqs. (57)–(60), whereby redundant coordinates are used. For a differentially flat crane Fumagalli et al. (2011) propose a solution based on backward differentiation formula.

In Fig. 11 the analytical solution for the desired trajectories of the generalized coordinates and its derivatives are shown. Since hardly any differences to the numerical solutions are visible, the presentation of the numerical solutions are omitted here. From Fig. 11, as well as from the control force plot, it is seen that the complete system is at rest after $t_f = 6\,\mathrm{s}$, indicating the final position of the output. This is due to the fact that for this system the output trajectory specifies the complete system behavior and no internal dynamics remain. This is typical for a differentially flat system.

5.2 Systems with stable internal dynamics

Underactuated multibody systems with stable internal dynamics can be solved by forward time integration. Therefore, the same numerical integrators as in the previous discussed differentially flat case might be used. This is demonstrated using the mass on car system with an inclination angle of $\alpha = 5°$, representing a non-ideal orthogonal realiza-

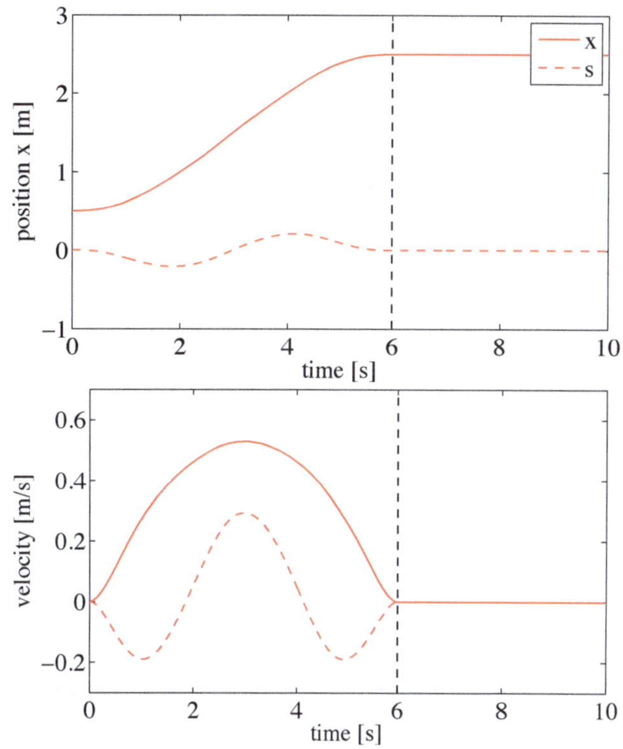

Figure 11. Mass on car – case 4, flat system: computed states $\boldsymbol{q}_d$, $\dot{\boldsymbol{q}}_d$.

tion. Thereby the case of a strongly damped system with $d = 1\,\mathrm{Ns\,m^{-1}}$ and the undamped system are considered.

By using the projection approach, index 1 DAEs arise for the orthogonal realization. As numerical solution schema for the projected servo-constraint Eqs. (15)–(18) the previously presented Euler backward schema with time step size 1ms and 10ms are used. In addition a numerical backward differentiation formula is used, as implemented in the Matlab function *ode15s*, see Shampine et al. (1999). This is capable of solving index 1 DAEs. Also the coordinate transformation approach is used. Hereby the internal dynamics is given explicitly by Eq. (36) and the control input follows from Eq. (35). For the numerical solution of the internal dynamics (Eq. 36) the Matlab *ode45* integrator is used, which is an explicit Runge-Kutta formula of order 4 and 5 using the Dormand-Prince pair.

The obtained control forces using the different solution methods are shown for the damped and undamped case in Fig. 12. The velocities of the generalized coordinates are presented in Fig. 13. Since differences between the solution methods are seen best in the control force, the velocity plot shows only the solution obtained using the coordinate transformation approach.

Due to the presence of the internal dynamics the output does not specify completely the motion of the system. After the output reaches its final position, it remains at rest. However, the system itself is at time $t_f = 6\,\mathrm{s}$ not at rest, as

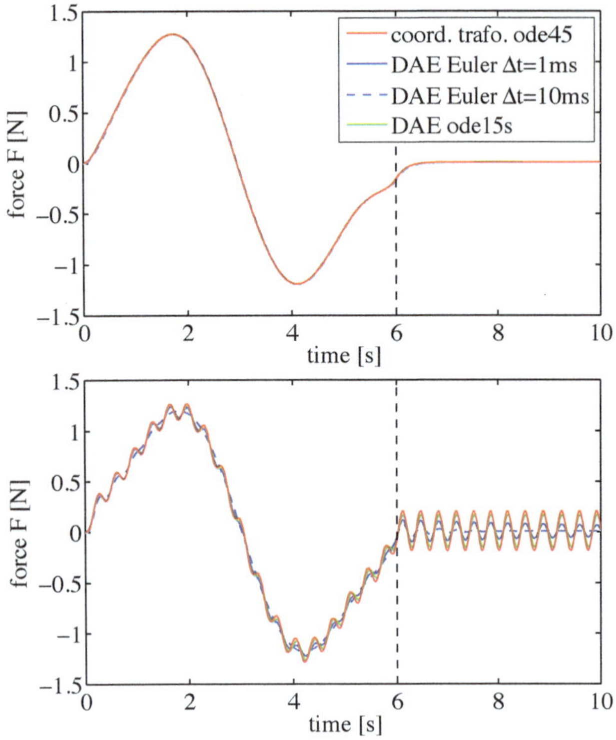

Figure 12. Mass on car – case 2, $\alpha = 5°$: computed control input $\boldsymbol{u}_d$. With damping $d = 1\,\mathrm{Ns\,m^{-1}}$ (top) and without damping (bottom).

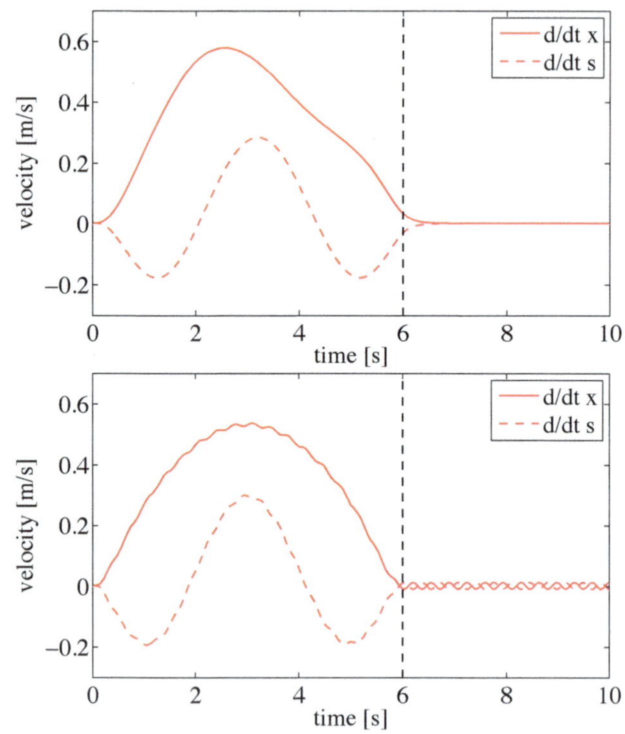

Figure 13. Mass on car – case 2, $\alpha = 5°$: generalized velocities $\dot{\boldsymbol{q}}_d$. With damping $d = 1\,\mathrm{Ns\,m^{-1}}$ (top) and without damping (bottom).

seen from the force and velocity plots. The remaining motion of the system is its zero dynamics. For the case with $d = 1\,\mathrm{Ns\,m^{-1}}$ an overdamped internal dynamics occurs, as also seen from the eigenvalue plot of the zero dynamics in Fig. 6. Thus, the internal dynamics decays rapidly. In contrast for the undamped case strong vibrations occur. These are best visible in the control force plot, whereby vibrations of the internal dynamics occur during trajectory tracking as well as after the output reaches its final position. It should be noted, that the eigenfrequency of the internal dynamics is much higher than the natural frequency of the uncontrolled system, see also Fig. 6.

For the damped case the numerical solutions using the different methods widely coincide, which is seen in the upper plot of Fig. 12. In contrast, for the undamped case some clear differences are observed, see the lower plot of Fig. 12. Here vibrations occur whose frequency is over 10 times higher than in the uncontrolled case. These high frequency vibrations of the internal dynamics are numerically damped using the Euler backward schema. This yields less accurate control inputs, deteriorating the performance of the feedforward control. However, using the more sophisticated methods, the control inputs computed with the projection approach and the coordinate transformation approach coincide.

5.3 Systems with unstable internal dynamics

Forward time integration of systems with unstable internal dynamics yield unbounded states and thus unbounded control inputs $\boldsymbol{u}_d$. This does not provide a feasible feedforward control. Therefore the previously presented solution schema for differentially flat systems and systems with stable internal dynamics cannot be used. Compared to the previous cases, there are much less approaches for the solution of inverse models with unstable internal dynamics.

In the following the so-called stable inversion approach is briefly presented, which is due to Devasia et al. (1996). This approach has been so far applied for solving the internal dynamics given as ODE, such as Eq. (25) for the orthogonal realization. Examples are the feedforward control design of flexible manipulators, see Seifried et al. (2011). With this approach bounded trajectories $\boldsymbol{q}_\mathrm{u}, \dot{\boldsymbol{q}}_u$ of the internal dynamics (Eq. 25) and thus bounded control inputs $\boldsymbol{u}_d$ are obtained. However, the solution might be non-causal, i.e. the trajectories depend on future states providing a so-called pre-actuation phase.

In stable inversion it is assumed that the starting and ending point of the desired system output trajectory $\boldsymbol{y}_d$ are equilibrium points of the multibody system. Further, it is required that the corresponding equilibrium points of the internal dynamics are hyperbolic such that stable manifold $\boldsymbol{W}^\mathrm{s}$ and unstable manifold $\boldsymbol{W}^\mathrm{u}$ exist at each equilibrium point, Sastry

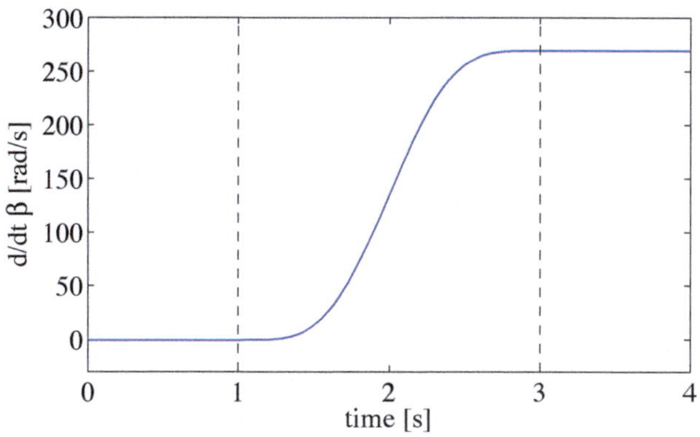

Figure 14. Desired output trajectory y_d for rotational manipulator.

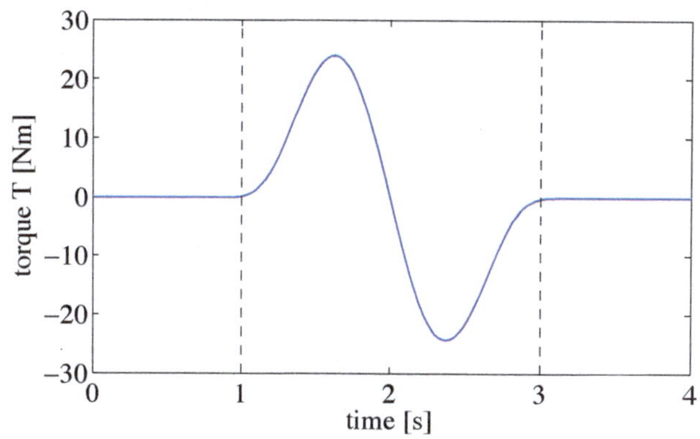

Figure 15. Rotational manipulator arm: computed control input $\boldsymbol{u}_d$.

(1999). Any trajectory starting on the stable manifold $\boldsymbol{W}^{\mathrm{s}}$ converges to the equilibrium point as time $t \to \infty$ and any trajectory starting on the unstable manifold $\boldsymbol{W}^{\mathrm{u}}$ converges to the equilibrium point as time $t \to -\infty$. The solution of the stable inversion is then formulated as a two-sided boundary value problem. The boundary conditions are described by the unstable and stable eigenspaces $\boldsymbol{E}_0^{\mathrm{u}}, \boldsymbol{E}_{\mathrm{f}}^{\mathrm{s}}$ at the corresponding equilibrium points, which are local approximations of the unstable manifold $\boldsymbol{W}_0^{\mathrm{u}}$ and stable manifold $\boldsymbol{W}_{\mathrm{f}}^{\mathrm{s}}$, respectively, Sastry (1999). This yields for the internal dynamics bounded trajectories $\boldsymbol{q}_{\mathrm{u}}, \dot{\boldsymbol{q}}_{\mathrm{u}}$ which start at time t_0 on the unstable manifold $\boldsymbol{W}_0^{\mathrm{u}}$ and reach the stable manifold $\boldsymbol{W}_{\mathrm{f}}^{\mathrm{s}}$ at time t_{f}. Thus, the initial conditions $\boldsymbol{q}_{u_0}, \dot{\boldsymbol{q}}_{u_0}$ at time t_0 cannot exactly be pre-designated. Therefore, a pre-actuation phase is necessary which drives the system along the unstable manifold to a particular initial condition $\boldsymbol{q}_{\mathrm{u}}(t_0), \dot{\boldsymbol{q}}_{\mathrm{u}}(t_0)$, while maintaining the constant output $\boldsymbol{y}_d = \boldsymbol{y}_d(t_0)$. Also a post-actuation phase is necessary to drive the internal dynamics along the stable manifold close to its resting position. The two-sided boundary value problem must be solved numerically, e.g. by a finite difference method as proposed by Taylor and Li (2002).

This stable inversion approach is applied to the rotational manipulator example. Thereby, the system output y should follow the trajectory shown in Fig. 14. The system output should be at rest for $t < t_0 = 1\mathrm{s}$ and $t > t_{\mathrm{f}} = 3\mathrm{s}$. For $1\mathrm{s} \leq t \leq 3\mathrm{s}$ the output should move form 0° to 270°, whereby the trajectory has the same form as Eq. (65).

For the model inversion the internal dynamics is derived using the coordinate transformation approach, and is given by Eq. (25). Thus for the internal dynamics a second order differential equation for β arises, which is the unactuated coordinate of this manipulator arm. The numerical solution of the stable inversion problem is computed in Matlab using the general boundary value solver *bvp5c*, see Kierzenka and Shampine (2008). In Fig. 15 the obtained control torque $u_d = T$ is shown. Figure 16 presents the trajectory for the unactuated coordinate β. It is clearly seen that the obtained solution for the inverse model is bounded. However, it turns out, that the computed solution is non-causal, i.e. trajectories

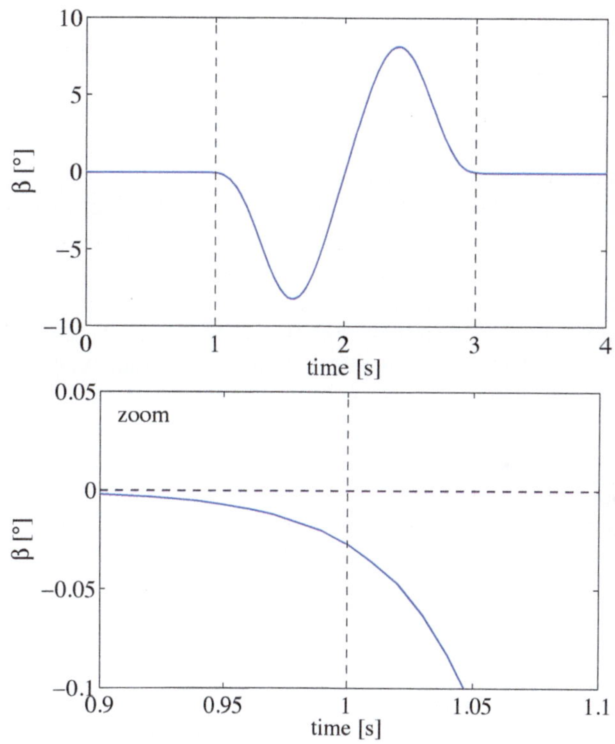

Figure 16. Rotational manipulator arm: computed trajectory of unactuated coordinate β.

for β as well as the control torque T start before $t = 1\mathrm{s}$, which indicates the start of the output trajectory. This is best seen in the enlargement of β around $t = 1\mathrm{s}$, which is also shown in Fig. 16.

Alternative methods for solving model inversion for systems with unstable internal dynamics have been recently proposed for the original servo-constraint formulation (Eqs. 8–9). In Seifried et al. (2013) the problem is projected numerically into the unconstrained subspace, from which the internal dynamics as ODE arises. This is then solved using the previously described stable inversion method. This approach has been applied to the feedforward control of a flexible multibody system with a kinematic loop, featuring 18 unactuated elastic degrees of freedom which represent the internal dynamics. Other solution approaches have been proposed, where the servo-constraint problem is reformulated into optimization problems, see Moberg and Hanssen (2009); Bastos et al. (2011). For example, in Bastos et al. (2011) the solution of the servo-constraint problem is stated as optimal control problem, where the objective function is the minimization of the trajectories of internal dynamics. Thus a bounded solution can be obtained. The stated problem is then solved by standard optimal control algorithms using a direct collocation method.

6 Conclusions

The servo-constraint approach is an efficient and straightforward way for deriving inverse models of underactuated multibody systems, yielding firstly a set of DAEs. In the servo-constraint approach the control inputs are used to meet the imposed servo-constraints. In the case of an orthogonal realization the control forces can directly regulate the constraint manifold, which is defined by the servo constraint. This provides a DAE system with differentiation index 3. In contrast for a tangential realization the control forces can only regulate the constraint manifold indirectly, yielding index larger 3. For the orthogonal realization of underactuated multibody systems internal dynamics remain in the inverse model. For a tangential realization, depending on the system parameters, internal dynamics might or might not exist. In the latter case the complete motion of the system is specified by the system output and a finite number of its derivatives. Such a system is called differentially flat. In this paper it is demonstrated that these different system properties, i.e. the existence of internal dynamics, might be changed by altering some system parameters. For the presented mass-spring-damper system on a car these system parameters are the inclination angle and the damping parameter.

In order to obtain a meaningful solution these different system properties must be investigated. Especially the internal dynamics might be very different form the dynamics of the system without servo-constraint. In the extreme case the internal dynamics might be unstable. The stability of the internal dynamics depends on physical system parameters and the selected system output. For example, the presented case of a passive joint manipulator shows, that the stability of the internal dynamics changes, if the output location is altered. Here, using the end-effector position of the manipulator as system output yields unstable internal dynamics. This can also be observed for many manipulators with flexible links.

In this paper two formulations have been presented for the analysis and solution of servo-constraint problems. The first approach is a projection approach which yields a DAE formulation with reduced differentiation index. The second approach is a coordinate transformation approach which yields, at least for the orthogonal realization, an ODE formulation. This coordinate transformation approach is inspired by nonlinear control theory. It is especially useful if the servo-constraint equations can be solved analytically. For the numerical solution of inverse models time integration schemas can be used for differentially flat systems or systems with stable internal dynamics. Hereby a variety of time-integration schema for DAEs or ODEs can be used. For the projection approach a simple Euler-backwards differentiation schema might be used. This provides accurate results for flat systems. In contrast, it might provide only rough accuracy for systems with internal dynamics, which is due to errors induced by numerical damping. In this case more sophisticated numerical methods must be used.

For systems with unstable internal dynamics forward time integration cannot be used to compute a meaningful feed-forward control. Therefore, the problem can be reformulate into a boundary value problem, which must be solved offline. This provides then a bounded but non-causal solution for the control inputs and the trajectories of all states.

Acknowledgements. The first author would like to thank the German Research Foundation (DFG) for financial support of the project within the Cluster of Excellence in Simulation Technology (EXC 310/1) at the University of Stuttgart.

Edited by: A. Tasora
Reviewed by: two anonymous referees

References

Bajodah, A., Hodges, H., and Chen, Y.-H.: Inverse Dynamics of Servo-Constraints Based on the Generalized Inverse, Nonlinear Dynam., 39, 179–196, 2005.

Bastos, G., Seifried, R., and Brüls, O.: Inverse Dynamics of Underactuated Multibody Systems using a DAE Optimal Control Approach, in: Proceedings of the ECCOMAS Thematic Conference Multibody Dynamics 2011, edited by: Samin, J. C. and Fisette, P., Brussels, Belgium, 4–7 July 2011.

Betsch, P., Uhlar, S., and Quasem, M.: On the Incorporation of Servo Constraints into a Rotationless Formulation of Flexible Multibody Dynamics, in: Proceedings of the ECCOMAS Thematic Conference Multibody Dynamics, Milano, Italy, 2007.

Betsch, P., Uhlar, S., and Quasem, M.: Numerical Integration of Discrete Mechanical Systems with Mixed Holonomic and Control Constraints, in: Proceedings of the 4th Asian Conference on Multibody Dynamics, Jeju, Korea, 2008.

Blajer, W.: Index of Differential-Algebraic Equations Governing the Dynamics of Constraint Mechanical Systems, Appl. Math. Model., 16, 70–77, 1992.

Blajer, W.: A Geometric Unification of Constrained System Dynamics, Multibody Syst. Dyn., 1, 3–21, 1997a.

Blajer, W.: Dynamics and Control of Mechanical Systems in Partly Specified Motion, J. Frankl. Inst., 334B, 407–426, 1997b.

Blajer, W.: A Geometric Interpretation and Uniform Matrix Formulation for Multibody System Dynamics, ZAMM, 81, 247–259, 2001.

Blajer, W. and Kolodziejczyk, K.: A Geometric Approach to Solving Problems of Control Constraints: Theory and a DAE Framework, Multibody Syst. Dyn., 11, 343–364, 2004.

Blajer, W. and Kolodziejczyk, K.: Control of Underactuated Mechanical Systems with Servo-Constraints, Nonlinear Dynam., 50, 781–791, 2007.

Blajer, W. and Kolodziejczyk, K.: Improved DAE Formulation for Inverse Dynamics Simulation of Cranes, Multibody Syst. Dyn., 25, 131–143, 2011.

Blajer, W. and Seifried, R.: From Orthogonal to Tangential Control of Underactuated Mechanical Systems in Partly Specified Motion – A Case Study Illustration, Int. J. Appl. Mech. Eng., 17, 689–706, 2012.

Blajer, W., Seifried, R., and Kolodziejczyk, K.: Diversity of Servo-Constraint Problems for Underactuated Mechanical Systems: A Case Study Illustration, Solid State Phenomena, Mechatronic Systems and Materials, IV, 473–482, 2013.

Campbell, S.: High-Index Differential Algebraic Equations, Mech. Based Des. Struc., 23, 199–222, 1995.

De Luca, A.: Trajectory Control of Flexible Manipulators, Control Problems in Robotics and Automation, in: Lecture Notes in Control and Information Sciences, edited by: Siciliano, B. and Valavanis, K. P., Springer, London, 230, 83–104, 1998.

Devasia, S., Chen, D., and Paden, B.: Nonlinear Inversion-Based Output Tracking, IEEE T. Automat. Contr., 41, 930–942, 1996.

Fliess, M., Lévine, J., Martin, P., and Rouchon, P.: Flatness and Defect of Nonlinear Systems: Introductory Theory and Examples, Int. J. Control, 61, 1327–1361, 1995.

Frye, J. and Fabien, B.: Control of Constrained Systems Described by Lagrangin DAEs, in: Proceedings of the ASME 2011 International Design Engineering Technical Conferences & Computers and Information in Engineering Conference, Washington, 28–31 August 2011.

Fumagalli, A., Masarati, P., Morandini, M., and Mantegazza, P.: Control Constraint Realization for Multibody Systems, J. Comput. Nonlin. Dyn., 6, 011002-1-8, 2011.

Hairer, E. and Wanner, G.: Solving Ordinary Differential Equations II – Stiff and Differential Algebraic Problems, Springer, Berlin, 2010.

Isidori, A.: Nonlinear Control Systems, Springer, London, 1995.

Isidori, A. and Moog, C.: On the Nonlinear Equivalent of the Notion of Transmission Zeros, Modelling and Adaptive Control, in: Lecture Notes in Control and Information Sciences, 105, Springer, 1988.

Kierzenka, J. and Shampine, L.: A BVP Solver that Controls Residual and Error, Journal of Numerical Analysis, Industrial and Applied Mathematics, 3, 27–41, 2008.

Khalil, H. K.: Nonlinear Systems, Prentice Hall, Upper Saddle River, 3rd Edn., 2002.

Kirgetov, V. I.: The Motion of Controlled Mechanical Systems with Prescribed Constraints (Servoconstraints), J. Appl. Math. Mech., 21, 433–466, 1967.

Kovács L., Köveces, J., Zelei, A., Bencsik, L., and Stépan, G.: Servo-Constraint Based Computed Torque Control of Underactuated Mechanical Systems, in: Proceedings of the ASME 2011 International Design Engineering Technical Conferences & Computers and Information in Engineering Conference, Washington, DETC2011-48533, 28–31 August 2011.

Masarati, P., Morandini, M., and Fumagalli, A.: Control Constraint Realization Applied to Underactuated Aerospace Systems, in: Proceedings of the ASME 2011 International Design Engineering Technical Conferences & Computers and Information in Engineering Conference, Washington, DETC2011-47276, 28–31 August 2011.

Moberg, S. and Hanssen, S.: A DAE Approach to Feedforward Control of Flexible Manipulators, in: Proceedings of the 2007 IEEE International Conference on Robotics and Automation, 3439–3444, 2007.

Moberg, S. and Hanssen, S.: Inverse Dynamics of Flexible Manipulators, in: Proceedings of the ECCOMAS Thematic Conference Multibody Dynamics, Warsaw, Poland, 2009.

Rosen, A.: Applying the Lagrange Method to Solve Problems of Control Constraints, J. Appl. Mech., 66, 1013–1015, 1999.

Rothfuss, R., Rudolph, J., and Zeitz, M.: Flachheit: Ein neuer Zugang zur Steuerung und Regelung nichtlinearer Systeme, at – Automatisierungstechnik, 45, 517–525, 1997 (in German).

Sastry, S.: Nonlinear Systems: Analysis, Stability and Control, Springer, New York, 1999.

Schiehlen, W., Guse, N., and Seifried, R.: Multibody Dynamics in Computational Mechanics and Engineering Applications, Comput. Meth. Appl. M., 195, 5509–5522, 2006.

Seifried, R.: Two Approaches for Feedforward Control and Optimal Design of Underactuated Multibody Systems, Multibody Syst. Dyn., 27, 75–93, 2012a.

Seifried, R.: Integrated Mechanical and Control Design of Underactuated Multibody Systems, Nonlinear Dynam., 67, 1539–1557, 2012b.

Seifried, R., Held, A., and Dietmann, F.: Analysis of Feed-Forward Control Design Approaches for Flexible Multibody Systems, Journal of System Design and Dynamics, 5, 429–440, 2011.

Seifried, R., Burkhardt, M., and Held, A.: Trajectory Control of Serial and Parallel Flexible Manipulators using Model Inversion, Multibody Dynamics: Computational Methods and Applications, in: Computational Methods in Applied Sciences, edited by: Samin, J. C. and Fisette, P., Springer, 28, 53–75, 2013.

Shampine, L., Reichelt, M., and Kierzenka, J.: Solving Index-1 DAEs in MATLAB and Simulink, SIAM Rev., 41, 538–552, 1999.

Siciliano, B., Sciavicco, L., Villani, L., and Oriolo, G.: Robotics – Modelling, Planning and Control, Springer, London, 2010.

Spong, M., Hutchinson, S., and Vidyasagar M.: Robot Modeling and Control, John Wiley & Sons, Hoboken, 2006.

Taylor, D. and Li, S.: Stable Inversion of Continuous-Time Nonlinear Systems by Finite-Difference Methods, IEEE T. Automat. Contr., 47, 537–542, 2002.

6

Concept of modular flexure-based mechanisms for ultra-high precision robot design

M. Richard and R. Clavel

Ecole Polytechnique Fédérale de Lausanne, Laboratoire de Systèmes Robotiques (LSRO), Station 9, 1015 Lausanne, Switzerland

Abstract. This paper introduces a new concept of modular flexure-based mechanisms to design industrial ultra-high precision robots, which aims at significantly reducing both the complexity of their design and their development time. This modular concept can be considered as a robotic Lego, where a finite number of building bricks is used to quickly build a high-precision robot. The core of the concept is the transformation of a 3-D design problem into several 2-D ones, which are simpler and well-mastered. This paper will first briefly present the theoretical bases of this methodology and the requirements of both types of building bricks: the active and the passive bricks. The section dedicated to the design of the active bricks will detail the current research directions, mainly the maximisation of the strokes and the development of an actuation sub-brick. As for the passive bricks, some examples will be presented, and a discussion regarding the establishment of a mechanical solution catalogue will conclude the section. Last, this modular concept will be illustrated with a practical example, consisting in the design of a 5-degree of freedom ultra-high precision robot.

1 Motivation

The current trend in numerous industrial domains is to miniaturise products, mainly microelectronic, optic and biomedical devices; the design of industrial robots capable of performing micromanipulation and micro-assembly tasks with a sub-micrometric precision is thus becoming a crucial need (Tolfree, 2006). As the use of flexure joints is compulsory to meet these precision requirements, a key aspect of the robot development consists in the design of the flexure-based mechanisms. Although this step is now widely investigated for planar and low-DOF (Degrees of Freedom) compliant structures, the development of a whole flexure-based robot is still infrequent, especially in the industrial context. Two examples of industrial prototypes are the Delta^3 robot family, which consists in an adaptation of the classical Delta kinematics for μ-Wire Electro-Discharge Machining (3 DOF in translation, strokes: ± 2 mm, resolution: 5 nm, see Fig. 1) (Bacher et al., 2002; Bottinelli et al., 2002), and the 6-DOF Sigma 6 robot, intended to perform active optical alignment (strokes: ± 4 mm and $\pm 4°$, resolution: 5 nm on the linear axes) (Helmer et al., 2004; Helmer, 2006). Nonetheless, the development time of such ultra-high precision machines, and thus their time-to-market, is considerably high. Furthermore, if the industrial specifications are modified, requiring for instance to add a degree of freedom or to change the position of a rotation centre, the whole design process has to be restarted from the beginning.

This paper thus presents a concept of modular flexure-based mechanisms, which allows to rapidly design an ultra-high precision robot and to easily change its mobility. This approach can be compared to a robotic Lego, where a finite number of building bricks can be chosen and assembled within a small amout of time to create the desired machine (Richard and Clavel, 2010). Besides the kinematical aspects of the robot, which will not be exposed here, the concept makes use of mechanical bricks, which consist in planar flexure-based mechanisms. This modular concept thus allows to transform a 3-D mechanical design problem into several 2-D ones: the flexure-based structures resulting from the 3-D conception are complex and scarcely flexible, whereas the 2-D solutions are simple, well-mastered and easy to machine monolithically. Likewise, both the kinematical and mechanical designs are considerably simplified thanks to the use of solution catalogues. The robots resulting from this methodology thus achieve similar high performances as more traditionally designed machines, with a significantly shortened development time.

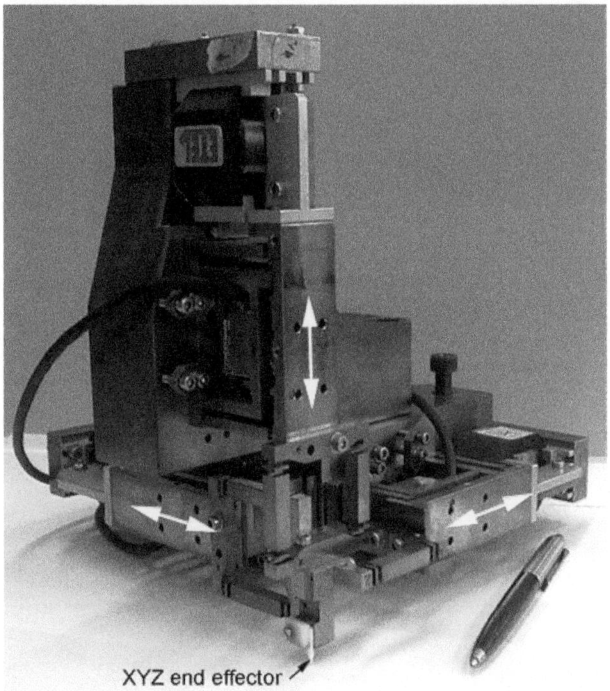

Figure 1. Delta[3] II robot (3 translational DOF, strokes: ±2 mm, resolution: 5 nm) (Bacher et al., 2002).

The current research on this modular concept will be presented in this paper: first, a brief description of the theoretical aspects will allow to establish the terminology and the chief specifications of the flexure-based building bricks. Then, the latest research directions will be presented, with a special focus on the maximisation of the flexure strokes and the development of an actuation sub-brick. Last, the application of this modular concept to design a 5-DOF ultra-high precision robot will illustrate the potentialities of this new approach.

2 Concept of modular flexure-based robot

A parallel robot resulting from this modular concept is composed of one to three kinematic chains and can be symbolised by a cube: the chains are disposed orthogonally along its faces, while the end-effector is located on one of its corners. The design of this robot makes use of a finite number of building bricks: these consist in 2-D flexure-based mechanisms and present two states: active and passive. As their name indicates, the role of the active bricks is to actuate from one to three degrees of freedom, making use of linear actuators only. The translations are obtained by the synchronous displacement of the actuators, whereas the rotations are performed by their differential motion. As for the passive bricks, their function is to link the output of the active bricks to the end-effector of the robot, thus performing a transmission of the actuated motion. Each kinematical chain of the modular robot therefore consists in the serial arrangement of an active brick with a passive one (see Fig. 2).

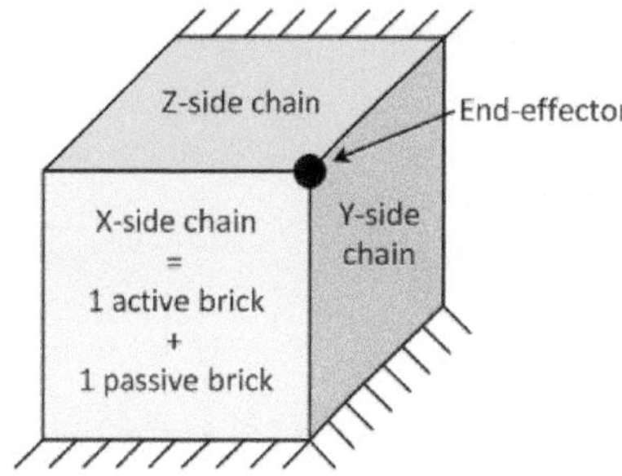

Figure 2. Symbolism of the parallel robot designed with the modular concept.

Briefly, the kinematical part of the modular concept uses as an input the desired mobility of the robot, i.e. the translational and rotational degrees of freedom which have to be performed (Tx, Ty, Tz, Rx, Ry, Rz). Then, with the help of a conceptual solution catalogue, this global requirement is transformed into an exhaustive list of all possible combinations of 2-D active and passive bricks fufilling it. At that point, the bricks are expressed only with their degrees of freedom arrangement, independently from any mechanical realisation. The kinematical part of this methodology consequently allows to design a large variety of machines, from machine-tools to microscale robots. The choice of the optimal solution in the exhaustive list is motivated by the specifications of the aimed robot.

The second part of the concept consists in mechanically designing these modular building bricks: for the high precision robots considered in this work, the use of flexure joints is compulsory. Consequently, the aim is to propose a mechanical solution catalogue which, for each brick, lists several possible flexure-based designs. The current research on this modular concept is thus focusing on the generation of this mechanical solution catalogue. The challenges encountered will be presented in the next paragraphs.

3 Active bricks

As aforementioned, the active bricks are planar flexure-based mechanisms whose role is to actuate one to three degrees of freedom. The motions which are not actuated are blocked, i.e. the stiffness of the structure along these movements should be at least 100 times higher than the stiffness of the actuated motions (Henein, 2001); practically, a factor of around 1000 is recommended to avoid low transverse eigenfrequencies. The planarity of the mechanisms consists in a noteworthy advantage to achieve this desideratum, as the maximisation of the transverse stiffnesses is much more straightforward than with 3-D structures. The thickness of the planar system can indeed be adjusted to fulfil this requirement without deteriorating the desired strokes.

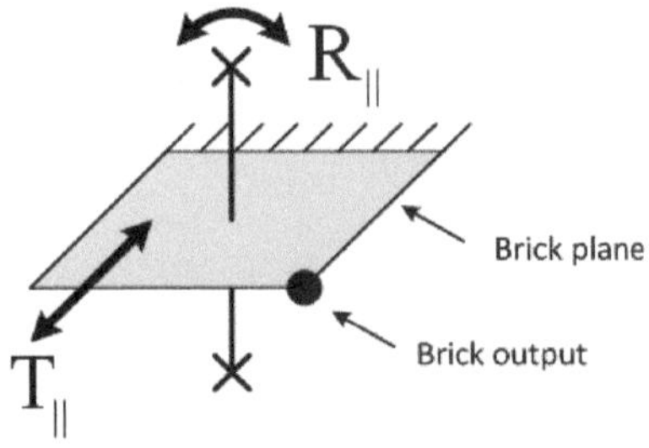

Figure 3. Mobility of the $T_{\parallel}R_{\perp}$ brick.

As the trend to miniaturise industrial products goes hand in hand with the miniaturisation of the production lines, and thus of the robots (Kobel and Clavel, 2011), the main limitation of the existing industrial high precision machines lies in their bulky volume (of some dm^3) for the required workspace (usually some mm^3). Consequently, the chief challenge of the active brick design consists in the maximisation of the ratio between the translational strokes and the characteristic dimension of the robot, which is defined as the highest dimension of the prism enclosing its mechanical parts (see Eq. 1).

$$R_s = \frac{\text{stroke [mm]}}{\text{characteristic dimension [mm]}} \tag{1}$$

For instance, the high-precision robots introduced in Sect. 1 present ratios of 0.02 for the Delta[3] II (±2 mm for a characteristic dimension of 200 mm, total volume: 3 dm^3, see Fig. 1) and of 0.032 for the Sigma 6 (±4 mm for a characteristic dimension of 250 mm, total volume: 6 dm^3). The ideal R_s value which this modular concept aims to achieve is around 0.1, which means performing strokes of ±5 mm with a cubic robot presenting a characteristic dimension of 100 mm.

3.1 Remote Centre of Motion (RCM)

Most high-precision robots and mechanisms performing both rotational and translational motions suffer from the same limitations: besides the low angular strokes, the maximum angle and the maximum translation cannot be achieved at the same time, as in the Sigma 6 robot; its maximal angle of 4° can indeed be obtained only with a translation of less than 1 mm. This phenomenon is first caused by the strokes of the actuators, which are not sufficient to allow the combined translational and rotational motions. Second, this effect is also induced by the high translational compensatory motions needed during the rotations. This is due to the fact that the centre of rotation does not coincide with the robot output; compensatory translations are thus necessary to keep its position constant during the rotation. A shrewd design making use of a Remote Centre of Motion (RCM) can overcome this problem by virtually locating the rotation centre on the system output, without adding any mechanical part at this point.

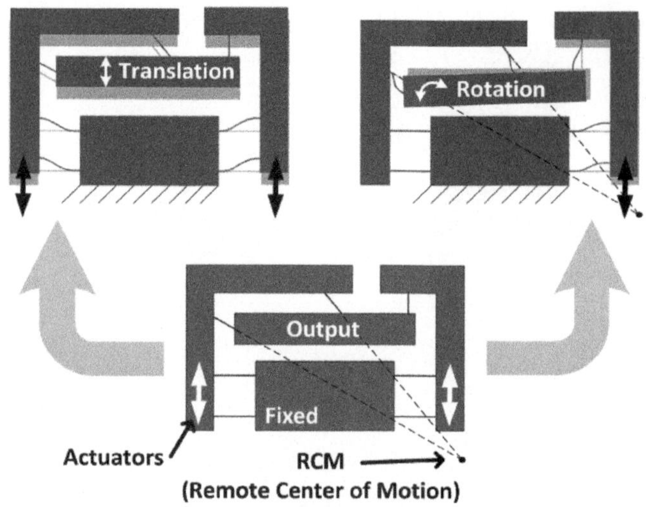

Figure 4. Principle of the designed active $T_{\parallel}R_{\perp}$ brick: the synchronous motion of the actuators performs the translation, while the rotation around the RCM results from their differential displacement.

To illustrate this solution, the case of the $T_{\parallel}R_{\perp}$ active brick is now detailed. This planar mechanism performs a translation and a rotation: its notation indicates that the rotation axis is orthogonal to the plane of the brick, whereas the translation is along one direction of that plan (see Fig. 3). According to the definition of the concept, the mechanism comprises two linear actuators: their synchronous motion performs the translation, whereas their differential motion achieves the rotation.

The principle of the mechanical flexure-based design making use of a RCM is shown in Fig. 4, and is illustrated with a machined mock-up with assembled flexure blades (see Fig. 5). The two lower structures are simple 2-blade parallel tables whose function is to guide the actuator moving part (this aspect will be further explained in Sect. 3.2). The position of the left actuator is kept constant during the rotation; both blades which are fixed on the corresponding L-shaped part define the position of the rotation centre, which is located at their virtual intersection. The angle between these two blades should be as near as 90° as possible in order to maximise the transverse stiffness. The linear motion of the right actuator performs the rotation: through the vertical blade, the force produced by the actuator is transformed into a moment, which causes the output to rotate. In this situation, the lever law $M = F \cdot d$ can be applied, where M is the moment, F is the force, and d is the perpendicular distance between the force and the rotation centre. Consequently, the actuation force which is necessary to rotate the output with a given angle decreases as the distance between the actuator and the RCM increases. Nonetheless, the linear stroke of the actuator which is needed to perform this same angle conversely increases. The distance between the actuator and the

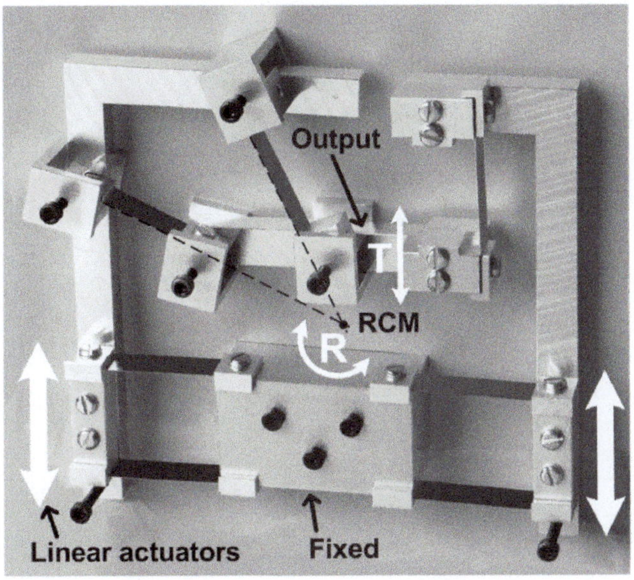

Figure 5. Mock-up of the designed active $T_{\|}R_{\perp}$ brick

RCM should thus be optimised in regard to the robot specifications, as well as to the actuator performances.

This design presents the noteworthy advantage that the location of the rotation centre can be defined at any point of the brick plan. When this brick is integrated into a modular high-precision robot, its RCM must be coincident with the robot output to completely avoid compensatory motions.

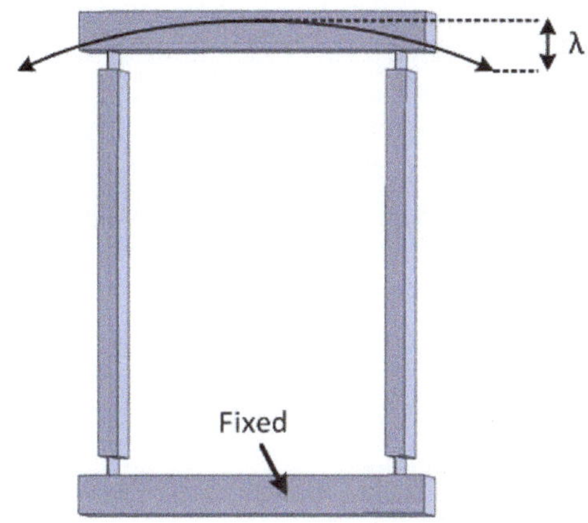

Figure 6. Sketch of the 4-prismatic hinge table trajectory.

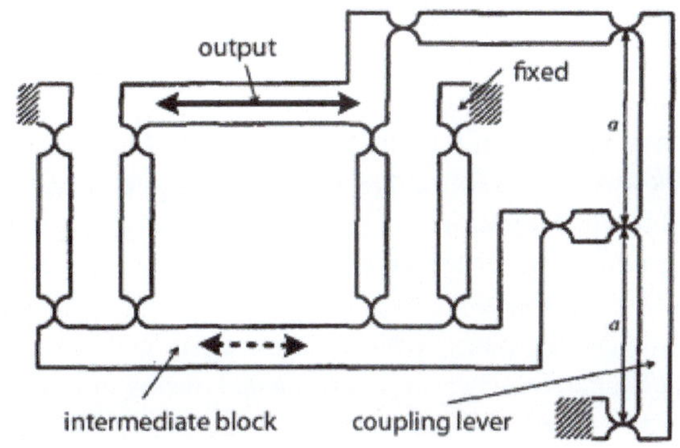

Figure 7. Sketch of the 13-hinge table principle.

3.2 Actuation sub-brick

In the frame of this concept of modular flexure-based mechanisms, the actuation of the bricks is always achieved with linear actuators only. Moreover, the guiding of the actuator moving part has to be performed without friction to fulfil the required sub-micrometric precision; a flexure-based structure is thus necessary. Consequently, each active brick comprises the same actuation system, which is composed of the actuator, the guiding structure of the actuator moving part, and a position sensor. This system can be considered as a lower level of modularity, thus defining an actuation sub-brick. The flexure-based mechanisms which transform the linear motion of the actuators into the desired degrees of freedom are then attached to this actuation sub-brick in order to create the desired building brick. The following paragraphs will detail the problematics linked with the guiding system, the choice of the actuators, as well as the mechanical design of the actuation sub-brick.

3.2.1 Actuator flexure-based guiding mechanism

A simple flexure-based mechanism performing a translational motion consists in the 4-hinge table (see Fig. 6) (Henein, 2001). Either elliptic, prismatic or truncated circular hinges are conceivable; however, the radius of the circular and elliptic joints necessary to achieve the required stroke is so high that the hinge amounts to a prismatic shape. The critical point of the latter is the stress concentration at the edge of the hinge, which can eventually lead to fracture. To prevent this effect, the edge is rounded: the chosen radius is approximately twice the radius of the wire used for machining.

The crucial issue of this guiding system is its non-rectilinear movement: it indeed performs a translation along a parabolic trajectory (see Fig. 6). For instance, the best designs of a 4-hinge table performing a ± 10 mm stroke within a volume of $10 \times 10\,\text{cm}^2$ achieve a parasitic translation, λ, of $500\,\mu\text{m}$ at both ends of the motion. This undesired movement thus becomes problematic for high strokes, as the actuator must withstand the parabolic trajectory of its moving part. The first option to solve this issue is to compensate for the parasitic motion in order to obtain a perfectly rectilinear displacement, making use of a 13-hinge table (see Fig. 7) (Henein, 2001). However, this mechanism is complex and bulky, especially for an actuation sub-brick. Indeed, as the active bricks can actuate up to three degrees of freedom, as many 13-hinge tables should be integrated into the same brick, besides the other mechanisms necessary to transform the actuator displacements into the desired motions. Moreover, the force which is necessary to achieve the desired stroke with the 13-hinge table is higher than the force needed with the 4-hinge table, as the number of hinges to deform is increased. This option is thus obviously unconceivable in the frame of this modular concept. The second solution to overcome this issue, which is more straightforward, is to find an actuation principle which is compatible with the parasitic motion of its moving part. The following paragraph will detail the investigation of this second option.

3.2.2 Actuator problematics

As the modular robots designed with this concept are intended for industrial use, electromagnetic actuators are selected for their high repetivity relatively to piezo-electric or ultrasonic actuators. Likewise, the moving coil/static magnets configuration is prefered to minimise both the moving mass and the hysteresis created by the moving magnets. Typical values of force requirements for high-precision tasks are continuous forces of 10 N, and peak forces of around 50 N to allow for high accelerations ($\sim 10\,g$). Thus, two types of actuators are conceivable: the voice coil and the ironless electronic commutation (EC) actuators.

The voice coil actuator, which is frequently encountered in loudspeakers, consists in a cylindrical arrangement of the moving coil and the static magnets. Like the classical brushed DC (direct current) motors, its control is simple, as the actuation force is directly proportional to the current injected into the coil. However, the value of the parasitic motion of the 4-hinge table is generally higher than the actuator air gap; nonetheless, a simple way to compensate for that limitation is to take advantage of the symmetrical behaviour of the parasitic motion, which is indeed minimal at the equilibrium position of the 4-hinge table (zero stroke) and maximal with the same absolute value at both ends of the stroke. Therefore, the solution consists in eccentrically mounting the coil at zero stroke to make a better use of the available air gap between the coil and the magnets (see Fig. 8).

As for the ironless electronic commutation actuator, it consists in a U-shape frame with fixed magnets and a prismatic moving coil (see Fig. 9). Similarly to brushless DC motors, this 2-phase or 3-phase actuator needs an integrated Hall sensor and a dedicated electronics in order to manage the phase commutation: its control is thus not straightforward. Nonetheless, the main advantage of this solution is its direct compatibility with the parasitic motion of the 4-hinge table. However, the standard EC actuators available on the market are designed for high-power applications, which makes them unsuited for small high-precision mechanisms.

To conclude this actuator problematics, the linear voice coil solution is obviously more adapted to high-precision applications. However, their requirements (strokes, forces) do not match most standard models. The scarce standard actuators which fulfil the requirements are bulky and thus difficult to combine with compact high-precision mechanisms, which tend to be as miniaturised as possible. This study of solutions to actuate high-precision systems highlights the lack of fully suitable off the shelf devices; the improvement of the compactness and miniaturisation of flexure-based active mechanisms is thus limited by the unsuitability of the available actuators for the specific requirements of such systems.

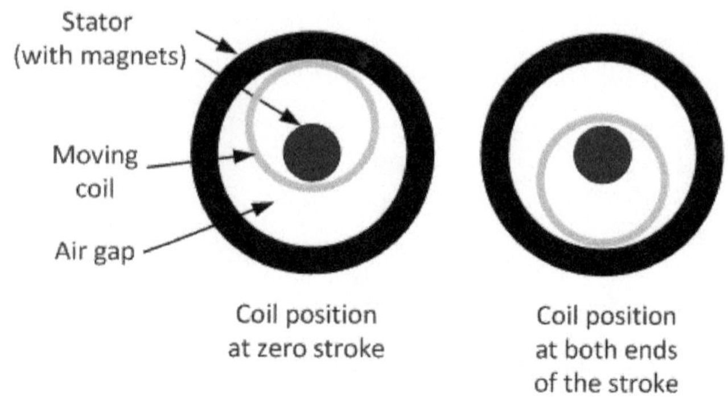

Figure 8. Principle of the eccentric coil mounting (transversal view; the actuation direction is orthogonal to the section).

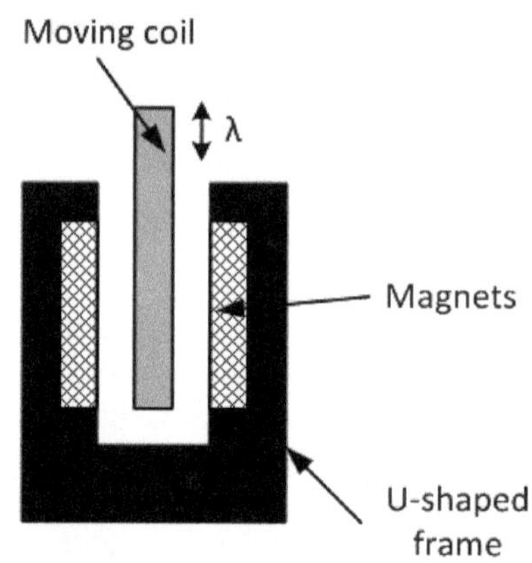

Figure 9. Transversal section of the linear EC actuator (the actuation direction is orthogonal to the section).

3.2.3 Design of the actuation sub-brick

A first prototype of the actuation sub-brick has been designed; the selected actuator is a cylindrical voice coil from BEI Kimco Motors (LA15-26-000A, stroke: ± 12.7 mm, continuous force: 11.57 N, peak force: 44.48 N), which presents an air gap of 390 μm on each side of the coil. With an eccentric mounting, as seen in Fig. 8, the parasitic motion of the guiding system must be below 780 μm. To perform that function, a 4-prismatic hinge table is dimensioned with the following requirements: the planar mechanism should not exceed $10 \times 10\,\text{cm}^2$, whereas the desired stroke is ± 10 mm. This value is the result of first estimations of the $T_{\parallel}R_{\perp}$ active

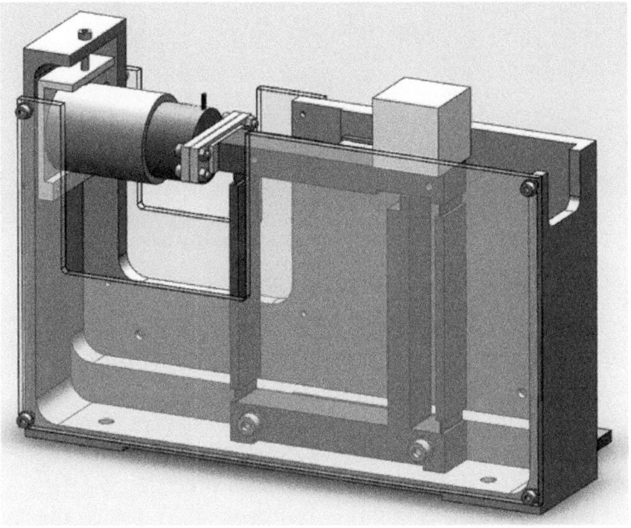

Figure 10. CAD image of the designed actuation sub-brick prototype.

brick described in Sect. 3.1: it allows to perform a translation of ±5 mm and a rotation of ±10°, the maximum angle being compatible with the maximum linear displacement. Last, the parasitic motion should be minimised. The optimal guiding mechanism design makes use of 250 µm-thick prismatic hinges, which creates a parasitic motion λ reaching 500 µm at both ends of the stroke. Finally, an optical encoder from Numerik Jena (LIK41, 100× interpolation, resolution: 50 nm) is also integrated. Figure 10 shows a CAD view of the final design.

In this design, thermal considerations have also been investigated, on the basis of the work presented in Lubrano and Clavel (2010): first, the conduction of the heat generated by the moving coil into the flexure-based mechanism should be minimised, in order not to deteriorate the repetitivity of the system. To do so, the parts which link the coil to the 4-hinge table have been machined in a material with poor thermal conductivity (namely POM, polyoxymethylene). Then, the guiding system should also be isolated from drafts: plexiglas plates have been designed to perform this function. It is to note that windows have been shrewdly integrated, in order to test a fan cooling around the actuator without influencing the flexure-based mechanism.

As for the feedback device, the aforementioned optical encoder, presenting a resolution of 50 nm, will be used to demonstrate the functionality of the actuation sub-brick. Nonetheless, its characterisation will be performed with more precise measurements: the feedback device will thus be a laser interferometer with a theoretical resolution of 1.25 nm (SIOS SP-2000 with IS-01 and interpolation modules). Consequently, the limiting resolution of the measurements will be the flatness of the optical surface of the mirror cube used to reflect the interferometer laser beam. The first tests of this prototype will allow to validate this approach and to study the need to integrate optical encoders with a higher resolution, which will obviously be more expensive and more complex to align in the flexure-based system.

As for the planned measurements, besides this validation of the feedback device, the relation between the current injected in the actuator coil and the position of the sub-brick output will be characterised. This will also allow to investigate the possible effects of the coil eccentricity on the voice coil behaviour.

4 Passive bricks

Some considerations concerning the design of the passive bricks will now be presented. As seen in Sect. 2, the passive bricks are transmission mechanisms which present from one to five degrees of freedom. These are passive, i.e. free to move but not actuated; the other movements are blocked (see definition in Sect. 3). Like the active bricks, the passive bricks are planar mechanisms; however, axisymmetrical structures, for example cylindrical, are also conceivable. In several cases, such as the mobility illustrated in Fig. 11, an arm-like system is simpler and more efficient than a planar mechanism. The chief challenge of the passive brick design also consists in the maximisation of the strokes: nonetheless, unlike the active case detailed in Sect. 3, this requirement is not necessary for all motions of the brick. Without detailing all kinematical aspects, the passive degrees of freedom can either allow a motion which is required at the robot output or suppress an overconstraint. In the first case, the strokes have to be maximised, because they are directly linked to the robot specifications. In other words, the passive motion must achieve the same displacement as the required stroke on the robot output. However, in the second case, the passive movement only suppresses an overconstraint, i.e. avoids that the position of a blocked degree of freedom is defined by two mechanisms, which would create internal stresses detrimental to the aimed precision. Consequently, the corresponding stroke does not need to be maximised to perform its function. Figures 11 and 12 present two possible designs of passive bricks.

This consideration leads to a reflection on the mechanical solution catalogue, to which the concept of modular flexure-based mechanisms aims. If its validity is undoubted for the active bricks, as the possible designs of these several mechanisms making use of the actuation sub-brick are limited, its usefulness for passive bricks is questionable. First, as these bricks perform from one to five degrees of freedom, the number of possible brick mobilities is higher than for the active bricks. Then, the previous discussion on the role of the passive motions leads to different potential mechanical designs, according to the function of each degree of freedom of the brick. Consequently, the solution catalogue of the passive bricks would present too many solutions to be of any practical use.

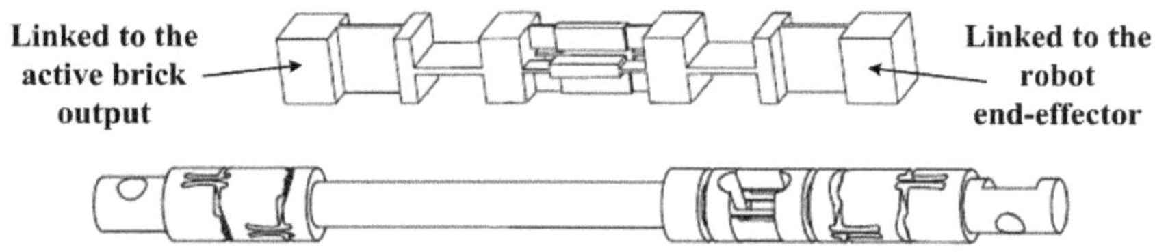

Figure 11. Arms of the Sigma 6 robot (Helmer, 2006), which can be used as passive bricks. They present 5 degrees of freedom; the only blocked motion is the translation along the arm axis.

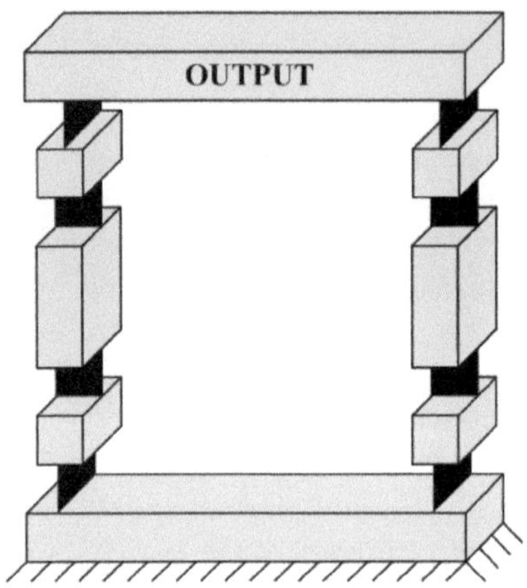

Figure 12. Principle of a possible design of a passive brick performing two rotations and two translations. This mechanism is often referred to as "space parallelogramm".

Finally, the definition of the modular concept states that the kinematic chains of the robot are composed of an active brick and a passive brick which are serially arranged. This point is crucial for the establishment of the conceptual solution catalogue, whose function is to transform a 3-D design problem into several 2-D ones. Nonetheless, when it comes to the mechanical design of the robot, this serial arrangement may not always consist in the optimal solution: both active and passive bricks could be merged in a unique flexure-based structure, which would perform actuated, passive and blocked degrees of freedom. Its design may be simpler and more efficient than the serial arrangement of two separate structures.

To conclude, a judicious mechanical solution catalogue consists in establishing a preferred design for each active brick, and several possibilities for the passive bricks. This catalogue can thus be considered as a useful tool to tackle ultra-high precision flexure-based design, without offering off the shelf solutions. The robot designer will still need to evaluate the validity of the proposed mechanims, and potentially to redesign active, passive, or unique bricks which optimally fulfil the requirements of his/her specific application.

5 Application of the concept to a 5-DOF robot

The last part of this paper will illustrate the concept of modular flexure-based mechanisms by detailing a practical example, namely the design of an ultra-high precision 5-DOF robot. The choice of this mobility has been motivated by the micromanipulation and micro-assembly tasks, which consist in typical industrial applications of high-precision robots. These tasks often require five degrees of freedom, namely the three translations (Tx, Ty, Tz) and two rotations (Rx, Ry). In other words, the only motion which is undesired is the rotation around the axis of the robot tool (which is most of the time vertical, thus Rz). Even though this kinematics is often required for industrial applications, it is complex to design and machine, which explains the scarcity of industrial prototypes.

The specifications of this 5-DOF robot are thus performing three translations and two rotations; the required strokes are ± 5 mm and $\pm 10°$, the maximum linear and rotational displacements being achievable at the same time. As for the resolutions, the ideal values are 5 nm and 0.1 µrad. The overall volume of the robot should not exceed 1 dm^3, thus achieving a R_s ratio of around 0.1 (see Sect. 3). Eigenfrequencies should also be maximised in order to allow for accelerations up to 10 g.

The first step of the modular concept consists in making use of the conceptual solution catalogue to transform the 3-D design problem into several 2-D brick designs. The kinematical aspects of the methodology being out of the frame of this paper, only the selected robot kinematics is presented (see Fig. 13). The criteria which have been used to choose this structure in the solution catalogue are the following: the number of different building bricks must be minimised, and the passage from this mobility to another one must be simple, i.e. this robot must be easily transformable into a 4-DOF or a 6-DOF one.

Figure 13 shows the 5-DOF kinematics, with the three different 2-D bricks necessary to build this robot. The first active brick, noted $T_{\|}$, consists in a translational motion; its mechanical design is a simple 4-prismatic hinge table. Then, the second active brick is the $T_{\|}R_{\perp}$ mechanism which has been detailed in Sect. 3.1: it performs a translation and a rotation, making use of a Remote Centre of Motion to avoid compensatory motions during the rotation. Last, only one passive

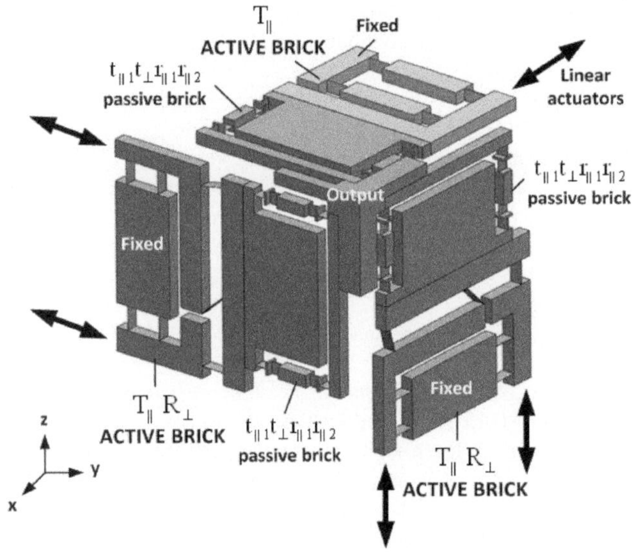

Figure 13. 5-DOF robot kinematics, with sketches of the brick design principles. The notation of the passive bricks is similar to the notation of the active bricks (see 3.1), with lowercase letters to indicate non-actuated motions.

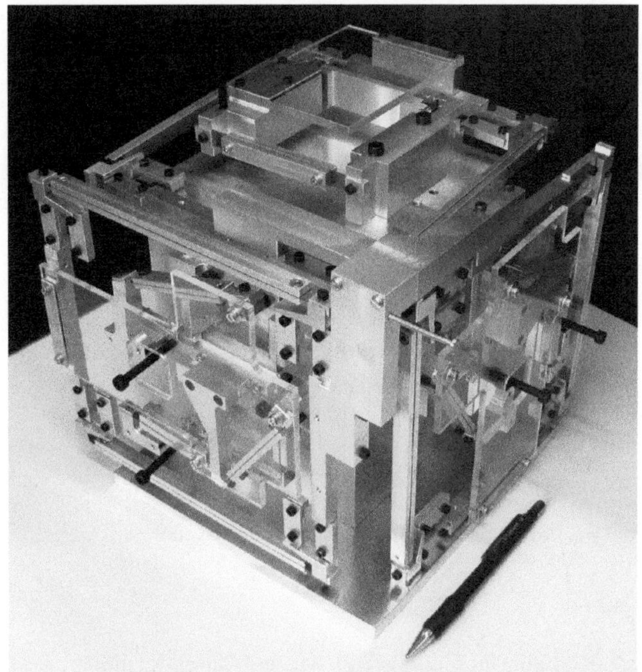

Figure 14. Double-scaled mock-up of the Legolas 5 robot.

brick ($t_{\|1}t_{\perp}r_{\|1}r_{\|2}$) is required, which performs two translations and two rotations. As its notation indicates, one translation is along an axis which is orthogonal to the brick plane, whereas the second is along one axis of the plane. As for the two rotations, their axes correspond to both directions of the brick plane. One possible mechanical design consists in the space parallelogramm illustrated in Fig. 12. A double-scaled mock-up of this 5-DOF robot, named Legolas 5, has been machined with assembled flexure blades in order to validate the design principles (see Fig. 14).

The next step of this robot development is the conception of an industrial prototype; as the bricks will be monolithically machined by W-EDM (Wire-Electro-Discharge Machining), the design must be reviewed in order to be more compact and adapted to this machining process; among other originalities, it includes the aforepresented actuation sub-brick, as well as gravity compensation mechanisms.

6 Summary and conclusions

A new concept of modular flexure-based mechanisms to design ultra-high precision robots has been presented in this paper. Its crucial advantage over other methodologies is the transformation of a 3-D design problem into several 2-D designs, which are simpler, well-mastered and easy to machine monolithically. A short description of the concept bases has allowed to introduce the representation of the modular parallel robot, as well as both types of planar flexure-based building bricks which are required for its design, namely the active and passive bricks.

The section dedicated to active bricks has detailed the main design challenge, which consists in the maximisation of the strokes. A ratio allowing the comparison between high-precision robots has been introduced, and a design making use of a Remote Centre of Motion has been proposed as a solution to avoid translational compensatory motions during the rotations. Moreover, an actuation sub-brick has been designed: some crucial issues, as the parasitic motion of the actuator moving part guiding system and the actuator choice have been detailed. The problematics of the actuator integration has highlighted the need for actuation solutions specifically intented for high-precision robots and mechanisms. The future work on this sub-brick is its experimental characterisation, which will allow to validate the measurement feedback system, the eccentric mounting of the actuator coil, as well as the thermal isolation of the flexure-based mechanism.

As for the passive bricks, the maximisation of some specific strokes, based on the role of the corresponding degrees of freedom, has been presented. The discussion on the validity of a mechanical solution catalogue has first highlighted that the serial arrangement of the active and passive bricks should be extended, permitting their merging into a unique mechanism. Then, it also allowed to conclude that the solution catalogue should include some possible designs, thus consisting in a useful tool for the designer, but cannot and should not be considered as exhaustive.

Last, this modular concept has been illustrated with a practical application, consisting in a high-precision 5-DOF robot. A double-scaled mock-up with assembled blades has validated the first design; the development of an industrial prototype, which would among other originalities include the actuation sub-brick, is currently in progress. Its machining and characterisation will allow to prove that the performances of a robot which has been designed with this new methodology are similar or better than the performances of robots resulting from more classical conception procedures.

Acknowledgements. The authors would like to thank the Laboratoire de Systèmes Robotiques collaborators, for their precious advice and the constructive brainstorming sessions, which contributed to the development of this work.

Edited by: A. Barari
Reviewed by: S. Henein and another anonymous referee

References

Bacher, J.-P., Bottinelli, S., Breguet, J.-M., and Clavel, R.: Delta^3: a New Ultra-high Precision Micro-Robot: Design and Control of a Flexure Mechanism, Journal Européen des Systèmes Automatisés, 36, 1263–1275, 2002.

Bottinelli, S., Henein, S., Aymon, C., and Clavel, R.: Movement transmission unit and movement transmission apparatus employing the same, US Patent US006453566B1, 24 September 2002.

Helmer, P.: Conception systématique de structures cinématiques orthogonales pour la micro-robotique, Ph. D. Thesis, No. 3365, Ecole Polytechnique Fédérale de Lausanne (EPFL), 2006.

Helmer, P., Mabillard, Y., Clavel, R., and Bottinelli, S.: High precision apparatus for imposing or measuring a position or a force, US Patent US2004025569A1, 23 December 2004.

Henein, S.: Conception des guidages flexibles, Presses Polytechniques et Universitaires Romandes, Collection Meta, 2001.

Kobel, P. and Clavel, R.: Miniaturization Challenges and Their Impact on the Micro-factory Concept and Manipulators, Japan Society of Precision Engineering, 77, 3, 263–268, 2011.

Lubrano, E. and Clavel, R.: Thermal Calibration of a 3 DOF Ultra High-Precision Robot Operating in Industrial Environment, Proceedings of the 2010 IEEE International Conference on Robotics and Automation, IEEE, 3692–3697, 2010.

Richard, M. and Clavel, R.: A new concept of modular kinematics to design ultra-high precision flexure-based robots, Proceedings for the joint conference of the 41st International Symposium of Robotics and the 6th German Conference on Robotics, Munich, 940–947, 2010.

Tolfree, D.: Commercialising Nanotechnology, concepts – products – markets, International Journal of Nanomanufacturing, 1, 1, 117–133, 2006.

7

Review article: locomotion systems for ground mobile robots in unstructured environments

L. Bruzzone[1] **and G. Quaglia**[2]

[1]University of Genoa, DIME, Via Opera Pia 15A, 16145 Genoa, Italy
[2]Politecnico di Torino, DIMEAS, Corso Duca degli Abruzzi, 24, 10129 Turin, Italy

Correspondence to: L. Bruzzone (bruzzone@dimec.unige.it)

Abstract. The world market of mobile robotics is expected to increase substantially in the next 20 yr, surpassing the market of industrial robotics in terms of units and sales. Important fields of application are homeland security, surveillance, demining, reconnaissance in dangerous situations, and agriculture. The design of the locomotion systems of mobile robots for unstructured environments is generally complex, particularly when they are required to move on uneven or soft terrains, or to climb obstacles. This paper sets out to analyse the state-of-the-art of locomotion mechanisms for ground mobile robots, focussing on solutions for unstructured environments, in order to help designers to select the optimal solution for specific operating requirements. The three main categories of locomotion systems (wheeled – W, tracked – T and legged – L) and the four hybrid categories that can be derived by combining these main locomotion systems are discussed with reference to maximum speed, obstacle-crossing capability, step/stair climbing capability, slope climbing capability, walking capability on soft terrains, walking capability on uneven terrains, energy efficiency, mechanical complexity, control complexity and technology readiness. The current and future trends of mobile robotics are also outlined.

1 Introduction

The forecasts of all the major robotics research institutions clearly indicate that the world market of service robotics is expected to increase dramatically over the next 20 yr, surpassing the market of industrial robotics in terms of units and sales (International Federation of Robotics, 2012; Prassler and Kosuge, 2008; EURON, 2009; CCC and CRA, 2009); in particular, ground mobile robots are the most widespread category of service robots; 75 % of total unit sales of professional service robots in 2010 were defence or field robots (International Federation of Robotics, 2012). Most mobile robots are designed to operate not only in structured environments but also in unstructured situations: important fields of application are homeland security (Murphy, 2004), surveillance (Quaglia et al., 2011), intervention in case of terrorist attacks (Birk and Carpin, 2006; Snyder, 2001), demining (Havlik, 2005; Cepolina and Hemapala, 2007), reconnaissance in dangerous situations (such as radioactive or chemical contamination; Hamel and Cress, 2001), agriculture (González et al., 2009; van Straten et al., 2002) and planetary exploration (Mishkin, 2004; Iagnemma and Dubowsky, 2004). Moreover, mobile robotics is a typical dual technology as it has significant military applications (Playter et al., 2006).

Several mechanical architectures of mobile robots have been proposed by academic and industrial researchers. These, of course, feature various combinations of advantages and drawbacks. Consequently, when conceiving a new mobile robot for a specific application, a designer has to evaluate an extensive range of possible technological solutions for its locomotion system, performing complex and time-consuming assessments. In the early design stages of a ground mobile robot, the expected operating environments must be analysed as these can belong to many different categories: indoor structured environments with flat and compact ground, with or without stairs; outdoor environments with differing terrain firmness, with or without obstacles, and so on. It is therefore useful to outline the variety of existing locomotion systems and to synthetically compare their strengths and weaknesses in different operating conditions.

This is the main aim of the work, which considers both research prototypes and commercially available industrial products. The analysis of state-of-the-art of locomotion systems for ground mobile robots has mainly focused on solutions for unstructured environments. As a matter of fact, many mobile robots for structured environments are quite simple from the mechanical point of view (in most cases, a layout with two active wheels featuring differential steering and one passive caster wheel is sufficient). Otherwise, when it is required to move on uneven or soft terrains or to climb obstacles, the mechanical design of the locomotion system is much more complex and many different schemes have been proposed; therefore, a synthetic comparison is more useful for designers.

It is very difficult to make a synthetic but exhaustive comparison of locomotion systems for ground mobile robots since the potential operating conditions are widely diversified (Siegwart and Nourbakhsh, 2004). An overview of legged robots is proposed in Machado and Silva (2006). In Seeni et al. (2008), a survey of the mobility concepts for extra-terrestrial surface exploration is proposed with reference to a selection of qualitative and quantitative parameters; the same topic is discussed in Fiorini (2000).

A similar comparison is made in this paper, but with two main differences: the analysis does not only focus on the field of application of spatial exploration, and the four hybrid categories (legs-wheels – LW, legs-tracks – LT, wheels-tracks – WT, and legs-wheels-tracks – LWT) are discussed separately. The evaluation considers the following features: maximum speed, obstacle crossing capability, step/stair climbing capability, slope climbing capability, walking capability on soft terrains, walking capability on uneven terrains, energy efficiency, mechanical complexity, control complexity, and technology readiness. These features are defined in Sect. 3.

It should be pointed out that there is a vast amount of scientific literature concerning mobile robotics which cannot be exhaustively quoted or included in the References for reasons of space. The works that have been quoted were selected to synthetically represent and compare all locomotion system categories. Broader bibliographies are available in Siegwart and Nourbakhsh (2004); Campion and Chung (2008); Kajita and Espiau (2008); Kemp et al. (2008); Meyer and Guillot (2008).

2 Classification of mobile robot locomotion systems and comparison methodology

This paper considers locomotion systems for robots moving predominantly on the ground, even if some of them are capable of travelling on water for short distances (Altendorfer et al., 2001). Furthermore, special-purpose locomotion

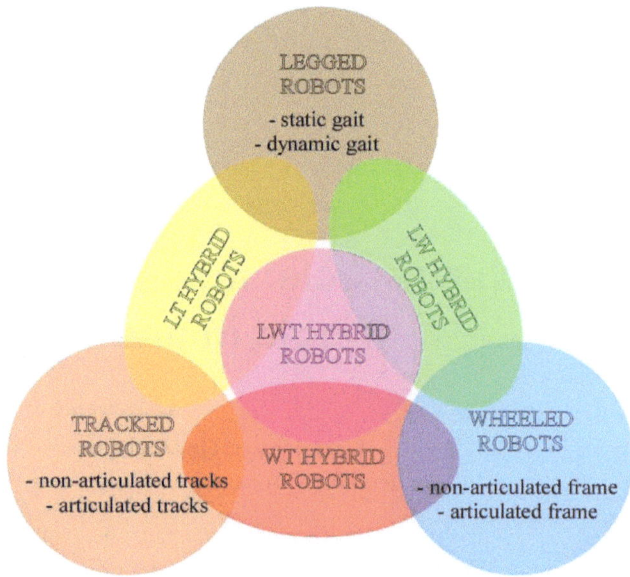

Figure 1. Ground mobile robot categories.

principles (e.g. jumping robots, snake-like slithering robots and wall-adhering robots) will not be discussed since they are devoted to very specific applications. Moreover, some researchers have proposed advanced locomotion strategies exploiting two or more co-operative mobile robots (Seeni et al., 2008); however, these co-operative approaches to robot mobility will not be considered since they require complex supervision systems and are rarely adopted; only locomotion principles for independent autonomous ground robots, not interacting with other vehicles or fixed devices, will be discussed.

On these premises, ground mobile robots can be classified into three main classes:

- wheeled robots (W)
- tracked robots (T)
- legged robots (L)

Moreover, there also exist robots featuring combinations of these locomotion principles, known as *hybrid robots*. In the scientific and industrial worlds, there are examples of the four possible combinations that lead to hybrid locomotion systems: legs-wheels (LW), legs-tracks (LT), wheels-tracks (WT), and legs-wheels-tracks (LWT). The three main categories and the four hybrid categories of ground mobile robots are summarised in Fig. 1.

For a detailed comparison of locomotion systems, a set of features that can be objectively evaluated must be defined; to this end, a set of ten features is proposed in Table 1 together with their corresponding definitions.

The first seven features (*maximum speed, obstacle crossing capability, step/stair climbing capability, slope climbing*

Table 1. Description of the features considered in the comparison of locomotion systems.

Feature	Definition
maximum speed	maximum speed on flat and compact surfaces in the absence of obstacles
obstacle crossing capability	capability of crossing obstacles with random shapes in unstructured environments (e.g. rocks)
step/stair climbing capability	capability of climbing up single steps and stairs in environments structured for humans
slope climbing capability	capability of climbing compact slopes with a sufficient friction coefficient (> 0.5)
walking capability on soft terrains	capability of walking on soft and yielding terrains (e.g. sand)
walking capability on uneven terrains	capability of walking on uneven terrains (e.g. grassy ground, rocky ground)
energy efficiency	energy efficiency in normal operating conditions, on flat and compact terrains
mechanical complexity	level of complexity of the mechanical architecture
control complexity	level of complexity of the control system (hardware and software)
technology readiness	level of maturity of the necessary enabling technologies

capability, walking capability on soft terrains, walking capability on uneven terrains, and energy efficiency) are related to pure mobility performance and are quantitatively measurable; the remaining three features (*mechanical complexity, control complexity, and technology readiness*) describe system complexity and influence other features, such as reliability, which are not expressly mentioned. Autonomy is not explicitly considered as it depends on energy efficiency.

3 Mobile robot locomotion systems

In this section, the main characteristics of the mobile robot locomotion system categories are summarised and compared with reference to the features defined in Table 1.

3.1 Wheeled locomotion systems

Wheeled robots can reach high speeds with low power consumption, and can be guided by controlling a few active degrees of freedom (Morin and Samson, 2008), but their ability to overcome obstacles is generally limited. Wheeled robots can be classified according to the number and position of the wheels. The minimum number of wheels to achieve static stability is three (the condition for static stability is that the vertical projection of the robot's centre of gravity on the ground must lie within the polygon formed by the wheel-ground contact points), through stability is improved with four or more wheels.

It should be noted that two-wheeled vehicles with an inverted pendulum layout have been developed for personal transportation (Liu and Parthasarathy, 2003), but in this case equilibrium is achieved in dynamic conditions by means of a complex control system and there is no substantial advantage with respect to three wheels if the overall height of vehicle and payload is not as relevant as in human transportation.

There are also examples of two-wheeled robots with inline configuration (robot motorcycles), but the mechanical model and the corresponding balance stabilisation and trajectory tracking control systems are even more complex than those governing the two-wheeled inverted pendulum scheme (Yi et al., 2006). Moreover, auxiliary legs or stabilisation rotors are required at zero-speed; for all of these reasons, unlike manned vehicles, this scheme is rarely adopted for autonomous robots.

Three-wheeled mobile robots with two differentially steered wheels and an idle caster wheel are widely used in structured environments with flat and even surfaces, for example, for transporting medicines in hospitals, because they require a simple control strategy and can spin around a vertical axis (Siegwart and Nourbakhsh, 2004). The same three-wheeled architecture is used in commercial vacuum-cleaner robots such as Roomba by iRobot (www.irobot.com). On the other hand, this locomotion architecture is not suitable for unstructured environments because of its poor stability in the presence of uneven and sloping surfaces.

To achieve improved stability, most wheeled robots for unstructured environments usually have 4, 6 or 8 wheels. These layouts, however, are hyperstatic and require an articulated frame or suspensions to ensure all the wheels make contact with the ground.

Regardless of the number of wheels, it is possible to distinguish two categories of wheeled robots: with non-articulated frames and with articulated fames.

It should be pointed out that the second category only includes robots with passively articulated frames. If the wheels are carried by members in relative motion with respect to the main robot body, and this motion is actuated, these members are considered as legs and the robot becomes a leg-wheel hybrid (see Sect. 3.4.1).

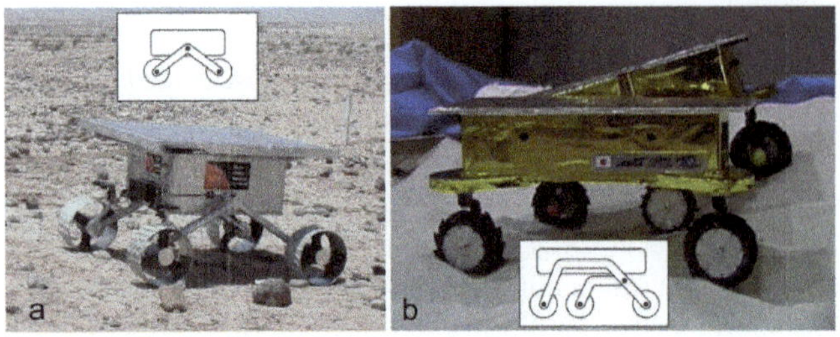

Figure 2. The four-wheeled SR2 rover (**a**, Miller et al., 2003) and the five-wheeled Micro5 rover (**b**, Kubotam et al., 2005).

In wheeled robots with non-articulated frames, the relative positions of the wheels are fixed, and with more than three wheels the configuration is hyperstatic and rarely adopted. Wheeled robots with articulated frames can adapt their configuration to the terrain, thanks to the mobility of the passive joints, and this reduces resistance while overcoming obstacles and irregularities.

The simplest possible mechanical design for wheeled robots for unstructured environments with four wheels is the $4\times4\times0$ wheel formula (*wheel formula:* total number of wheels $\times$ number of actuated wheels $\times$ number of actively steered wheels); the two wheels on each side are actuated by the same motor; steering is differential, achieved by setting different speeds on the two actuators. An example of this scheme is the SR2 rover developed by the University of Oklahoma for Malin Space Science Systems (Miller et al., 2003). The rover chassis is articulated in order to provide equal load distribution on the ground (Fig. 2a); this scheme allows pivotal turning, but the wheels slip on the ground while steering, thus reducing energy efficiency.

Where high speed is required, the robot locomotion system can be derived from automotive technology, adopting car suspensions and Ackerman steering geometry. Depending on the required mobile robot size, an unmanned vehicle can be created by fitting a car or another vehicle for human transport (usually four-wheeled) with an autonomous or remotely-controlled navigation system. This is the approach used in the DARPA Grand Challenge, funded by the United States Department of Defense (Thrun et al., 2006). With this approach, of course, robot speed is very high but motion control accuracy and manoeuvrability in small areas is worse with respect to other electrically-actuated wheeled robots.

Mobile robots with five wheels are rare; the Micro5 rover, developed by the Japanese Developer Aerospace Exploration Agency (JAXA), features the $5\times5\times0$ wheel formula (Fig. 2b). The load is distributed by the suspension system known as the *Pentad Grade Assist Suspension (PEGASUS)* (Kubotam et al., 2005); its fifth wheel is connected centrally to the main body by a passive revolute joint. While climbing obstacles, the traction of the fifth wheel produces a nose-dive momentum which increases the load on the front wheels and improves traction. This five-wheeled architecture was conceived to improve traction in applications with very low speeds (about $3\,\mathrm{cm\,s^{-1}}$) and low energy consumption, while it is not suitable for high speed operation due to the absence of suspensions.

Examples of six-wheeled robots with articulated frames are the rocker-bogie type rovers developed by NASA for Mars exploration, Spirit/Opportunity and Sojourner (Lindemann and Voorhees, 2005). These rovers feature the $6\times6\times4$ wheel formula (the front and rear wheels are independently steered). The two rocker-bogie mechanisms on each side of the vehicle (Fig. 3a) keep all the wheels in contact with the ground, even in presence of uneven terrain and obstacles. The average pressure on the wheels is passively equilibrated and this improves motion capability on soft terrains.

Other examples of wheeled robots with articulated frames are the SOLERO and CRAB rovers developed by the European Space Agency for the exploration of Mars (Michaud et al., 2002; Thueer et al., 2006a, b). SOLERO (Fig. 3b), based on the Shrimp mechanical design (Siegwart et al., 2002), features a $6\times6\times2$ wheel formula. The chassis has a bogie with two wheels on each side; the front wheel has a suspension system; the front and rear wheels are steered. CRAB II (Fig. 3c) is a rover with a $6\times6\times4$ wheel formula, comprising two symmetrical bogies on each side (the central wheel is connected to both bogies).

Eight-wheeled robots are quite rare. This scheme was used for the early lunar explorations (Lunokhod mission), but the latest space rovers are generally six-wheeled. Most robots with more than six wheels have an articulated frame with actively controlled degrees of freedom; therefore, they are considered as leg-wheel hybrid robots (see Sect. 3.4.1).

For the sake of completeness, it should be stated that most wheeled robots are fitted with deformable tyres (the larger sizes are usually inflatable while the smaller sizes are not) in order to reduce shocks while avoiding suspension systems. In some cases, omni-directional wheels (also known as Swedish wheels) are used to enhance robot mobility by reducing kinematic constraints while steering (Diegel et al., 2002). Nevertheless, these wheels are never adopted for unstructured environments as their mechanics can be damaged by shocks and dirt; moreover, conventional tyres give a better grip.

3.2 Tracked locomotion systems

Tracked robots are well suited to move on uneven and soft terrains and overcome obstacles, thanks to their large ground contact surface, but they move more slowly and consume more energy than wheeled robots: as a matter of fact, tracked robots are subject to vibrations as the lateral track profile is a polygon with moving vertices, and they are rarely fitted with damping systems. This limits maximum speed and reduces mechanical efficiency.

Tracked robots can be classified according to the number and layout of the tracks. First of all, robots can have non-articulated tracks or articulated tracks. If the relative

Figure 3. Six-wheeled rovers: rocker-bogie type (**a**, Lindemann and Voorhees, 2005), SOLERO (**b**, Michaud et al., 2002), CRAB II (**c**, Thueer et al., 2006a).

Figure 4. Tracked robots: Nanokhod (**a**, Klinker et al., 2007), Robhaz DT3 (**b**, Woosub et al., 2004) and Gunryu (**c**, Hirose et al., 1996).

mobility of the tracks is actuated, they are considered as leg-track hybrid robots (see Sect. 3.4.2). Robots with non-articulated tracks feature very simple mechanics and controls. They generally have two parallel tracks with differential steering. In spite of their simplicity, they move well on uneven and soft terrains and over small obstacles. An example of this scheme is Nanokhod, the miniaturised space exploration tracker (Fig. 4a) based on Russian technology (Klinker et al., 2007). An example of an agricultural robot with two parallel tracks is Fitorobot (González et al., 2009), which is used for spraying in greenhouses.

To improve the capacity to handle uneven terrains and climb obstacles, more than two tracks with relative passive mobility can be adopted. For example, the Robhaz DT3 and DT5 are four-tracked robots developed by the KIST Intelligent Robotics Research Center (Woosub et al., 2004). They feature two parallel tracks with common actuation on each side, independently articulated with respect to the frame by revolute joints. The triangular shape of the forward tracks is designed to facilitate the approach to high obstacles (Fig. 4b).

To further enhance adaptability to uneven terrains and obstacles, other articulated tracked robots feature more complex mechanics. For example, Gunryu (Fig. 4c), developed by the Hirose-Fukushima Robotics Lab, has four independent tracks; the front and rear tracks are connected to two independent main bodies by revolute joints. Moreover, the two main bodies are connected by a passively articulated arm (Hirose et al., 1996). This configuration allows superior relative track mobility and, consequently, a high capability of overcoming obstacles.

3.3 Legged locomotion systems

Legged robots have a broad mobility which makes them suitable for applications both in structured environments and on uneven terrains. However, they are relatively slow and consume a lot of energy. Generally speaking, legged robots have a large number of actuators and a complex control system.

The development of legged mobile robots is evidently biologically inspired: there exist robots with two legs (inspired by humans), with four legs (inspired by quadrupeds) and with more than four legs (inspired by insects). Though the most evident classification of legged robots is on the basis of the number of legs, their most important characteristic is the type of gait, which can either be *static* or *dynamic*.

Robots with static gait are always balanced, similarly to slow insects: their gait is planned so that the vertical projection of the robot's centre of gravity always remains within the polygon formed by the contact points of the legs with the ground. The gait is designed by assessing the quality of the robot's stability using proper stability indices (McGhee, 1985).

On the contrary, with a dynamic gait, the robot is not always balanced, similarly to fast walking, running, trotting or galloping animals (Raibert, 1986). Evidently, this type of locomotion requires a much more complex control system:

Figure 5. Legged robots: BigDog (**a**, Playter et al., 2006), RHex (**b**, Altendorfer et al., 2001), Titan XI (**c**, Hodoshima et al., 2007).

- the static gait design can only be based on the kinematic model, while the dynamic gait design must also be based on the dynamic model;
- with the static gait, different robot speeds can be achieved with the same leg motions performed at different speeds, while with the dynamic gait, leg trajectories must be modified as a function of the robot's speed;
- with the static gait, the robot can stop in any position and remain balanced, while with the dynamic gait, a proper movement must be performed in order to reach a rest position;
- with the dynamic gait, it is complex to compensate external disturbance forces and requires complete motion replanning.

Moreover, the increased complexity of the dynamic gait is not only related to control, but also to the mechanical architecture: for efficient dynamic walking, each robot leg should have several actuated degrees of freedom and, preferably, force sensors on its feet.

On the other hand, dynamic walking brings remarkable advantages. First of all, the energy efficiency is higher: with the static gait, the robot continuously uses power to change its configuration, but kinetic energy is lost at every step, while with the dynamic gait, most of the kinetic energy is conserved. Moreover, the dynamic gait can efficiently isolate the body from terrain irregularities and compensate external disturbance forces.

There are two main classes of model-based dynamic gait control strategies: the approaches based on the *zero-moment point (ZMP) principle* (Vukobratovic and Borovac, 2004) and those based on *passive dynamic walking* and *limit cycle walking* (McGeer, 1990; Manchester et al., 2011).

The basic idea of the ZMP approach is that the pressure centre remains within the polygon of the foot in contact with the terrain. The movements are conservative and stable but neither natural looking nor highly energy efficient. On the contrary, passive-dynamic and limit-cycle walkers are more life-like and energy efficient, as gravity and inertial effects have an important part to play in the generation of motion.

Moreover, strict ZMP walking does not admit a flight phase which characterises many dynamic movements, such as running or hopping. For all these reasons, many researchers consider the ZMP walking to be a quasi-static gait. On the contrary, passive-dynamic and limit-cycle walkers can perform motions with flying phases. In fact, they can be considered as biped or quadruped evolutions of the one-legged hopping robots by Raibert (1986) and Raibert et al. (1984).

Though discussion of these control techniques is out of the scope of this paper, it is important to note that the real-time implementation of these algorithms is the main obstacle to the realisation of legged robots with dynamic gait. Nevertheless, thanks to the progress made in microprocessor technology, legged mobile robots with advanced dynamic walking capability, agility and equilibrium comparable to human beings and animals have been developed during the last decade.

As regards biped locomotion, in the most advanced humanoid robots, the upper body and arms contribute to the movement, achieving a human-like dynamic gait with extended legs, while the first biped robots featured an unnatural gait with semi-bent legs. These robots can also soften impacts and stand up autonomously in case of tumbles due to strong unplanned external forces or slippery terrains (Hirukawa et al., 2005). They can also perform complex tasks such as dancing (Aucouturier et al., 2008).

Up until now, these state-of-the-art humanoid biped robots have mainly been developed and tested for research purposes, and their cost limits the range of possible applications. There is a growing number of commercially available small-scale humanoid robots, such as Nao by Aldebaran Robotics and Robonova by Hitec (http://www.aldebaran-robotics.com; http://www.robonova.de). Nevertheless, these robots are designed for edutainment and not for outdoor operations. On the contrary, quadruped robots with advanced dynamic walking capabilities, suitable for operations in highly unstructured environments, are already available. Bigdog by Boston Dynamics is a quadruped robot (height: 1 m; length: 1 m; weight: 90 kg) powered by a diesel engine which drives a hydraulic actuation system (Fig. 5a). Each leg features three active degrees of freedom and one passive linear pneumatic

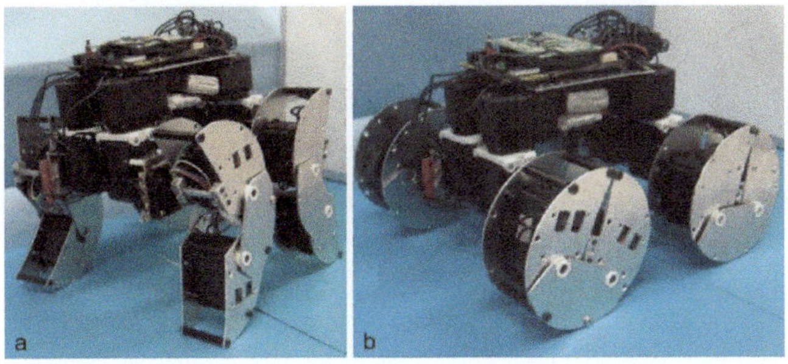

Figure 6. Leg-wheel hybrid robot with retractable locomotion modules: legged mode **(a)** and wheeled mode **(b)** (Tadakuma et al., 2010).

compliance in the lower leg. BigDog can walk and trot at speeds of up to $0.8\,\mathrm{m\,s^{-1}}$, and move on rough terrains and slopes up to 35°, while carrying a 50 kg payload (Playter et al., 2006).

In applications with small payloads, e.g. surveillance, robot size can be decreased. On this smaller scale, the complexity of the leg architecture and, consequently, control complexity can be reduced whilst preserving good motion performance. For example, RHex is a cockroach-inspired hexapod robot with compliant legs providing a self-stabilised gait, developed by a consortium of six American and Canadian universities (Fig. 5b). In spite of its simple mechanical design with one actuator per leg, RHex is capable of performing a wide variety of tasks, such as walking, running, leaping over obstacles and climbing stairs (Altendorfer et al., 2001). Another biologically inspired legged robot with simplified leg design is Whegs, developed by the Case Western Reserve University, which is fitted with three-spoke locomotion units (Quinn et al., 2003).

On the other hand, large legged robots for heavy payloads are usually controlled with static gait and move slowly in order to reduce the structural stresses caused by inertial effects. In this case, the most popular mechanical architecture is the quadruped. One example is Titan XI, developed by the Hirose-Fukushima Robotics Lab (Fig. 5c) for consolidating rocky slopes (mass: 7000 kg; length of the hydraulically actuated legs: 3.7 m; Hodoshima et al., 2007).

3.4 Hybrid locomotion systems

Hybrid locomotion systems are probably the most interesting solutions for mobile robots as they combine the advantages of the various classes while attempting to avoid the drawbacks. In fact, legged locomotion is the most suitable solution in unstructured environments or in environments structured for humans (in presence of stairs, for example). On the other hand, humans often use wheeled or tracked extensions to increase their speed and energy efficiency. Unlike nature, in the design of a mobile robot, legs, wheels and tracks can be freely integrated. The four categories of hybrid mobile robots (LT, LW, WT, and LWT) are now discussed.

Figure 7. The leg-wheel hybrid Octopus robot (Lauria et al., 2002).

3.4.1 Leg-wheel hybrid locomotion systems

Leg-wheel robots combine the energy efficiency of wheels with the operative flexibility of legs. Legs and wheels can be basically combined in three ways (Tadakuma et al., 2010):

- fitting a wheeled robot with additional legs connected to the robot body;
- using retractable modules that can be used as wheels or legs;
- placing the wheels on the leg links (usually, but not always, at the ends of the legs).

The first approach is rarely adopted. Robot design is conceptually simple if legs and wheels are used alternatively, depending on terrain conditions. The main drawback is that the mass of the robot is usually high as the vehicle is fitted with two separate locomotion systems. Furthermore, if wheels and legs are used simultaneously, the advantage in terms of energy efficiency due to the wheels is significantly reduced.

The second approach is very interesting. For example in Tadakuma et al. (2010), a mobile robot with retractable leg-wheel modules is proposed (Fig. 6). The main drawback of this design is the mechanical complexity of the retractable modules and their low reliability in dirty environments or in case of shocks.

The third way of combining legs and wheels is probably the most effective. An example of a hybrid leg-wheel robot is Octopus (Fig. 7), developed by Ecole Polytechnique Fédérale de Lausanne (EPFL) (Lauria et al., 2002). It is fitted with tilt sensors and tactile wheels, while its sophisticated locomotion system has 8 motorised wheels and a total of 15 degrees

Figure 8. Stepping triple wheel hybrid robots: Spacecat (**a**, Siegwart et al., 1998); Epi.q (**b**, Quaglia et al., 2011).

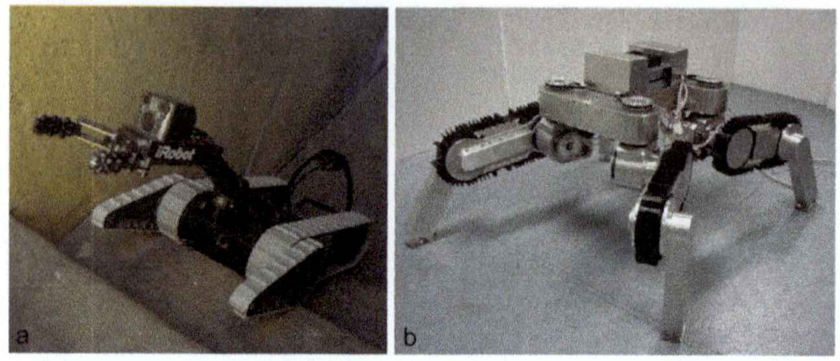

Figure 9. Leg-track hybrid robots: iRobot SUGV (**a**, http://www.irobot.com/gi/ground/) and Titan X (**b**, Hirose et al., 2009).

of freedom which ensure great adaptability to uneven terrain and obstacle-climbing capability.

Another example of the third approach is the *stepping triple wheel* concept, introduced in the Spacecat microrover by EPFL (Siegwart et al., 1998): two three-wheeled locomotion modules can rotate independently with respect to the main body and allow the rover to actively lift one wheel to climb obstacles (Fig. 8a). Eight independent actuators are required for motion control: two for the locomotion modules and six for the wheels.

The same three-wheeled locomotion unit geometry, which assures good motion performance on uneven terrains and obstacle-climbing capability, is adopted in the Epi.q mobile robots family, developed in co-operation between the Polytechnic of Turin and the University of Genoa (Fig. 8b). Each Epi.q locomotion unit is underactuated by a single gearmotor through an epicyclical mechanism which automatically switches between legged locomotion and wheeled locomotion depending on dynamic and friction conditions, without control intervention. This significantly reduces the complexity of the control system (Quaglia et al., 2011).

3.4.2 Leg-track hybrid locomotion systems

Hybrid mobile robots with legs and tracks are fairly popular for difficult environments, provided that speed and energy efficiency are not crucial.

There are many ways of combining legs with tracks. The simplest approach is to use more than two tracks (usually four), which are in relative mobility with respect to the robot frame, in order to achieve legged locomotion. Examples of this approach are the commercially available mobile robots developed by iRobot for homeland security, surveillance, inspection, and explosive detection (Fig. 9a). These robots are extremely sturdy and reliable, with motion performance similar to tracked robots, but a greater capability of overcoming obstacles.

On the other hand, leg-track hybrid robots with more complex leg architectures have more extended mobility features and gait capabilities. Titan X, developed by the Hirose-Fukushima Robotics Lab (Fig. 9b), is a quadruped mobile robot with three degrees of freedom per leg; the four belts have a double function: mechanical transmission for actuation of the knee joints during legged locomotion, and tracks during tracked locomotion (Hirose et al., 2009). Another example of a leg-track hybrid robot is the one developed by a research group led by Yokota et al. (2006).

3.4.3 Wheel-track hybrid locomotion systems

The combination of wheels and tracks is very effective when good motion performance on soft and uneven terrains is required to be combined with energy efficiency on flat and compact artificial ground.

In wheel-track hybrid robots, the relative position of the tracks and wheels or the track shape can usually be changed to enable or disable wheel contact with the ground. For example, the Daegu Gyeongbuk Institute of Science & Technology has developed a hybrid robot, featuring variable shape tracks (Fig. 10a) (Kim et al., 2010). In the configuration shown in Fig. 10, locomotion is achieved by the tracks; on flat compact ground, the tracks can be folded and locomotion obtained by the wheels.

On the contrary, in the Helios VI tracked carrier by the Hirose-Fukushima Robotics Lab (Fig. 10b), locomotion is never exclusively performed by the wheels; the two front wheels are added not to improve energy efficiency, but to increase stair climbing capabilities (Hirose et al., 2001).

Another interesting approach is the Galileo wheel patented system which, for example, is adopted in the VIPeR, a mobile platform for surveillance tasks developed by Elbit Systems (Fig. 10c). Thanks to extensible crawlers, the Galileo wheel combines wheels and tracks in a single group.

3.4.4 Leg-wheel-track hybrid locomotion systems

Azimuth is an example of a robotic platform that combines all three types of locomotion (Fig. 11); it is fitted with four independent leg-track-wheel articulations which can generate a wide variety of locomotion modes (holonomic and omni-directional motions, climbing of obstacles and stairs) (Michaud et al., 2005). Azimuth was mainly designed to perform in indoor environments, though the concept can be

Figure 10. Wheel-track hybrid robots: with variable-shape tracks (**a**, Kim et al., 2010); Helios VI (**b**, Hirose et al., 2001); with Galileo wheel (**c**, http://www.galileomobility.com).

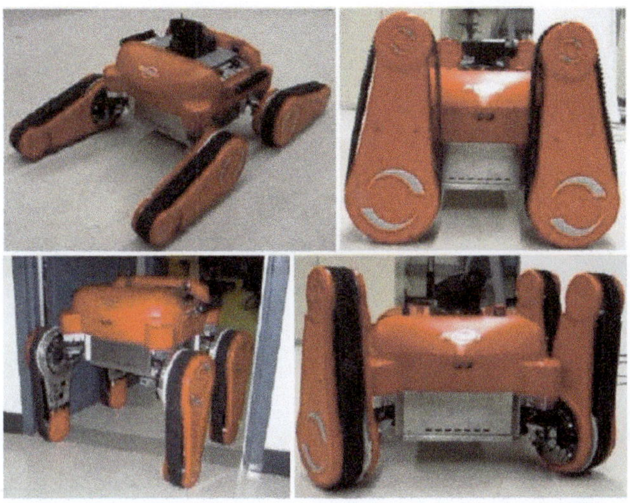

Figure 11. The hybrid leg-wheel-track Azimuth robot (Michaud et al., 2005).

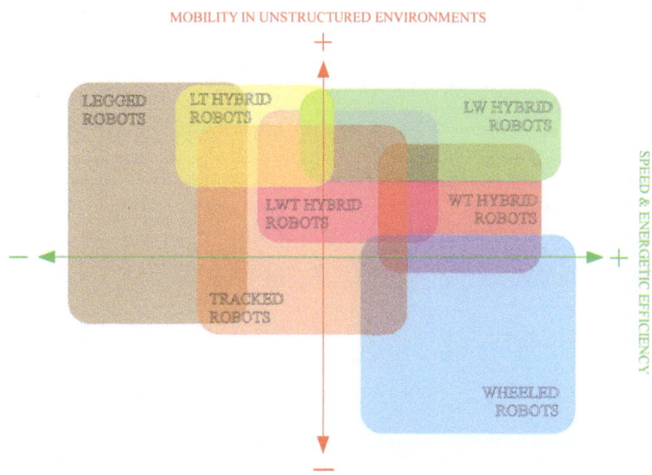

Figure 12. Comparison of locomotion systems.

adapted to outdoor settings. The main drawback is its high mechanical complexity.

3.5 Comparison of locomotion system features

With reference to the three main categories and the four hybrid categories of locomotion systems (Fig. 1), a very synthetic outline of locomotion system features is shown in Fig. 12, where mobility in unstructured environments (y) vs. speed and energy efficiency (x) is qualitatively graphed. Wheeled robots are in the right lower zone (better x, worst y); legged robots are in the left upper zone (worst y, better x); tracked robots are in the middle; hybrid combinations tend towards the right upper zone (combination of the benefits).

Along the x-axis,

- legged robots are in the left zone as this type of locomotion implies impacts between feet and ground. Moreover, it requires torque delivery in the actuated joints, also in static conditions (unless non-reversible transmissions or elastic preload elements are used to balance the forces of gravity).

- tracked robots are in the middle as the impacts with the ground are less significant but not absent. Moreover, the track elements are connected by several revolute joints with simple constructions and medium to poor energy efficiency.

- wheeled robots are in the right zone as they have few revolute joints (one per wheel), usually realised with ball bearings and therefore with a high energy efficiency.

Along the y-axis,

- wheeled robots are in the lower zone, because a wheel without additional devices cannot overcome obstacles if initial contact with the obstacle profile is not significantly lower than the wheel axis.

- tracked robots are in the middle as the large ground contact surface allows terrain unevenness to be filtered and contact pressure reduced.

- legged robots are in the upper zone, because this locomotion provides many degrees of freedom during motion in unstructured environments, since it allows the

Table 2. Synthetic comparison of locomotion system features.

locomotion → feature ↓	Wheeled	Tracked	Legged	LW hybrid	LT hybrid	WT hybrid	LWT hybrid
maximum speed	high	medium/ high	low (static walking) to medium (dynamic walking)	medium/ high	medium	medium/ high	medium/ high
obstacle crossing capability	low	medium/ high	high	medium/ high	high	medium	high
step/stair climbing capability	low	medium	high	high	high	medium	high
slope climbing capability	low/ medium	high	medium/ high	medium/ high	high	medium/ high	high
walking capability on soft terrains	low	high	low/ medium	low/ medium	medium/ high	high	medium/ high
walking capability on uneven terrains	low	medium/ high	high	high	high	medium/ high	high
energy efficiency	high	medium	low (static walking) to medium (dynamic walking)	medium/ high	medium	medium/ high	medium/ high
mechanical complexity	low	low	high	medium/ high	medium/ high	low/ medium	high
control complexity	low	low	high	medium/ high	medium/ high	medium	high
technology readiness	full	full	full for static walking; in progress for dynamic walking	full (usually static walking)	full (usually static walking)	full	full (usually static walking)

contact points between feet and terrain/obstacles to be selected.

In general, hybrid locomotion systems should combine the benefits of the different categories from which they derive. Therefore, in the graph in Fig. 12, each hybrid category should be located in an area characterised by the maximum values of the original categories along each axis. Nevertheless, this does not occur completely (and the graph is drawn accordingly), because the combination of alternative locomotion devices lowers overall performance: in fact, a non-active locomotion device is a non-negligible payload. From this point of view, hybrid solutions in which different locomotion systems share common members (for example, leg-wheel robots based on the stepping triple-wheel concept; see Sect. 3.4.1) are better, while hybrid solutions in which two locomotion systems are never used in combination (such as the wheel-track robot in Fig. 10) are worse.

Table 2 compares the various locomotion systems with reference to the ten features discussed in Sect. 2. The first seven features describe mobility performance and are qualitatively measurable, while the last three features describe system complexity. The proposed criteria for evaluating the first seven features are shown in Table 3 with ranges corresponding to the different ranks. These criteria have been applied in Table 2.

Since mobile robots can have different sizes, some features (maximum speed, obstacle crossing capability, step/stair climbing capability, walking capability on uneven terrains) have been normalised with respect to the robot dimensions.

Table 3. Evaluation criteria.

Feature	Evaluation criteria	Ranges		
		Low	Medium	High
maximum speed	ratio between maximum speed and overall length of robot	$< 0.5\,s^{-1}$	$0.5–3\,s^{-1}$	$> 3\,s^{-1}$
obstacle crossing capability	ratio between the maximum height of an obstacle with a semi-circular lateral profile which can be crossed and the height of the robot	< 0.25	0.25–0.5	> 0.5
step/stair climbing capability	ratio between the maximum height of the square step which can be crossed and the height of the robot	< 0.15	0.15–0.35	> 0.35
slope climbing capability	maximum slope that can be climbed (compact surface, friction coefficient > 0.5)	< 15°	15–30°	> 30°
walking capability on soft terrains	capability of walking on:	compact soil	medium to soft soil	soft sand
walking capability on uneven terrains	ratio between the maximum ground roughness that can be crossed (rocky ground with random profile) and the height of the robot	< 0.1	0.1–0.2	> 0.2
energy efficiency	ratio between the final gravitational potential energy acquired while climbing a slope at low speed and the energy supplied to the actuators	< 0.1	0.1–0.35	> 0.35

The obstacle crossing capability is defined as the capability of crossing obstacles with random shapes in unstructured environments (Table 1). Nevertheless, a semi-circular shape was selected for quantitative comparison.

As previously mentioned, locomotion systems including wheels enjoy high speed and energy efficiency. In particular, wheeled robots derived from automotive technology, featuring car-like suspensions and Ackermann steering, maximise these parameters. If not only these features were of primary importance, but obstacle crossing and climbing capabilities would also be required, wheels could be properly combined with legs in hybrid solutions. On the other hand, if task requirements are focussed on soft and yielding terrains, it is preferable to enhance robot mobility by adding tracks rather than legs, because tracks have a larger contact surface, while legs are more efficient for rigid obstacles with complex shapes.

It is important to note that slippage in the presence of yielding terrains, slopes or obstacles, which is one of the main hindrances to robot mobility, is a complex phenomenon that is remarkably influenced by the control approach (Wong, 2001). Suitable visual-based and/or adaptive control strategies can enhance the mobility performance of wheeled and tracked systems (Angelova et al., 2007; González et al., 2010). As regards legged locomotion, balance control in the presence of slippage is achieved by state-of-the-art robots with dynamic gait, such as Bigdog.

Fundamental design choice criteria are mechanical complexity and control complexity (Dudek and Jenkin, 2010). In particular, mechanical complexity has a significant influence on robot operating dependability, in terms of both mean time between failure and mean time to repair. In this sense, pure tracked and wheeled robots are evidently simple and robust, while robots with complex mechanical designs, especially as regards the linkages in contact with the terrain (such as the leg-wheel robot in Fig. 6) should be avoided in heavy-duty applications. As regards control complexity, this is evidently higher for solutions involving legs due to gait planning requirements. Nevertheless, only these architectures, especially if controlled by model-based dynamic algorithms, can provide top performance in obstacle crossing and climbing in unstructured environments.

Starting from these general considerations, the proper mix of locomotion systems for hybrid solutions can be selected on the basis of specific task requirements.

4 Discussion about future trends in mobile robotics

The global market of service robotics is expected to increase substantially during the next 20 yr, surpassing the market of industrial robotics in terms of both units and sales. In this scenario, the development of efficient and flexible locomotion systems for ground mobile robots is a fundamental research issue.

As regards the future trends of mobile robotics, the most significant enabling conditions that can lead to the spread of legged mobile robots, with a level of operating flexibility that is comparable to humans and quadrupeds, concern the development of model-based and adaptive dynamic gait control methodologies, capable of faithfully reproducing animal behaviour. Moreover, from the hardware point of view, other important enabling conditions are the ongoing increase in the computational power of microprocessors and in the energy density of batteries, and the availability of actuators with highly specific power ratings.

Probably one of the most important technological challenges for robotics during next 10–15 yr will be to transfer high-performance legged robots from the world of research to the world of industry and commerce by means of a difficult but necessary cost reduction process.

On the other hand, when some specific sets of features are of primary importance in the mix of operating requirements (e.g. speed, energy efficiency or mobility on yielding terrains) or when robot cost and complexity must be limited due to the economic feasibility of the application, hybrid locomotion systems are often the most suitable solutions.

For all of these reasons, the strong diversification in the current mobile robotics scenario will probably persist in the future, with a wide range of technological approaches and different levels of mechanical and control complexity.

5 Conclusions

The state-of-the-art of locomotion systems for mobile robots has been analysed in this paper, considering both research prototypes and industrial products, in order to provide useful indications for the early design stages, when the type of locomotion mechanism has to be selected on the basis of operating requirements.

To this end, the three main categories (wheeled, tracked and legged) and the four hybrid categories (legged-wheeled, legged-tracked, wheeled-tracked, and legged-wheeled-tracked) of locomotion systems have been compared with reference to a selection of ten features, in order to assess not only mobility performance in the most significant conditions, but also overall system complexity and reliability.

Future trends in mobile robotics are also briefly discussed. It can be said that two of the most promising lines of research are quadruped legged locomotion with dynamic gait for high-end applications in completely unstructured outdoor environments (such as military and homeland security tasks) and hybrid locomotion, especially leg-wheel and leg-track, for compact and low-cost mobile robotic platforms.

Edited by: J. Schmiedele
Reviewed by: T. Brown and another anonymous referee

References

Altendorfer, R., Moore, N., Komsuoglu, H., Buehler, M., Brown Jr., H. B., McMordie, D., Saranli, U., Full, R., and Koditschek, D. E.: RHex: a biologically inspired hexapod runner, Auton. Robot., 11, 207–213, 2001.

Angelova, A., Matthies, L., Helmick, D., and Perona, P.: Learning and prediction of slip from visual information, J. Field Robot., 24, 205–231, 2007.

Aucouturier, J. J., Ikeuchi, K., Hirukawa, H., Nakaoka, S., Shiratori, T., Kudoh, S., Kanehiro, F., Ogata, T., Kozima, H., Okuno, H. G., Michalowski, M. P., Ogai, Y., Ikegami, T., Kosuge, K., Takeda, T., and Hirata, Y.: Cheek to Chip: Dancing Robots and AI's Future, IEEE Intell. Syst., 23, 74–84, 2008.

Birk, A. and Carpin, S.: Rescue robotics: a crucial milestone on the road to autonomous systems, Adv. Robotics, 20, 596–605, 2006.

Campion, G. and Chung, W.: Wheeled robots, in: Springer Handbook of Robotics, edited by: Siciliano, B. and Khatib, O., Springer, Heidelberg, 391–410, 2008.

CCC (Computing Community Consortium) and CRA (Computing Research Association), A Roadmap for US Robotics. From Internet to Robotics, Technical report, http://www.us-robotics.us/, 2009.

Cepolina, E. E. and Hemapala, M. U.: Power tillers for demining: blast test, International Journal of Advanced Robotic Systems, 4, 253–257, 2007.

Diegel, O., Badve, A., Bright, G., Potgieter, J., and Tlale, S.: Improved Mecanum Wheel Design for Omni-directional Robots, Engineering , Proceedings of the Australasian Conference on Robotics and Automation, 117–121, 2002.

Dudek, G. and Jenkin, M.: Computational Principles of Mobile Robotics, Cambridge University Press, 2010.

EURON (EUropean RObotics research Network) and EUROP (EUropean RObotics technology Platform), The strategic research agenda for robotics in Europe, Technical report, http://www.robotics-platform.eu/sra, 2009.

Fiorini, P.: Ground Mobility Systems for Planetary Exploration, Proceedings of the IEEE International Conference on Robotics & Automation, 908–913, 2000.

González, R., Rodríguez, F., Sánchez-Hermosilla, J., and Donaire, J. G.: Navigation Techniques for Mobile Robots in Greenhouses, Appl. Eng. Agric., 25, 153–165, 2009.

González, R., Fiacchini, M., Alamo, T., Guzman, J. L., and Rodríguez, F.: Adaptive Control for a Mobile Robot under Slip Conditions using an LMI-based Approach, Eur. J. Control, 16, 144–155, 2010.

Hamel, W. R. and Cress, R. C.: Elements of telerobotics necessary for waste clean up automation, Proc. IEEE International Conference on Robotics and Automation, 393–400, 2001.

Havlik, S.: A modular concept of the robotic vehicle for demining operations, Auton. Robot., 18, 253–262, 2005.

Hirose, S., Shirasu, T., and Fukushima, E. F.: Proposal For Cooperative Robot Gunryu Composed of Autonomous Segments, Robot. Auton. Syst., 17, 107–118, 1996.

Hirose, S., Fukushima, E., Damoto, R., and Nakamoto, H.: Design of terrain adaptive versatile crawler vehicle HELIOS-VI , Proceedings of the IEEE International Conference on Intelligent Robots and Systems, 1540–1545, 2001.

Hirose, S., Fukuda, Y., Yoneda, K., Nagakubo, A., Tsukagoshi, H., Arikawa, K., Endo, G., Doi, T., and Hodoshima, R.: Quadruped Walking Robots at Tokyo Institute of Technology, IEEE Robotics and Automation Magazine, 16, 104–114, 2009.

Hirukawa, H., Kajita, S., Kanehiro, F., Kaneko, K., and Isozumi, T.: The Human-size Humanoid Robot That Can Walk, Lie Down and Get Up, International Journal of Robotic Research, 24, 755–769, 2005.

Hodoshima, R., Doi, T., Fukuda, Y., Hirose, S., Okamoto, T., and Mori, J.: Development of a Quadruped Walking Robot TITAN XI for Steep Slope Operation – Step Over Gait to Concrete Frames on Steep Slopes, Journal of Robotics and Mechatronics, 19, 13–26, 2007.

Iagnemma, K. and Dubowsky, S.: Mobile Robots in Rough Terrain. Estimation, Motion Planning, and Control with Application to Planetary Rovers, Springer, 2004.

International Federation of Robotics, http://www.ifr.org, last access: 11 July 2012.

Kajita, S. and Espiau, B.: Legged robots, in: Springer Handbook of Robotics, edited by: Siciliano B. and Khatib, O., Springer, Heidelberg, 361–389, 2008.

Kemp, C. C., Fitzpatrick, P., Hirukawa, H., Yokoi, K., Harada, K., and Matsumoto, Y.: Humanoids, in: Springer Handbook of Robotics, edited by: Siciliano, B. and Khatib, O., Springer, Heidelberg, 1307–1333, 2008.

Kim, J., Kim, Y.-G., Kwak, J.-H., Hong, D.-H., and An, J.: Wheel & Track Hybrid Robot Platform for Optimal Navigation in an Urban Environment, Proceedings of the SICE Annual Conference, 881–884, 2010.

Klinker, S., Lee, C. G. Y., Wagner, C., Hlawatsch, W., Schreyer, A. M., and Roser, H. P.: Destination Moon and beyond for the Micro rover Nanokhod, Proceedings of the DGLR International Symposium To Moon and beyond, 2007.

Kubotam, T., Kunii, Y., Kuroda, Y., and Working Group: Japanese lunar robotics exploration by co-operation with lander and rover, J. Earth Syst. Sci., 114, 777–785, 2005.

Lauria, M., Piquet, Y., and Siegwart, R.: OCTOPUS: an autonomous wheeled climbing robot, Proceedings of the 5th International Conference on Climbing and Walking Robots (CLAWAR), 315–322, 2002.

Lindemann, R. and Voorhees, C.: Mars exploration rover mobility assembly design, test, and performance , Proceedings of the 2005 IEEE Conference on Systems, Man and Cybernetics, 450–455, 2005.

Liu, R. and Parthasarathy, R.: Segway Human Transporter (HT): potential opportunities and challenges for transportation systems, presented at the 82nd Annual Meeting of the Transportation Research Board, Washington, DC, 2003.

Machado, J. A. T. and Silva, M. F.: An Overview of Legged Robots, Proceedings of the MME International Symposium on Mathematical Methods in Engineering, 2006.

Manchester, I. R., Mettin, U., Iida, F., and Tedrake, R.: Stable dynamic walking over uneven terrain, Int. J. Robot. Res., 30, 265–279, 2011.

McGeer, T.: Passive dynamic walking, Int. J. Robot. Res., 9, 62–82, 1990.

McGhee, R. B.: Vehicular legged locomotion, in: Advances in Automation and Robotics, edited by: Saridis, G. N., JAI Press, New York, 1, 259–284, 1985.

Meyer, J.-A. and Guillot, A.: Biologically inspired robots, in: Springer Handbook of Robotics, edited by: Siciliano, B. and Khatib, O., Springer, Heidelberg, 1395–1422, 2008.

Michaud, F., Letourneau, D., Arsenault, M., Bergeron, Y., Cadrin, R., Gagnon, F., Legault, M. A., Millette, M., Pare, J. F., Remblay, M. C., Lepage, P., Morin, Y., Bisson, J., and Caron, S.: Multi-Modal Locomotion Robotic Platform Using Leg-Track-Wheel Articulations, Auton. Robot., 18, 137–156, 2005.

Michaud, S., Schneider, A., Bertrand, R., Lamon, P., Siegwart, R., van Winnendael, M., and Schiele, A.: SOLERO: Solar Powered Exploration Rover, Proceedings of the 7th ESA Workshop on Advanced Space Technologies for Robotics and Automation, The Netherlands, 2002.

Miller, D. P., Hunt, T., Roman, M., Swindell, S., Tan, L., and Wainterholler, A.: Experiments with a Long-Range Planetary Rover, Proceedings of the International Symposium on Artificial Intelligence, Robotics and Automation in Space, 2003.

Mishkin, A.: Sojourner: an insider's view of the Mars Pathfinder mission, Berkley Books, 2004.

Morin, P. and Samson, C.: Motion control of wheeled mobile robots, in: Springer Handbook of Robotics, edited by: Siciliano, B. and Khatib, O., Springer, Heidelberg, 799–826, 2008.

Murphy, R. R.: Rescue robotics for homeland security, Communications of the ACM – Homeland security, 47, 66–68, 2004.

Playter, R., Buehler, M., and Raibert, M.: Bigdog, Proceedings of the SPIE Defense&Security Symposium, Unmanned Systems Technology, 2006.

Prassler, E. and Kosuge, K.: Domestic robotics, in: Springer Handbook of Robotics, edited by: Siciliano, B. and Khatib, O., Springer, Heidelberg, 1253–1281, 2008.

Quaglia, G., Bruzzone, L., Bozzini, G., Oderio, R., and Razzoli, R.: Epi.q-TG: mobile robot for surveillance, Industrial Robot, 38, 282–291, 2011.

Quinn, R. D., Nelson, G. M., Bachmann, R. J., Kingsley, D. A., Offi, J. T., Allen, T. J., and Ritzmann, R. E.: Parallel complementary strategies for implementing biological principles into mobile robots, International Journal of Robotics Research, 22, 169–186, 2003.

Raibert, M. H.: Legged Robots That Balance, MIT Press, Cambridge, Mass, 1986.

Raibert, M. H., Brown, H. B., and Chepponis, M.: Experiments in balance with a 3D one-legged hopping machine, Int. J. Robot. Res., 3, 75–92, 1984.

Seeni, A., Schafer, B., Rebele, B., and Tolyarenko, N.: Robot mobility concepts for extraterrestrial surface exploration, Proceedings of the IEEE Aerospace Conference, 1–14, 2008.

Siegwart, R. and Nourbakhsh, I. R.: Introduction to autonomous mobile robots, MIT Press, Cambridge, 2004.

Siegwart, R., Lauria, M., Maeusli, P. A., and Van Winnendael, M.: Design and implementation of an innovative micro rover, Proceedings of Robotics 98, the 3rd Conference and Exposition on Robotics in Challenging Environments, 1998.

Siegwart, R., Lamon, P., Estier, T., Lauria, M., and Piguet, R.: Innovative Design for Wheeled Locomotion in Rough Terrain, Robot. Auton. Syst., 40, 151–162, 2002.

Snyder, R. G.: Robots assist in search and rescue efforts at WTC, IEEE Robotics and Automation Magazine, 8, 26–28, 2001.

Tadakuma, K., Tadakuma, R., Maruyama, A., Rohmer, E., Nagatani, K., Yoshida, K., Ming, A., Shimojo, M., Higashimori, M., and Kaneko, M.: Mechanical Design of the Wheel-Leg Hybrid Mobile Robot to Realize a Large Wheel Diameter, Proceedings of the IEEE/RSJ International Conference on Intelligent Robots and Systems, 3358–3365, 2010.

Thueer, T., Lamon P., Krebs, A., and Siegwart, R.: CRAB-Exploration rover with advanced obstacle negotiation capabilities, Proceedings of the 9th ESA Workshop on Advanced Space Technologies for Robotics and Automation, Noordwijk, The Netherlands, 2006a.

Thueer, T., Krebs, A., and Siegwart, R.: Comprehensive Locomotion Performance Evaluation of All-Terrain Robots, Proceedings of the IEEE/RSJ International Conference on Intelligent Robots and Systems, 2006b.

Thrun, S., Montemerlo M., Dahlkamp, H., Stavens, D., Aron, A., Diebel, J., Fong, P., Gale, J., Halpenny, M., Hoffmann, G., Lau, K., Oakley, C., Palatucci, M., Pratt, V., Stang, P., Strohband, S., Dupont, C., Jendrossek, L.-E., Koelen, C., Markey, C., Rummel, C., van Niekerk, J., Jensen, E., Alessandrini, P., Bradski, G., Davies, B., Ettinger, S., Kaehler, A., Nefian, A., and Mahoney, P.: STANLEY, the robot that won the DARPA Gran Challenge, J. Field Robot., 23, 661–692, 2006.

Van Straten, G., van Willigenburg, L. G., and Tap, R. F.: The Significance of Crop Co-States for Receding Horizon Optimal Control of Greenhouse Climate, Control Eng. Pract., 10, 625–632, 2002.

Vukobratovic, M. and Borovac, B.: Zero-moment point- thirty five years of its life, Int. J. Hum. Robot., 1, 157–173, 2004.

Wong, J. Y.: Theory of Ground Vehicles, John Wiley & Sons, 2001.

Woosub, L., Sungchul, K., Munsang, K., and Mignon, P.: ROBHAZ-DT3: teleoperated mobile platform with passively adaptive double-track for hazardous environment applications, Proceedings of the IEEE/RSJ International Conference on Intelligent Robots and Systems, 1, 33–38, 2004.

Yi, J., Song, D., Levandowski, A., and Jayasuriya, S.: Trajectory tracking and balance stabilization control of autonomous motorcycles, Proceedings of the IEEE International Conference on Robotics and Automation ICRA 2006, 2583–2589, 2006.

Yokota, S., Kawabata, K., Blazevic, P., and Kobayashi, H.: Control Law for Rough Terrain Robot with Leg-type Crawler, Proceedings of the IEEE International Conference on Mechatronics and Automation, 2006.

8

Compliant space mechanisms: a new frontier for compliant mechanisms

R. M. Fowler, L. L. Howell, and S. P. Magleby

Department of Mechanical Engineering, Brigham Young University, Provo, UT 84602, USA

Abstract. Compliant mechanisms offer distinct advantages for use in space that can address many of the issues encountered with current rigid-link space mechanisms. Compliant space mechanisms are defined as moveable mechanical assemblies that achieve their desired motion, force, or displacement by means of the deflection of flexible members and can perform a necessary function in the environments of launch and space. Many current space mechanisms are already highly optimized, yet they still experience inherent challenges, and it is unclear if significant improvements in performance can be made by continuing to refine current designs. Compliant space mechanisms offer a promising opportunity to change the fundamental approach to achieving controlled motion in space systems and have potential for dramatic increases in mechanism performance given the constraints of the space environment. This paper proposes the merger of the fields of compliant mechanisms and space mechanisms as a future direction of research in compliant mechanisms, discusses in detail the motivation to do so, and addresses the key factors of applying compliant mechanism technology to space mechanisms.

1 Introduction

Satellites, rovers, the international space station, and other space vehicles require mechanisms to perform mechanical tasks. These space mechanisms have been designed to perform in the demanding environments of space and launch. Efforts are continually being made to improve their performance and reliability while considering cost (Boesiger, 2008). Many current mechanisms are already highly optimized and it is unclear if significant improvements in performance can be made by continuing to refine current designs. Compliant mechanisms offer a promising opportunity to change the fundamental approach to achieving controlled motion in space systems and have potential for dramatic increases in mechanism performance given the constraints of the space environment. Compliant mechanisms gain motion from the elastic deflection of flexible components. Advantages of compliant mechanisms over traditional mechanisms include increased performance through reduced weight, increased precision, reduced friction, elimination of lubricants, ease of miniaturization, and integration of functions into fewer components. Figure 1 shows a few examples of compliant mechanisms designed by the Compliant Mechanisms Research Group (CMR) at Brigham Young University. The field of compliant mechanisms has matured to the point that design and analysis methods, and the increase in commercial compliant mechanisms, makes it possible to apply them in critical applications.

This paper proposes Compliant Space Mechanisms as a new research direction in compliant mechanisms, provides the motivation to do so, shows the application and benefits of this research, and discusses the crucial factors in beginning to understand this field.

2 Proposed research direction

A *compliant space mechanism* is a moveable mechanical assembly that achieves its desired motion, force, or displacement by means of the deflection of flexible members and that can perform a necessary function in the environments of launch and space.

The field of compliant space mechanisms has the potential for significant impact on the performance of space mechanisms because compliant mechanisms offer distinct advantages that can address many of the issues encountered in current rigid-link mechanisms. The advent of design and analysis methods for compliant mechanisms allows for design engineers to accurately model flexible segments and to

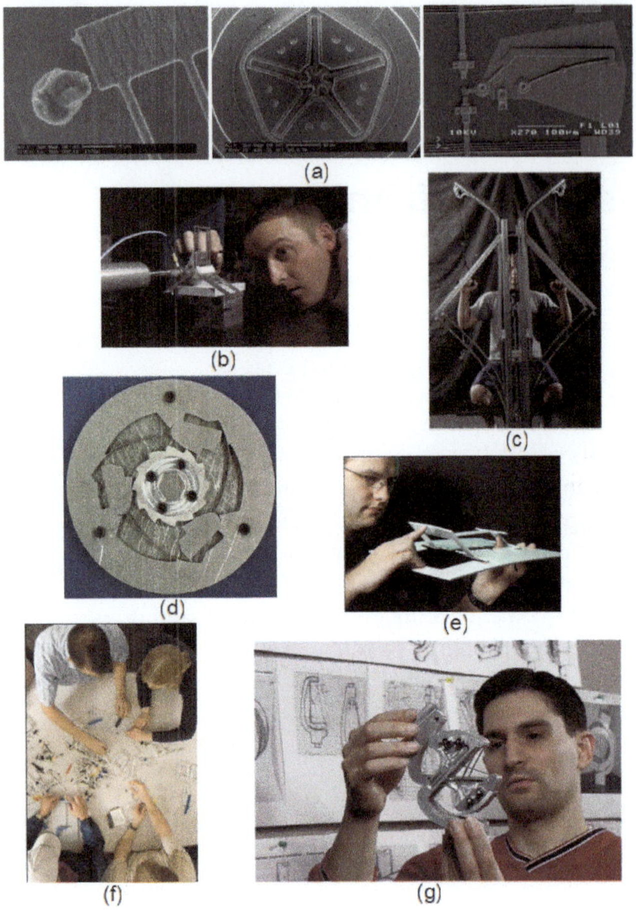

Figure 1. Examples of compliant mechanisms designed by the CMR: **(a)** scanning electron micrographs of compliant micro mechanism next to a white blood cell, a micro compliant joint for high g-loads, and a micro bistable mechanism, **(b)** a high precision device for nanoscribing, **(c)** an exercise device with specified force-displacement characteristics, **(d)** a compliant overrunning (one-way) clutch, **(e)** a lamina emergent mechanism (LEM), **(f)** a custom prototyping kit, compatible with Legos, to facilitate quick design and testing of compliant mechanisms, and **(g)** a compliant prosthetic knee.

design precision devices for specific tasks. The pseudo-rigid-body model (Howell, 2001) enables flexures to be modeled as rigid-link assemblies with representative link lengths and torsional spring constants. It also enables the use of well-known rigid-link kinematic and dynamic analysis software packages in analyzing compliant mechanisms. The advent of this approach has brought about many advances in numerous fields and lends itself particularly well to the space industry. Topology optimization (Sigmund, 1997; Frecker et al., 1997; Kim et al., 2006; Saxena and Ananthasuresh, 2000), another method of compliant mechanism design, can be used in the conceptual phase of design to arrive at the optimal geometry for specific loading and boundary conditions. These and other tools, further discussed in Sect. 4.2, provide the ability to merge compliant mechanism technology and space mechanisms so it can be readily used and implemented by designers and analysts in the space industry.

3 Motivation

This section addresses the motivation for creating compliant space mechanisms, including the challenges faced by current space mechanisms, the potential advantages of compliant space mechanisms, and limiting factors. The known challenges inherent in compliant mechanisms are discussed, as are the lessons learned from space mechanism failures. The relationship between compliant mechanism research and NASA's Technology Roadmaps is summarized.

3.1 State of current space mechanisms

Current space mechanisms are almost entirely composed of traditional rigid-link assemblies. These mechanisms perform a variety of functions. However, the harsh environments of space impose demanding requirements and rigid-link mechanisms can experience a variety of issues, including lubrication outgassing, friction and binding of joints, and inadequate force or torque margin of safety. The mission objectives or desired mechanism functionality also have demanding requirements and rigid-link mechanisms are naturally prone to issues concerning size, weight, and accuracy of motion. Continual effort is put into improving the performance and reliability of space mechanisms, yet many of the inherent challenges still remain and lead to compromises in performance and reliability. Since mechanisms often perform functions that are singularly vital for mission success, a failure could be catastrophic to the mission. Many of the failures of space mechanisms have been documented (Shapiro et al., 1995; Fusaro, 1999) and occur because of the design tradeoffs and inherent challenges.

3.2 Possible advantages of compliant mechanisms in space applications

The application of compliant mechanism technology could prove vital in overcoming some of the difficult challenges that current space mechanisms face when put in the space environment. Table 1 shows key challenges of space mechanisms and which advantages of compliant mechanisms could possibly address each challenge.

These advantages eliminate or reduce many of the disadvantages inherent in rigid-link space mechanisms. Compliant mechanisms also offer an increased number of mechanism designs, joints, and configurations. This provides more options in finding an optimized, low-cost mechanism design. The distributed compliance of some compliant mechanisms could be particularly useful in robotic grasping, sample collection, landing platforms, and rover suspensions. Resulting

Table 1. Current space mechanism challenges and the advantages of compliant space mechanisms that overcome those challenges.

Challenges of Current Space Mechanism	Advantages of Compliant Space Mechanisms
outgassing of lubrication, friction, wear and binding of joints	elimination of joints requiring lubrication or friction
large mass/weight (and accompanying cost)	significant part count reduction; miniaturization possible
large size/volume	significant part count reduction, increased number of possible joints and design configurations, integration of multiple functions into one mechanism, simpler geometries can lead to a reduction in material and assemblies needed to achieve the required motion
incorrect mechanism stiffness (components assumed rigid are not entirely rigid)	ability to accurately model and predict joint and mechanism stiffnesses; distributed compliance is more accurate for analysis than lumped compliance and it can be accurately calculated
complex or costly to manufacture and integrate	less expensive manufacturing methods possible (e.g. planar); simpler, more integrated geometries can achieve the required motion; significant part count reduction; little or no assembly required
feedback noise in attitude control system due to mechanism dynamics	precision motion provides increased predictability and control over mechanism mode shapes, natural frequencies, and component stiffnesses; can improve isolation from deployment dynamics with distributed compliance
single point failure modes	redundancy in actuation and motion possible; elimination of lubrication and friction dependent joints
inadequate force or torque margin of safety	accurate analysis methods; redundancy in actuation and motion possible
lack of accurate modeling and analysis methods for flexible and large-displacement segments	proven design and analysis methods
reduced reliability in off-nominal conditions	analysis methods provide increased predictability of behavior in off-nominal conditions; reduced susceptibility to foreign objects during testing and operation
thermal gradients cause joint binding or misalignment	mechanisms constructed of a single continuous material
backlash, hysteresis, and joint misalignment	monolithic (single piece) nature of compliant joints eliminates backlash, makes hysteresis predictable, and reduces the need for assembly, thus reducing the possibility of joint misalignment

from reduced part count and simpler topologies is the possible effect on a program level of fewer schedule delays during design, procurement, or manufacturing.

Overall, the advantages of compliant space mechanisms provide the opportunity to design simpler, more reliable, better performing, and more cost-effective solutions for many space applications.

3.3 Limiting factors

3.3.1 The space environment

Much is known about the environment of space and the challenges it presents in designing mechanisms (Sellers, 2004; Sarafin, 1995; Conley, 1998). The basics of understanding the space environment include the effects of the vacuum, electromagnetic radiation, and charged particles.

The vacuum of space creates three potential problems

- Outgassing: the release of gasses from spacecraft materials, often seen in composites and joint lubrication.
- Cold welding: fusing together of metal components since the vacuum eliminates small air gaps.
- Limited means of heat transfer: heat can only be transfered by radiation between components and space, and by conduction between components in contact.

The effects of these three problems can be mitigated by employing compliant mechanisms.

Outgassing and cold welding can be avoided by designing joints requiring neither lubrication nor contact surfaces. The often monolithic nature of compliant mechanisms lends itself well to designing joints that overcome these challenges.

The limited means of heat transfer is a problem inherent to the space environment. Compliant space mechanisms need to meet the thermal requirements of their specific mission and must be able to handle the thermal loading or be properly isolated from it. The reduction in number of parts and contact surfaces of compliant mechanisms provides increased simplicity in predicting the temperatures and thermal paths in the mechanism. The thermal properties of space vehicles are important and are carefully analyzed and monitored.

The thermal environment is determined by the temperature and radiation interactions of all the components of the space vehicle and the sun or planets. Convection is not a means of heat transfer in space and excess heat must be properly dissipated from critical components using radiation, or conduction into a neighboring component. A monolithic compliant joint has the potential for significantly better heat transport capability than a multi-element ball bearing.

Electromagnetic radiation from the sun heats up surfaces exposed to it. Whether the mechanism is exposed to the sun or in its shadow can significantly change its temperature and thus its material properties. The resulting thermal extremes can reduce the accuracy and reliability of mechanism performance and can cause damage to surfaces and electronic components. The performance of compliant mechanisms designed for use in space applications must not be adversely affected by the extreme range of temperature. The temperature dependent performance of flexures will need to be better understood.

Charged particles are encountered in the space environment and come from different sources such as the solar wind and flares on the sun, galactic cosmic rays from outside the solar system, or charged particles from the Van Allen radiation belts. They can cause spacecraft charging and possible discharges, sputtering, and single event phenomenon (Sellers, 2004).

Research will be vital in evaluating how these challenges affect compliant space mechanisms and how those differ from the challenges affecting current space mechanisms.

3.3.2 The launch environment

The launch environment subjects mechanisms to some of the most intense loads and vibrations experienced on a mission. The launch loads often range from 25 to 100 g's. The vibration profile is random and can excite many modes. Understanding these loads and the vibrational response of the mechanisms requires extensive analysis and testing. The stresses, deflections, natural frequencies, mode shapes, and other crucial characteristics will need to be found for any compliant space mechanism.

Compliant mechanisms lend themselves well to vibration isolation applications. A satellite and all its appendages have many needs for vibration isolation. Compliant mechanisms offer the potential to design custom vibration isolation management systems. The design and analysis of large-displacement flexures for space applications is a matter of energy management. The understanding and control of strain energy in compliant mechanisms can be applied to address this critical need. Compliant mechanisms provide the ability to safely manage the vibrations from launch, separation, or deployment events.

3.3.3 Performance requirements

The NASA Space Mechanisms Handbook (Fusaro, 1999) identifies distinct measurable quantities as mechanism performance metrics. They are:

- Range of Motion
- Torque (or Force) Margin
- Operating Speed
- Operating Life
- Pointing Accuracy
- Slew (or Scan) Rate
- Deployment Time
- Restow Capability
- Stowage Duration

Sarafin (1995) also has a listing of similar metrics.

3.3.4 System requirements

Aside from mechanism performance, a mechanism must be able to be integrated into the other subsystems and become part of a whole space vehicle. These considerations (Fusaro, 1999) affect not only the mechanism but the entire space vehicle.

- Weight
- Stiffness
- Envelope
- Clearance
- Alignment
- Interface
- Environments (temperature, vibration, shock, vacuum, transportation, and storage)

These are often vital factors that influence the design of mechanisms and will become avenues of opportunity for innovation as compliant space mechanism design research advances.

3.3.5 Possible challenges of compliant mechanisms in space applications

The numerous advantages of compliant mechanisms also come with some challenges. Two distinct challenges are:

- The coupling of motion and forces in compliant mechanisms creates a more complex design situation
- Off-axis stiffness and motion are possible

Typically, a mechanism is designed so it can achieve a certain motion, then, the forces in the joints and links are determined. In compliant mechanisms, the kinematics and dynamics are coupled.

In current space mechanisms, off-axis stiffnesses are significantly larger than the stiffness in the desired direction/axis and are often neglected. In compliant mechanisms, off-axis stiffness can be lower and could create undesirable parasitic motion.

Other known challenges associated with compliant mechanisms are:

- Designing joints or flexures capable of large deflection without failure due to stress, fatigue, or vibration
- Stress relaxation or creep (e.g. under preload, particularly at elevated temperatures or extended stowage duration)
- Designing thin flexible segments to withstand the vacuum and thermal extremes of space (Sellers, 2004)
- Continuously rotating joints often require a hybrid, or partially compliant design
- Designing complex compliant mechanisms that can be manufactured from a planar (or similar) state
- Stored strain energy (can be an advantage or drawback depending on the application)

Fortunately, these challenges are surmountable if proper attention is given to already established compliant mechanism design and analysis guidelines. It will remain the mechanism designer's responsibility to be vigilant in accounting for these challenges. Some aspects, such as manufacturing and testing, will involve a certain level of difficulty due to the stringent requirements for space applications.

3.4 Documented failures: Space Mechanisms Lessons Learned Study

The Space Mechanisms Lessons Learned Study performed by NASA (Shapiro et al., 1995) is an extensive collection of knowledge learned from past space mechanism failures, and is evidence that many of the failures in space mechanisms are due to problems for which compliant space mechanisms may provide solutions. The Lessons Learned Study investigated available literature on mechanism failures as well as the research being performed to eliminate those failures. It included a review of the first 28 Annual Proceedings of the Aerospace Mechanism Symposium, documents on deployable appendages from NASA Goddard, an industrial survey that yielded meaningful anomaly reports from companies in the space industry, and a review of the European literature contributed by the European Space Tribology Lab. The "Needs Analysis" section lists future needs for space mechanisms and will be useful in guiding compliant space mechanism research. It begins by stating, "A review of the information compiled for the Lessons Learned study reveals that bearing and lubrication problems are the most prevalent and, thus, improved technologies are most needed in these areas". It then details the specific needs for each of the three categories: Deployable Appendages, Rotating Systems, and Oscillating Systems.

Compliant mechanisms show potential for overcoming lubrication and friction issues caused by traditional mechanisms with lubricated contact surfaces, e.g. ball bearings, and present an opportunity to meet the needs defined by past experience and summarized in the NASA Lessons Learned Study.

3.5 A call for future direction: NASA Technology Roadmaps

The NASA Technology Roadmaps provide specific areas where compliant mechanisms may make an immediate impact and help to show how research in this field contributes to technological goals on a national level. The Roadmaps can be found at: http://www.nasa.gov/offices/oct/home/roadmaps/.

Technology Roadmaps 12 and 9 are most applicable to compliant space mechanisms. Technology Roadmap 12, "Materials, Structures, Mechanical Systems, and Manufacturing" (Piascik et al., 2010) provides detailed technologies that are priority for research and development in the areas most related to compliant mechanisms. Several of the identified technologies are listed in Table 2.

Compliant mechanisms show great promise in these areas.

Technology Roadmap 9 "Entry, Descent, and Landing" (Adler et al., 2010) also provides areas for application of compliant space mechanisms, such as: flexible thermal protection systems for entry; mechanical deployments for attached deployable decelerators for descent; and anchoring,

Table 2. Technologies Identified for Future Development by NASA Technology Roadmap 12 (Piascik et al., 2010).

Roadmap Section	Subsection	Specific Technologies Identified
12.1 Materials	Flexible Material Systems	Expandable Habitat; Flexible EDL Materials; Solar Sail; Shape Morphing Materials; Advanced Expandable Materials
12.2 Structures	Lightweight Concepts	Composite/Inflatable Habitats; Expandable Structures (Precision Mirrors and Solar/Antenna Arrays); Landers
12.2 Structures	Innovative, Multifunctional Concepts	Reusable Modular Components
12.3 Mechanical Systems	Deployables, Docking and Interfaces	Common Universal Interchangeable Interfaces; Restraint/Release Devices; Deployment of Flex Materials; Large Lightweight Stiff Deployable; Precision Structure Deploy Mechanism
12.3 Mechanical Systems	Electro-mechanical, Mechanical and Micromechanisms	Active Landing Attenuation System; New Concepts
12.3 Mechanical Systems	Design and Analysis Tools and Methods	Kinematics and Rotor Dynamics Analysis

touchdown, and extreme terrain suspension systems for landing.

Some of the areas listed are large scope goals for the future and it is understood that compliant mechanisms would be an integral part of the solutions.

3.6 Building upon, not replacing, heritage

From the Russian's Sputnik I (Fig. 2), to present day, an impressive amount has been learned about space mechanisms. Many mechanisms have performed vital functions, such as the sample collection mechanism on the Phoenix Mars Lander (Fig. 3) that uncovered ice on Mars.

This rich heritage is not to be replaced or forgotten; it is to be continued and built upon to take space mechanisms to the next level of performance and reliability.

4 Merging compliant mechanisms and space mechanisms

The merger of compliant mechanism technology to space mechanisms is addressed in this section by discussing the current state of space mechanisms, the current state of compliant mechanism research, and the proposed merger of these two fields. Possible technologies to develop and suggested areas to begin research are summarized.

4.1 Current space mechanisms

Many earth-sensing satellites, planetary rovers and orbiters, and manned spacecraft have been developed that require mechanisms to perform specific tasks, such as deployments, instrument pointing, stage separations, dockings, sample return, landings, retention and release, attitude stability, etc.

4.1.1 Space mechanism design rules

Since the general paradigm for designing space mechanisms has remained fairly consistent over the years, design rules exist for space mechanisms and have been generally well established by industry. A good example of a list of design rules is shown on page 499 in Space Vehicle Mechanisms – Elements of Successful Design (Conley, 1998). The NASA Space Mechanisms Handbook (Fusaro, 1999) is the space industry's authoritative document on space mechanism design and contains guidelines and details for space mechanisms of all types. It is available for free to approved US citizens at www.grc.nasa.gov/WWW/spacemech/ (with updates through 2009). The AIAA has a similar document that standardizes how to, for example, calculate margins of safety for a space mechanism (AIAA, 2005) and is commonly used in the space industry. Research needs to be performed in collaboration with these knowledge standards.

In addition to design rules, NASA's Space Mechanisms Handbook (Fusaro, 1999) cites general design objectives for space mechanisms:

Figure 2. Sputnik I. Courtesy of NASA.

- Identify and eliminate failure modes
- Increase mechanism robustness (performance in off-nominal conditions)
- Provide adequate force or torque margin
- Provide redundancy
- Design for producibility/manufacturability (without compromising performance)
- Design mechanisms for test and analysis
- Provide access for repair
- Provide adequate instrumentation
- Design for storage (stress relaxation, lubrication evaporation)

Many of these design objectives can be met by the advantages of compliant space mechanisms (Table 1).

4.1.2 Space mechanism types and mechanism components

Table 3 lists different types of mechanisms needed in space, most of which are identified in the NASA Space Mechanisms Handbook (Fusaro, 1999).

Common deployable mechanisms include: hinges, booms, antennas, covers, solar arrays, trusses, and dampers. Common rotating mechanisms include control moment gyroscopes, momentum wheels, reaction wheels, slip rings, and solar array drives. Common pointing/oscillating mechanisms include gimbals, harmonic drives, bearings, swash plates, pivot joints, antennas, telescopes, and scanning or pointing

Figure 3. Example of a current space mechanism. Phoenix Mars Lander sample collection shovel. Courtesy of NASA/JPL-CalTech-University of Arizona.

Table 3. Space Mechanism Types.

Space Mechanism Types		
Deployable	Rotating	Suspension
Restraint and Release	Latches and Stops	Drive
Vibration Isolation	Separation	Landing
Sample Return	Docking	Shape Control
Pointing/Oscillating		

mirrors. Precision pointing mirrors are common in space astronomy and can often have tight requirements for pointing, even needing to have a resolution of 1 microradian. A common example of a restraint and release mechanism is a pyrotechnic separation nut used to disconnect the solar arrays from the structure at deployment. A common example of a separation mechanism is known as a Lightband, which releases the payload from the launch vehicle at the separation stage. An example of a mobility suspension mechanism is the Mars Rover rocker-bogie suspension or the ATHLETE rover multi-DOF legs.

4.2 Current state of compliant mechanisms

Significant work has already been done in the field of compliant mechanisms. Many compliant mechanisms have become commercially available and research is ongoing in many areas of application. A complete review of the research and publications in compliant mechanisms is not feasible to present here. Compliant mechanisms symposia are held annually as part of the ASME Mechanisms & Robotics Conference, and the International Symposium on Compliant Mechanisms has been held every four years. Significant advances have been made in compliant mechanism:

- analysis and design methods (e.g. *the pseudo-rigid-body model* (Howell, 2001), *topology optimization*

(Sigmund, 1997, Frecker et al., 1997; Kim et al., 2006; Saxena and Ananthasuresh, 2000), and *FACT* (Hopkins and Culpepper, 2010a,b; Chen and Culpepper, 2006))

- advanced applications (e.g. *lamina emergent mechanisms* (Jacobsen et al., 2010; Winder et al., 2008; Albrechtsen et al., 2010; Gollnick, 2010), *bistable mechanisms* (Jensen and Howell, 2003; Howell, 2001; Qiu et al., 2004; Lusk and Howell, 2008a), *constant-force mechanisms* (Weight, 2001; Howell, 2001), *metamorphic mechanisms* (Dai and Jones, 2005, 1999; Wuxiang and Xilun, 2009), *large-displacement mechanisms* (Trease et al., 2005; Howell, 2001), *contact-aided mechanisms* (Mankame and Ananthasuresh, 2007, 2004; Cannon and Howell, 2005; Halverson et al., 2010; Mehta et al., 2009), *spherical mechanisms* (Wilding et al., 2011; Lusk and Howell, 2008b), *embedded actuators and sensors* (Trease and Kota, 2009), and *statically balanced mechanisms* (Hoetmer et al., 2010, 2009)).

A Handbook of Compliant Mechanisms will be published by John Wiley & Sons in 2012 and will include a library to illustrate a few hundred examples.

The research in compliant mechanisms and their resulting components and systems are available to address many of the issues related to space mechanisms. An active compliant mechanisms research community exists with applications ranging from microelectromechanical systems to shape morphing wings to spinal implants.

4.3 Applying compliant mechanism technology to space mechanisms

The types of space mechanisms listed in Table 3 can benefit from the application of compliant mechanism technology. Design and analysis methods, such as the pseudo-rigid-body model, topology optimization and FACT, allow for the design of joints or linkages using flexible segments of all types of boundary conditions. Simple flexures, four bar mechanisms, spherical linkages, and many other types of kinematic chains can be modeled and designed. Revolute, torsional, spherical, planar, and many other types of joints can be designed. Examples could include a fixed-guided flexure for attaching a thermally expanding-contracting telecom waveguide to the structural panels or a one-piece bistable compliant four-bar deployment hinge.

The different types of current space mechanisms (Fusaro, 1999; Conley, 1998; Sarafin, 1995) (also shown in Table 3) can be evaluated for their potential for being replaced by compliant mechanisms. Some types will be better suited for the merger than others. In doing this, the distinct motions and functions (Shean-Juinn and Kota, 1999; Fusaro, 1999) that are commonly required by space mechanisms need to be defined. Working from the undertstanding of current space mechanism functions and requirements, compliant space mechanisms can be created that are either compliant redesigns of current space mechanisms or are novel mechanism designs that fulfill a required function or motion. This can be done using a toolbox of compliant mechanism joint types (Olsen et al., 2009; Trease et al., 2005; Winder et al., 2008; Howell, 2001) and design techniques (Howell, 2001; Shean-Juinn and Kota, 1999; Jacobsen et al., 2010; Trease et al., 2005; Berglund et al., 2000; Guerinot et al., 2005). Design guidelines and rules need to be developed on how to apply compliant mechanism technology to space mechanisms so that the process is standardized and widely usable. The breadth of possible applications and configurations of compliant mechanisms in space applications should also be explored.

Candidate designs can be evaluated against the design criteria that has been developed. Promising candidate designs will be selected for detailed analysis. Analysis of the mechanisms will be performed to better understand the behavior of the mechanisms, especially in terms of their desired functionality in the space environments. The performance of the mechanisms should be quantified in several categories, which were outlined earlier as the Performance and System Requirements of the NASA Space Mechanisms Handbook or similarly in Conley (1998); Sarafin (1995).

Testing will be performed following the same standards that are used to qualify flight hardware. The testing results will be compared to the analytical results and models can be verified or corrected, and added to the pool of design knowledge. Efforts should be taken to not only evaluate the performance of the mechanism but to optimize its design and performance. The lessons learned through this process of design, analysis, and testing will be documented as initial guidelines for this field and act as a stepping stone for continued advancements.

4.3.1 Example of a compliant space mechanism

Although not yet widespread, flexible links have been successfully used on space mechanisms (Conley, 1998) and the field is just beginning to be explored (Gore et al., 2006). Compliant vibration isolation systems such as SoftRide are produced by CSA Engineering (a division of Moog) and have been used on numerous satellite missions. Figures 4 and 5 show these compliant mechanisms being effectively used to isolate a satellite from launch vehicle vibrations. Similar mechanisms have been used on the Hubble Space Telescope for on-orbit jitter reduction and solar array vibration damping. Propellant tank tab flexures and compliant universal joints are other common examples.

4.3.2 Possible concepts to develop for compliant space mechanisms

In addition to converting current space mechanisms to compliant mechanisms, innovative compliant mechanism designs

Figure 4. Example of a compliant space mechanism. SoftRide vibration isolation mechanism. Compliant segments are labeled. Courtesy of U.C. Berkeley.

may be developed. This could include the development of the concepts listed here:

- Compliant joints capable of 90 and even 180 degree deflection; multiple segment joints
- Statically balanced hingeline joints
- Compliant joints with improved off-axis properties
- Temperature insensitive joints
- Modular vibration isolation mechanisms
- Compliant bi/tri-stable hinges
- Creasing joints for deployable arrays and masts
- Integrated deployment springs
- Redundant compliant latches
- Multi-Layer lamina emergent thermal radiators
- Pointing array mechanisms
- Two/three axis compliant gimbal platforms
- Compliant joints with integrated damping

4.4 Areas of research and questions to answer

Merging the fields of compliant and space mechanisms opens up many possible areas of research. Research can be started in a variety of ways. Each of the space mechanism types should be looked at for their compliant potential. A desired mechanism function and its requirements could be the starting point. Replacing a current mechanism component, such as a ball bearing, would yield valuable results. Designing to overcome a reoccurring issue or failure in a current mechanism could guide the research. Guidelines, methods, and tools for the design, analysis, and testing of compliant space mechanisms need to be developed and established for use in industry.

Figure 5. Example of a compliant space mechanism. SoftRide vibration isolation mechanism. The compliant segments separate the launch vehicle adapter and the satellite WISE that was launched on a Delta II in December 2009. Courtesy of U.C. Berkeley.

This research hopes to yield ground-breaking answers to detailed questions such as this one: how do compliant segments, especially large-displacement, behave in the thermal extremes and the vacuum of space? There are many fascinating questions and there will be even more as this field unfolds.

Compliant space mechanism research will help fulfill the needs expressed in the NASA Technology Roadmaps and Lessons Learned Study, while also creating next generation space mechanisms.

5 Conclusions

Failures in space can be costly, dangerous, and can hinder progress of technology development and space exploration. The field of compliant space mechanisms has the potential for significant impact on the performance of future space mechanisms. Research in compliant space mechanisms will create innovative new space technologies. It could initiate critical advances that would lead to more affordable and more capable space mechanism solutions. This research cross-cuts several technologies and applies to a large range of different space mechanism types.

Compliant space mechanism research hopes to inspire collaboration between government space agencies (such as NASA), companies in the space industry, and academia. Overall, it would present opportunities for game-changing technologies and lead to enhanced capabilities in space exploration.

Acknowledgements. This material is based upon work supported by the National Science Foundation under Grant No. CMMI-0800606. Any opinions, findings, and conclusions or recommendations expressed in this material are those of the authors and do not necessarily reflect the views of the National Science Foundation.

Edited by: J. Herder
Reviewed by: B. Trease and another anonymous referee

References

AIAA: AIAA S-114, Moving Mechanical Assemblies for Space and Launch Vehicles, 2005.

Albrechtsen, N. B., Magleby, S. P., and Howell, L. L.: Using Technology Push Product Development to Discover Potential Applications and Categories of Enabled Opportunities for Lamina Emergent Mechanisms, in: ASME 2010 International Design Engineering Technical Conference, DETC2010-28531, ASME, Montreal, Quebec, Canada, 2010.

Adler, M., Wright, M., Campbell, C., Clark, I., Engelund, W., and Rivellini, T.: Entry, Descent, and Landing Roadmap – Technology Area 09, NASA, draft edn., 2010.

Berglund, M. D., Magleby, S. P., and Howell, L. L.: Design Rules for Selecting and Designing Compliant Mechanisms for Rigid-Body Replacement Synthesis, in: ASME 2000 Design Engineering Technical Conference, DETC00/DAC-14225, ASME, Baltimore, Maryland, 2000.

Boesiger, E. A. (Ed.): 39th Aerospace Mechanisms Symposium, NASA/CP-2008-215252, NASA and Lockheed Martin Space Systems Company, 2008.

Cannon, J. R. and Howell, L. L.: A compliant contact-aided revolute joint, Mech. Mach. Theory, 40, 1273–1293, 2005.

Chen, S.-C. and Culpepper, M. L.: Design of a six-axis micro-scale nanopositioner – HexFlex, Precis. Eng., 30, 314–324, 2006.

Conley, P. L. (Ed.): Space Vehicle Mechanisms – Elements of Successful Design, John Wiley & Sons, Inc., 1998.

Dai, J. S. and Jones, J. R.: Mobility in metamorphic mechanisms of foldable/erectable kinds, J. Mech. Des.-T. ASME, 121, 375–382, 1999.

Dai, J. S. and Jones, J. R.: Matrix representation of topological changes in metamorphic mechanisms, J. Mech. Des.-T. ASME, 127, 837–840, 2005.

Frecker, M., Ananthasuresh, G., Nishiwaki, S., Kikuchi, N., and Kota, S.: Topological synthesis of compliant mechanisms using multi-criteria optimization, J. Mech. Des.-T. ASME, 119, 238–245, 1997.

Fusaro, R. L. (Ed.): NASA Space Mechanisms Handbook, Glenn Research Center NASA/TP 206988, Cleveland, OH, 1999.

Gollnick, P. S.: Design Principles and Preliminary Actuation Approaches for Novel Multiple-Layer Lamina Emergent Mechanisms, Master's thesis, Brigham Young University, 2010.

Gore, B., Didziulis, S., and Hilton, M.: Space Vehicle Mechanisms, Crosslink Magazine (Aerospace Corporation), vol. 12, available at: http://www.aero.org/publications/crosslink/fall2006/03.html, 2006.

Guerinot, A. E., Magleby, S. P., Howell, L. L., and Todd, R. H.: Compliant Joint Design Principles for High Compressive Load Situations, J. Mech. Design, 127, 774–781, 2005.

Halverson, P. A., Howell, L. L., and Magleby, S. P.: Tension-Based Multi-stable Compliant Rolling-contact Elements, Mech. Mach. Theory, 45, 147–156, 2010.

Hoetmer, K., Herder, J., and Kim, C.: A building block approach for the design of statically balanced compliant mechanisms, in: ASME International Design Engineering Technical Conferences and Computers and Information in Engineering Conference, vol. 7, 313–323, ASME, dETC2009, 2009.

Hoetmer, K., Woo, G., Kim, C., and Herder, J.: Negative Stiffness Building Blocks for Statically Balanced Compliant Mechanisms: Design and Testing, Journal of Mechanisms and Robotics, 2, 041007-1–041007-7, 2010.

Hopkins, J. B. and Culpepper, M. L.: Synthesis of multi-degree of freedom, parallel flexure system concepts via Freedom and Constraint Topology (FACT) – Part I: Principles, Precision Engineering, 34, 259–270, 2010a.

Hopkins, J. B. and Culpepper, M. L.: Synthesis of multi-degree of freedom, parallel flexure system concepts via freedom and constraint topology (FACT). Part II: Practice, Precis. Eng., 34, 271–278, 2010b.

Howell, L. L.: Compliant Mechanisms, John Wiley & Sons, Inc., 2001.

Jacobsen, J. O., Winder, B. G., Howell, L. L., and Magleby, S. P.: Lamina Emergent Mechanisms and Their Basic Elements, Journal of Mechanisms and Robotics, 2, 1–9, 2010.

Jensen, B. D. and Howell, L. L.: Identification of Compliant Pseudo-Rigid-Body Four-Link Mechanism Configurations Resulting in Bistable Behavior, J. Mech. Design, 125, 701–708, 2003.

Kim, J. C., Kota, S., and Moon, Y.: An instant center approach toward the conceptual design of compliant mechanisms, J. Mech. Des.-T. ASME, 128, 542–550, 2006.

Lusk, C. P. and Howell, L. L.: Spherical Bistable Micromechanism, J. Mech. Design, 130, 1–6, 2008a.

Lusk, C. P. and Howell, L. L.: Components, Building Blocks, and Demonstrations of Spherical Mechanisms in Microelectromechanical Systems, J. Mech. Design, 130, 034503-1–034503-4,

2008b.

Mankame, N. D. and Ananthasuresh, G. K.: Topology optimization for synthesis of contact-aided compliant mechanisms using regularized contact modeling, Comput. Struct., 82, 1267–1290, 2004.

Mankame, N. D. and Ananthasuresh, G. K.: Synthesis of contact-aided compliant mechanisms for non-smooth path generation, Int. J. Numer. Meth. Eng., 69, 2564–2605, 2007.

Mehta, V., Frecker, M., and Lesieutre, G. A.: Stress relief in contact-aided compliant cellular mechanisms, J. Mech. Des.-T. ASME, 131, 091009-1–091009-11, 2009.

Olsen, B. M., Hopkins, J. B., Howell, L. L., Magleby, S. P., and Culpepper, M. L.: A Proposed Extendable Classification Scheme for Compliant Mechanisms, Proceedings of the ASME International Design Engineering Technical Conferences, San Diego, CA, 30 August–2 September, DETC2009-87290, 2009.

Piascik, B., Vickers, J., Lowry, D., Scotti, S., Stewart, J., and Calomino, A.: Materials, Structures, Mechanical Systems, and Manufacturing Roadmap – Technology Area 12, NASA, draft edn., 2010.

Qiu, J., Lang, J. H., and Slocum, A. H.: A Curved-Beam Bistable Mechanism, J. Microelectromech. S., 13, 137–146, 2004.

Sarafin, T. P. (Ed.): Spacecraft Structures and Mechanisms: From Concept to Launch, Microcosm Press and Springer, 1995.

Saxena, A. and Ananthasuresh, G. K.: On an optimal property of compliant topologies, Struct. Optimization, 19, 36–49, 2000.

Sellers, J. J.: Understanding Space – An Introduction to Astronautics, McGraw Hill Custom Publishing, revised second edn., 2004.

Shapiro, W., Murray, F., and Howarth, R.: NASA Space Mechanisms Lessons Learned Study, vol. 1-Summary, NASA Lewis Research Center Cleveland, Ohio TM 107046, 1995.

Shean-Juinn, C. and Kota, S.: Automated conceptual design of mechanisms, Mech. Mach. Theory, 34, 467–495, 1999.

Sigmund, O.: On the Design of Compliant Mechanisms Using Topology Optimization, Mechanics Based Design of Structures and Machines, MARCEL DEKKER INC, 25, 493–524, 1997.

Trease, B. and Kota, S.: Design of adaptive and controllable compliant systems with embedded actuators and sensors, J. Mech. Des.-T. ASME, 131, 111001-1–111001-12, 2009.

Trease, B. P., Moon, Y.-M., and Kota, S.: Design of Large-Displacement Compliant Joints, J. Mech. Design, 127, 788–798, 2005.

Weight, B. L.: Development and Design of Constant-Force Mechanisms, Master's thesis, Brigham Young University, 2001.

Wilding, S. E., Howell, L. L., and Magleby, S. P.: Spherical Lamina Emergent Mechanisms, Mech. Mach. Theory, in press, 2011.

Winder, B. G., Magleby, S. P., and Howell, L. L.: A Study of Joints Suitable for Lamina Emergent Mechanisms, in: ASME 2008 International Design Engineering Technical Conference, DETC2008-49914, ASME, Brooklyn, New York, 2008.

Wuxiang, Z. and Xilun, D.: A method for designing metamorphic mechanisms and its application, in: ASME International Conference on Reconfigurable Mechanisms and Robots, London, 2009.

9

The use of underactuation in prosthetic grasping

P. J. Kyberd, A. Clawson, and B. Jones

Institute of Biomedical Engineering, University of New Brunswick, Fredericton, New Brunswick, E3B 4A3, Canada

Abstract. Underactuation as a method of driving prosthetic hands has a long history. The pragmatic requirements of such a device to be light enough to be worn and used regularly have meant that any multi degree of freedom prosthetic hand must have fewer actuators than the usable degrees of freedom. Aesthetics ensures that while the hand needs five fingers, five actuators have considerable mass, and only in recent years has it even been possible to construct a practical anthropomorphic hand with five motors. Thus there is an important trade off as to which fingers are driven, and which joints on which fingers are actuated, and how the forces are distributed to create a functional device. This paper outlines some of the historical solutions created for this problem and includes those designs of recent years that are now beginning to be used in the commercial environment.

This paper was presented at the IFToMM/ASME International Workshop on Underactuated Grasping (UG2010), 19 August 2010, Montréal, Canada.

1 Introduction

The loss of a hand can be a significant impairment to any individual. In a society where independence and self determination is important, the inability to perform two handed tasks can reduce a person's capacity to function easily in daily life. This alone is sufficient motivation to create effective prosthetic hands to replace the loss. However, while the adaptability of homo sapiens means that a loss of a single hand below the elbow can be less a handicap than an inconvenience, it is the methods by which the person compensates for the loss than have a significant impact on their lives. To place a prosthetic hand, or the end of a remnant limb, in the correct place to manipulate an object a person may have to, flex, bend or stretch further than they would if they did not have the loss (Zinck, 2008). With these poorer biomechanics the person may have to use greater forces and larger ranges of motions, these actions are associated with overuse injuries (Kidd et al., 2000) which are observed in long term users of prosthetic arms (Jones and Davidson, 1999) and this ultimately may disable the person more comprehensively. Thus the design of an effective prosthetic hand is an important challenge for engineers, prosthetists and therapists.

The requirement that the prosthesis can be derived from questionnaires of the general population (Atkins et al., 1996; Fraser, 1998; Heger et al., 1985; Kyberd et al., 2007) among others. What is observed it that it must be a *practical* device. This creates some severe constraints on the design: It has to be light enough to be worn all day, easy to control, cheap enough to be readily available, robust enough to operate in the field for months at a time without the need for servicing. Additionally, it must take high levels of abuse.

While the natural hand has its drives in the forearm, for a prosthesis this is impractical. The level at which the loss occurs is unique to the individual, and the numbers of persons with a particular level of loss increase the further down the arm (Kyberd et al., 1998a). So that any design that uses any part of the forearm for drives of the hand limits its potential market considerably. Any device constructed in this way is not a prosthesis (despite any claims to the contrary), but an anthropomorphic robot, and of little clinical use or significance.

The result of these requirements is that the number of actuators is limited to the space in the hand and by the mass of the motors. Hence there is a need to gain function from limited actuation, as a result, the use of underactuation in prosthetics has long been considered a solution (Borchardt et al., 1919). Recent years have seen a renewed interest in multi-articulated prosthetic hands and rise in the complexity of commercial devices. This paper outlines some of the solutions that have been suggested in the past and relates them to designs that are being proposed currently. It starts with a review of the constraints on the design.

2 Solutions

2.1 Appearance and the choice of the number of digits

While appearance is important to some users, it is not essential (Kyberd et al., 1998a), but the majority of effort towards multifunctional terminal devices has been towards fully anthropomorphic hands. This means that the device will possess five digits, but the choice on how they are driven will dictate how functionally effective the device is.

The principle role of the prosthetic hand is to hold objects stably, and the aim of a complex hand is to be more functionally capable. The hand must therefore adapt to a wide range of target objects of different size, shape, mass and surface finishes. Additionally the world it is used is the one that was developed by human beings, so the majority of objects have been made to a particular size, shape and mass to be easily manipulated by the human hand. Thus even an ideal prosthesis shape, without any reference to appearance, may well be similar to a natural hand.

The natural hand achieves its breadth of capability through the use of multiple fingers which are loosely coupled together. The fingers are able to adapt their posture in response to the way the held object is going to be used/carried. To keep the physical effort to a minimum, the hand conforms round the object and spreads the grip forces across as large an area as possible, and so the grip force required is as low as practical (Schulz et al., 2005).

2.2 Flexible actuation

While cables create a flexible drive that will fit round corners and allow the actuators to be distant from the motion, it remains difficult to create robust solutions with cables. Strong termination is complex, cables tend to stretch and require constant maintenance. So, for example, although the MIT/Utah dextrous hand (Jaconsen et al., 1986) was derived from a prosthetics research program (Sears et al., 1988), its use of cables meant that it was in constant need of servicing and adjusting, which would preclude it from a prosthetic solution. Instead rigid links are generally used. The flexibility is then created using some form of differential. One solution is to use links in the form of a whiffle tree. All elements are moved by a single drive, when one link is stopped the others continue to move until they all are halted. A similar process can be achieved using a geared differential, where the relative proportions are those of the gear ratios (see Sect. 3). The biggest disadvantage to this form of actuation is to easily return of the system to its start point. While motion in one direction (the "forward" direction) is balanced between the different branches, all elements move together, but the reverse direction tends to channel all the motion from the drive through the branch that is easiest to move. Once it has reached the end the second easiest will move and so on. Thus for a hand, instead of all the fingers extending together, one moves quickly to fully extended, before the second, etcetera. In order to balance the branches on the return leg, an external return force is required, such as springs, which would add mass and complexity to the hand. The spring also opposes the flexion drive, hence weakening the grip.

2.3 Digit selection

While there are theoretical considerations as to the minimum number of digits necessary to stably hold an abstract object, other matters dominate with a prosthesis. For example; the requirement that the device is natural in appearance ensures that there are five curling fingers. Effective manipulation can be conducted with three digits, but a fourth and fifth increases the chances of a stable contact. When a hand is used in an unstructured environment this level of redundancy and flexibility is important. Having to adjust the grip or change the posture to get sufficient contact is not readily tolerated by prosthesis users.

2.3.1 Classes of grip

With many fingers different grip forms can be created. The hand has a single opposing digit (the thumb) and most grips are controlled by its use. There are numerous different ways to classify grasp types. Napier (1956) suggested two: Precision (where the tips of the fingers oppose each other) and Power (where the fingers thumb wrap around the object and the hand is able to impart more force). Others have seen six broad grasps (Kamakura et al., 1980; MacKenzie and Iberall, 1994) (tripod, cylinder, extension, tip, lateral and spherical). While these summarise the grip shapes the hand must perform, Iberall simplified the matter, considering only what part of the hand the tips of the fingers oppose on to: So Precision grips are where the finger tips opposed the thumb tip, Power-like grips, when the finger tips oppose the palm, or Lateral grips, when the thumb tip opposes the side of the flexed index finger (Iberall, 1997). This final form of classification gives the engineer clear indications as to the required path the digit tips must move through to create as wider a range of grips (and so function) for the hand. To achieve all these grasps the fingers need to be driven separately to the thumb and they need to curl to reach the palm. A prosthesis should be able to perform all of these grips to be truly practical. When compromises are made it might be imagined that the most common grips are those which are considered first. However, the differences in the way that a prosthesis is used, compared with a natural hand means that it is still uncertain which grips are most useful. The latest generation of commercial hands use the lateral grip more often than the natural hand does, (proportions according to Sollerman and Sperling (1978)) and also for gripping objects not considered for this grip in natural prehension.

2.3.2 Precision grips

An important distinction are the two and three jaw precision grips (tripod and tip). Two jaw opposition dictates which fingers are linked to the same drive. Two general solutions have been investigated; each splits the fingers into two groups: Either two fingers in each group or three in one group and the index driven separately. Early examples of these two solutions are the Belgrade Hand (Bekey et al., 1990; Iberall et al., 1996) and the Southampton Hand (Todd, 1969; Kyberd et al., 2001). The latter split is similar to the way work gloves are designed for use in cold climates, there the three fingers are set in a mitt for warmth and the index is kept separate for ease of manipulation. In both of these hands the drives passed through whiffle trees to the fingers, with two motors in the palm for four fingers. Both formats allow balancing of forces across the surface of the held object with the grasp curving round the outside of an object.

Both also allowed for two point grips between the index finger and the thumb. In the case of the Belgrade Hand the middle finger needs to close fully once the thumb was opposed by the index, to allow the index to grip the object. With the Southampton Hand the three fingers could be held in the palm (or extended out of the way) and a two point grip maintained (Moore, 1980). A second advantage is that repeated grips with the Belgrade format force both of the fingers to fully open and close again, while with the Southampton Hand the index finger is fully isolated, so the index only needs be opened to reapply the grip.

A third hand with whiffle trees across the palm are the Sven hand series (Almström, 1977). All four fingers were driven through a single whiffle tree, providing grip flexibility from one motor. Later this gave rise to the ES Hand (Almström et al., 1981), which is the only underactutated prosthetic hand to be used in the field (albeit in limited numbers).

2.4 Thumb actuation

In multifunction prosthetic hands the role of the thumb can vary considerably. The natural thumb can move in and out from the fingers (flex and extend) as well as across the hand (abduct and adduct). If the thumb can abduct a small amount it can provide both two and three jaw opposition, if it is allowed to move round to the side then the lateral grip is also possible. Some hands require two motors to provide this extra motion (Kyberd, 1990) and the Otto Bock Michelangelo hand. Others, such as the TouchBionics and RSL Steeper hands require external forces from the user to move the thumb across the palm (Newman, 2008). A more sophisticated design is to use opposite ends of a more complex range of motion of the thumb to create motion in one plane or the other (Clawson et al., 2010), or Pons et al. (2005) who used a Geneva mechanism. Thus getting the two motions from a single actuator.

Additionally, in the natural hand the motion of fingers and thumb are different (Wing et al., 1986), with the thumb moving more slowly and allowing the fingers to close on to the thumb, which acts as a reference for the person grasping. So powerful is this need, that in simple prostheses where both fingers and thumb are linked and move with the same angular velocity, users move their forearm in such a way that the relative motions of finger and thumb are closer to the natural response (Wing and Fraser, 1987). Using this knowledge it is clear that if the speed of the fingers and thumb are under the control of the designer, that it is appropriate to allow the thumb to move more slowly to get closer to the natural solution. This also allows the thumb to be geared differently so it can create the power in the grip, in a similar manner to the natural solution (Kyberd et al., 1998b). Most conventional prosthetic hands have the same angular velocity for fingers and thumb.

2.5 Finger construction

The natural finger has three joints outside the palm. This allows the finger to wrap around an object and spread the grip across a larger surface area (reducing the need for a greater grip force for a stable grip). The tip of the finger can then oppose the palm at the base of the finger, allowing power grip on a small object. Each natural segment consists of a phalangial bone, linked by a rolling, sliding joint. In a prosthetic replacement uses a link between successive phalanges to give a similar motion of the tip of the finger, Fig. 1. This increase in complexity from rigid fingers to three separate elements, may not provide sufficient functional increase to justify the increase in complexity, for example; The Montreal Hand (Vinet et al., 1995) chose to fix the top *two* phalanxes in slight flexion and altered the ratios of the lengths of the phalanxes from roughly equal, in the anthropomorphic case, to the distal pair being roughly the same length as the proximal segment, so that with two elements the tips of the fingers still opposed the base of the finger. There is as yet no evidence that this compromise effects the functional capabilities.

Alternatively increased grip flexibility can be achieved if the top joint and the lower joints are linked via a whiffle tree across the finger. In this case the tips of the fingers will continue to close onto an object even when the base of the finger is in contact with the object. This means that the contact area is spread more widely (Kyberd et al., 2001), Fig. 2.

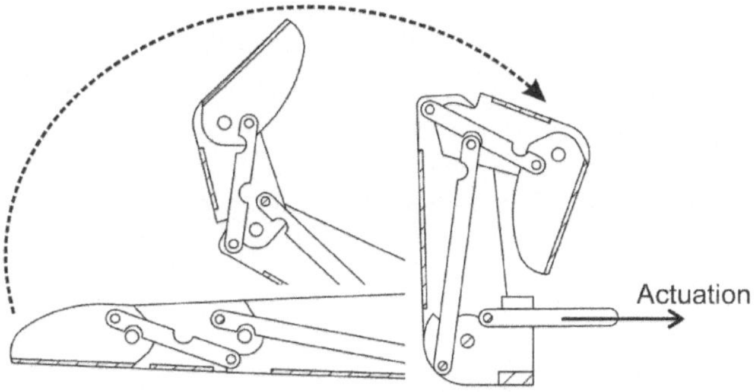

Figure 1. Links between the interphalangial joints cause the finger to flex as it closes. the finger closes progressively, but if the lower phalanx is halted by an external object the tips cannot continue to move.

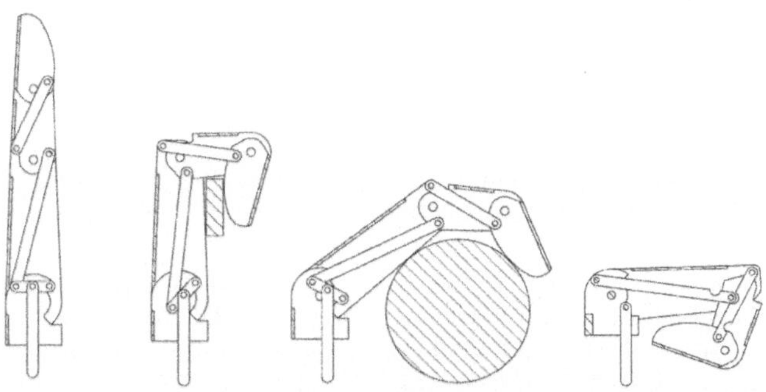

Figure 2. Use of the equalising mechanism across the finger to allow an adaptive grip. if the proximal phalanx is stopped by an external object. the distal phalanx closes onto the object. for a larger object the finger curls similar to the design in Fig. 1.if unobstructed, the tip wraps round to the base of the finger.

2.6 Power sources

The overwhelming choice for power sources has been the permanent magnet electric motor. This is again for practical reasons. For example; batteries are easy to recharge in any domestic situation, while the accumulators or pressure vessels needed for pneumatic and hydraulic systems require considerable specialist equipment which is unlikely to be available to the average user, making the maintenance of a device using such sources very much more difficult. Indeed this advantage has increased over the past 20 years as the technology has continued to advance, with increasing capacity and peak power output for simple motors, so that alternative systems remain less practical. One drawback with motors is that they are most efficient when they are moving at a constant rate and in one direction of rotation. Biological systems tend to move slowly, cyclicly and start and stop often. Hence a motor is the poorest choice from the perspective of efficiency. One group in Germany (Schulz et al., 2005) has taken a different approach. They generate pressure for a lightweight hydraulic hand using an electric motor turning at a constant rate. The power source can still be a battery, with all the advantages of their ability to be easily recharged, but the actuator is hydraulic, which is better at low speeds and from rest. Each finger can be linked via an equalising mechanism, but from a single source, with the joints flexing along the fingers and across the hand. This device was undergoing clinical assessment (Schulz et al., 2008).

3 Current designs

Historically, numerous solutions have been proposed for prosthetic hands. Some have been used in the field, but the limitations on technology has meant that the mass of the designs has restricted their application. With the launch of the TouchBionics Hand in 2005, a new generation of multifunction hands with flexible fingers is beginning to enjoy genuine clinical use. The TouchBionics hand is fully actuated (Newman, 2008). The next device to reach the market, was the RSL Steeper, "BeBionic" hand in spring 2010. This too has separately driven fingers which curl as they flex and a thumb with unpowered abduction. However neither of these hands are underactuated. Yet to reach the market, is the Otto Bock Michaleangelo hand (Puchhammer, 2008). The main flexion power comes from a single brushless motor in the palm, and a combination of a cam at the opposite ends of the motion and a smaller flexion motor for the thumb, allows the hand to adopt tip and lateral prehension. The fingers do not curl, merely flexing from their base and their motions are linked. The hand is statically very anthropomorphic in appearance, but without some independent clinical testing, it is unclear if these changes give it any functional advantages over the more conventional single degree of freedom designs. Further from the market is the Vicent hand from Schutz's team (Schulz, 2010), with individually driven fingers.

The team at the Institute of Biomedical Engineering at the University of New Brunswick have many years of experience with novel means to control prosthetic systems, most recently they have begun to design a hand which will incorporate their knowledge of pattern recognition with the collected ideas concerning hand design to product a lightweight compact anthropomorphic hand. It has a three and one finger split with the three fingers driven through two stages of differential gears. Each finger is jointed at the lower interphalangial joints. The thumb has a single motor and creates lateral or tip pinch at opposite ends of its travel. Each finger is instrumented so that contact points are recorded and the most appropriate grip for the target object can be automatically selected and the grip force applied will be sufficient to hold the object stably, Fig. 3. If the object slips within the grip it will be detected and the hand automatically grips tighter to arrest any slide. Using this design the hand can match all six of the basic grip classes.

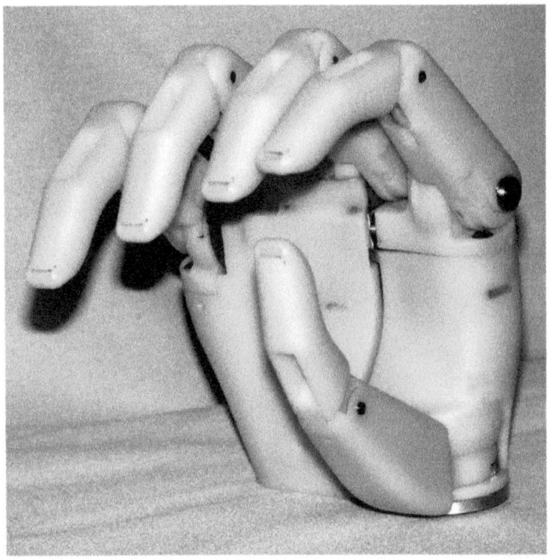

Figure 3. The design of the UNB hand. it incorporates three motors to actuate five fingers, the thumb with two separate motions.

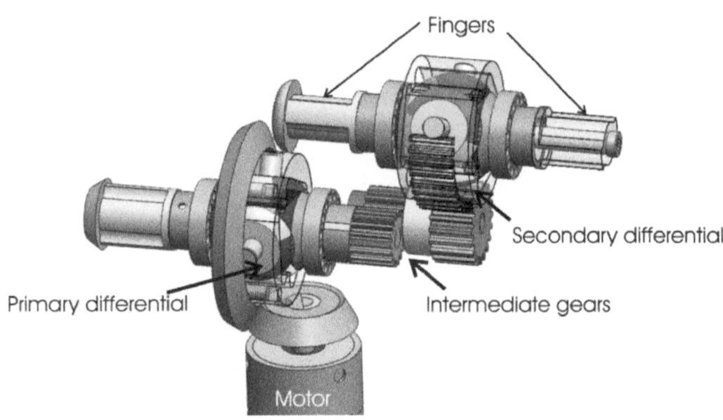

Figure 4. The use of a pair of differentials to create equalising mechanism across two fingers.

The finger drive uses two differentials with three internal crown gears, Fig. 4. The motor drives the housing of the first gear, the output of that gear passes through an intermediate shaft before driving the second differential via spur gearing. This allows there to be a different ratio between the fingers to equalise the forces transmitted. It also means that the base of the fingers is stepped in an anthropomorphic manner. This is different to the the DEKA hand which places the differentials in series and so the bases of the fingers are all in one line (Langefeld et al., 2008).

4 Conclusions

The use of underactuation in prosthetics is the result of pragmatic choices to create functional and adaptive hand mechanisms, while reducing size and mass a level that is practical. This has resulted in different numbers of drives and degrees of freedom. To obtain a flexible grip the use of underactuation allows greater flexibility with fewer drives. However few such hands have achieved significant clinical use so far.

Acknowledgements. The authors wish to thank their support, from the Natural Sciences and Engineering Research Council of Canada, the Atlantic Innovation Fund and the New Brunswick Innovation Foundation.

Edited by: J. L. Herder
Reviewed by: two anonymous referees

References

Almström, C.: An Electronic Control System for a Prosthetic Hand with Six Degrees of Freedom, Tech. Rep. 1:77, Research Laboratory of Medical Electronics, Chalmers University of Technology, 1977.

Almström, C., Herberts, P., and Körner, L.: Experience with Swedish Multifunctional Prosthetic Hands Controlled by Pattern Recognition of Multiple Myoelectric Signals, International Orthopaedics (SICOT), 5, 15–21, 1981.

Atkins, D., Heard, D., and Donovan, D.: Epidemiologic Overview of idividuals with Upper-Limb Loss and their Reported Research Priorities, J. Prosthet. Orthot., 8, 2–11, 1996.

Bekey, G., Tomovic, R., and Zeljkovic, I.: Dextrous Robot Hands, chap. Control Architecture for the Belgrade USC hand, Springer-Verlag, 1990.

Borchardt, Hartmann, Leymann, Radike, Schlesinger, and Schwiening: Ersatzglieder und Arbeitshilfen fr Kriegsbeschädigte und Unfallverletzte, Springer Verglag, 1919.

Clawson, A., Kyberd, P., Jones, B., Losier, Y., Wilson, A., Segil, J., and Weir, R.: The UNB Hand – Multifunction Hand Prosthesis System, in: 13th World Congress of the International Society of Prosthetics and Orthotics, International Society of Prosthetics and Orthotics, 2010.

Fraser, C. M.: Laterality, gender and age differences in estimated frequency and actual registration of people with congenital upper limb absences, Prosthet. Orthot. Int., 22, 224–229, 1998.

Heger, H., Milstien, S., and Hunter, G. A.: Electrically powered prostheses for the adult with an Upper Limb Amputation, J. Bone Joint Surg., 67-B, 278–281, 1985.

Iberall, T.: Human Prehension and Dextrous Robot Hands, Int. J. Robot. Res., 16, 285–299, 1997.

Iberall, T., Sukhatme, G., Beattie, D., and Bekey, G.: Intelligent Control Systems, chap. Control philosophy and simulation of a robotic hand as a model for prosthetic hands, IEEE Press, 682–701, 1996.

Jaconsen, S., Iversen, E., Knutti, D., Johnson, R., and Biggers, K.: Design of the Utah/MIT Dexterous Hand, in: IEEE international conferance on robotics and automation, San Francisco, California, IEEE, 1986.

Jones, L. and Davidson, J.: Save that arm: a study of problems in the remaining arm of unilateral upper limb amputees, Prosthet. Orthot. Int., 23, 55–58, 1999.

Kamakura, N., Matsuo, M., Harumi, I., Mitsuboshi, F., and Miura, Y.: Patterns of static prehension in normal hands, Am. J. Occup. Ther., 34, 437–445, 1980.

Kidd, P., MoCoy, C., and Steenbergen, L.: Repetitive Strain Injuries in Youths, Jo. Am. Acad. Nurse Prac., 12, 413–426, 2000.

Kyberd, P.: The Algorithmic Control of a multifunction hand prosthesis, Ph.D. thesis, Electrical Engineering Department, Southampton University, 1990.

Kyberd, P., Beard, D., Davey, J., and Morrison, D.: Survey of upper limb prostheses users in Oxfordshire, J. Prosthet. Orthot., 10, 85–91, 1998a.

Kyberd, P., Evans, M., and te Winkel, S.: An Intelligent Anthropomorphic Hand, with Automatic Grasp, Robotica, 16, 531–536, 1998b.

Kyberd, P., Light, C., Nightlingale, J., Whatley, D., and Evans, M.: The Design of Anthropomorphic Prosthetic Hands, a study of the Southampton Hand, Robotica, 19, 593–600, 2001.

Kyberd, P. J., Wartenberg, C., Sandsjö, L., Jönsson, S., Gow, D., Frid, J., Almström, C., and Sperling, L.: Survey of upper extremity prosthesis users in Sweden and the United Kingdom, J. Prosthet. Orthot., 19, 55–62, 2007.

Langefeld, C., Evans, C., Smith, S., Muller, A., Kerwin, J., Schnellinger, T., Guay, G., and der Merwe, D. V.: Patent number US2008/0288088 – Arm Prosthetic Device, 2008.

MacKenzie, C. and Iberall, T.: The Grasping Hand, North-Holland, 1994.

Moore, D.: Development of multifunctional adaptive hand prosthesis, Ph.D. thesis, Electrical Engineering Department, University of Southampton, 1980.

Napier, J.: The prehensile movements of the human hand, J. Bone Joint Surg., 38B, 902–913, 1956.

Newman, P.: Multi-articulating Hands and Fingers – Technical Strategies for Improved Function and Myoelectric Uptake(Abstract), in: MEC '08 Measuring Success in Upper Limb Prosthetics, UNB, IBME, Fredericton, New Brunswick, Canada, p. 183, 2008.

Pons, J., Ceres, R., Reynaerts, D., Saro, B., Levin, S., and Moorleghem, D. V.: Objectives and technological approach to the development of the multifunctional MANUS upper limb prosthesis, Robotica, 23, 301–310, 2005.

Puchhammer, G.: Michelangelo 03 – A versatile hand prosthesis, featuring superb controllability and sophisticated bio mimicry (Abstract), in: MEC '08 Measuring Success in Upper Limb Prosthetics, UNB, IBME, Fredericton, New Brunswick, Canada, 162–163, 2008.

Schulz, S.: Introducing a New Multiarticulating Myoelectric Hand System, in: 13th World Congress, 10–15 May 2010, Leipzig, Germany, International Society of Prosthetics and Orthotics, ISPO, 598–599, 2010.

Schulz, S., Pylatiuk, C., Reischl, M., Martin, J., and Bretthauer, R. M. G.: A hydraulically driven multifunctional prosthetic hand, Robotica, 23, 293–300, 2005.

Schulz, S., Pylatiuk, C., Kargov, A., Werner, T., Gaiser, I., Reischl, M., and Oberle, R.: Design and Preliminary Experience with Fludihand MK III, in: MEC '08 Measuring Success in Upper Limb Prosthetics, UNB, IBME, Fredericton, New Brunswick, Canada, 197–199, 2008.

Sears, H., Andrew, J., and Jacobsen, S.: Experience with the Utah arm, Hand and Terminal Device, chap. 18, 194–210, Springer-Verlag, 1988.

Sollerman, C. and Sperling, L.: Evaluation of Activities of Daily Living – Especially hand function, Scand. J. Rehabil. Med., 6, 139–143, 1978.

Todd, R.: Adaptive control of a hand prosthesis, Ph.D. thesis, Electrical Engineering Department, University of Southampton, 1969.

Vinet, R., Lozac'h, Y., Beaudry, N., and Drouin, G.: Design methodology for a multifunctional hand prosthesis, J. Rehabil. Res. Dev., 32, 316–324, 1995.

Wing, A. M. and Fraser, C.: The contribution of the thumb to reaching movements, Q. J. Exp. Psychol., 35A, 297–309, 1983.

Wing, A., Turton, A., and Fraser, C.: Grasp size and accuracy of approach in reaching, J. Motor Behav., 18, 245–260, 1986.

Zinck, A.: The Investigation of Compensatory Movements in Prosthesis Users and the Design of a Novel Wrist, Masters, Department of Mechanical Engineering, University of New Brunswick, 2008.

10

Towards the design of monolithic decoupled XYZ compliant parallel mechanisms for multi-function applications

G. Hao

Department of Electrical and Electronic Engineering, School of Engineering, University College Cork, Cork, Ireland

Correspondence to: G. Hao (g.hao@ucc.ie)

Abstract. This paper deals with the monolithic decoupled XYZ compliant parallel mechanisms (CPMs) for multi-function applications, which can be fabricated monolithically without assembly and has the capability of kinetostatic decoupling. At first, the conceptual design of monolithic decoupled XYZ CPMs is presented using identical spatial compliant multi-beam modules based on a decoupled 3-PPPR parallel kinematic mechanism. Three types of applications: motion/positioning stages, force/acceleration sensors and energy harvesting devices are described in principle. The kinetostatic and dynamic modelling is then conducted to capture the displacements of any stage under loads acting at any stage and the natural frequency with the comparisons with FEA results. Finally, performance characteristics analysis for motion stage applications is detailed investigated to show how the change of the geometrical parameters can affect the performance characteristics, which provides initial optimal estimations. Results show that the smaller thickness of beams and larger dimension of cubic stages can improve the performance characteristics excluding natural frequency under allowable conditions. In order to improve the natural frequency characteristic, a monolithic decoupled configuration that is achieved through employing more beams in the spatial modules or reducing the mass of each cubic stage mass can be adopted. In addition, an isotropic variation with different motion range along each axis and same payload in each leg is proposed. The redundant design for monolithic fabrication is introduced in this paper, which can overcome the drawback of monolithic fabrication that the failed compliant beam is difficult to replace, and extend the CPM's life.

1 Introduction

Compliant parallel mechanisms (CPMs) transmit motion/loads by deformation of their compliant links (namely jointless), and belong to a class of parallel-type mechanisms. They aim to utilize the material compliance/flexibility instead of only analyzing/suppressing the negative flexibility effect like those initial works in the area of kinematics of mechanisms with elasticity (Howell, 2001; Hao, 2011). This revolutionary change leads to many potential merits such as reduced part count (monolithic), zero backlashes, no need for lubrication, reduced wear, high precision and compact configuration in comparison with the rigid-body counterparts. CPMs with multiple DoF (degrees of freedom) have drawn more attentions from academia and industries due to their extensive applications such as motion/positioning stages (Awtar and Slocum, 2007; Dong et al., 2007; Hao and Kong, 2012a, b), acceleration/force sensors (Gao and Zhang, 2010; Hansen et al., 2007; Cappelleri et al., 2010) and energy harvesting devices (Rupp et al., 2009; Ando et al., 2010).

For a planar multi-DoF CPM such as the XY CPM, it is always easy to fabricate towards a monolithic configuration using existing well-developed planar manufacturing technologies such as wire EDM, water jet, and laser cutting (Awtar, 2004). However, these manufacturing technologies usually fail to satisfy the needs of fabricating most spatial multi-DoF CPMs (such as XYZ CPMs) monolithically, and therefore assembly has to be passively applied as shown in (Dong et al., 2007; Hao and Kong, 2012b; Gao and Zhang, 2010), which leads to some issues such as assembly error, increased number of parts, reduced stiffness (by about 30 % by bolted joints), and increased cost (Hao and Kong, 2012b). Over recent years, 3-D printing technology has been developed rapidly. Various base/substrate materials, such as engineering plastics, ceramics and metal, can be used for fabrication for a variety of applications. But the emerging 3-D printing technology may lead to limited or undesired performance characteristics of material due to no traditional heat treatment applied and inherent layer-by-layer fabrication. This shortcoming has been proved by testing our initial prototype, made of engineering plastic, obtained using a 3-D printer. Therefore, better manufacturing approaches/strategies for spatial multi-DoF CPMs are potentially needed. Averting the manufacturing issue on spatial CPMs, one can design a type of spatial multi-DoF CPMs that are possible to be fabricated monolithically using the above planar manufacturing technologies without bringing any assembly issues.

In this paper, we will mainly deal with the XYZ CPMs with (kinetostatically) decoupled configuration. Kinetostatic decoupling means that one primary output translational displacement is only affected by the actuation force along the same direction, which describes the relationship between the input force and output motion. This decoupling (not absolute) is also called the output-decoupling/minimal cross-axis coupling in CPMs. Kinetostatic coupling may lead to complicated motion control, which is the sufficient condition of kinematic decoupling. A number of literatures have reported the design of decoupled XYZ CPMs for motion/positioning stage and sensing applications (Hao and Kong, 2009, 2012b; Gao and Zhang, 2010; Li and Xu, 2011; Pham et al., 2006; Yue et al., 2010; Tang et al., 2006) using kinematics design methods (Hao, 2011). Here, each of the three kinematic legs/chains, which are coupled in parallel, is individually a serial-parallel hybrid arrangement. But none of them have showed the possibility for monolithic fabrication. Also, these designs have their own limitations such small motion range (Li and Xu, 2011; Pham et al., 2006; Yue et al., 2010), bulky and complex configuration (due to the serial-parallel hybrid arrangement) (Li and Xu, 2011; Pham et al., 2006; Hao and Kong, 2009; Tang et al., 2006), and poor out-of-plane stiffness of the PP plane in each leg (Li and Xu, 2011; Pham et al., 2006; Hao and Kong, 2009; Yue et al., 2010; Tang et al., 2006). Recently, Awtar et al. (2013) proposed a novel XYZ parallel kinematic flexure mechanism with geometrically decoupled DoF using identical flexure plates, which has a more compact and simpler construction and has the possibility to be fabricated monolithically. However, this design suffers from complicated modelling, bad out-of-plane stiffness and big lost motion, especially its three actuation directions are skew and cannot intersect at the center of the primary motion stage so that its applications are limited in low payload, and/or low speed. Hao and Kong (2012b) reported a decoupled XYZ CPM composed of identical spatial modules, but still has the challenging issue on fabrication.

This work builds on the above advances on decoupled XYZ CPMs towards a monolithic configuration (also compact and simple) for manufacturing purpose. It also stresses the potential extensive applications in multi-axis motion/positioning stages, multi-axis sensors (acceleration/force), and energy harvesting devices using the present monolithic decoupled XYZ CPMs.

This paper is organized as follows. Section 2 proposes the conceptual design of monolithic decoupled XYZ CPMs for three-type applications: motion/positioning stages, acceleration/force sensors, and energy harvesting devices at first. Then the analytical kinetostatic and dynamic modelling is undertaken in Sect. 3. In Sect. 4, the performance characteristics analysis to reflect the change of performance characteristics with that of the geometrical parameters is investigated. Section 5 discusses the thermal stability. Conclusions are drawn in Sect. 6.

2 Conceptual design of monolithic decoupled XYZ CPMs for multi-function applications

2.1 Monolithic decoupled XYZ CPMs

A decoupled XYZ CPM for the motion/positioning stage can be generated using a design approach proposed in the references (Hao, 2011; Hao and Kong, 2012b). This design is demonstrated in Fig. 1, which is based on a decoupled XYZ parallel kinematic mechanism (PKM) whose three planes associated with the passive PPR kinematic chains are orthogonal. It is obtained by replacing the actuated P joint and the passive PPR chain in each leg of the 3-PPPR XYZ PKM (Fig. 1a) with the compliant P joint in Fig. 1c and a compliant PPR joint in Fig. 1b, respectively, and making appropriate arrangement for the identical building blocks (spatial four-beam modules).

In order to facilitate the monolithic fabrication, an improved design of the decoupled XYZ CPMs (Fig. 2) is adopted in terms of the proposed decoupled XYZ CPM (Fig. 1d), which can be fabricated monolithically from a cubic material by three orthogonal directions' cutting using EDM/milling machining for a macro-version, or lithography/DRIE for an MEMS version with same masks on three surfaces of the cube. The improved design is composed of eight rigid cubic stages organically connected by twelve identical spatial multi-beam modules with planar motion to form a monolithic and compact cubic confirguration. When four adjacent rigid stages are fixed in the undeformed configuration (and therefore three spatial modules are inactive), the other four rigid stages act as the mobiles stages (X-, Y-, Z-, and XYZ-stages), displaced by the deformation of the nine spatial modules, to achieve the function of XYZ CPMs.

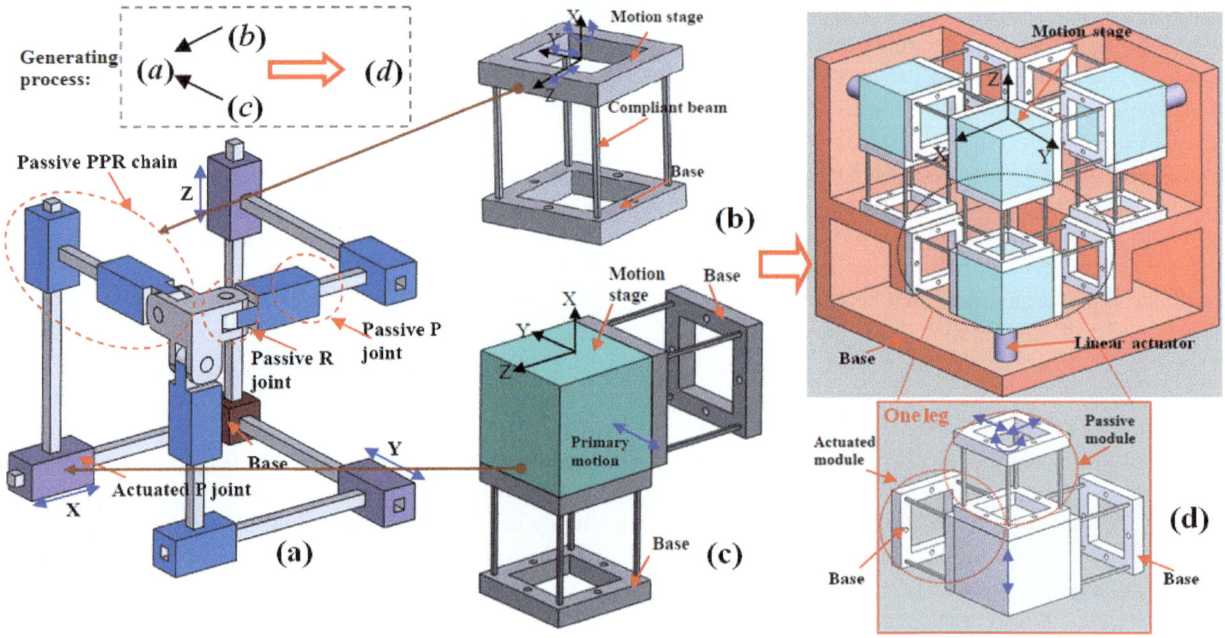

Figure 1. The generating process of a decoupled XYZ CPM: **(a)** A decoupled 3-PPPR XYZ PKM (Hao, 2011; Hao and Kong, 2012b) with three planes associated with the passive PPR kinematic chains orthogonal; **(b)** A compliant planar-motion PPR joint (Hao, 2011; Hao et al., 2011; Hao and Kong, 2013): spatial four-beam module that is composed four identical symmetrical square wire beams spaced around a circle uniformly; **(c)** A compliant P joint (Hao, 2011; Hao and Kong, 2012b): two spatial four-beam modules connected in parallel, two planes associated two PPR joints of which are orthogonal; **(d)** A decoupled XYZ CPM using identical spatial four-beam modules.

Three different monolithic forms are shown in Fig. 2a, b and c, respectively. As mentioned earlier, the monolithic decoupled XYZ CPM (for instance, Fig. 2a) has mainly three types of applications: motion/positioning stages, acceleration/force sensors, and energy harvesting devices, which is detailed as follows.

2.1.1 High-precision motion/positioning stages

For the applications as high-precision motion/positioning stages, the proposed monolithic decoupled XYZ CPM (Fig. 2a) can be actuated by three linear Voice Coil (VC) actuators for large motion range or by three PZT actuators for small motion range as shown in Fig. 1d. In addition three optical linear encoders (input sensing) and three capacitive measuring systems (output sensing) are required.

2.1.2 Acceleration/force sensors

The presented monolithic decoupled XYZ CPM (Fig. 2a) can be used as the 3-axis acceleration/force sensor without using linear actuators. For the use as the force sensor, any external force exerted at the XYZ-stage along each axis can be sensed by measuring the displacements of the X-, Y- and Z- stages along the X-, Y-, and Z-axes, respectively, by piezoresistors or other types of sensors. When used as the acceleration sensor, the XYZ-, X-, Y- and Z-stages are served as the inertial mass of the sensor. If there is a acceleration along certain resultant direction, the components of the resultant iniertial force along the X-, Y-, and Z-axes will result from the contribution of the combined mass of the XYZ- and X-stages, that of the XYZ- and Y-stages, and that of the of the XYZ- and Z-stages, respectively. Then by sensing displacements of the X-, Y- and Z-stages along the X-, Y-, and Z-axes, respectively, one can obtain the inertial force along each axis and then work out the acceleration along each axis.

2.1.3 Energy harvesting devices

Coupling with magnets and coils, an example 3-axis energy harvesting device (Fig. 3) can be obtained based on the monolithic decoupled XYZ CPM (Fig. 2a). It harvests energy in three axes through the well-known eletromagnetic induction. If there is an external exitation along a certain axis, the inertial mass (including the magnet) coupled with the compliant members (spring) along this axis will produce a vabriation to make the magnet to go through the coil, and then magnetic flux changes with time, which induces the electricity production for harvesting.

2.2 Redundant design for monolithic fabrication

A major drawback of the monolithic fabrication is that the failure (yield/fraction) of certain compliant beam(s) can cause the whole system's permanent strike due to the fact that the failed wire beam is difficult to replace. However, the

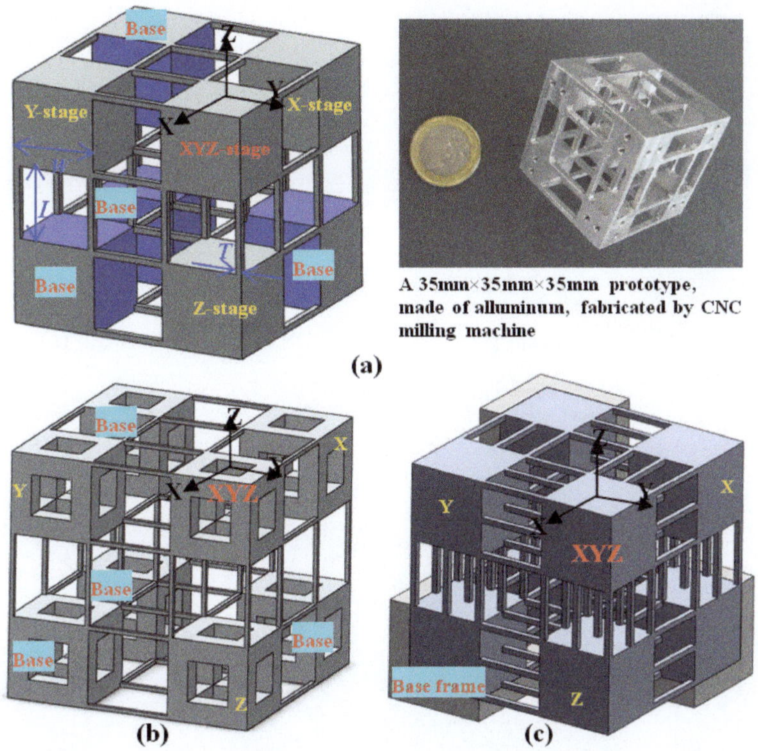

Figure 2. Monolithic decoupled XYZ CPMs: **(a)** a monolithic decoupled XYZ CPM with three geometrical parameters; **(b)** a monolithic decoupled XYZ CPM with improved natural frequency via reducing the cubic stage mass; and **(c)** a monolithic decoupled XYZ CPM with improved natural frequency via increasing the beam number (i.e. stiffness enhanced) (all configurations have same motion range).

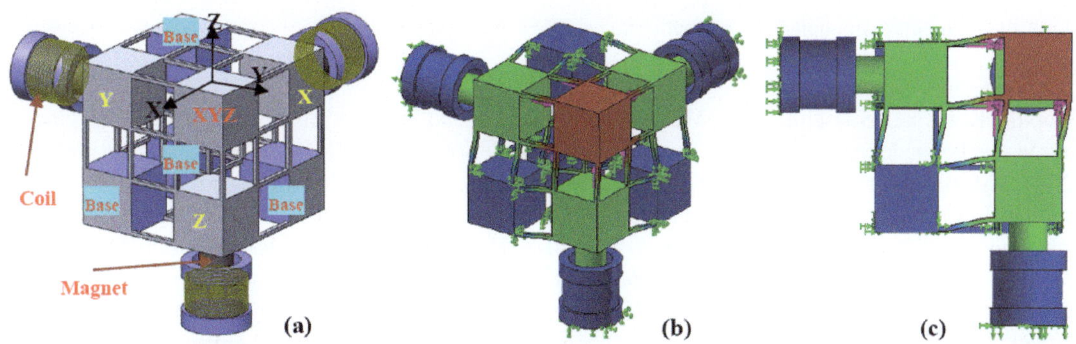

Figure 3. Energy harvesting devices: **(a)** Energy harvesting device based on a monolithic decoupled XYZ CPM; **(b)** Perspective view of FEA resutls in deformation; **(c)** Top view of FEA results in deformation.

present monolithic decoupled XYZ CPM in this paper is a redundant design with three redundant spatial four-beam modules inactive (or four cubic stages fixed), and therefore the redundant building blocks (or fixed cubic stages) can swap the functions with certain failure's mobile building blocks (or mobile cubic stages) to extend the system life.

In our design, each of three passive spatial four-beam modules undergoes two translations, and is prone to fail compared to other spatial modules to produce only one translation. If either of the three passive spatial four-beam modules fails, the base frame originally connecting the four fixed cubic stages can be moved to connect with the four originally mobile cubic stages in their initially undeformed configuration. Such a way, the life of the XYZ CPM is retrieved. A more clear illustration is shown in Fig. 4.

2.3 Variation for the monolithic decoupled XYZ CPM

It is easy to understand that the above proposed monolithic decoupled XYZ CPM have same motion range along each axis, i.e. a cubic workspace. However, for the purposes that the motion range in a certain direction is required larger/smaller than the other(s) and also that each leg has

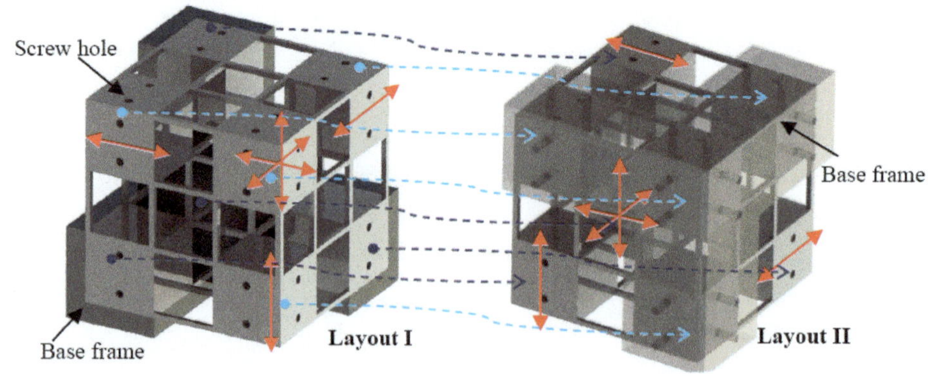

Figure 4. Demonstration of redundant design (rendered).

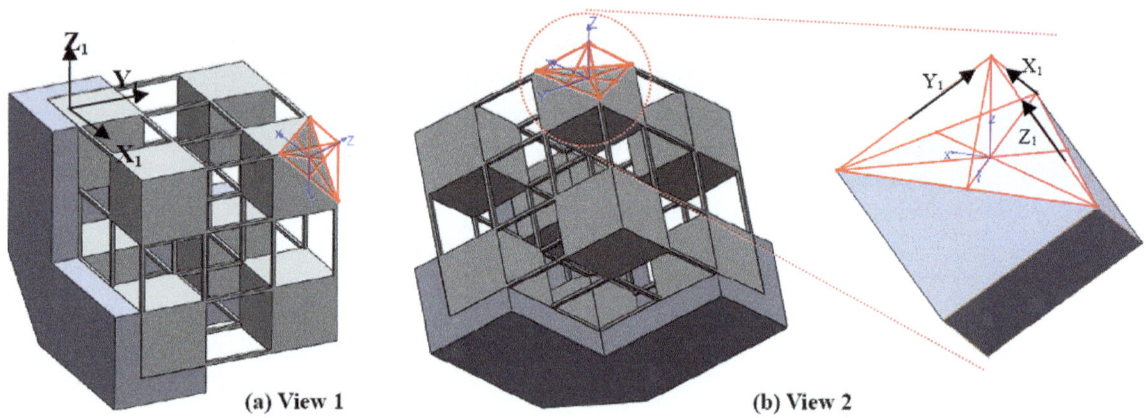

Figure 5. A variation for the monolithic decoupled XYZ CPM.

same payload (isotropic) (Kong and Gosselin, 2002; Werner et al., 2010), a variation can be made from the present monolithic decoupled XYZ CPM as shown in Fig. 5.

This variation has the decoupling property with regard to the original coordinate system $X_1Y_1Z_1$. But in the new coordinate system XYZ, the motion is not decoupled. The motion relationships between the two coordinate systems will be followed in the subsequent section.

3 Modelling of the monolithic decoupled XYZ CPM

In order to analyzing the performance characteristics of the monolithic decoupled XYZ CPM (Fig. 2a), it is essential to carry out the kinetostatic modelling and dynamic modelling.

3.1 Kinetostatic modelling

References (Hao, 2011; Hao and Kong, 2012b) have given the detailed analytical modelling derivation for an XYZ CPM using identical spatial double four-beam modules, therefore, in this paper, we will directly use the associated linear equations from these references with differenct geometrical parameters substitution. The purpose for linear kinetostatic modelling is to approximately estimate the displacements of the centers of the XYZ-, X-, Y-, and Z-stages under the action of loads at the centers of those stages to suit different applications. Here, the normalization-based strategy (Hao et al., 2011; Hao and Kong, 2013) is also adopted to represent loads and displacements using the corresponding lower-case letters, which refers to that all translational displacements and length parameters are normalized by the beam actual length L, forces by EI/L^2, and moments by EI/L. Here, E and I denotes the Young's modulus and the second moment of the area of a symmetrical cross-section, respectively.

The compliance matrix for the CPM system (Fig. 2a) with regard to the center of the XYZ-stage in the global coordinate system XYZ, i.e. the loads and displacements are both defined at the center, is obtained as

$$\mathbf{C}_{\rm cpm}=\mathbf{K}_{\rm cpm}^{-1}=(\mathbf{R}_{\rm leg1}\mathbf{K}_{\rm leg2}\mathbf{R}_{\rm leg1}^{-1}+\mathbf{K}_{\rm leg2}+\mathbf{R}_{\rm leg3}\mathbf{K}_{\rm leg2}\mathbf{R}_{\rm leg3}^{-1})^{-1} \quad (1)$$

where $\mathbf{C}_{\rm cpm}$ and $\mathbf{K}_{\rm cpm}$ are the compliance and stiffness matrices of the CPM system, respectively; $\mathbf{K}_{\rm leg2}$ is the stiffness matrix of Leg 2 (with the Y-stage) defined at the center of XYZ-stage in the global coordinate system, which is also a reference based on which the stiffness (or compliance) matrices for Legs 1 and 3 can be obtained by appropriate coordinate transformation so that $\mathbf{K}_{\text{leg1}} = \mathbf{R}_{\text{leg1}}\mathbf{K}_{\text{leg2}}\mathbf{R}_{\text{leg1}}^{-1}$, and $\mathbf{K}_{\text{leg3}} = \mathbf{R}_{\text{leg3}}\mathbf{K}_{\text{leg2}}\mathbf{R}_{\text{leg3}}^{-1}$; $\mathbf{R}_{\text{leg1}}$ and $\mathbf{R}_{\text{leg3}}$ are the rotation transformation matrices for Legs 1 and 3, respectively, which are given as

$$\mathbf{R}_{\text{leg1}} = \begin{bmatrix} \mathbf{R}_Z(-\pi/2)_{3\times3} & \mathbf{0}_{3\times3} \\ \mathbf{0}_{3\times3} & \mathbf{R}_Z(-\pi/2)_{3\times3} \end{bmatrix} \times \begin{bmatrix} \mathbf{R}_Y(-\pi/2)_{3\times3} & \mathbf{0}_{3\times3} \\ \mathbf{0}_{3\times3} & \mathbf{R}_Y(-\pi/2)_{3\times3} \end{bmatrix},$$

and

$$\mathbf{R}_{\text{leg3}} = \begin{bmatrix} \mathbf{R}_Y(\pi/2)_{3\times3} & \mathbf{0}_{3\times3} \\ \mathbf{0}_{3\times3} & \mathbf{R}_Y(\pi/2)_{3\times3} \end{bmatrix} \times \begin{bmatrix} \mathbf{R}_Z(\pi/2)_{3\times3} & \mathbf{0}_{3\times3} \\ \mathbf{0}_{3\times3} & \mathbf{R}_Z(\pi/2)_{3\times3} \end{bmatrix}.$$

The stiffness matrix of Leg 2 can be represented as

$$\mathbf{K}_{\text{leg2}} = [\mathbf{J}_{\text{m3}}(\mathbf{R}_{\text{pp}}\mathbf{C}_{\text{m}}\mathbf{R}_{\text{pp}}^{-1})\mathbf{J}_{\text{m3}}^{\text{T}} + \mathbf{J}_{\text{p}}\mathbf{C}_{\text{p}}\mathbf{J}_{\text{p}}^{\text{T}}]^{-1} \tag{2}$$

where $\mathbf{C}_{\text{m}}$ is the compliance matrix of the spatial four-beam module (compliant PPR joint, Fig. 1b) in Leg 2 with regard to the center of the bottom-plane of its own motion stage in its own local coordinate system; $\mathbf{C}_{\text{p}}$ is the compliance matrix of the compliant P joint (Fig. 1c) in Leg 2 with regard to the Y-stage center in the global coordinate system; $\mathbf{J}_{\text{m3}}$ and $\mathbf{J}_{\text{p}}$ are the position transformation matrices, and $\mathbf{R}_{\text{pp}}$ is the rotation transformation matrix, which are detailed below:

$$\mathbf{J}_{\text{m3}} = \begin{bmatrix} \mathbf{I}_{3\times3} & \begin{bmatrix} 0 & 0 & -w/2 \\ 0 & 0 & 0 \\ w/2 & 0 & 0 \end{bmatrix} \\ \mathbf{0}_{3\times3} & \mathbf{I}_{3\times3} \end{bmatrix},$$

$$\mathbf{J}_{\text{p}} = \begin{bmatrix} \mathbf{I}_{3\times3} & \begin{bmatrix} 0 & 0 & -1-w \\ 0 & 0 & 0 \\ 1+w & 0 & 0 \end{bmatrix} \\ \mathbf{0}_{3\times3} & \mathbf{I}_{3\times3} \end{bmatrix}, \text{ and}$$

$$\mathbf{R}_{\text{pp}} = \begin{bmatrix} \mathbf{R}_Z(\pi/2)_{3\times3} & \mathbf{0}_{3\times3} \\ \mathbf{0}_{3\times3} & \mathbf{R}_Z(\pi/2)_{3\times3} \end{bmatrix}.$$

The compliance matrix of the compliant P joint is further represented as

$$\mathbf{C}_{\text{p}} = \{(\mathbf{J}_{\text{m1}}\mathbf{C}_{\text{m}}\mathbf{J}_{\text{m1}}^{\text{T}})^{-1} + [\mathbf{J}_{\text{m2}}(\mathbf{R}_{\text{m}}\mathbf{C}_{\text{m}}\mathbf{R}_{\text{m}}^{-1})\mathbf{J}_{\text{m2}}^{\text{T}}]^{-1}\}^{-1} \tag{3}$$

where $\mathbf{J}_{\text{m1}}$ and $\mathbf{J}_{\text{m2}}$ are the position transformation matrices, and $\mathbf{R}_{\text{m}}$ is the rotation transformation matrix, which are expressed below:

$$\mathbf{J}_{\text{m1}} = \begin{bmatrix} \mathbf{I}_{3\times3} & \begin{bmatrix} 0 & 0 & 0 \\ 0 & 0 & w/2 \\ 0 & -w/2 & 0 \end{bmatrix} \\ \mathbf{0}_{3\times3} & \mathbf{I}_{3\times3} \end{bmatrix},$$

$$\mathbf{J}_{\text{m2}} = \begin{bmatrix} \mathbf{I}_{3\times3} & \begin{bmatrix} 0 & w/2 & 0 \\ -w/2 & 0 & 0 \\ 0 & 0 & 0 \end{bmatrix} \\ \mathbf{0}_{3\times3} & \mathbf{I}_{3\times3} \end{bmatrix}, \text{ and}$$

$$\mathbf{R}_{\text{m}} = \begin{bmatrix} \mathbf{R}_Y(\pi/2)_{3\times3} & \mathbf{0}_{3\times3} \\ \mathbf{0}_{3\times3} & \mathbf{R}_Y(\pi/2)_{3\times3} \end{bmatrix} \times \begin{bmatrix} \mathbf{R}_Z(\pi)_{3\times3} & \mathbf{0}_{3\times3} \\ \mathbf{0}_{3\times3} & \mathbf{R}_Z(\pi)_{3\times3} \end{bmatrix}.$$

The compliance matrix of the spatial four-beam module is further derived as

$$\mathbf{C}_{\text{m}} = \left(\sum_{i=0}^{4} \mathbf{D}_i^{\text{T}}\mathbf{K}\mathbf{D}_i\right)^{-1} \tag{4}$$

where

$$\mathbf{D}_i = \begin{bmatrix} \mathbf{I}_{3\times3} & \begin{bmatrix} 0 & z_i' & -y_i' \\ -z_i' & 0 & x_i' \\ y_i' & -x_i' & 0 \end{bmatrix} \\ \mathbf{0}_{3\times3} & \mathbf{I}_{3\times3} \end{bmatrix}, \text{ and}$$

$$\mathbf{K} = \begin{bmatrix} d & 0 & 0 & 0 & 0 & 0 \\ 0 & 12 & 0 & 0 & 0 & -6 \\ 0 & 0 & 12 & 0 & 6 & 0 \\ 0 & 0 & 0 & 1/(1+v) & 0 & 0 \\ 0 & 0 & 6 & 0 & 4 & 0 \\ 0 & -6 & 0 & 0 & 0 & 4 \end{bmatrix}.$$

Here, $x_1' = 0$, $y_1' = (w-t)/2$, and $z_1' = (w-t)/2$; $x_2' = 0$, $y_2' = (w-t)/2$, and $z_2' = -(w-t)/2$; $x_3' = 0$, $y_3' = -(w-t)/2$, and $z_3' = -(w-t)/2$; $x_4' = 0$, $y_4' = -(w-t)/2$, and $z_4' = (w-t)/2$. $d = 12/(t)^2$ for square cross-section with normalized thickness t. v is the Poisson's ratio of the material.

Based on the above modelling results, the relationships between the displacements at the center of the XYZ-stage and the loads acting at both the actuation points and the centers of the XYZ-, X-, Y-, and Z-stages are derived as

$$\boldsymbol{F} = \mathbf{K}_{\text{cpm}}\boldsymbol{X}_{\text{s}} - \mathbf{K}_{\text{leg1}}\mathbf{R}_{\text{leg1}}(\mathbf{J}_{\text{p}}\mathbf{C}_{\text{p}}\mathbf{J}_{\text{p}}^{\text{T}})\mathbf{R}_{\text{leg1}}^{-1}\mathbf{J}_{\text{pa1}}^{\text{T}}\boldsymbol{F}_{\text{ax}} - \mathbf{K}_{\text{leg2}}(\mathbf{J}_{\text{p}}\mathbf{C}_{\text{p}}\mathbf{J}_{\text{p}}^{\text{T}})\mathbf{J}_{\text{pa2}}^{\text{T}}\boldsymbol{F}_{\text{ay}} - \mathbf{K}_{\text{leg3}}\mathbf{R}_{\text{leg3}}(\mathbf{J}_{\text{p}}\mathbf{C}_{\text{p}}\mathbf{J}_{\text{p}}^{\text{T}})\mathbf{R}_{\text{leg3}}^{-1}\mathbf{J}_{\text{pa3}}^{\text{T}}\boldsymbol{F}_{\text{az}},$$

i.e.

$$\boldsymbol{X}_{\text{s}} = \mathbf{C}_{\text{cpm}}\Big[\boldsymbol{F} + \mathbf{K}_{\text{leg1}}\mathbf{R}_{\text{leg1}}(\mathbf{J}_{\text{p}}\mathbf{C}_{\text{p}}\mathbf{J}_{\text{p}}^{\text{T}})\mathbf{R}_{\text{leg1}}^{-1}\mathbf{J}_{\text{pa1}}^{\text{T}}\boldsymbol{F}_{\text{ax}} + \mathbf{K}_{\text{leg2}} \times (\mathbf{J}_{\text{p}}\mathbf{C}_{\text{p}}\mathbf{J}_{\text{p}}^{\text{T}})\mathbf{J}_{\text{pa2}}^{\text{T}}\boldsymbol{F}_{\text{ay}} + \mathbf{K}_{\text{leg3}}\mathbf{R}_{\text{leg3}}(\mathbf{J}_{\text{p}}\mathbf{C}_{\text{p}}\mathbf{J}_{\text{p}}^{\text{T}})\mathbf{R}_{\text{leg3}}^{-1}\mathbf{J}_{\text{pa3}}^{\text{T}}\boldsymbol{F}_{\text{az}}\Big] \tag{5}$$

where $\boldsymbol{F} = [f_x, f_y, f_z, m_x, m_y, m_z]^{\text{T}}$ and $\boldsymbol{X}_{\text{s}} = [x_s, y_s, z_s, \theta_{sx}, \theta_{sy}, \theta_{sz}]^{\text{T}}$ which are the load-vector and displacement-vector are first defined at the center of the XYZ-stage;

$\boldsymbol{F}_{ax} = [f_{ax-x}, f_{ax-y}, f_{ax-z}, m_{ax-x}, m_{ax-y}, m_{ax-z}]^T$, $\boldsymbol{F}_{ay} = [f_{ay-x}, f_{ay-y}, f_{ay-z}, m_{ay-x}, m_{ay-y}, m_{ay-z}]^T$, and $\boldsymbol{F}_{az} = [f_{az-x}, f_{az-y}, f_{az-z}, m_{az-x}, m_{az-y}, m_{az-z}]^T$, which denote the load vectors at the centers of the X-, Y- and Z-stages, respectively. As an example for explaining the force symbols, f_{ax-x}, f_{ax-y} and f_{ax-z} denote the forces acting at the center of the X-stage along the X-, Y-, and Z-axes, respectively;

$\mathbf{J}_{pai}$ $(i = 1, 2, 3)$ in Eq. (5) is the position transformation matrix for loads in each leg, which is shown below

$$\mathbf{J}_{pai} = \begin{bmatrix} \mathbf{I}_{3\times3} & \begin{bmatrix} 0 & z_i' & -y_i' \\ -z_i' & 0 & x_i' \\ y_i' & -x_i' & 0 \end{bmatrix} \\ \mathbf{0}_{3\times3} & \mathbf{I}_{3\times3} \end{bmatrix}.$$

Where, $x_1' = -(1+w), y_1' = 0$, and $z_1' = 0$; $x_2' = 0, y_2' = -(1+w)$, and $z_2' = 0$; $x_3' = 0, y_3' = 0$, and $z_3' = -(1+w)$.

We can then obtain the displacements at the centers of the X-, Y-, and Z-stages as

$$\boldsymbol{X}_{ax} = \mathbf{R}_{leg1}\mathbf{C}_p\mathbf{R}_{leg1}^{-1}\mathbf{J}_{p1}^T\left[\mathbf{K}_{leg1}\left(\boldsymbol{X}_s - \mathbf{R}_{leg1}(\mathbf{J}_p\mathbf{C}_p\mathbf{J}_p^T)\right.\right.$$
$$\left.\left.\times\, \mathbf{R}_{leg1}^{-1}\mathbf{J}_{pa1}^T\boldsymbol{F}_{ax}\right) + \mathbf{J}_{pa1}^T\boldsymbol{F}_{ax}\right],$$

$$\boldsymbol{X}_{ay} = \mathbf{C}_p\mathbf{J}_{p2}^T[\mathbf{K}_{leg2}(\boldsymbol{X}_s - \mathbf{J}_p\mathbf{C}_p\mathbf{J}_p^T\mathbf{J}_{pa2}^T\boldsymbol{F}_{ay}) + \mathbf{J}_{pa2}^T\boldsymbol{F}_{ay}],$$

$$\boldsymbol{X}_{az} = \mathbf{R}_{leg3}\mathbf{C}_p\mathbf{R}_{leg3}^{-1}\left[\mathbf{K}_{leg3}\left(\boldsymbol{X}_s - \mathbf{R}_{leg3}\right.\right.$$
$$\left.\left.\times\,(\mathbf{J}_p\mathbf{C}_p\mathbf{J}_p^T)\mathbf{R}_{leg3}^{-1}\mathbf{J}_{pa3}^T\boldsymbol{F}_{az}\right) + \mathbf{J}_{pa3}^T\boldsymbol{F}_{az}\right] \tag{6}$$

where $\boldsymbol{X}_{ax} = [x_{ax}, y_{ax}, z_{ax}, \theta_{ax-x}, \theta_{ax-y}, \theta_{ax-z}]^T$,

$\boldsymbol{X}_{ay} = [x_{ay}, y_{ay}, z_{ay}, \theta_{ay-x}, \theta_{ay-y}, \theta_{ay-z}]^T$, and

$\boldsymbol{X}_{az} = [x_{az}, y_{az}, z_{az}, \theta_{az-x}, \theta_{az-y}, \theta_{az-z}]^T$, which denote the displacement vectors at the centers of the X-, Y- and Z-stages. As an example for explaining the displacement symbols, x_{ay}, y_{ay}, and z_{ay} denote the translational displacements of the Y-stage center along the X-, Y- and Z-axes, respectively; $\mathbf{J}_{pi}$ $(i = 1, 2, 3)$ is the position transformation matrix for loads in each leg as

$$\mathbf{J}_{pi} = \begin{bmatrix} \mathbf{I}_{3\times3} & \begin{bmatrix} 0 & z_i' & -y_i' \\ -z_i' & 0 & x_i' \\ y_i' & -x_i' & 0 \end{bmatrix} \\ \mathbf{0}_{3\times3} & \mathbf{I}_{3\times3} \end{bmatrix}.$$

Here, $x_1' = (1+w), y_1' = 0$, and $z_1' = 0$; $x_2' = 0, y_2' = (1+w)$, and $z_2' = 0$; $x_3' = 0, y_3' = 0$, and $z_3' = (1+w)$.

It should be noted that the above derived analytical models are capable of comprehensively reflecting the displacements of any stage under loading at any stage, and are accurate enough under small motion range. The proposed analytical models can be used for quick design synthesis, and also offer a reference for further nonlinear kinetostatic analysis and optimization.

3.2 Dynamic modelling

Accurate dynamic equations can be obtained from the classical Lagrange equation building on the above kinetostatic modelling results. However, we only give approximate estimatations of the natural frequencies of the monolithic decoupled XYZ CPM using simplified stiffness models.

The actual primary translational stiffness along each axis can be simplied as

$$K = 16\frac{12EI}{L^3} = 16\frac{ET^4}{L^3}. \tag{7}$$

Then the equal first, second or third-order natural frequency is derived as

$$f_1 = \sqrt{K/(M)}/(2\pi) \tag{8}$$

where M is the actual motion mass along each axis only considering two stages neglecting the compliant beams' mass.

Given that all stages are identical, M is double of mass of each (cubic) stage, which is equal to $2\rho W^3$ with a density of ρ. So Eq. (8) is rewritten as

$$f_1 = \sqrt{8ET^4/(\rho W^3 L^3)}/(2\pi) = \sqrt{8Et^4/(\rho w^3 L^2)}/(2\pi). \tag{9}$$

3.3 Resutls analysis

Let the material be a standard aluminum alloy AL 6061-T651 with Young's Modules $E = 69$ Gpa and Poisson's ratio $v = 0.33$. Also, we define the geometrical parameters as: $L = 20$ mm (beam length), $W = 20$ mm (cubic stage's side-length), $T = 1$ mm (beam thickness) (i.e. a system dimension of 60 mm × 60 mm × 60 mm) for the initial performance analysis.

Given an example of only three forces acting at the centers of the X-, Y- and Z-stages for the motion stage applications, the normalized load-displacement relationships based on Eqs. (5) and (6) can be expressed as

$$\begin{bmatrix} x_{\mathrm{s}} \\ y_{\mathrm{s}} \\ z_{\mathrm{s}} \\ \theta_{\mathrm{sx}} \\ \theta_{\mathrm{sy}} \\ \theta_{\mathrm{sz}} \end{bmatrix} = \mathbf{C}_{\mathrm{a-s}} \begin{bmatrix} f_{\mathrm{ax-x}} \\ f_{\mathrm{ay-y}} \\ f_{\mathrm{az-z}} \\ f_{\mathrm{ax-x}} \\ f_{\mathrm{ay-y}} \\ f_{\mathrm{az-z}} \end{bmatrix} = \begin{bmatrix} 0.005341 & -0.00006975 & -0.00006975 & 0 & 0 & 0 \\ -0.00006975 & 0.005341 & -0.00006975 & 0 & 0 & 0 \\ -0.00006975 & -0.00006975 & 0.005341 & 0 & 0 & 0 \\ 0 & 0 & 0 & 0 & -0.0001112 & 0.0001112 \\ 0 & 0 & 0 & 0.0001112 & 0 & -0.0001112 \\ 0 & 0 & 0 & -0.0001112 & 0.0001112 & 0 \end{bmatrix} \begin{bmatrix} f_{\mathrm{ax-x}} \\ f_{\mathrm{ay-y}} \\ f_{\mathrm{az-z}} \\ f_{\mathrm{ax-x}} \\ f_{\mathrm{ay-y}} \\ f_{\mathrm{az-z}} \end{bmatrix}. \tag{10a}$$

$$\begin{bmatrix} x_{\mathrm{ay}} \\ y_{\mathrm{ay}} \\ z_{\mathrm{ay}} \\ \theta_{\mathrm{ay-x}} \\ \theta_{\mathrm{ay-y}} \\ \theta_{\mathrm{ay-z}} \end{bmatrix} = \mathbf{C}_{\mathrm{a-a}} \begin{bmatrix} f_{\mathrm{ax-x}} \\ f_{\mathrm{ay-y}} \\ f_{\mathrm{az-z}} \\ f_{\mathrm{ax-x}} \\ f_{\mathrm{ay-y}} \\ f_{\mathrm{az-z}} \end{bmatrix} = \begin{bmatrix} 1.280\times10^{-5} & 3.1193\times10^{-7} & -1.7446\times10^{-7} & 0 & 0 & 0 \\ -0.00006961 & 0.005367 & -0.00006961 & 0 & 0 & 0 \\ -1.7446\times10^{-7} & 3.1193\times10^{-7} & 1.280\times10^{-5} & 0 & 0 & 0 \\ 0 & 0 & 0 & 0 & -0.00008392 & 0.00008371 \\ 0 & 0 & 0 & 3.8288\times10^{-7} & 0 & -3.8288\times10^{-7} \\ 0 & 0 & 0 & -0.00008371 & 0.00008392 & 0 \end{bmatrix} \begin{bmatrix} f_{\mathrm{ax-x}} \\ f_{\mathrm{ay-y}} \\ f_{\mathrm{az-z}} \\ f_{\mathrm{ax-x}} \\ f_{\mathrm{ay-y}} \\ f_{\mathrm{az-z}} \end{bmatrix}. \tag{10b}$$

The above equations can clearly show the performance characteristics for any three-axis loading. For example, under a single force, the ratio of magnitude of the parasitic rotation about the X- (or Z-axis) to the primary translation along the Y-axis can be obtained based on Eq. (10a) as

$$\left|\frac{\mathbf{C}_{\mathrm{a-s}}(4,5)}{\mathbf{C}_{\mathrm{a-s}}(2,2)}\right| = \frac{0.0001112}{0.005341} = 2.08\,\%. \tag{11}$$

The cross-axis coupling effect, for example the effect of $f_{\mathrm{ay-y}}$ upon x_{s}, is then determined using Eq. (10a) by

$$\left|\frac{\mathbf{C}_{\mathrm{a-s}}(1,2)}{\mathbf{C}_{\mathrm{a-s}}(2,2)}\right| = \frac{0.00006975}{0.005341} = 1.31\,\%. \tag{12}$$

Moreover, the lost motion percentage under a single loading can be written using the results in Eqs. (10a) and (10b) as

$$\frac{\mathbf{C}_{\mathrm{a-a}}(2,2)-\mathbf{C}_{\mathrm{a-s}}(2,2)}{\mathbf{C}_{\mathrm{a-s}}(2,2)} = \frac{0.005367-0.005341}{0.005341} = 0.49\,\%. \tag{13}$$

Equation (10a) shows that the force applied along an axis cannot produce the parasitic rotation about this axis. And two equal forces applied along two out of three axes, respectively, cannot cause the parasitic rotation about the third axis either.

In addition, simple comparisons, including kinetostatic and dynamic, between the analytical results and the FEA results (using Solidworks) are given in the Table 1, which shows good agreements in primary motion displacement, parasitic rotational angles, and modal frequency. There is relatively large difference only in cross-axis coupling motion, which may result from the error of the linear analytical modeling (or that of the FEA results), but have reasonable estimation in the changing trends.

4 Performance characteristics analysis

In this section, performance characteristics analysis for the monolithic decouple XYZ CPM as the motion stage (Fig. 2a) is conducted to see how the geometrical parameters' change can affect the performance characteristics and which performance characteristic is most sensitive to a geometrical parameter. This analysis will provide an initial optimization.

Figures 6, 7, and 8 illustrate the performance characteristics, defined in Eqs. (11), (12) and (13), against the normalized beam thickness, t, and the side-length, w, of the (cubic) stage under the conditions of specified material and beam length. Some key findings are summarized as follows.

- The parasitic rotation (Fig. 6) is influenced by both w and t, and increases with the increase of t and decreases with the increase of w. It is more sensitive to t compared with w. The smaller w is, the larger the effect of t on the parasitic rotation is. Similarly, the larger t is, the larger the effect of w on the parasitic rotation is.
- The cross-coupling (Fig. 7) is also affected by both w and t, and increases with the increase of t or with the decrease of w. It is also more sensitive to t in comparison with w. The smaller w is, the larger the effect of t on the cross-coupling is, and the larger t is, the larger the effect of w on the cross-coupling is.
- The lost motion (Fig. 8) is insensitive to w, and only is dominated by t. the smaller t, the smaller the lost motion is.

In addition, under the specified material, the translational motion range (same motion range along each axis) is only affected by the normalized beam thickness, t, and insensitive to the side-length, w, of the stage. The decrease of t can improve the motion range.

We can conclude from the above results that desired performance characteristics (large-range motion, minimized parasitic rotation, minimized cross-coupling and minimal lost motion) can be achieved by employing smaller t and appropriately larger w under the allowable conditions such as the minimum fabrication thickness, the overall system size, and stiffness/frequency requirements.

It is observed from Fig. 9 (or Eq. 9) that the first natural frequency goes up with the increase of t and/or the decrease of w, and is a little more sensitive to t. Therefore, in order to improve the natural frequency characteristic under small t and/or large w, we can enhance the system stiffness through using a better elasticity-average configuration with

Table 1. Comparisons of the analytical results and the FEA results.

Methods	Displacements under single loading: $F_y = 50$ N						Modal frequency (Hz)		
	X_s (mm)	Y_s (mm)	Z_s (mm)	θ_{sx} (rad)	θ_{sy} (rad)	θ_{sz} (rad)	First mode	Second mode	Third mode
Analytical results	−0.0049	0.3715	−0.0049	-3.87×10^{-4}	0	3.87×10^{-4}	284	284	284
FEA results	−0.0081	0.3884	−0.0077	-4.11×10^{-4}	-3×10^{-7}	4.20×10^{-4}	273	273	281
Difference |(Analytical-FEA)/FEA|	39.51 %	4.35 %	36.36 %	5.84 %	negligible	7.86 %	4.03 %	4.03 %	1.07 %

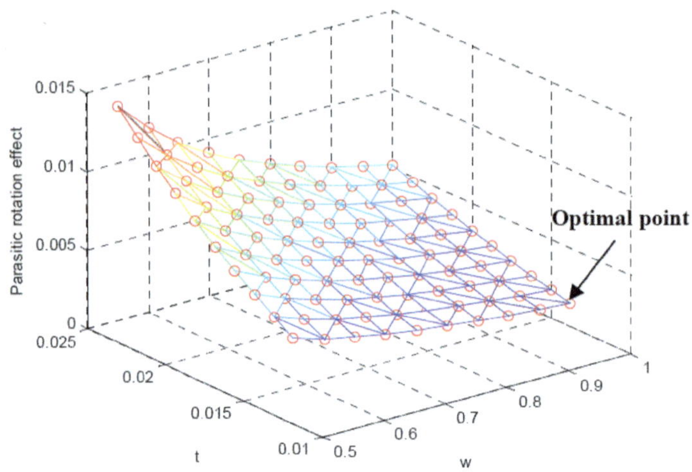

Figure 6. Parasitic rotation effect.

more beams in each spatial module as indicated in Fig. 2c, and/or reducing the mass of the cubic stages as indicated in Fig. 2b.

The motion of the variation (Fig. 5) in the coordinate system XYZ can be further determined by the following equations

$$Z=(X_{1z}^2+Y_{1z}^2+Z_{1z}^2)^{0.5}$$
subject to $X_{1z}=Y_{1z}=Z_{1z}$ with all positive values;
$$Y=(X_{1y}^2+Y_{1y}^2+Z_{1y}^2)^{0.5}$$
subject to $Y_{1y}=Z_{1y}$ with both negtive values and $|Y_{1y}|=0.5X_{1y}$;
$$X=(Y_{1x}^2+Z_{1x}^2)^{0.5}$$
subject to $|Y_{1x}|=|Z_{1x}|$
with a positive Z_{1x} and a negtive Y_{1x} (14)

where X, Y, and Z is the positive resultant motion in the coordinate system XYZ. X_{1z}, Y_{1z}, and Z_{1z} are the motion along the X_1-, Y_1-, and Z_1-axes, respectively, in the coordinate system $X_1Y_1Z_1$ contributing to Z. Y_{1y}, and Z_{1y} are the motion along the Y_1-, and Z_1-axes, respectively, in the coordinate system $X_1Y_1Z_1$ contributing to Y. X_{1x}, Y_{1x}, and Z_{1x} are the motion along the X_1-, Y_1-, and Z_1-axes, respectively, in the coordinate system $X_1Y_1Z_1$ contributing to X.

Let the motion range along each positive axis in the coordinate system $X_1Y_1Z_1$ be δ, the following constraint conditions should be met:

$$0 <= X_1 = X_{1y} + X_{1z} <= \delta;$$
$$-\delta <= Y_1 = Y_{1x} + Y_{1y} + Y_{1z} <= \delta;$$
$$-\delta <= Z_1 = Z_{1x} + Z_{1y} + Z_{1z} <= \delta. \quad (15)$$

From the above resutls, it is directly obtained that the maximal single-axis motion along the positive X-axis is $2^{0.5}\delta$, the maximal single-axis motion along the positive Y-axis is $1.5^{0.5}\delta$, and the maximal single-axis motion along the positive Z-axis is $3^{0.5}\delta$.

The workspace for this variation in Octant I of the coordinate system XYZ can be obtained using numerical approach based on the above results (Eqs. 14 and 15), and is shown in Fig. 10. Note that the number of points in Fig. 10 depends on the set-up step size in the numerical approach.

5 Discussions

The large motion range requires a large-range linear actuator, which cannot be a PZT actuator. Although displacement amplifiers as actuated compliant P joints can be combined with the PZT actuators to enlarge the motion range, adversely, they lead to relatively low off-axis stiffness and augment the minimum incremental motion of the actuators, i.e. poor resolution. Thus, one needs to use the linear VC actuator for millimeter-level actuation range, which will generate heat to the mechanism.

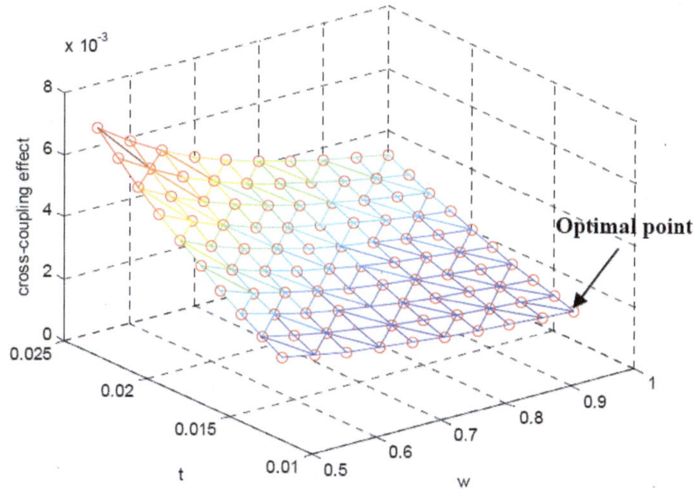

Figure 7. Cross-coupling effect.

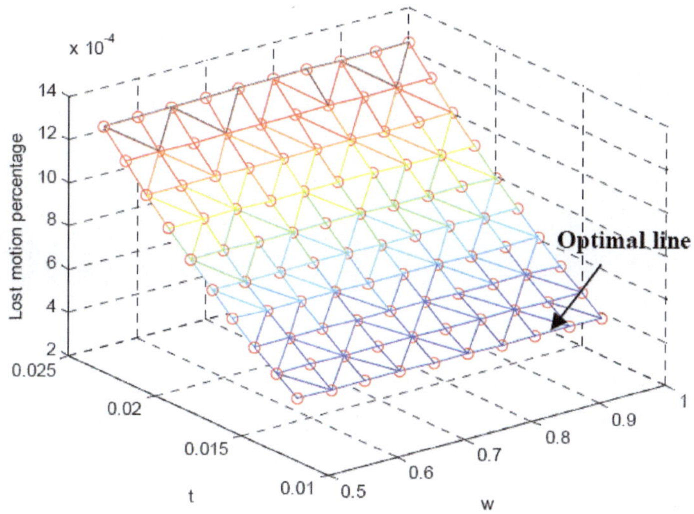

Figure 8. Lost motion percentage.

The monolithic design proposed in this paper is an over-constraint design in both the individual spatial multi-beam modules and the three identical legs. Although the problems of over-constraint are largely mitigated by the fact that the mechanism is monolithic and requires no assembly, there are still problems with its over-constraint. For instance, when VC actuators are used, the temperature will vary over the mechanism producing stress building up that can be problematic for precision performance. Moreover, when the mechanism heats up, the stage will drift as the design is not fully symmetric and thus not considered thermally stable.

6 Conclusions

This paper has presented and modelled a monolithic decoupled XYZ CPM (Fig. 2a) for multi-function applications: motion/positioning stages, acceleration/force sensors, and energy harvesting devices. The proposed monolithic decoupled XYZ CPM uses only identical spatial multi-beam modules as the building blocks involving three geometrical parameters and can be fabricated by the planar manufacturing technologies (such as EDM) without assembly as the 2-D compliant mechanisms.

In addition, the monolithic decoupled XYZ CPM with improved natural frequency (Fig. 2b and c) and the variation with different motion range in each axis and same payload in each leg (Fig. 5) have been proposed. Redundant design for monolithic fabrication has been discussed in this paper, which can be used to extend the CPM's life.

The derived analytical kinetostatic models can capture the displacements of any stage under loading at any stage. The performance characteristics analysis for the motion stage application has been implemented to identify the optimal geometrical parameters for beam thickness and stage dimension.

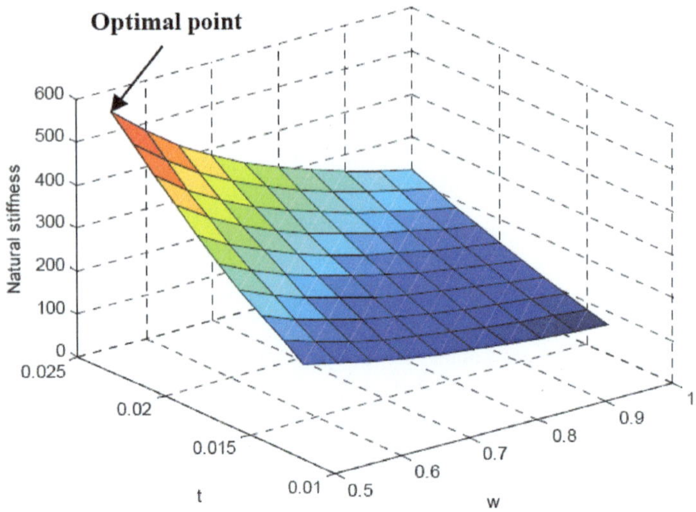

Figure 9. Natural frequency.

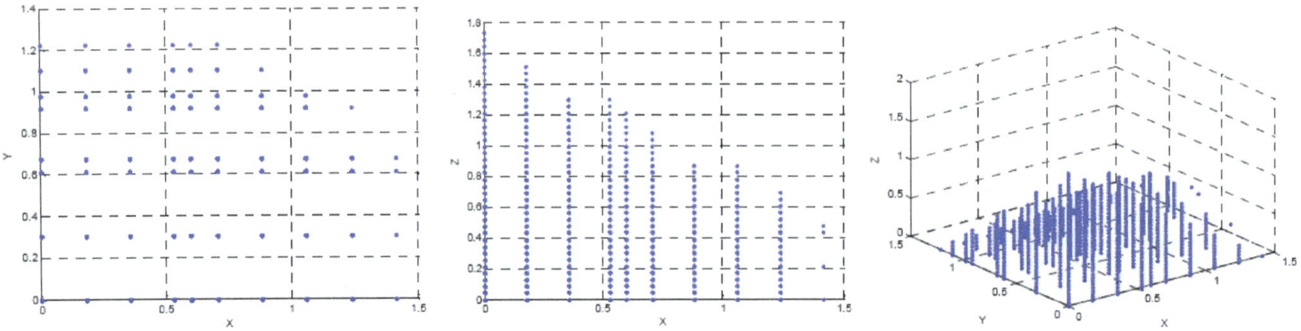

(a) Top view (b) Front view (c) Side view

Figure 10. Workspace of the variation in Octant I of the coordinate system XYZ.

It is noted that the proposed design may promote the fabrication using the carbon nanotubes or carbon fibers, which may lead to novel compliant mechanisms used in the emerging MEMS or nano-electro-mechanical-systems (NEMS). Experiment verification, nonlinear modelling, fatigue analysis, and optimization deserve further investigation in the near future.

Acknowledgements. The author would like to sincerely appreciate the reviewers' valuable comments for improving the quality of this paper. Special thanks goes to Tim Power and Michael O'Shea for their excellent technical knowledge in fabricating the prototype in UCC.

Edited by: Y. Li
Reviewed by: H. Tang and one anonymous referee

References

Ando, B., Baglio, S., Trigona, C., Dumas, N., Latorre, L., and Nouet, P.: Nonlinear mechanism in MEMS devices for energy harvesting applications, J. Micromech. Microeng., 20, 125020, doi:10.1088/0960-1317/20/12/125020, 2010.

Awtar, S.: Analysis and Synthesis of Planar Kinematic XY Mechanisms, Sc.D. thesis, Massachusetts Institute of Technology, Cambridge, MA, 2004.

Awtar, S. and Slocum, A. H.: Constraint-Based Design of Parallel Kinematic XY Flexure Mechanisms, J. Mech. Design, 129, 816–830, 2007.

Awtar, S., Ustick, J., and Sen, S.: An XYZ Parallel Kinematic Flexure Mechanism with Geometrically Decoupled Degrees of Freedom, Journal of Mechanisms and Robotics, 5, 015001, doi:10.1115/1.4007768, 2013.

Cappelleri, D., Krishnan, G., Kim, C., Kumar, and V., Kota, S.: Toward the Design of a Decoupled, Two-Dimensional, Vision-Based uN Force Sensor, Journal of Mechanisms and Robotics, 2, 021010, doi:10.1115/1.4001093, 2010.

Dong, W., Sun, L. N., and Du, Z. J.: Design of a Precision Compliant Parallel Positioner Driven by Dual Piezoelectric Actuators, Sensor. Actuat. A-Phys., 135, 250–256, 2007.

Gao, Z. and Zhang, D.: Design, Analysis and Fabrication of a Multidimensional Acceleration Sensor Based on Fully Decoupled Compliant Parallel Mechanism, Sensor. Actuat. A-Phys., 163,

418–427, 2010.

Hansen, B. J., Carron, C. J., Jensen, B. D., Hawkins, A. R., and Schultz, S. M.: Plastic Latching Accelerometer Based on Bistable Compliant Mechanisms, Smart Mater. Struct., 16, 1967–1972, 2007.

Hao, G.: Creative Design and Modelling of Large-Range Translational Compliant Parallel Manipulators, Ph.D. thesis, Heriot-Watt University, Edinburgh, UK, 2011.

Hao, G. and Kong, X.: A 3-DOF Translational Compliant Parallel Manipulator Based on Flexure Motion, in: Proceedings of 2009 ASME International Design Engineering Technical Conferences & Computers and Information in Engineering Conference, San Diego, CA, USA, 30 August–2 September, DETC2009-86075, 2009.

Hao, G. and Kong, X.: Novel XY Compliant Parallel Manipulators for Large Translation with Enhanced Out-of-Plane Stiffness, J. Mech. Design, 134, 061009, doi:10.1115/1.4006653, 2012a.

Hao, G. and Kong, X.: Design and Modelling of a Large-Range Modular XYZ Compliant Parallel Manipulators Using Identical Spatial Modules, Journal of Mechanisms and Robotics, 4, 021009, doi:10.1115/1.4006188, 2012b.

Hao, G. and Kong, X.: A Normalization-Based Approach to the Mobility Analysis of Spatial Compliant Multi-Beam Modules, Mech. Mach. Theory, 59, 1–19, 2013.

Hao, G., Kong, X., and Reuben, R. L.: A Nonlinear Analysis of Spatial Compliant Parallel Modules: Multi-beam Modules, Mech. Mach. Theory, 46, 680–706, 2011.

Howell, L. L.: Compliant Mechanisms, Wiley, New York, 2001.

Kong, X. W. and Gosselin, C. M.: Kinematics and Singularity Analysis of a Novel Type of 3-CRR 3-DOF Translational Parallel Manipulator, I. J. Robot. Res., 21, 791–798, 2002.

Li, Y. and Xu, Q.: A Totally Decoupled Piezo-Driven XYZ Flexure Parallel Micropositioning Stage for Micro/Nanomanipulation, IEEE T. Autom. Sci. Eng., 8, 265–279, 2011.

Pham, H.-H., Yeh, H. C., and Chen, I.-M.: Micromanipulation System Design Based on Selective Actuation Mechanisms, Int. J. Robot. Res., 25, 171–185, 2006.

Rupp, C. J., Evgrafov, A., Maute K., and Dunn, M. L.: Design of Piezoelectric Energy Harvesting Systems: A Topology Optimization Approach Based on Multilayer Plates and Shells, J. Intel. Mat. Syst. Str., 20, 1923–1939, 2009.

Tang, X., Chen, I.-M., and Li, Q.: Design and Nonlinear Modeling of a Large-Displacement XYZ Flexure Parallel Mechanism with Decoupled Kinematics Structure, Rev. Sci. Instrum., 77, 115101, doi:10.1063/1.2364132, 2006.

Werner, C., Rosielle, P. C. J. N., and Steinbuch, M.: Design of a Long Stroke Translation Stage for AFM, International Journal Machine Tools and Manufacture, 50, 183–190, 2010.

Yue, Y., Gao, F., Zhao, X., and Ge, Q.: Relationship among Input-Force, Payload, Stiffness and Displacement of a 3-DOF Perpendicular Parallel Micro-Manipulator, Mech. Mach. Theory, 45, 756–771, 2010.

11

The kinematic preshaping of triggered self-adaptive linkage-driven robotic fingers

L. Birglen

Ecole Polytechnique of Montreal, Department of Mechanical Engineering, Montreal, QC, H3T 1J4, Canada

Abstract. In this paper, the issue of the kinematic – as opposed to dynamic – preshaping of self-adaptive robotic fingers driven by linkages is discussed. A method to obtain designs of these fingers capable of various behaviours during their closing motions is presented. The method is based on using triggered passive elements in carefully selected joints of the finger and the selection or optimization of geometric parameters to obtain particular kinematic relationships between the motions of the phalanges. This method is very general and can be applied to any self-adaptive robotic finger in order to obtain many different types of closing motions. Examples given in this paper are focusing on two different preshaping motions, the first one aims at allowing pinch grasps while the second mimics a human finger. The fundamental aim of this paper is to show that various preshapings of self-adaptive fingers are possible, not just one, and to give two step-by-step examples.

This paper was presented at the IFToMM/ASME International Workshop on Underactuated Grasping (UG2010), 19 August 2010, Montréal, Canada.

1 Introduction

1.1 Self-adaptive fingers

In the past few years, a significant increase in the development of innovative technologies has tried to address the lack of commercial success of complex robotic hands. Significant efforts have been made to find designs simple enough to be easily built and controlled (Bicchi, 2000), particularly in human prosthetics. A special emphasis has been placed on the reduction of the number of degrees of freedom (DOF), thereby decreasing the number of required actuators. In particular, the SSL hand (Akin et al., 2002), the DIES-DIEM hand (Biagiotti et al., 2001) and the Cassino finger (Figliolini and Ceccarelli, 2002) have followed this path. On the other hand, a rapidly growing number of prototypes involve a smaller number of actuators without decreasing the number of DOF by taking advantages of self-adaptive, a.k.a. underactuated, mechanisms. These prototypes are usually either driven by tendons (Hirose and Umetani, 1978; Crisman et al., 1996; Massa et al., 2002; Higashimori et al., 2005a) or by linkages (Bekey et al., 1999; Ulrich, 1988; Laliberté and Gosselin, 1998; Kennedy, 2001; Ceccarelli et al., 2006b). This approach leads to the automatic/mechanical adaptation of the robotic finger to the shape of the object seized. A classic example of a self-adaptive two-DOF finger driven by linkages and its closing sequence is illustrated in Fig. 1.

Self-adaptive mechanisms offer tremendous possibilities and have been known for decades but they have been kept in the shadows of the exponentially growing capability of digital architectures. These mechanical systems have been superseded by their electronic counterpart which provided better, more accurate, cheaper and faster means of performing computing tasks (Gosselin, 2006). However, the confinement of mechanical systems to the execution of motion under the control of a higher-level controller has to be considered as a design choice rather than an obligation. In this paper:

- the kinematic preshaping of self-adaptive fingers is defined,
- two examples are given: pinch and anthropomorphic with the two-phalanx class S finger,
- geometric parameters to achieve these preshapings are either selected or optimized,
- limitations and extensions of the proposed methodology are discussed.

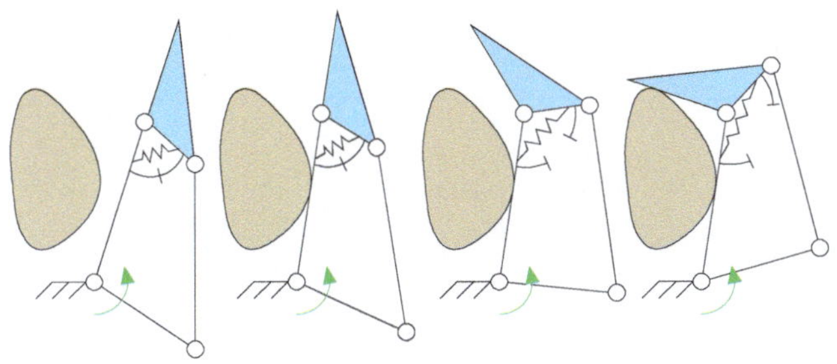

Figure 1. Closing sequence of a two-DOF self-adaptive finger.

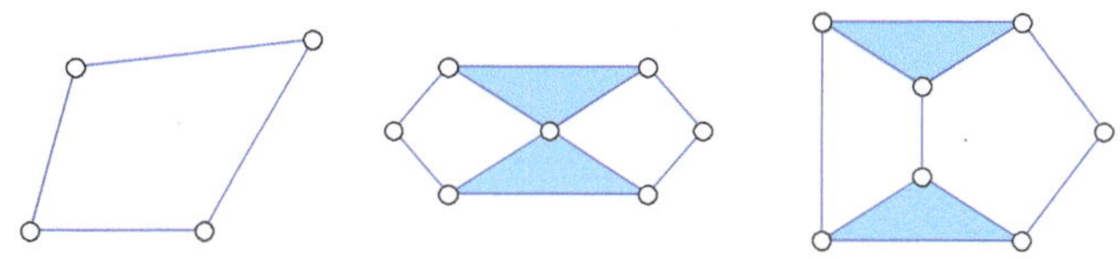

Figure 2. Four-bar, Watt's and Stephenson's one-DOF linkages.

2 Kinematics of triggered self-adaptive fingers

2.1 Introduction

The essence of self-adaptive fingers is twofold: a transmission linkage used to transmit actuation to the phalanges, and passive elements used to constrain the finger. In this paper, it is assumed that the finger is driven by a single actuator and unless otherwise specified the latter is located between the ground and the transmission linkage. Passive elements, as illustrated in Fig. 1, are key to prevent any undesired motions of the finger since the latter has n-DOF while having as few as one actuator. It is the aim of this paper to present how, by carefully selecting the type and location of these passive elements, several very interesting behaviours can be achieved. Many different types of passive elements may be considered (Birglen, 2009). Each of them applies a constraint (i.e. *a torque/force*) to a selected joint in the linkage. Springs and triggered springs are by far the most popular choice to achieve this goal. Using a triggered element in one of the joints of a self-adaptive finger, one locks this joint when no external constraint if applied to the mechanism, i.e. when no contact has yet occurred. Therefore, with a n-phalanx self-adaptive finger when $n-1$ triggered elements are used, one obtains a single DOF linkage. If a reasonable number (e.g. less than 6) of binary/ternary links is considered, the only possible one-DOF linkages (Birglen, 2009) are the four-bar, Watt's and Stephenson's linkages, illustrated in Fig. 2.

Hence, when no external contact has yet occurred, the self-adaptive finger considered behaves as one of these three linkages. Therefore, if triggered elements are used, a purely kinematic relationship exists between the angles of the phalanges θ_i and actuation, i.e. one has

$$g_i(\theta_i, \theta_a) = 0 \quad \text{with} \quad i = 1, \ldots, n \tag{1}$$

with a n-phalanx finger where θ_a is the angle associated with actuation. A trivial case occurs when the triggered element(s) is (are) located between the phalanges j and $j-1$ or between the ground and the proximal phalanx. Both cases respectively yields $\theta_j = K$ or $\theta_1 = L$ where K and L are constants. If a triggered element is located elsewhere in the mechanism, i.e. in the transmission linkage, the functions g_i can be obtained by writing the loop-closure equations of the linkage and a relationship $h(\theta_1, \ldots, \theta_n)$ can be obtained. With Watt's and Stephenson's linkages, this might be challenging as these equations are highly nonlinear and closed-form solutions have not been reported in the literature to the best of the author's knowledge (except in few cases with simplified geometries). Nevertheless, numerical values can be obtained. On the other hand, if the finger is reduced to a four-bar linkage, a closed-form solution exists. The position analysis of a four-bar linkage is actually a well-known textbook problem (Norton, 1992; McCarthy, 2000) and the relationships between the angles of the driven/driving cranks and the coupler can be obtained. Even if only the four-bar case is considered, a wide range of behaviours can be obtained as will be shown.

2.2 Kinematics of the class S

As an example, let us consider the linkage-driven two-phalanx class S finger (Birglen, 2009). A class S finger with n phalanges is constituted by a single loop linkage with $n+3$ joints. The phalanges and the ground to which they are attached are chosen as consecutive links of the mechanism and connected by revolute joints. Previously, only the case where this finger presented the particularity of having a ground link reduced to a single point was usually considered (cf. Fig. 1). This choice considerably simplifies mechanical design and the finger based on this simplification has been extensively investigated (Birglen et al., 2008). However, the more general case where two distinct joints constitute the ground link of the mechanism allows for much more freedom. It is worth mentioning that this class can be easily extended to an arbitrary number of phalanges. Indeed, the definition of this class is that the transmission mechanism is a simple planar motion generator (e.g. a RRR or RPR chain).

Triggered elements placed in any joint(s) of such a three-phalanx finger to reduce its DOF during the closing motion results in a four-bar linkage, as illustrated in Fig. 3. In this figure, the "lock" symbol has been used to indicate which joint holds the triggered element and the resulting four-bar linkage is highlighted in dashed lines. The kinematic reduction of the finger to a four-bar is true for any finger of this class when an appropriate number of triggered elements is used. A maximum number of two passive elements can be located in the joints of the transmission linkage since the latter is constituted by only three joints and the actuated joint

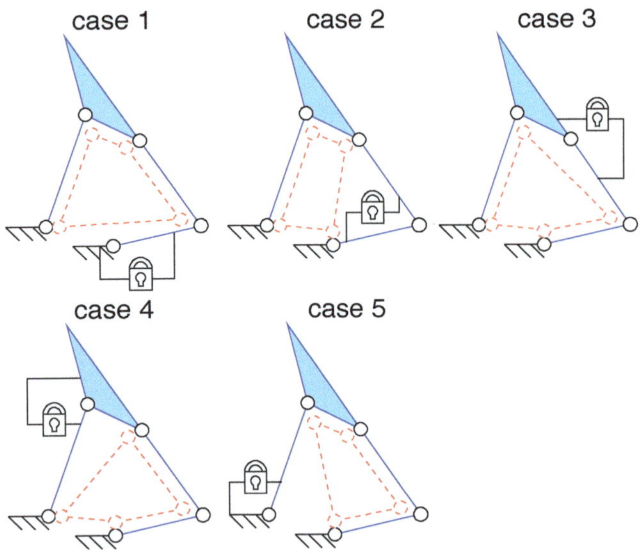

Figure 3. Kinematic reductions of the two-phalanx class S finger with one triggered element.

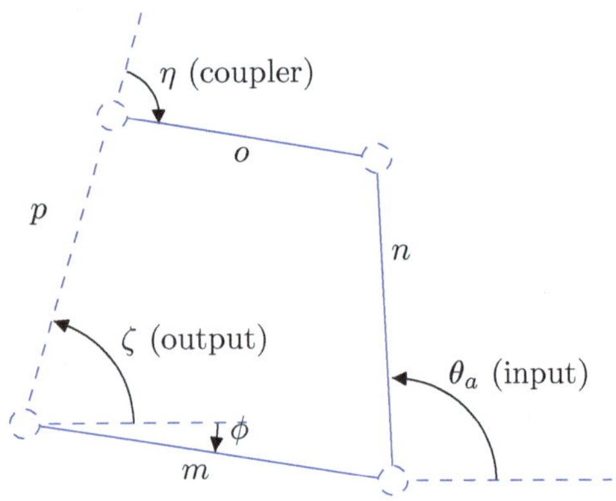

Figure 4. Modeling of a four-bar linkage.

is not a proper choice (although in certain cases it can hold a passive element, e.g. with compliant fingers). Therefore, if more passive elements are required they are necessarily located between the phalanges. Yet, the final one-DOF linkage remains a four-bar.

Let us consider the kinematic modeling of a general four-bar linkage illustrated in Fig. 4. Using the loop-closure equations, the relationships between the output/coupler, output/input and coupler/input angles are obtained. One has respectively:

$$p^2+o^2+m^2-n^2+2op\cos\eta-2mp\cos(\zeta-\phi) -2mo\cos(\phi-\zeta-\eta)=0, \quad (2)$$

$$m^2+n^2+p^2-o^2+2mn\cos(\phi-\theta_a) -2mp\cos(\phi-\zeta)-2np\cos(\theta_a-\zeta)=0, \quad (3)$$

$$m^2+n^2-o^2-p^2+2mn\cos(\theta_a-\phi)-2op\cos\eta=0. \quad (4)$$

With the two-phalanx finger and disregarding the trivial cases

Table 1. Associations of geometric parameters.

	m	n	o	p	ϕ	ζ	η
Case 1	r	b	c	l_1	$\alpha+\beta$	θ_1	$\theta_2-\psi$
Case 2	d	s	c	l_1	α	θ_1	$\theta_2-\psi$
Case 3	d	a	t	l_1	α	θ_1	$\theta_2-\psi-\gamma$

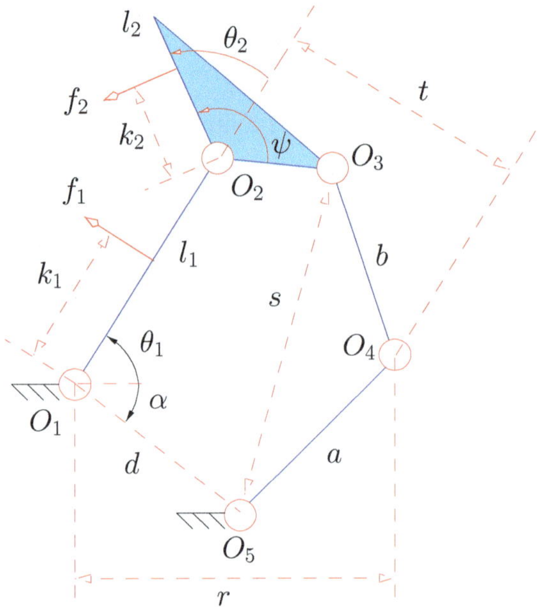

Figure 5. Details of the class S two-phalanx finger.

4 and 5, each location for the passive element yields a different association between the geometric parameters of the four-bar with these of the finger (illustrated in Fig. 5) as described in Table 1.

In this table, r is the distance between O_1 and O_4, s is the distance between O_3 and O_5 and t is the distance between O_2 and O_4. Each one of these distances is kept constant in its respective case. The angles β and γ are defined by

$$\beta=\arccos\left(\frac{r^2+d^2-a^2}{2rd}\right) \quad \text{and} \quad \gamma=\arccos\left(\frac{c^2+t^2-b^2}{2ct}\right) \quad (5)$$

Note that in case 1, actuation is assumed to be located in O_4.

Equation (2) is very important since it describes the relationship between the output phalanx angles of the mechanism directly, i.e. without θ_a. By carefully choosing the geometric parameters of the fingers and therefore the coefficients of this equation, it is possible to approximate several very interesting relationships between θ_1 and θ_2 for the finger, as will be shown.

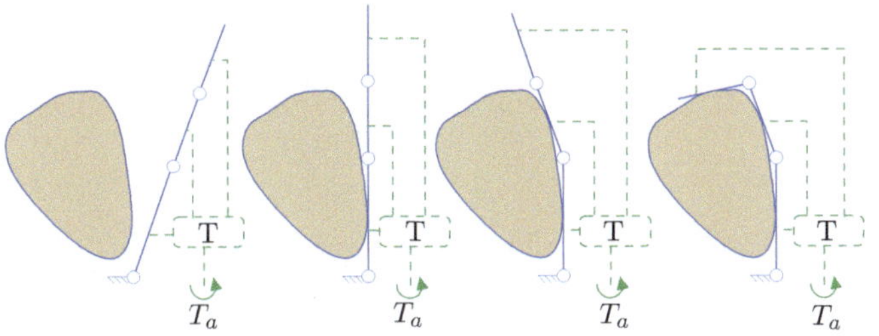

Figure 6. Closing sequence of a three-phalanx finger keeping its phalanges aligned.

3 Kinematic preshaping

3.1 Introduction and straight line preshaping

First, let us briefly recall the definition of preshaping: the preshaping of a finger – artificial or not – is defined as the set of geometric configurations this finger undertakes during the pregrasping phase, when no contact has yet occurred (Zhiwei et al., 2002; Supuk et al., 2005). With self-adaptive fingers, this issue has been overlooked in the past because of the limited set of available kinematic architectures known to be able to achieve self-adaption. Indeed, to the best of the author's knowledge, most prototypes of self-adaptive robotic fingers have springs located between their phalanges (or a kinematically equivalent design) in order for the latter to stay aligned during the pregrasping phase (as illustrated in Figs. 1 and 6). This preshaping can be done with any architectures of self-adaptive robotic finger since by definition, the phalanges are consecutive links of the mechanism and therefore, it is possible to locate spring(s) and mechanical limit(s) between them.

Recently, the Robotics Laboratory of Hiroshima University has proposed a method to optimize the preshaping of one self-adaptive robotic finger originally proposed in Hirose and Umetani (1978). This method was dealing with *dynamic preshaping* (Higashimori et al., 2005b), a particular aspect of the preshaping issue based on a dynamic analysis. Indeed, the proposed finger was subjected to very high accelerations (the authors claim to approach 100 g). With such high accelerations, dynamics becomes a critical issue and a specific dynamic preshaping, optimized to increase the probability for the finger to make contact with a spherical object using all its phalanges, was proposed based on a refined dynamic modeling. However, this method requires a thorough knowledge of the dynamics of the system which yields a significant burden to the designer and is unnecessarily complicated when low accelerations are considered. Furthermore, although this method is very accurate, results obtained are difficult to extend to other designs, for instance fingers based on linkages as in this paper.

On the other hand, *kinematic preshaping* as proposed here is defined as the preshaping of a self-adaptive finger from a quasi-static perspective, i.e. when dynamics is neglected in favor of kinematics. This approach does not require any knowledge of the dynamic of the system. Triggered elements are key to this method since they prevent motions under the assumption that dynamic forces are small. Continuous elements on the other hand are much more sensitive to these forces. Furthermore, triggered passive elements allow to consider simpler one-DOF linkages as discussed in Sect. 2 and illustrated in Fig. 2 instead of the complete mechanism. For instance, the preshaping of two-phalanx fingers such as the one described in Fig. 5 is the set of both phalanx joint angles when it is reduced to one of the linkages in Fig. 2. Disregarding again trivial cases where the triggered element is located in one of the phalanx joints, the preshaping can be described by a set of two equations, namely

$$\begin{cases} \theta_1 = g_1(\theta_a) \\ \theta_2 = g_2(\theta_a) \end{cases} . \tag{6}$$

Therefore, providing that g_1 and g_2 are invertible (at least locally), the preshaping can be described by a single equation, namely

$$g_1^{-1}(\theta_1) = g_2^{-1}(\theta_2) \Leftrightarrow h(\theta_1, \theta_2) = 0. \tag{7}$$

When the finger can be reduced to a four-bar linkage, Eq. (7) simply becomes Eq. (2). Following the same method, the preshaping of an n-phalanx finger can be described by a set of $n-1$ equations. This set of equations usually cannot match a particular desired preshaping except in some simple cases (e.g. in Sect. 3.2). However, the difference between the sets of equations describing the actual and desired preshaping can be characterized by a performance index and an optimization procedure can be used to maximize this index.

3.2 Pinch preshaping

An interesting feature of a few self-adaptive robotic fingers proposed in the literature (Gosselin and Laliberté, 1996; Bartholet, 1992; Bégoc et al., 2004) is the ability to maintain their distal phalanx perpendicular to the palm during their closing sequence. This kinematic property allows pinch grasps as illustrated in Fig. 7. This is very useful when the finger is seizing objects with linear edges. It should be noted that the shape adaptation capability must be kept to perform enveloping grasps if required (upper part of Fig. 7). The preshaping equation characteristic of a finger capable of pinch grasps is

$$\sum_{i=1}^{n} \theta_i = \frac{\pi}{2}. \tag{8}$$

In this case, this preshaping can be described by a single (and simple) equation. To achieve this feature, prototypes have been developped using various solutions such as a complex cam-linkage mechanism that can be disengaged (Bartholet, 1992), an additional linkage (Gosselin and Laliberté, 1996), or more recently, a combination of two interconnected

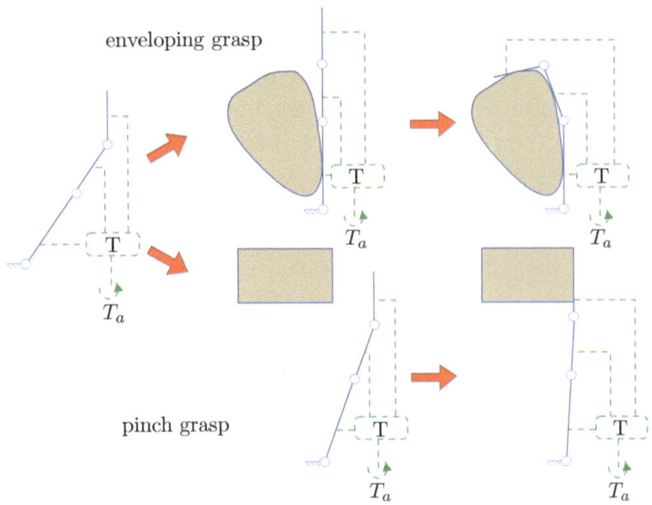

Figure 7. Two possible closing sequences of a three-phalanx finger designed to allow pinch grasps.

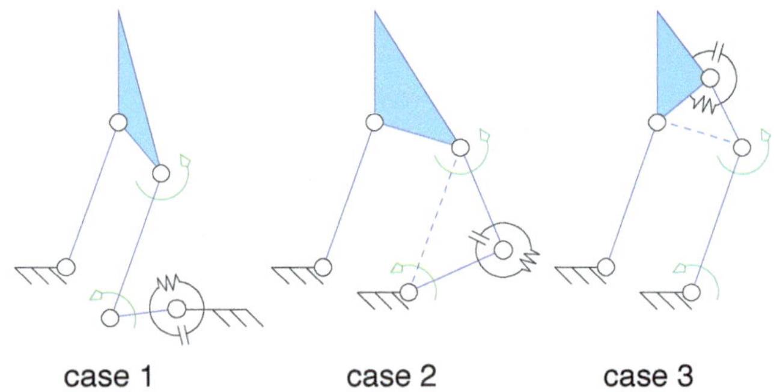

Figure 8. Designs of the two-phalanx class S architecture allowing pinch preshaping.

struts with non-reversible valves dubbed the *air parallelogram* (Bégoc et al., 2004). These solutions have been reported to work well but they all require an addition to the transmission linkage used for self-adaptation. However, if one carefully uses triggered elements in certain joint(s) of the mechanism, this behaviour can be closely approached without requiring any addition.

It is well known that a parallelogram linkage maintains its coupler link parallel to the ground. This property can be used in the design of self-adaptive robotic fingers with pinch preshaping. For instance, let us consider the mechanical architectures presented in Fig. 3. Non-trivial cases $(1-3)$ can be very easily kinematically reduced to a parallelogram if simple geometric conditions on the lengths of the links are satisfied. Namely, in case 1 they are $b = l_1$ and $r = c$. In case 2, they become $c = d$ and $s = l_1$. Finally, in case 3 one has $a = l_1$ and $t = d$. If these conditions are satisfied, Eq. (2) becomes

$$\theta_1 + \theta_2 = \phi - \eta + \theta_2. \tag{9}$$

This equation is valid with any of the three cases considered and it is trivial to show that the right hand side of the previous equation is constant, i.e. θ_2 is canceled in all cases by η and only the design parameters are left. Therefore, it is possible to choose these parameters to exactly match Eq. (8). Hence, for the distal phalanx of a two-phalanx class S finger to maintain a constant orientation with respect to the ground, one simply has to place a triggered element in one joint of the transmission linkage, adjust the lengths of the latter in order for the resulting one-DOF mechanism to be a parallelogram and choose the remaining geometric parameters to ensure that Eq. (9) matches Eq. (8). The method is illustrated in Fig. 8, the green circular arrows indicate possible actuation location. With these designs, the distal phalanx will remain perpendicular to the palm until a contact occurs. When this contact is established, the actuation torque will overcome the preloading of the triggered element and initiate the closing of the other phalanx. Hence, pinch preshaping is achieved without any additional mechanisms.

Nevertheless, if the contact occurs with the distal phalanx during this pregrasping phase, the resulting grasp stability must be studied to determine if the linear contact can be maintained. Indeed, with a contact on only one phalanx, i.e. one constraint, a two-DOF finger without additional mechanisms is not necessarily in static equilibrium as discussed in Birglen and Gosselin (2004). If this is not the case, contact with the whole line will be lost and a sliding motion of the finger in contact with only one vertex of the object will occur. This contact situation can lead to either a stable grasp or ejection (Birglen and Gosselin, 2006b). It should be noted that the parallelogram linkages proposed in Fig. 8 are known to achieve very poorly with respect to grasp stability. However, this is not an issue here because the parallelogram shape is only maintained during the preshaping phase and pinch grasps. It is automatically disengaged when a contact does not occur with the distal phalanx.

Let us consider the case 2 of Fig. 8. The equilibrium equation is defined as a set of contact locations on the distal phalanx leading to a static equilibrium of the finger with no contact on the proximal phalanx, i.e. the set of k_2 such as $f_1 = 0$ (cf. Fig 5). Note that this situation corresponds to the desired pinch grasp. The equilibrium equation of the finger discussed here is

$$e(\theta_1, \theta_2) = k_2 = h_4 \cos\theta_2 \tag{10}$$

where h_4 is the signed distance between O_2 and the intersection of lines (O_1O_2) and (O_3O_4). The equilibrium equation value e is a particular location of the contact force on the distal phalanx (not the value of the force but its location) and a function of h_4 which is itself a complex function of θ_1 and θ_2. The equilibrium equation defines a set of contact locations (a surface) in the grasp-state space $(\theta_1, \theta_2, k_2)$. Note that its value can be smaller than zero or greater than l_2, i.e. outside the physical limits of the phalanx. The equilibrium equation

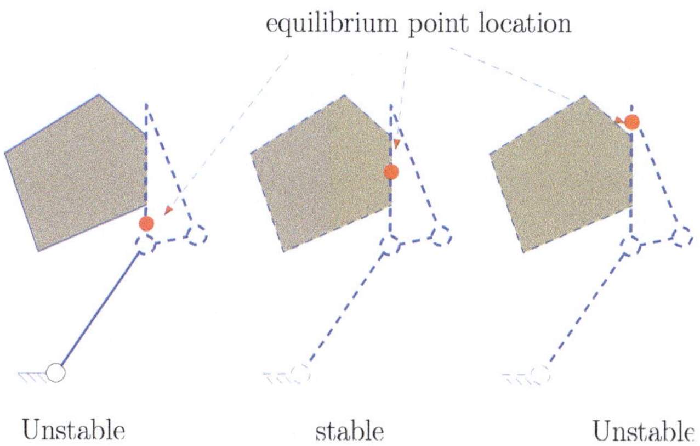

Figure 9. Stability of a planar pinch grasp depending on the equilibrium location.

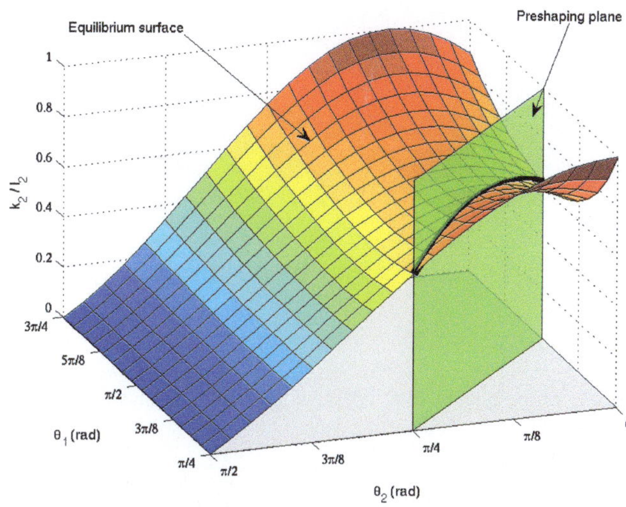

Figure 10. Equilibrium surface of the optimal design of two-phalanx finger with pinch preshaping.

is critical with respect to the grasp stability of a linear contact. It has also been demonstrated in Birglen and Gosselin (2006b) that a linear contact with the distal phalanx of a two-phalanx self-adaptive finger is stable if and only if the location of the equilibrium position is located between both vertices of the line (cf. Fig. 9). However, it should be noted that if this condition cannot be met, ejection will not necessarily occur. Indeed, the grasp can still be stable if $e(\theta_1,\theta_2) < l_2$ but with a contact maintained only with one of the edges of the object.

Therefore, the stability of a linear grasp cannot be guaranteed in all cases, since the object cannot be assumed to have a linear contact with the entire distal phalanx. However, in most common cases, the linear contact is long enough to exceed the physical limit of the distal phalanx, i.e. the rightmost case of Fig. 9 can be disregarded, and a compromise can be found. This compromise lies between the avoidance of ejection, favored by a small value of the equilibrium location, and the ability to perform pinch grasp, favored by a large value of the equilibrium location. This compromise can be part of the optimization procedure characterized by an index I_e^*:

$$I_e^* = \begin{cases} 0 & \text{if} \quad \max(e)_W > l_2 \\ \min(e/l_2)_{W^*} & \text{otherwise} \end{cases} \tag{11}$$

where W^* is the set of phalanx angles during the preshaping, i.e. the locus described by $\theta_1 + \theta_2 = \pi/2$. This performance index is to be maximized. It ensures that the equilibrium value e is the closest possible to l_2 over the workspace W^* (its minimal value is maximized) without being greater than l_2 (the score drops to zero).

If the lengths of the proximal and distal phalanges are respectively chosen as 1 and $l_1/\sqrt{2} = 0.707$. The first step of the optimization is to find out an initial estimate of the geometric parameters. This initial choice can be done very easily by choosing the parameters to form a parallelogram. Then a numerical optimization toolbox can be used to further refine the design. Taking into account the constraints mentioned

Table 2. Two-phalanx finger optimal geometric parameters.

ψ	a	b	c	d	α
95°	0.84	1.32	0.39	0.39	–5°

above, one can reduce the original set of 8 geometric parameters to only four: α, a, b and c in case 2 (since $\psi = \pi/2 - \alpha$ and $d = c$) to speed up calculations. An example of the result of such an optimization procedure is presented in Table 2 corresponding to the equilibrium locus illustrated in Fig. 10. Note that the set of angles θ_1 and θ_2 during the preshaping phase, denoted W^*, corresponds to a plane in the grasp-state space and the associated values of the equilibrium locations – maximized through the index I_e^* – are outlined in black in Fig. 10. The maximal value of the equilibrium equation over the workspace W does not exceed 1. Hence, no ejection occurs and the finger is always stable. The minimal value of the equilibrium equation in the preshaping phase W^* is 62.1% of l_2. This means that any linear contact with one vertex below $0.621 l_2$ and the other beyond the limit the phalanx is stable. This behaviour has been verified with a virtual prototype using a dynamic simulation package.

The procedure leading to this two-phalanx design can be adapted to a finger with three phalanges by taking into account the grasp stability and the expressions of the contact forces of three-phalanx fingers (Birglen and Gosselin, 2006a). Furthermore, it should be emphasized that this method is actually not limited to one class of kinematic architectures (the class S in this example). In order to maintain the desired orientation of the distal phalanx, one can use cascaded parallelograms as illustrated in Fig. 11. In this figure, three architectures from Birglen (2009) are illustrated

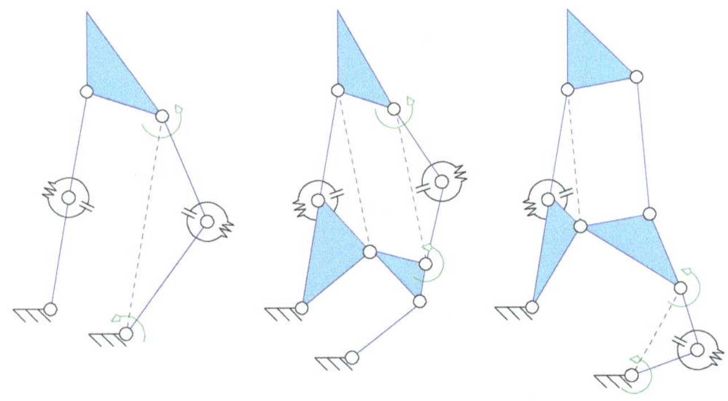

Figure 11. Three-phalanx self-adaptive fingers with pinch preshaping.

where two triggered elements are used to maintain the orientation of the distal phalanx during the closing preshaping. The green circular arrows indicate possible actuation location, but it should be noted that only one location has to be chosen, i.e. all three fingers have three DOF but can be driven by a single actuator.

3.3 Anthropomorphic preshaping

When designing prosthetic fingers, two approaches are possible. Either attention is paid to the closing motion of the prosthesis to mimic a human finger, i.e. an *anthropomorphic preshaping* (Figliolini and Ceccarelli, 2002; Ceccarelli et al., 2006b,a), or to the forces generated by the phalanges (de Visser and Herder, 2000). Since the latter approach does not require a particular closing motion, it has lead to the introduction of self-adaptive fingers in prosthetics (de Visser and Herder, 2000; Lotti and Vassura, 2002; Carrozza et al., 2004). However, it is possible to design an artificial finger that has an anthropomorphic closing motion while keeping the self-adaptation property as also very recently proposed in Wu et al. (2009). In Figliolini and Ceccarelli (2002), it is illustrated that in the absence of obstacles, the closing motion of a human finger exhibits approximately linear relationships between the rotations of the individual phalanges. Namely, one has

$$\theta_i = m_1(\theta_p - \pi/2) \quad \text{and} \quad \theta_d = m_2(\theta_p - \pi/2), \tag{12}$$

where $(\theta_p, \theta_i, \theta_d)$ are respectively the proximal, intermediate and distal relative angles between the phalanges. Numerical values reported in Figliolini and Ceccarelli (2002) are used in this paper, i.e. $(m_1, m_2) = (3.96, 2.37)$ with $90° \leq \theta_p \leq 197°$. Using again the two-phalanx finger illustrated in Fig. 5, the mechanism is reduced to a four-bar linkage by locking one joint to approximate the motion described by Eq. (12). If the resulting four-bar linkage is used to drive the last two phalanges of a prosthetic finger, i.e. $\theta_1 = \theta_i$ and $\theta_2 = \theta_d$, the preshaping equation is

$$\theta_2 = m\theta_1 \quad \text{with} \quad m = m_2/m_1. \tag{13}$$

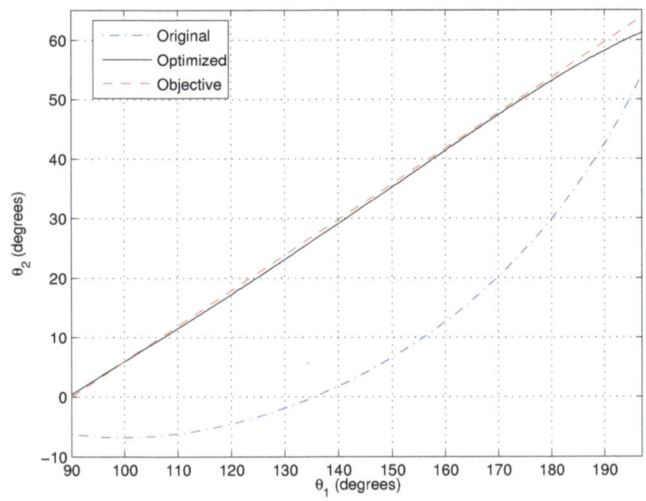

Figure 12. Comparison of preshaping motions.

Table 3. Two-phalanx finger second optimal geometric parameters.

ψ	a	b	c	d	α
119°	0.94	0.81	0.29	0.20	−149°

It is obvious that Eq. (2) is highly nonlinear and cannot usually be reduced into Eq. (13). However, a performance index measuring the closeness of the desired preshaping to the actual motion can be used, e.g.

$$I_a = \frac{1}{W^*} \int_{W^*} \frac{1}{1 + \left(\frac{f^*(\theta_1)}{m\theta_1} - 1\right)^2} d\theta_1 \tag{14}$$

where $f^*(\theta_1)$ is the solution of Eq. (2) for θ_2. Again, in the case of a four-bar linkage it is interesting to note that a closed-form expression of $f^*(\theta_1)$ exists (McCarthy, 2000) although its expression is impractical. The workspace of the finger is defined here as $W = \{90 < \theta_1 < 197°, 0 < \theta_2 < 64°\}$ to match the workspace proposed in Figliolini and Ceccarelli (2002) and I_a is therefore evaluated for $W^* \equiv 90 < \theta_1 < 197°$. If the joint locked during the preshaping is O_4 (case 2 in Fig. 8) the same optimization procedure discussed in the previous section yields the results presented in Table 3. Again, these parameters have been obtained with $l_1 = 1$ and $l_2 = l_1/\sqrt{2} = 0.707$. This time, the initial estimate has been chosen geometrically to have approximately the desired initial and final position, disregarding intermediate behaviour. A comparison of the initial, optimal and desired preshaping motions is illustrated in Fig. 12.

The equilibrium surface associated with this optimal set of geometric parameters is illustrated in Fig. 13. Equation (2) for any value of k_2 defines in this grasp-state space a preshaping surface illustrated in Fig. 13. Note that ideally, this surface should be a plane as the objective of the optimization

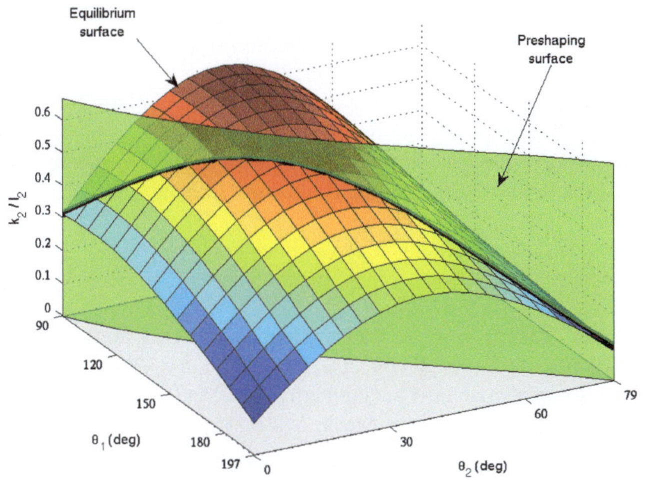

Figure 13. Equilibrium surface of the optimal design of two-phalanx finger with anthropomorphic preshaping.

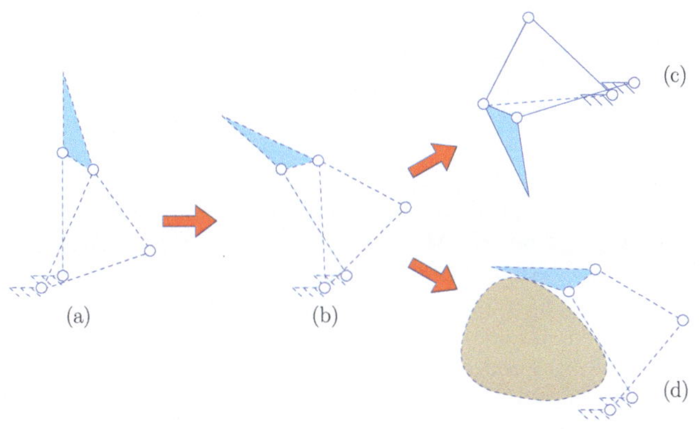

Figure 14. Possible closing motions of the optimized finger.

procedure was to approach Eq. (13). The finger built with these parameters and with a triggered passive element in O_4 will have a preshaping close to a human finger while keeping the self-adaptation property as illustrated in Fig. 14.

4 Conclusions

The issue of the kinematic preshaping of linkage-driven self-adaptive finger was discussed in this paper. A general method was proposed to achieve various preshapings based on using triggered passive elements. These elements reduce the number of DOF of self-adaptive fingers to one and allow to study the motion undergone by these fingers when no contact has occurred. Based on basic kinematic properties of four-bar linkages, designs of self-adaptive fingers were proposed that allow pinch grasps without requiring any additional mechanisms. The method was subsequently used in another example aiming at mimicking the closing motion of a human finger. These two examples of preshapings were demonstrated experimentally during the 2010 IFToMM/ASME International Workshop on Underactuated Grasping using fused deposition models illustrated in Fig. 15. Finally, only two different types of preshaping were discussed in this paper and with only one simple class of self-adaptive fingers. However, it is probable that any kind of motion is achievable using other architectures of self-adaptive fingers and that the presented preshapings can also be achieved by most of these other architectures.

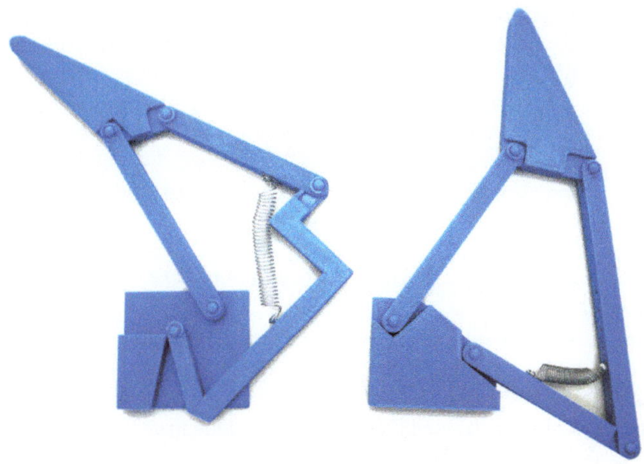

Figure 15. Prototypes of self-adaptive fingers with the discussed preshapings.

Acknowledgements. The financial support of the Natural Sciences and Engineering Research Council of Canada (NSERC) and the *Fonds québécois de la recherche sur la nature et les technologies* (FQRNT) is gratefully acknowledged.

Edited by: J. L. Herder
Reviewed by: two anonymous referees

References

Akin, D. L., Carignan, C. R., and Foster, A. W.: Development of a Four-Fingered Dexterous Robot End Effector For Space Operations, Proceedings of the 2002 IEEE International Conference on Robotics and Automation, Washington, DC, USA, 2302–2308, 2002.

Bartholet, S. J.: Reconfigurable End Effector, US Patent No. 5108140, 1992.

Bégoc, V., Krut, S., Dombre, E., Durand, C., and Pierrot, F.: Mechanical Design of a New Pneumatically Driven Underactuated Hand, Proceedings of the IEEE International Conference on Robotics and Automation, 927–933, Roma, Italy, 2004.

Bekey, G. A., Tomovic, R., and Zeljkovic, I.: Control Architecture for the Belgrade/USC Hand in Dextrous Robot Hands, Springer-Verlag, New-York, 1999.

Biagiotti, L., Melchiorri, C., and Vassura, G.: Control of a Robotic Gripper for Grasping Objects in No-Gravity Conditions, Proceedings of the 2001 IEEE International Conference on Robotics and Automation, 1427–1432, Seoul, Korea, 2001.

Bicchi, A.: Hands for Dexterous Manipulation and Powerful Grasping: A Difficult Road Towards Simplicity, IEEE T. Robotic. Autom., 16, 652–662, 2000.

Birglen, L.: The Synthesis of Linkage-Driven Self-Adaptive Fingers, ASME Journal of Mechanisms and Robotics, 1, 809–816, 2009.

Birglen, L. and Gosselin, C.: Kinetostatic Analysis of Underactuated Fingers, IEEE T. Robotic. Autom., 20, 211–221, 2004.

Birglen, L. and Gosselin, C.: Geometric Design of Three-Phalanx Underactuated Fingers, J. Mech. Des.-T. ASME, 128, 356–364, 2006a.

Birglen, L. and Gosselin, C.: Grasp-State Plane Analysis of Two-Phalanx Underactuated Fingers, Mech. Mach. Theory, 41, 807–822, 2006b.

Birglen, L., Laliberté, T., and Gosselin, C.: Underactuated Robotic Hands, Springer, New-York, 2008.

Carrozza, M. C., Suppo, C., Sebastiani, F., Massa, B., Vecchi, F., Lazzarini, R., Cutkosky, M. R., and Dario, P.: The SPRING Hand: Development of a Self-Adaptive Prosthesis for Restoring Natural Grasping, Auton. Robot., 16, 125–141, 2004.

Ceccarelli, M., Rodriguez, N. E. N., and Carbone, G.: Design and tests of a three finger hand with 1-DOF articulated fingers, Robotica, 24, 183–196, 2006a.

Ceccarelli, M., Tavolieri, C., and Lu, Z.: Design Considerations for Underactuated Grasp with a one D.O.F. Anthropomorphic Finger Mechanism, 2006 IEEE/RSJ International Conference on Intelligent Robots and Systems, 1611–1616, 2006b.

Crisman, J. D., Kanojia, C., and Zeid, I.: Graspar: A Flexible, Easily Controllable Robotic Hand, IEEE Robot. Autom. Mag., 32–38, 1996.

de Visser, H. and Herder, J. L.: Force Directed Design of a Voluntary Closing Hand Prosthesis, Proceedings of the 2000 ASME Design Engineering Technical Conferences, Baltimore, MD, USA, 1–8, 2000.

Figliolini, G. and Ceccarelli, M.: A Novel Articulated Mechanism Mimicking the Motion of Index Fingers, Robotica, 20, 13–22, 2002.

Gosselin, C.: Adaptive Robotic Mechanical Systems: A Design Paradigm, J. Mech. Des.-T. ASME, 128, 192–198, 2006.

Gosselin, C. and Laliberté, T.: Underactuated mechanical finger with return actuation, US Patent No. 5762390, 1996.

Higashimori, M., Kaneko, M., Namiki, A., and Ishikawa, M.: Design of the 100G Capturing Robot Based on Dynamic Preshaping, Int. Jo. Robot. Res., 24, 743–753, 2005a.

Higashimori, M., Nishio, S., and Kaneko, M.: Dynamic Preshaping Based Optimum Design for High Speed Capturing Robots, Proceedings of the 2005 IEEE/RSJ International Conference on Intelligent Robots and Systems, Edmonton, Canada, 913–918, 2005b.

Hirose, S. and Umetani, Y.: The Development of Soft Gripper for the Versatile Robot Hand, Mech. Mach. Theory, 13, 351–358, 1978.

Kennedy, B.: Three-Fingered Robot Hand With Self-Adjusting Grip, Nasa Tech Briefs, 25, 59, 2001.

Laliberté, T. and Gosselin, C.: Simulation and Design of Underactuated Mechanical Hands, Mech. Mach. Theory, 33, 39–57, 1998.

Lotti, F. and Vassura, G.: A Novel Approach to Mechanical Design of Articulated Fingers for Robotic Hands, Proceedings of the 2002 IEEE/RSJ International Conference on Intelligent Robots and System, Vol. 2, 1687–1692, 2002.

Massa, B., Roccella, S., Carrozza, M. C., and Dario, P.: Design and Development of an Underactuated Prosthetic Hand, Proceedings of the 2002 IEEE International Conference on Robotics and Automation, Washington, DC, USA, 3374–3379, 2002.

McCarthy, J. M.: Geometric Design of Linkages, Springer-Verlag, 2000.

Norton, R. L.: Design of Machinery, McGraw-Hill, Inc., 1992.

Supuk, T., Kodek, T., and Bajd, T.: Estimation of Hand Preshaping During Human Grasping, Med. Eng. Phys., 27, 790–797, 2005.

Ulrich, N. T.: Methods and Apparatus for Mechanically Intelligent Grasping, US Patent No. 4957320, 1988.

Wu, L., Carbone, G., and Ceccarelli, M.: Designing an underactuated mechanism for a 1 active DOF finger operation, Mech. Mach. Theory, 44, 336–348, doi:10.1016/j.mechmachtheory.2008.03.011, 2009.

Zhiwei, L., Ito, T., Sugimoto, N., Odashima, T., and Hosoe, S.: Virtual impedance control for preshaping of a robot hand, Proceedings of the 41st SICE Annual Conference, Osaka, Japan, 2058–2059, 2002.

12

Quasi-static modelling of compliant mechanisms: application to a 2-DOF underactuated finger

C. Quennouelle and C. Gosselin

Laboratoire de robotique, Département de génie mécanique, Université Laval, Québec, QC, Canada

Abstract. In this paper, the kinematostatic and the quasi-static models of parallel mechanisms are applied to underactuated mechanisms. Both models are extended to the cases for which the actuated joints are not kinematically independent, and for which the external loads are function of the configuration of the mechanism, the grasped object being considered as not perfectly rigid. An application to a 2-DOF underactuated compliant finger is then presented with details about the implementation of the kinematostatic and the quasi-static models. Finally, some numerical results are given that illustrate possible contributions of these models for the analysis and the control of underactuated mechanisms.

This paper was presented at the IFToMM/ASME International Workshop on Underactuated Grasping (UG2010), 19 August 2010, Montréal, Canada.

1 Introduction

Compliant mechanisms, i.e., mechanisms built with elastic joints, offer several advantages compared to conventional mechanisms: reduction of wear, clearance and backlash, compactness, no need for lubrication, simplified assembly, etc. However, due to their particular behaviour – namely the coupling between applied forces/torques and motion, and the parasitic deformations (a single joint can have several degrees of freedom (Su, 2009), the classical kinematic models are not sufficient to accurately describe them. Therefore, some models taking into account their specific behaviour have recently been developed. These models describe the configuration of the mechanisms (represented by a minimal set of kinematic parameters, i.e., the generalized coordinates) as a function of some external static parameters, such as the external loads and the commanded positions of the actuators (Sun et al., 1997; Carricato, 1998; Quennouelle and Gosselin, 2009). Due to the multiple degrees of freedom that can be required for an accurate modelling of a single compliant joint, any compliant mechanism can be considered as an underactuated mechanism, i.e., with fewer actuators than degrees of mobility. Therefore, the *kinematostatic model* (KSM) and the *quasi-static model* (QSM) of compliant mechanisms appear to be useful tools for the study of underactuated mechanisms (UMs) in general.

On the other hand, in the recent years, research initiatives on underactuated robotic hands led to a miniaturization of these systems and several interesting compliant prototypes have been proposed (Lotti et al., 2005; Boudreault and Gosselin, 2006; Doria and Birglen, 2009). For example, Fig. 1 shows a compliant gripper designed for surgical applications.

Indeed, the use of compliant joints is a very promising avenue for the delevopment of compact underactuated grippers, even if their theoretical analysis is still in its infancy (Birglen, 2006). Important challenges can be mentioned like stability of grasping (Birglen et al., 2008; Kragten et al., 2008) or force-isotropic properties (Krut, 2005), and the use of compliant joints greatly increases the complexity of the corresponding analyses.

Yet, the present paper does not address these issues. Its objective is merely to illustrate the use of the KSM and the QSM in the modelling of UMs – and in particular underactuated fingers, in order to show the possibilities offered by these models. It is argued that, with further investigations, these models – especially the generalized stiffness matrix (generalized SM) (Cho et al., 1989; Griffis and Duffy, 1991; Quennouelle and Gosselin, 2008) and the transmission matrix – can become valuable tools for the analysis and control of underactuated robotic hands.

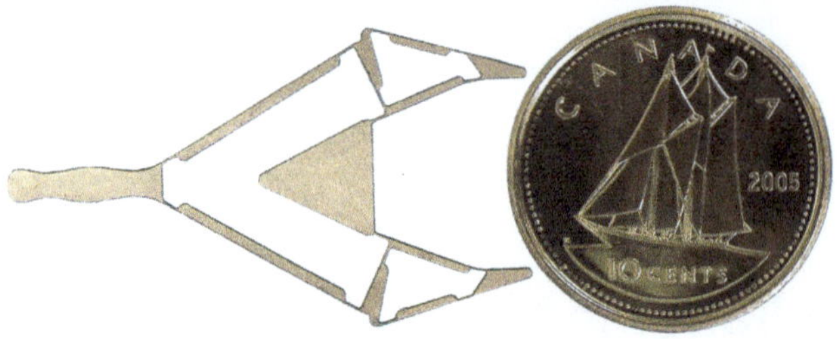

Figure 1. Sub-centimetre underactuated compliant gripper (Boudreault and Gosselin, 2006).

The first section of this paper deals with the theoretical formulation of the KSM and the QSM. The equivalence between the general formulation of the static equilibrium applied to UMs and the kinetostatic relation between the actuation torques and the contact forces used in the underactuation theory is demonstrated. For that, the models have been extended to the case of dependent actuated joints. Then, when all external forces are function of the configuration of the mechanism, the system {UM + grasped object} becomes equivalent to a fully actuated mechanism. In the second section, the KSM and the QSM are applied to a compliant 2-DOF underactuated finger. The implementation of the simulation is presented and some results are discussed. The contributions of the KSM and the QSM to this kind of application are underlined for analysis, design and control, respectively.

2 Kinematostatic modelling of underactuated mechanisms

2.1 Concept of kinematostatic modelling

2.1.1 Kinematic model

The kinematic constraints between the dependent coordinates λ, the pose of the end-effector platform x and the independent generalized coordinates ψ are written as $\mathcal{K}(\lambda,x,\psi)=\mathbf{0}$. By solving the latter, the kinematic model can be written as a function of the independent generalized coordinates: $x=\mathcal{P}(\psi)$ for the Cartesian pose and $\lambda=\mathcal{G}(\psi)$ for the dependent coordinates. The degree of mobility of the mechanism is noted $\mathfrak{M}$ and the number of actuators is noted $\mathfrak{A}$.

2.1.2 Static model

This model is less commonly used, (it is sometimes called *kinetostatic*). The static constraints between the external loads f, the commanded position of the actuators ϕ_0 and the configuration of the mechanism, represented by the generalized coordinates are written as $\mathcal{S}(\psi,f,\phi_0)=\mathbf{0}$. By solving these static constraints, the configuration of the mechanism can be written as a function of the external loads and the position of the actuators: $\psi=\mathcal{F}(f,\phi_0)$.

2.1.3 Kinematostatic model

By solving simultaneously the kinematic and the static constraints, the pose of the end-effector of a mechanism can be written as a function of the external parameters only. This is written as

$$x=\mathcal{M}(f,\phi_0)=\mathcal{P}(\mathcal{F}(f,\phi_0)). \tag{1}$$

2.2 General formulation of the static equilibrium

When the gravitational loads are not taken into account, the equation describing the static equilibrium in a parallel mechanism is written as

$$\mathcal{S}(\psi,f,\phi_0)=\boldsymbol{\tau}_\psi+\mathbf{G}^T\boldsymbol{\tau}_\lambda-\mathbf{J}^T f=\mathbf{0}, \tag{2}$$

where $\boldsymbol{\tau}_\psi$ and $\boldsymbol{\tau}_\lambda$ are the forces applied on the corresponding joints and $\mathbf{G}=d\lambda/d\psi$ and $\mathbf{J}=d\lambda/d\psi$ are the Jacobian matrices of the vectors of the dependent coordinates and of the pose of the end-effector with respect to the generalized coordinates.

2.3 Theory of underactuation

In Birglen et al. (2008), the relation between the contact force f_c of a grasping UM and the torque $\boldsymbol{\tau}_\phi$ provided by its actuators is expressed as

$$\mathbf{J}_c^\star f_c=\mathbf{T}_u^\star \boldsymbol{t}, \tag{3}$$

with $\mathbf{J}_c^\star=\mathbf{J}_c^T$, such that $\mathbf{J}_c^\star f_c$ corresponds to the projection of the effect of the contact forces, into the generalized coordinates domain; and $\mathbf{T}_u^\star=\mathbf{T}_u^{-T}$, such that $\mathbf{T}_u^\star \boldsymbol{t}$ corresponds to the projection of the torques provided by the actuator and the springs in the independent joints, into the generalized coordinates domain.

With the following assumptions, the general formulation of the static equilibrium (Eq. 2) can be brought to the particular case given in Eq. (3):

- The vector containing the position of all contact points is noted $x_c^T=[x_p^T;x_q^T;\ldots]$, and the corresponding Jacobian matrix is : $\mathbf{J}_c^T=[\mathbf{J}_p^T;\mathbf{J}_q^T;\ldots]$.
- The vector containing the components of all contact forces is $f_c^T=[f_p^T;f_q^T;\ldots]$. Therefore, the generalized forces due to the contact forces are written as $-\mathbf{J}_c^T f_c = -\mathbf{J}_p^T f_p-\mathbf{J}_q^T f_q-\ldots$
- The actuated joints are not considered as independent coordinates, i.e., the corresponding coordinates $\boldsymbol{\phi}$ are components of vector λ. The unactuated dependent coordinates are noted λ_N. They have zero stiffness and therefore cannot resist or transmit any force in a static configuration. One has $\boldsymbol{\tau}_\lambda^T=[\boldsymbol{\tau}_\phi^T;\mathbf{0}^T]$.

- Matrix $\mathbf{H}$ is composed of the rows of $\mathbf{G}$ corresponding to the actuated dependent joints and $\mathbf{N}$ is composed of the rows of $\mathbf{G}$ corresponding to the unactuated dependent joints, such that $\mathbf{G}^T\boldsymbol{\tau}_\lambda = \mathbf{H}^T\boldsymbol{\tau}_\phi + \mathbf{N}^T\mathbf{0} = \mathbf{H}^T\boldsymbol{\tau}_\phi$.
- Among the $\mathfrak{M}$ generalized coordinates, $\mathfrak{A}$ have a zero stiffness and the $\mathfrak{L}$ others, noted $\boldsymbol{\psi}_K$, have a finite stiffness ($\mathfrak{L} = \mathfrak{M} - \mathfrak{A}$). By defining $\mathbf{L} = [\mathbf{0}_{\mathfrak{L}\times\mathfrak{A}}; \mathbf{1}_{\mathfrak{L}\times\mathfrak{L}}]$, the torques applied to the independent joints are written as $\boldsymbol{\tau}_\psi = [\mathbf{0}_{\mathfrak{A}}^T; \boldsymbol{\tau}_{\psi_K}^T]^T = \mathbf{L}^T\boldsymbol{\tau}_{\psi_K}$.
- The vector of joint torques, used in Birglen et al. (2008), is $\boldsymbol{t}^T = [\boldsymbol{\tau}_\phi^T; \boldsymbol{\tau}_{\psi_K}^T]$.
- In Birglen et al. (2008), matrix $\mathbf{T}_u$ is the Jacobian of vector $\boldsymbol{\psi}$ with respect to vector $[\boldsymbol{\phi}^T; \boldsymbol{\psi}_K^T]^T$. It inverse is given by

$$\mathbf{T}_u^{-T} = [(\frac{d\boldsymbol{\phi}}{d\boldsymbol{\psi}})^T; (\frac{d\boldsymbol{\psi}_K}{d\boldsymbol{\psi}})^T] = \left[\mathbf{H}^T; \mathbf{L}^T\right]. \tag{4}$$

Matrices $\mathbf{H}$ and $\mathbf{l}$ being respectively of dimension $\mathfrak{M}\times\mathfrak{A}$ and $\mathfrak{M}\times\mathfrak{L}$, matrix $\mathbf{T}_u^{-1}$ is a square and invertible $\mathfrak{M}\times\mathfrak{M}$ matrix.

Thus, the static equilibrium (Eq. 2) is equivalent to

$$\mathcal{S} = \mathbf{L}^T\boldsymbol{\tau}_{\psi_K} + \mathbf{H}^T\boldsymbol{\tau}_\phi - \mathbf{J}_c^T\boldsymbol{f}_c = \mathbf{T}_u^{-T}\boldsymbol{t} - \mathbf{J}_c^T\boldsymbol{f}_c = \mathbf{0}. \tag{5}$$

This confirms that Eq. (3) is a particular case of the general formulation of the static equilibrium of a parallel mechanism.

2.4 Concept of underactuation

For the reasons mentioned above, the KSM of an UM can be easily established. In the case of a grasping mechanism, there is no pose of end-effector platform to be calculated, but a configuration $\boldsymbol{\psi}$ to be determined as a function of the positions of the actuators and of the contact forces, written as $\boldsymbol{\psi} = \mathcal{F}(\boldsymbol{\phi}_0, \boldsymbol{f}_c)$.

In practice, such a mechanism is called *underactuated* because there are $\mathfrak{M}$ degrees of mobility for only $\mathfrak{A}$ actuators. But in this theoretical KSM, it can be seen that, not only the actuators have an impact on the configuration of the mechanism, but also the $3\times n$ components of the contact forces.

If these external forces are considered as additional actuators acting on the mechanism, and if the $3\times n$ components of the external forces are reduced to only $\mathfrak{L}$ independent components chosen among them and noted $\boldsymbol{f}$, the KSM becomes the one of a classical fully actuated mechanism: $\boldsymbol{\psi} = \mathcal{F}(\boldsymbol{\phi}_0, \boldsymbol{f})$.

2.5 Quasi-static modelling

2.5.1 Linear relationship

The QSM of a parallel mechanism provides linear relationships between the infinitesimal variation of the configuration $\boldsymbol{\psi}$ and that of the external parameters ($\boldsymbol{\phi}_0$ and $\boldsymbol{f}$) (Quennouelle and Gosselin, 2009). This is written as

$$d\boldsymbol{\psi} = \mathbf{T}_0 d\boldsymbol{\phi}_0 + \mathbf{K}_{M_0}^{-1}\mathbf{J}^T d\boldsymbol{f}, \tag{6}$$

where $\mathbf{T}_0$ is the transmission matrix and $\mathbf{K}_{M_0}$ is the generalized SM where the external loads are independent from the configuration of the mechanism (Quennouelle and Gosselin, 2008). One has

$$\mathbf{K}_{M_0} = \mathbf{K}_\psi + \mathbf{K}_I + \mathbf{K}_{E_0},$$

with $\mathbf{K}_\psi$ the SM due to the independent joints, $\mathbf{K}_I$ the internal SM (due to the stiffness of the dependent joints and the internal forces) and $\mathbf{K}_{E_0}$ is the SM due to the external loads (independent from $\boldsymbol{\psi}$).

2.5.2 Transmission matrix

The transmission matrix is a $\mathfrak{M}\times\mathfrak{A}$ matrix that maps the effects of a variation in the commanded value of the actuators onto the generalized coordinates, taking into account the effects of the compliance due to the contact forces. It is different from the kinematic "transmission matrix" used in the theory of underactuation (Birglen et al., 2008), because only the former considers the compliance of the joints.

In Quennouelle and Gosselin (2009), only the case for which the actuated coordinates are independent is addressed. In the present case, the actuated joints are considered as dependent coordinates and the transmission matrix $\mathbf{T}_0$ is then defined as

$$\mathbf{T}_0 = \frac{\partial\boldsymbol{\psi}}{\partial\boldsymbol{\phi}_0} = -\left(\frac{d\mathcal{S}}{d\boldsymbol{\psi}}\right)^{-1}\frac{d\mathcal{S}}{d\boldsymbol{\phi}_0} = \mathbf{K}_{M_0}^{-1}\mathbf{H}^T\mathbf{K}_\phi. \tag{7}$$

When the actuated joints are dependent in Eq. 2), only the term $\mathbf{G}^T\boldsymbol{\tau}_\lambda$, and in particular $\mathbf{H}^T\boldsymbol{\tau}_\phi$, are function of $\boldsymbol{\phi}_0$. One has

$$\frac{d\mathcal{S}}{d\boldsymbol{\phi}_0} = \frac{d\left(\mathbf{H}^T\boldsymbol{\tau}_\phi\right)}{d\boldsymbol{\phi}_0} = \left(\frac{d\mathbf{H}^T}{d\boldsymbol{\phi}_0}\boldsymbol{\tau}_\phi\right) - \mathbf{H}^T\mathbf{K}_\phi = -\mathbf{H}^T\mathbf{K}_\phi, \tag{8}$$

with $(d\mathbf{H}^T/d\boldsymbol{\phi}_0)\boldsymbol{\tau}_\phi = \mathbf{0}$ because $\boldsymbol{\psi}$, the configuration of the mechanism, is not directly (kinematically) function of $\boldsymbol{\phi}_0$.

2.5.3 Effect of the external loads

Independent: In an UM used as a gripper, the external loads are the contact forces. When these forces are considered as independent from the configuration of the mechanism, they are taken into account through the $\mathbf{K}_{E_0}$ matrix as follows:

$$\mathbf{K}_{E_0} = -(\frac{d\mathbf{J}^T}{d\boldsymbol{\psi}})\boldsymbol{f}. \tag{9}$$

Therefore a contact force on one phalanx can modify the stiffness of the generalized coordinates and, in turn, the variation of the configuration of the mechanism due to a variation of the external parameters ($\boldsymbol{\phi}_0$ and $\boldsymbol{f}$).

Dependent: However, in a context of grasping, the contact forces are generally not independent from the configuration of the mechanism, since they are mainly generated by the pressure of the phalanges on the grasped object. Therefore, these contact forces are in turn, function of the only truly independent external parameters, namely, the commanded position of the actuators. The contact force applied on an object is related to the position of the phalanx through a SM, $\mathbf{K}_c = \text{diag}(\mathbf{K}_p;\mathbf{K}_q;\ldots)$, each $\mathbf{K}_i$ being the SM of the object at the ith contact point. The variation of the force is given by $d\boldsymbol{f} = \mathbf{K}_c d\boldsymbol{x} = \mathbf{K}_c\mathbf{J}d\boldsymbol{\psi}$.

Then, the external SM $\mathbf{K}_E$ is expressed as

$$\mathbf{K}_E = -(\frac{d\mathbf{J}^T}{d\boldsymbol{\psi}})\boldsymbol{f} - \mathbf{J}^T\mathbf{K}_c\mathbf{J} = \mathbf{K}_{E_0} - \mathbf{J}^T\mathbf{K}_c\mathbf{J}. \quad (10)$$

A generalized SM $\mathbf{K}_M$ can be defined that includes this stiffness due to the variation of the external loads. One has

$$\mathbf{K}_M = \mathbf{K}_{M0} - \mathbf{J}^T\mathbf{K}_c\mathbf{J}.$$

Thus, the QSM becomes simply function of the $\mathfrak{A}$ commanded positions of the actuators. One has

$$d\boldsymbol{\psi} = \mathbf{K}_M^{-1}\mathbf{K}_\phi^\star d\boldsymbol{\phi}_0 = \mathbf{T}d\boldsymbol{\phi}_0. \quad (11)$$

In the latter equation, $\mathbf{T}(= d\boldsymbol{\psi}/d\boldsymbol{\phi}_0)$ is no longer a partial derivative. This is equivalent to saying that when the mechanism is grasping a compliant object, the degree of mobility is reduced from $\mathfrak{M}$ to $\mathfrak{A}$.

3 Application to a 2-DOF compliant finger

3.1 2-DOF compliant underactuated finger

The KSM and the QSM of compliant mechanisms are applied to a compliant finger shown in Fig. 2a (more details about its design are presented in Boudreault and Gosselin (2006)).

A schematic representation of the mechanism is given in Fig. 2b. The lengths (in mm) of the links are $e = 6$, $h = 54$, $g = 10$, $b = 8$, $c = 54$, $n = 30$, $l_x = 25$, $l_y = 20$, $d_1 = 10$ and $d_2 = 9$. The angle $\phi = 134.5$ deg. The actuator is not modelled as a simple torque T_ρ as it is often the case in the literature, but – according to the reality – as a prismatic compliant actuator of variable commanded length ρ_0 and of actual length ρ. The contact forces $\boldsymbol{f}_p$ and $\boldsymbol{f}_q$ on the two phalanges are respectively applied at a distance l_p and l_q of the revolute joints θ_p and θ_q.

3.2 Symbolic modelling

The mechanism has eight joints with coordinates: $\theta_p, \theta_1, \theta_2, \theta_3, \theta_4, \theta_5, \theta_6$ and ρ (θ_q being rigidly linked to θ_2) and two planar kinematic loops, thus it has two degrees of mobility. The chosen independent generalized coordinates are $\boldsymbol{\psi}^T = \left[\theta_p;\theta_1\right]$ and the dependent ones are assembled in $\boldsymbol{\lambda}^T = [\theta_2;\theta_3;\theta_4]$ for the four-bar linkage and ρ for the actuated loop (actually joints θ_5 and θ_6 are assumed to be conventional, i.e., with a zero stiffness, and therefore their determination is not useful in the KSM). The position of the final phalanx is given by $\theta_q = \theta_2 + \phi - \pi$.

(a) Prototype with joints made of nitinol (Boudreault and Gosselin, 2006).

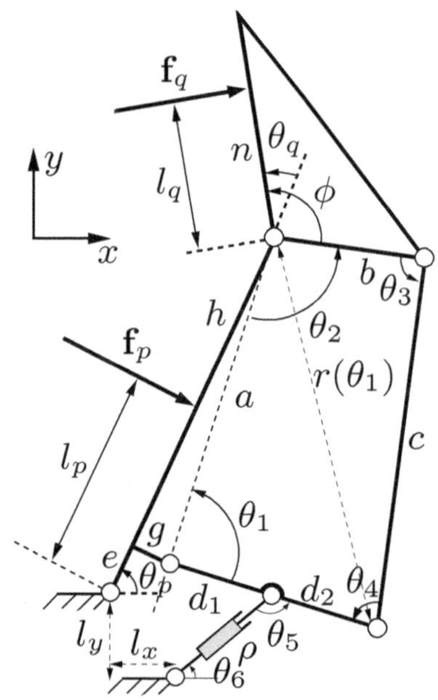

(b) Geometry of the 2-DOF underactuated finger.

Figure 2. 2-DOF compliant underactuated finger.

3.2.1 Kinematic model of the four-bar linkage

This passive loop defines three kinematic constraints that have to be satisfied by the dependent coordinates. Using two constant lengths $a = \sqrt{h^2 + g^2}$ and $d = d_1 + d_2$ and a variable length $r = \sqrt{a^2 + d^2 - 2ad\cos\theta_1}$, the solution to these constraints, corresponding to the configuration shown in Fig. 2b, is

$$\begin{cases} \theta_2 = \arccos\left(\dfrac{r^2+a^2-d^2}{2ar}\right) + \arccos\left(\dfrac{b^2+r^2-c^2}{2br}\right), \\ \theta_3 = \arccos\left(\dfrac{b^2+c^2-r^2}{2bc}\right), \\ \theta_4 = \arccos\left(\dfrac{r^2+c^2-b^2}{2cr}\right) + \arccos\left(\dfrac{d^2+r^2-a^2}{2dr}\right). \end{cases} \quad (12)$$

A simple analysis of the workspace of the four-bar mechanism, in the configuration shown in Fig. 2b, reveals that these relations are only valid for $0.9818 \le \theta_1 \le 1.8177$.

Instantaneous kinematic constraints: The differentiation of the kinematic constraints with respect to the joint coordinates can be written as $\mathbf{S}_{\theta 1} d\theta_1 = -\mathbf{S}_\lambda d\boldsymbol{\lambda}$, with

$$\mathbf{S}_\psi = \begin{bmatrix} 0 & -as_1+bs_{12}-cs_{123} \\ 0 & ac_1-bc_{12}+cc_{123} \\ 0 & 1 \end{bmatrix}; \mathbf{S}_\lambda = \begin{bmatrix} bs_{12}-cs_{123} & -cs_{123} & 0 \\ -bc_{12}+cc_{123} & cc_{123} & 0 \\ 1 & 1 & 1 \end{bmatrix} \quad (13)$$

where $s_{i...k}$ stands for $\sin(\theta_i+\cdots+\theta_k)$ (c for cos respectively). From the latter equation, matrix $\mathbf{G}$ is obtained as $\mathbf{G} = -\mathbf{S}_\lambda^{-1}\mathbf{S}_\psi$.

3.2.2 Kinematic model of the actuation loop

The actuated loop is shown in details in Fig. 3a and the following constant geometric parameters are introduced: $l = \sqrt{l_x^2+l_y^2}$, $r_1 = \sqrt{e^2+g^2}$, $\alpha = \arctan(g/e)$, $\beta = \pi - \alpha - \arctan(g/h)$, $\gamma = (\pi/2) - \arctan(l_x/l_y)$. The finger is actuated by a prismatic actuator that moves the link d. This actuator has a finite stiffness, therefore it is modelled as an ideal prismatic actuator ρ_0 connected in series with a linear spring of stiffness k_ρ. The commanded position is ρ_0 and the actual position of this compliant actuator is ρ.

By satisfying the loop-closure constraints and introducing $\theta'_p = \theta_p - \alpha + \gamma$ and $\theta'_1 = \theta_1 - \beta$, the actuated coordinate ρ can be written as a function of the generalized coordinates:

$$\rho = \sqrt{r_1^2 + d_1^2 + l^2 + 2d_1 r_1 c'_1 - 2r_1 l c'_p - 2d_1 l \cos\left(\theta'_p - \theta'_1\right)}. \quad (14)$$

And the Jacobian matrix $\mathbf{H}$ can be calculated as

$$\mathbf{H}^T = \begin{bmatrix} \dfrac{\partial \rho}{\partial \theta_p} \\ \dfrac{\partial \rho}{\partial \theta_1} \end{bmatrix} = \begin{bmatrix} \dfrac{l}{\rho}\left(r_1 \sin\theta'_p + d_1 \sin\left(\theta'_p - \theta'_1\right)\right) \\ -\dfrac{d_1}{\rho}\left(r_1 \sin\theta'_1 + l \sin\left(\theta'_p - \theta'_1\right)\right) \end{bmatrix}. \quad (15)$$

3.2.3 Contact points

In this paper, it is assumed that the grasped object is fixed, such that the effect of the force applied by the finger is to deform the object and not to displace it. It is also assumed that the first contact between the object and the finger is on the first phalanx, and then the second contact point is on the second phalanx (see Fig. 3b). The design of the finger and the position of the second contact point are chosen to prevent any ejection, i.e., the first contact cannot be lost when the second contact is established.

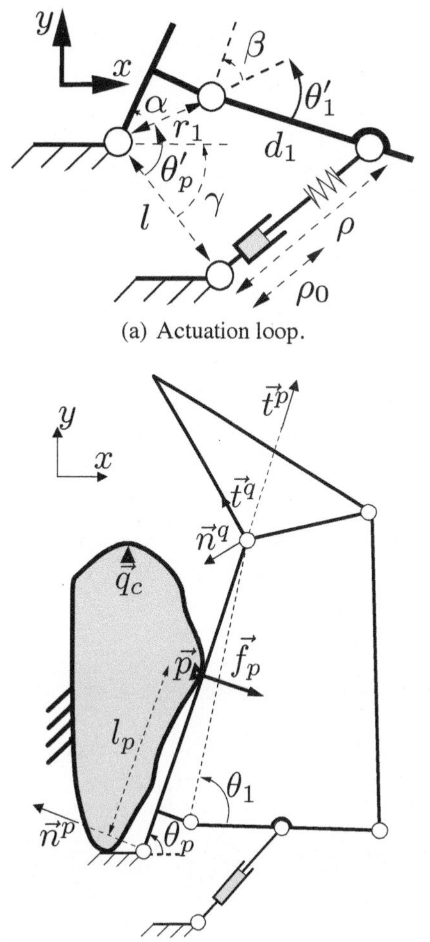
(a) Actuation loop.

(b) Finger grasping a compliant object.

Figure 3. 2-DOF compliant underactuated finger.

The first contact is established in $\boldsymbol{p}_c = \left[x_{pc}; y_{pc}\right]^T$ and the second one in $\boldsymbol{q}_c = \left[x_{qc}; y_{qc}\right]^T$. After the deformation of the object, the actual positions of the contact points become $\boldsymbol{x}_p$ and $\boldsymbol{x}_q$ and the corresponding deformations are noted $\delta\boldsymbol{x}_p = \boldsymbol{x}_p - \boldsymbol{p}_c$ and $\delta\boldsymbol{x}_q = \boldsymbol{x}_q - \boldsymbol{q}_c$. The contact points correspond to the following particular configurations of the mechanism:

- the first contact is established when θ_p is greater than $\theta_{pc} = \arctan\left(y_{pc}, x_{pc}\right)$,
- then, the second contact is established when θ_q is greater than $\theta_{qc} = \arctan\left(y_{qc} - ms_p, x_{qc} - mc_p\right)$.

Finally, $\delta\boldsymbol{x}_p$ and $\delta\boldsymbol{x}_q$ are equal to $\mathbf{0}$ when the corresponding contact is not established.

Assuming that the position of a contact point relatively to the corresponding phalanx is constant, i.e., l_p and l_q are constant, the position of the contact points are given by

$$\boldsymbol{x}_p = \begin{bmatrix} l_p c_p \\ l_p s_p \end{bmatrix} \text{ and } \boldsymbol{x}_q = \begin{bmatrix} mc_p + l_q c_{pq} \\ ms_p + l_q s_{pq} \end{bmatrix}, \tag{16}$$

with $m = h + e$, $l_q = \sqrt{\left(x_{qc} - mc_p^q\right)^2 + \left(y_{qc} - ms_p^q\right)^2}$ and $l_p = \sqrt{x_{pc}^2 + y_{pc}^2}$, θ_p^q is the value of θ_p when the second contact is established. The mapping between the global frame and the local frames $(\boldsymbol{t}^i, \boldsymbol{n}^i)$ attached to each phalanx is performed through rotation matrices $\mathbf{Q}_i$, written as

$$\mathbf{Q}_p = \begin{bmatrix} c_p & s_p \\ -s_p & c_p \end{bmatrix} \text{ and } \mathbf{Q}_q = \begin{bmatrix} c_{pq} & s_{pq} \\ -s_{pq} & c_{pq} \end{bmatrix} \tag{17}$$

In these frames, the contact points are given by $\boldsymbol{x}_p^p = \mathbf{Q}_p \boldsymbol{x}_p = [l_p; 0]^T$ and $\boldsymbol{x}_q^q = [mc_p + l_q; -ms_p]^T$, and their corresponding Jacobian matrices are given by

$$\mathbf{J}_p^p = \mathbf{Q}_p \frac{d\boldsymbol{x}_p}{d\psi} = \begin{bmatrix} 0 & 0 \\ l_p & 0 \end{bmatrix} \text{ and } \mathbf{J}_q^q = \begin{bmatrix} ms_q & 0 \\ mc_q + l_q & l_q\eta \end{bmatrix}, \tag{18}$$

where η relates the variation of angle θ_q to the variation of θ_1. It is written as $\eta = d\theta_q / d\theta_1 = (as_{23} - bs_3)/bs_3$.

3.2.4 Static model

Torques/forces in the joints: In this compliant mechanism actuated by a compliant actuator, all the joints can be subjected to a torque/force, even in a static configuration. Since, the stiffness of the joints is supposed to be constant, these torques/forces are written $\boldsymbol{\tau}_\nu = \mathbf{K}_\nu(\nu - \nu_0)$, (for $\nu = \psi, \lambda, \rho$). Matrices $\mathbf{K}_\nu$ is diagonal and contains the stiffnesses of the corresponding joints, ν_0 is the vector composed by the undeformed value of the joints.

Contact forces: Since the ceased object is elastic, the contact forces can be written as a function of the contact points and in turn, of the mechanism configuration. Both contact forces are assumed to be proportional to the deformation of the object, i.e., the latter is supposed to have a constant and isotropic SM (this assumption being reasonable for relatively small deformations). Obviously the contact force is zero when there is no contact (so is $\delta\boldsymbol{x}_i$). Each force is expressed in the local frame as

$$\boldsymbol{f}_i^i = -\mathbf{K}_i \delta \boldsymbol{x}_i^i = -\mathbf{K}_i \mathbf{Q}_i \delta \boldsymbol{x}_i = \begin{bmatrix} -k_i(\; c_i\delta x_i + s_i \delta y_i) \\ -k_i(-s_i \delta x_i + c_i \delta y_i) \end{bmatrix} = \begin{bmatrix} f_{ti} \\ f_{ni} \end{bmatrix}. \tag{19}$$

In order to ease a future study of the stability of the grasp, the tangential components of all forces can be assembled in a vector $\boldsymbol{f}_t$ and the normal components in a vector $\boldsymbol{f}_n$ (see Birglen et al., 2008). One has $\boldsymbol{f}_t = \begin{bmatrix} f_{tp}; & f_{tq} \end{bmatrix}^T$ and $\boldsymbol{f}_n = \begin{bmatrix} f_{np}; & f_{nq} \end{bmatrix}^T$.

The rows of the Jacobian matrices $\mathbf{J}_p^p$ and $\mathbf{J}_q^q$ can be reassembled into two matrices $\mathbf{J}_t$ and $\mathbf{J}_n$ (respectively named $\mathbf{J}_2$ and $\mathbf{J}_1$ in Birglen et al. (2008)). These matrices are

$$\mathbf{J}_t = \begin{bmatrix} 0 & 0 \\ ms_q & 0 \end{bmatrix} \text{ and } \mathbf{J}_n = \begin{bmatrix} l_p & 0 \\ mc_q + l_q & l_q\eta \end{bmatrix}. \tag{20}$$

Static equilibrium: In this finger, the static equilibrium of the mechanism is written as

$$\mathcal{S}(\psi, \rho_0) = \boldsymbol{\tau}_\psi + \mathbf{H}^T \boldsymbol{\tau}_\rho + \mathbf{G}^T \boldsymbol{\tau}_\lambda - \mathbf{J}_n^T \boldsymbol{f}_n - \mathbf{J}_t^T \boldsymbol{f}_t = \mathbf{0}. \tag{21}$$

Ejection of the finger: The static equilibrium given in Eq. (21) can only be reached for positive values of δn_i, i.e., negatives values of the normal contact forces ($f_{ni} = -k_i \delta n_i$). Therefore, the following conditions have to be satisfied:

$$\begin{cases} f_{np} \geq 0 \\ f_{nq} \geq 0 \end{cases} \text{with} \begin{bmatrix} f_{np} \\ f_{nq} \end{bmatrix} = \mathbf{J}_n^{-T}\left(\boldsymbol{\tau}_\psi + \mathbf{H}^T \boldsymbol{\tau}_\rho + \mathbf{G}^T \boldsymbol{\tau}_\lambda - \mathbf{J}_t^T \boldsymbol{f}_t\right). \tag{22}$$

3.2.5 Stiffness matrices

Joint stiffness: The SM due to the dependent coordinates in the generalized space is given by

$$\mathbf{K}_I = \frac{d(\mathbf{G}^T \boldsymbol{\tau}_\lambda)}{d\psi} = \frac{d(\mathbf{G}^T \boldsymbol{\tau}_\lambda)}{d\lambda} \mathbf{G}, \text{ with } \boldsymbol{\tau}_\lambda = \mathbf{K}_\lambda (\lambda - \lambda_0). \tag{23}$$

In this matrix, only the bottom-right coefficient is not zero, i.e., the stiffness of the dependent joints affects only the position of the second phalanx θ_q.

Active stiffness due to contact forces: When the corresponding contact is established, its effects on the stiffness of the mechanism are given by

$$\mathbf{K}_{Et}^T = \begin{bmatrix} k_q m s_q\left(s_{pq} x_{pc} - c_{pq} y_{pc}\right) & 0 \\ \left(-mc_q f_{tq} + k_q m s_q\left(-ms_q + s_{pq} x_{pc} - c_{pq} y_{pc}\right)\right)\eta & 0 \end{bmatrix}$$

$$\mathbf{K}_{En} = \begin{bmatrix} D & E \\ F & G \end{bmatrix}, \text{ where the components are}$$

$$\begin{aligned} D &= l_p k_p \left(c_p x_{pc} + s_p y_{pc}\right) + k_q\left(mc_q + l_q\right)\left(c_{pq} x_{qc} + s_{pq} y_{qc}\right), \\ E &= -\left(-ms_q f_{nq} + k_q\left(mc_q + l_q\right)\left(mc_q - c_{pq} x_{qc} - s_{pq} y_{qc}\right)\right)\eta, \\ F &= l_q k_q \left(c_{pq} x_{qc} + s_{pq} y_{qc}\right)\eta, \\ G &= -l_q k_q \left(mc_q - c_{pq} x_{qc} - s_{pq} y_{qc}\right)\eta^2 - l_q f_{nq} \frac{as_2(c + ac_{23} - bc_3)}{bcs_3^2} \end{aligned} \tag{24}$$

Active stiffness due to the compliant actuator:

$$\mathbf{K}_\rho = \left(\frac{d(\mathbf{H}^T \boldsymbol{\tau}_\rho)}{d\psi}\right) = \left[\frac{\partial^2 \rho}{\partial \psi^2}\right] \boldsymbol{\tau}_\rho + \mathbf{H}^T k_\rho \mathbf{H}. \tag{25}$$

3.2.6 Quasi-static model

The generalized SM is given by $\mathbf{K}_M = \mathbf{K}_\psi + \mathbf{K}_I + \mathbf{K}_\rho + \mathbf{K}_{Et} + \mathbf{K}_{En}$ and the matrix of transmission ratio is calculated as $\mathbf{T} = \mathbf{K}_M^{-1} \mathbf{K}_{\rho 0}^\star$.

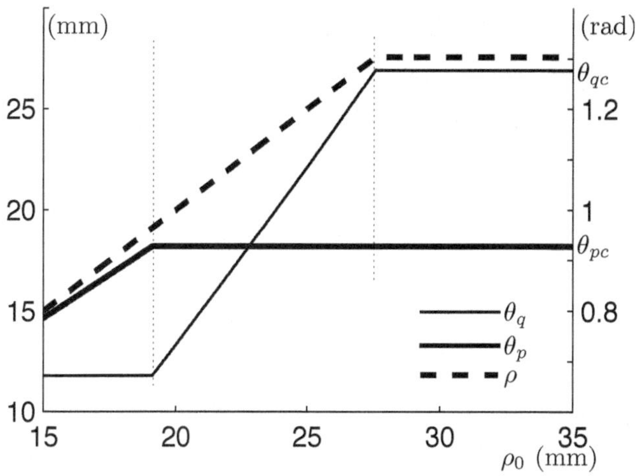

Figure 4. Configuration of the mechanism as a function of ρ_0.

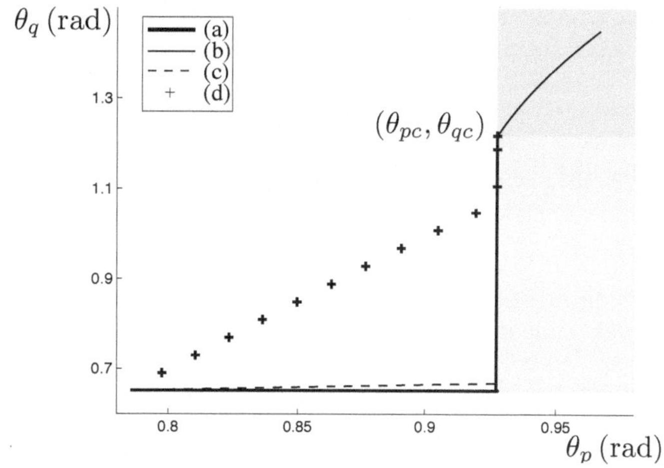

Figure 5. Configurations of several mechanisms in the (θ_p, θ_q) plane.

3.3 Numerical simulations

3.3.1 Management of the contact

The simulation (using the KSM or the QSM) starts with the finger open and the contact with the object not established, then the value of the only commandable parameter, i.e., ρ_0, is modified. At each step of the simulation, the corresponding values of the stiffness used for the computation are determined as follows: if $\theta_p \geq \theta_{pc}$ (first contact established), $k_p \leftarrow k_c$ else $k_p \leftarrow 0$. And if $\theta_p \geq \theta_{pc}$ and $\theta_q \geq \theta_{qc}$ (both contacts established), then $k_q \leftarrow k_c$ otherwise $k_q \leftarrow 0$. It is also assumed that the increments are small enough to neglect the effect of a contact established between two steps of computation.

Since when $k_i = 0$, the corresponding contact force $\boldsymbol{f}_i$ and SM $\mathbf{K}_{Ei}$ are zero for any value of $\delta \boldsymbol{x}_i$, the calculation can be directly performed with the relations given in the previous section.

3.3.2 Kinematostatic simulation

Simulation: By solving numerically the static constraints, the configuration of the mechanism can be determined as a function of the commanded value of the actuator ρ_0 only. Even if the computational time of the KSM depends on the gap between the initial configuration (used to start the numerical method) and the found configuration, it is relatively fast, about 10 ms in our Matlab program. This model provides exact results and is suitable for the analysis and design of a UM.

The undeformed angles of the compliant joints are (in rad): $\theta_{p0} = \pi/4$, $\theta_{10} = \pi/2$, $\theta_{20} = 1.437$, $\theta_{30} = 1.819$, $\theta_{40} = 1.457$. The positions of the objects are (in mm) $p_x = 15, p_y = 20, q_x = 15, q_y = 65$.

Results: Figure 4 shows the configuration (θ_p, θ_q) and the actual length of the actuator for ρ_0 between 15 mm and 35 mm. The stiffness coefficients used for this first simulation correspond to those of a conventional UM with compliant actuator grasping a rigid object, namely (in N.m^{-1}): $k_p = 0$, $k_2 = k_3 = k_4 = 0, k_1 = 1, k_\rho = 10^3$ and $k_c = 10^6$.

The three stages of the grasping process can be clearly seen. When there is no contact, the actual length ρ of the actuator is equal to its commanded value ρ_0 and the first phalanx (angle θ_p) is closing while the second one remains in its initial position. Then, when the first contact is established, θ_p remains equal to θ_{pc} and the second phalanx is closing (θ_q increases). Finally, when the object is grasped the configuration is set, the actual length of the actuator becomes different from ρ_0 and a force equal to $k_\rho(\rho - \rho_0)$ is applied on the object. It is important to mention that the different sections of the curves are not lines, the relation between ρ_0 and (θ_p, θ_q) being nonlinear.

Figure 5 shows the possible configurations taken by three different mechanisms grasping a rigid object and by a mechanism grasping a compliant object.

a. represents the mechanism used in Fig. 4 grasping a rigid object (its stiffness coefficient is $k_c = 10^6$ N.m^{-1}). It can be observed that the mechanism is closing or opening one phalanx at a time. When both contacts are established (i.e., when the point $(\theta_{pc}, \theta_{qc})$ is reached), the configuration is completely set.

b. still represents the same mechanism, but the grasped object is now compliant, its stiffness being half of that of the actuator ($k_c = 500$N.m^{-1} and $k_\rho = 10^3$N.m^{-1}). The curve corresponds to (a) when no contact is established (white area) and when only the first contact is established (pale gray area). However, when the second contact is set, the mechanism can enter in the dark gray area, i.e., the mechanism can keep on closing on itself by squeezing the object.

c. corresponds to a compliant mechanism. Each passive joint has a finite stiffness, however the stiffness related to the first phalanx ($k_p = 1$ n.rad^{-1}) is lower that the stiffness related to the second phalanx ($k_1 = 2$ N.rad^{-1} and $k_{2,3,4} = 1$ N.rad^{-1}). Even if with these values, the behaviour of the compliant mechanism is close to that of the conventional mechanism, it can be observed, that a slight coupling between the motions of the two phalanges appears when no contact is established (The first section of the dashed line is not exactly horizontal).

d. also corresponds to a compliant mechanism. In this case, the design of the gripper is poor because the stiffness related to the first phalanx ($k_p = 4$ N.rad^{-1}) is too high compared to the stiffness related to the second phalanx ($k_1 = 1$ N.rad^{-1} and $k_{2,3,4} = 0$ N.rad^{-1}) such that both phalanges are closing simultaneously. With this kind of mechanism, a risk exists that the second phalanx establishes the contact first; and even pokes the object. Graphically in this case, the line composed with crosses would not pass through point (θ_{pc},θ_{qc}).

3.3.3 Quasi-static simulation

Simulation: At each step, the configuration of the mechanism is calculated as a function of the previous configuration and the variation of the commanded value of the actuator $\Delta\rho_O$. This is written as $\boldsymbol{\psi} \leftarrow \boldsymbol{\psi} + \mathbf{T}\Delta\rho_O$.

The computation of this model is very fast, less than 0.3 ms in our Matlab program, however the accuracy depends on the magnitude of the increments $\Delta\rho_O$. In a context of simulation, these round-off errors are added step after step, but in a control application, since the parameters can be reinitialized using some sensors, the use of the QSM is suitable. Hence, it is important to keep in mind that the dynamic effects are neglected, therefore the variations have to remain small.

Results: The results provided by the QSM are similar to those of the KSM, except for the round-off errors. Figures 6 shows the magnitude of these errors for the computation of θ_p for different magnitudes of $\Delta\rho_O$ ($\Delta\rho_O = 0.2$ mm (100 iterations, 30 ms), $\Delta\rho_O = 20$ µm (1000 iterations, 0.20 s), $\Delta\rho_O = 2$ µm (10^4 iterations, 2 s) and $\Delta\rho_O = 0.20$ µm (10^5 iterations, 20 s).

First, it is interesting to notice that the error in θ_p is growing faster in the first part of the graph corresponding to the non-contact section. Indeed, in this part of the curves, θ_p is varying and therefore the error (that is related to the magnitude of $\Delta\theta_p$) is larger. Then, when θ_p is set (one and two contacts), $\Delta\theta_p$ is close to zero and thus, the growth of the error is smaller.

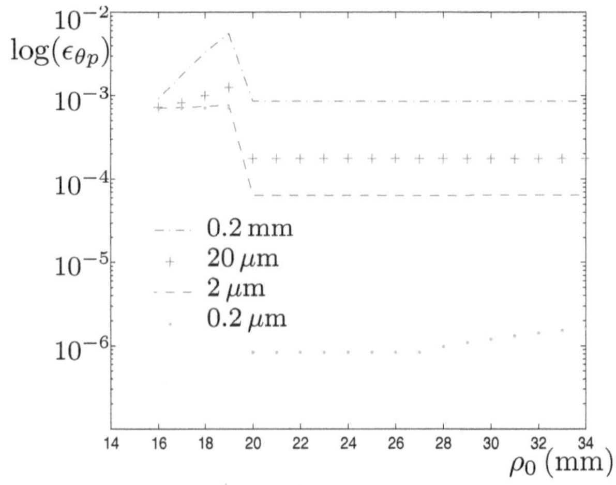

Figure 6. Discrepancy in θ_p between KSM and QSM for different $\Delta\rho_O$.

The second and the major source of inaccuracy in the use of the QSM comes from the step where the contact is established since the test has to be performed at the end of the step, so that the object can be squeezed up to $\approx l_i\Delta\theta_i$; which is another obvious advantage of using small increments. On the other hand, the fact that the contact is established for a certain value of θ_i has the advantage of "reinitializing" the round-off errors at this step.

3.4 Conclusion about the simulation

The comparison of the KSM and the QSM has shown their respective advantages and drawbacks. The constant and small computational time of the QSM makes it highly suitable in a context of real-time control of a manipulator. Hence, with the use of some sensors on this manipulator, the problem of the round-off errors disappears. The only remaining problem is the establishment of the contact. However, it is also interesting to notice that the presence of sensors would make the knowledge of the contact points $\boldsymbol{p}$ and $\boldsymbol{q}$ not necessary and therefore, would enable a simpler and more general use of the models. In practice depending on the relative stiffness of the actuator and the object, the finger may penetrate into the latter at a maximal depth of the increment.

The KSM, with kinematic parameters as well as static parameters that enable a detailled modelling of the mechanism, offers very accurate results and can be useful for the analysis and the design of an UM.

4 Conclusions

In this paper, the kinematostatic and quasi-static models of compliant mechanisms are applied to underactuated mechanisms. This application brings some interesting contributions to the modelling of compliant mechanisms. First, the theoretical models are extended to the case of actuated dependent joints. Second, the case of external loads that are function of the configuration of the mechanism has been studied and has shown that it was equivalent to a reduction of degree of mobility in the system {mechanism + grasped object}. On the other hand, since the equivalence between the general formulation of the static equilibrium of a compliant parallel mechanism and the kinetostatic relation of the underaction theory was proven, the former can as well be used as the starting point for the study of the grasp stability, the force isotropy or any other important property of grasping hands.

Then, in the application, a new possibility offered by these models is illustrated: the configuration of an underactuated finger can be described even when it is not in contact with the object to be grasped[1]. All degrees of mobility become function of the single commanded value of the actuator.

Acknowledgements. The authors would like to acknowledge the financial support of the Natural Sciences and Engineering Research Council of Canada (NSERC) as well as the Canada Research Chair (CRC) Program.

Edited by: J. L. Herder
Reviewed by: two anonymous referees

References

Birglen, L.: An Introduction to the Analysis of Linkage-driven Compliant Underactuated Fingers, in: Proceedings of the ASME Design Engineering Technical Conferences, 2006.

Birglen, L., Laliberté, T., and Gosselin, C. M.: Underactuated Robotic Hands, Spr. Tra. Adv. Robot., 40, 2008.

Boudreault, E. and Gosselin, C. M.: Design of Sub-Centimetre Underactuated Compliant Grippers, in: Proceedings of the ASME Design Engineering Technical Conferences, Philadelphia, PA, USA, 10–13 September 2006.

Carricato, M.: Contributi teorici per l'analisi statica dei meccanismi con elementi elastici, Master's Thesis, Università degli Studi di Bologna, Italy, 1998.

Cho, W., Tesar, D., and Freeman, R. A.: The Dynamic and Stiffness Modeling of General Robotic Manipulator Systems with Antagonistic Actuation, IEEE Int. Conference on Robotics and Automation, 1380–1387, 1989.

Doria, M. and Birglen, L.: Design of an Underactuated Compliant Gripper for Surgery Using Nitinol, Journal of Medical Devices, 3(1), 011007, 2009.

Griffis, M. W. and Duffy, J.: Kinestatic Control: a Novel Theory for Simultaneously Regulating Force and Displacement, J. Mech. Des.-T. ASME, 113(4), 508–515, 1991.

Kragten, G. A., Herder, J. L., and Schwab, A. L.: On the Influence of Contact Geometry on Grasp Stability, in: ASME Design Engineering Technical Conferences and Computer and Information in Engineering Conference, p. 49400, Brooklyn, New York, USA, 3–6 August 2008.

Krut, S.: A Force-Isotropic Underactuated Finger, in: IEEE International Conference on Robotics and Automation, 2325–2330, Barcelona, Spain, 18–22 April 2005.

Lotti, F., Tiezzi, P., Vassura, G., Biagiotti, L., Palli, G., and Melchiorri, C.: Development of UB hand 3: Early results, in: IEEE International Conference on Robotics and Automation, 4488–4493, Barcelona, Spain, 18–22 April 2005.

Quennouelle, C. and Gosselin, C. M.: Stiffness Matrix of Compliant Parallel Mechanisms, Advances in Robot Kinematics: Analysis and Design, 331–341, 2008.

Quennouelle, C. and Gosselin, C. M.: A Quasi-Static Model for Planar Compliant Parallel Mechanisms, Journal of Mechanisms and Robotics, 1(2), 021012, 2009.

Su, H.-J.: A Pseudorigid-Body 3R Model for Determining Large Deflection of Cantilever Beams Subject to Tip Loads, Journal of Mechanisms and Robotics, 1(2), 021008, 2009.

Sun, L., Liang, C. G., and Liao, Q. Z.: An inverse Static Force Analysis of a Special Planar Three-Spring System, Mech. Mach. Theory, 32(5), 609–615, 1997.

[1]Even in a conventional n-DOF underactuated mechanism, at least $n-1$ springs have to be added in order to set the configuration of the mechanism when no contact is established.

13

Torque control of underactuated tendon-driven fingers

M. E. Abdallah and C. W. Wampler

General Motors R&D, Manufacturing Systems Research Lab, Warren, MI 48090, USA

Abstract. Given an underactuated tendon-driven finger, the finger posture is underdetermined and can move freely ("flop") in a region of slack tendons. This work shows that such an underactuated finger can be operated in tendon force control (rather than position control) with effective performance. The force control eliminates the indeterminate slack while commanding a parameterized space of desired torques. The torque will either push the finger to the joint limits or wrap around an external object with variable torque – behavior that is sufficient for primarily gripping fingers. In addition, introducing asymmetric joint radii to the design allows the finger to command an expanded range of joint torques and to scan an expanded set of external surfaces. This study is motivated by the design and control of the secondary fingers of the NASA-GM R2 humanoid hand.

This paper was presented at the IFToMM/ASME International Workshop on Underactuated Grasping (UG2010), 19 August 2010, Montréal, Canada.

1 Introduction

Tendon or cable transmission systems are often used in the actuation of fingers in high degree of freedom (DOF) hands. The remote actuation allows for significant reductions to the size and weight of the fingers, features that are important for dexterous manipulation. Since the tendons can only transmit forces in tension, the number of actuators must exceed the DOF's to achieve fully determined control of the finger. It turns out that only *one* tendon more than the number of DOF's is needed (Murray et al., 1994). If arranged correctly, the $n+1$ tendons can independently control the n DOF's while always maintaining positive tensions.

In this sense, an n-DOF finger with only n tendons is underactuated, and the finger posture is underdetermined. This creates a slack space within which the finger posture is uncontrolled. In other words, the finger cannot hold a desired position and will be free to flop in some region where the tendons go slack.

Despite this handicap, having a reduced number of actuators is still very attractive. Space or power limitations can be significant in high DOF hands. Each extra actuator and tendon transmission system increases the demands on space and maintenance. This work shows that an underactuated finger with n tendons can be operated in tendon force control (rather than position control) with effective performance. Desired joint torques can be commanded to the finger in a feasible torque space. The slack space is eliminated as the torques either push the finger to the joint limits or wrap around an external object. Others in the robotics community have applied tension control to tendon-driven hands, most notably for the Stanford/JPL, Utah/MIT, POSTECH, and Robonaut-2 hands (Salisbury and Craig, 1982; Biggers et al., 1986; Lee et al., 1994; Abdallah et al., 2010). In these cases, however, only fully actuated control strategies have been presented. Others have presented underactuated grippers, though they do not use force control (Hirose and Umetani, 1978) or spring load the joints to eliminate the slack space (Dollar and Howe, 2006).

The ability to independently control the joint torques depends on a finger design that has asymmetric joint radii. We shall discuss how the radii shape the feasible torque space. Furthermore, the asymmetric radii allow the finger to scan external surfaces whose target directions fall within a feasible set.

Figure 1. The secondary fingers of R2 are underactuated.

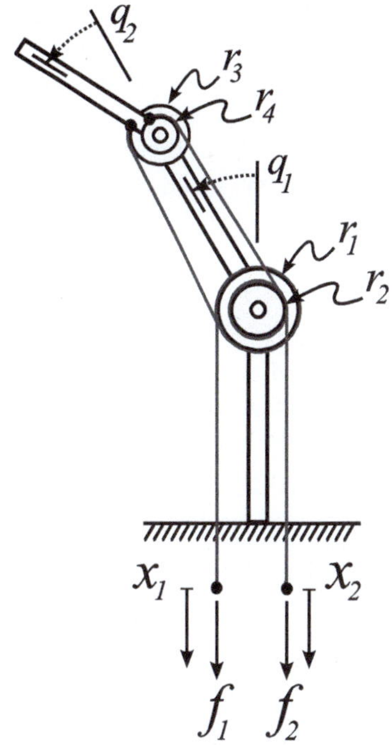

Figure 2. A 2 DOF finger with two tendons. x_1 represents the flexor tendon and x_2 the extensor.

In short, in asymmetric underactuated fingers, the torques are fully determined functions of the tendon tensions, whereas joint positions are not fully determined by the tendon positions. Therefore, we propose to control such fingers by controlling the tendon tensions instead of positions. This work starts with a thorough analysis of a two-DOF finger before generalizing the conditions to an n-DOF system.

This study was motivated by the design and control of the NASA-GM Robonaut-2 humanoid hand, shown in Fig. 1. The secondary fingers of the hand, i.e. the ring and little fingers, are underactuated to reduce the number of actuators and allow for a human-sized form factor. The primary objective of these fingers is to support the dexterous set of fingers in gripping larger objects.

2 Tension control

Suppose that we control the tensions in a tendon-actuated mechanism. For an open chain linkage with concentric circular pulleys and negligible friction, the relation between the actuated joint torques, $\boldsymbol{\tau}$, and tendon tensions, $\boldsymbol{f}$, is

$$\boldsymbol{\tau} = \mathbf{R}\boldsymbol{f}. \tag{1}$$

Matrix $\mathbf{R}$ is known as the tendon map matrix. If the pulleys are noncircular, the same relation holds, but $\mathbf{R}$ then varies with the configuration of the linkage. That is, if $\boldsymbol{q}$ is the set of joint angles for the linkage, we have $\boldsymbol{\tau} = \mathbf{R}(\boldsymbol{q})\boldsymbol{f}$. Although this relation is true for all tension values, one must remember that only non-negative tensions are possible. This limitation means that it is only feasible to access a limited portion of the joint torque space, as we will discuss further below.

As an illustration, consider the 2 DOF finger shown in Fig. 2. The tendons wrap around idler pulleys of radii r_1 and r_2 at the first joint, and radii of r_3 and r_4 at the second joint, terminating on the distal link. r_i are all positive scalars. In this case, the tendon map matrix becomes

$$\mathbf{R} = \begin{bmatrix} r_1 & -r_2 \\ r_3 & -r_4 \end{bmatrix}. \tag{2}$$

The dimensionality of the set of feasible torques depends on the design of the pulleys. For a symmetrical choice of pulley radii, $r_1 = r_2$ and $r_3 = r_4$, matrix $\mathbf{R}$ is singular. No matter what tensions are applied, the joint torques stay in a constant ratio $\tau_1/\tau_2 = r_1/r_3$. Thus, only a one-dimensional line in torque space is feasible, and this will be limited to just a line segment due to actuator saturation limits. In contrast, for a more general choice of radii, $\mathbf{R}$ becomes nonsingular, and we may invert Eq. (1) to obtain the tensions $\boldsymbol{f}$ that will produce a given joint torque $\boldsymbol{\tau}$:

$$\boldsymbol{f} = \mathbf{R}^{-1}\boldsymbol{\tau}. \tag{3}$$

Again, the limits on the tendon tensions give limits on what torques can be commanded. If $\mathbf{R}$ is nearly singular, even moderate torque commands in the singular direction will require large tensions. Nevertheless, there is a two-dimensional region of the torque space that is feasible.

3 Position control

Suppose that we control position of the tendons instead of their tensions. As long as the tendons remain taut, a standard virtual work argument implies that the relation between joint and actuator motion is governed by the same matrix $\mathbf{R}$ involved in the torque-tension relation (Eq. 1):

$$\dot{x} = \mathbf{R}^T \dot{q}, \tag{4}$$

where x is the set of tendon actuator positions. In the more general case where the tendons may become lax, it is helpful to introduce an intermediate variable y that represents the tendon extension that would keep the tendons taut, while x is the actual extension of the tendon actuators. Then, starting from any configuration in which the tendons are initially taut, i.e., $x = y$, we have

$$\dot{x} \le \dot{y} = \mathbf{R}^T \dot{q}. \tag{5}$$

(By this notation, we mean that the inequality holds for each row of the matrix expression.) Even if the actuators are held stationary, $\dot{x} = 0$, the finger can move with $\dot{y}$ in the positive quadrant: $\dot{y}_1 \ge 0$, $\dot{y}_2 \ge 0$. Such motions enter the *slack region*, a bounded region in which the finger may move freely even though the actuators are held stationary.

The slack region is described by inequalities at the position level. We assume that all quantities are measured from an initial position $x = y = q = 0$ in which the tendons are taut. Assuming inelastic tendons, the joint motion is constrained by the length of tendons:

$$x \le y = \mathbf{R}^T q. \tag{6}$$

In particular, for the finger of Fig. 2, we have

$$\begin{aligned} x_1 &\le r_1 q_1 + r_3 q_2 \\ x_2 &\le -r_2 q_1 - r_4 q_2 \end{aligned} \tag{7}$$

In general, the intersection of these two inequalities consists of a wedge that defines the slack region. As the input side of the tendons (x_1 and x_2) are displaced, these inequalities shift and change the slack region. An example is shown in Fig. 3. In the interior of the slack region, the tendons lose tension, while on either boundary, one tendon is taut, the other slack. For symmetric designs, the constraints become parallel as shown in Fig. 4. In this case, the tendons perfectly oppose each other, so they can be drawn taut, at which point their constraints in joint space collapse onto each other into a single line that matches the null-space of $\mathbf{R}^T$. Even though the tendons will remain taut, they cannot resist motion along this line.

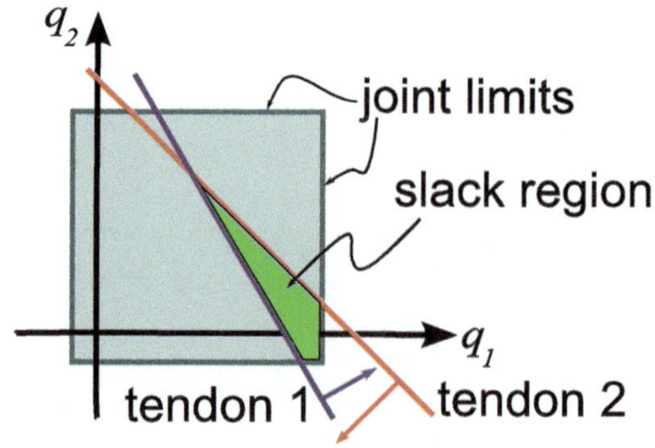

Figure 3. The slack space (shaded region) is bound by the two constraints (lines) and the joint limits (box).

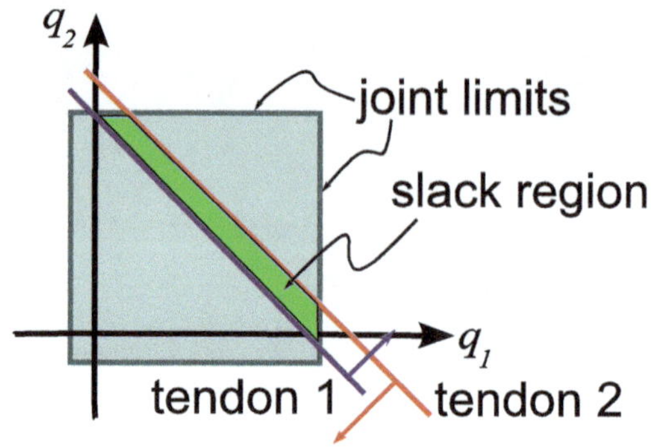

Figure 4. The slack space for a symmetric pulley design.

4 Feasible torques

Not all joint torques are possible due to the nature of both the tendons and actuators. Due to the tendons, each tendon tension $f_i, i = 1, \ldots, n$, must be nonnegative. Due to the actuators, each tendon tension also has an upper limit, say f_i^+. In short,

$$0 \le f_i \le f_i^+, \qquad i = 1, \ldots, n. \tag{8}$$

This rectangular region in tension space gets mapped by $\mathbf{R}$ into an n-parallelotope (the generalization to n dimensions of a parallelogram.) The 2^n vertices of this feasible torque region are given by evaluating $\mathbf{R}f$ for every combination of tendon tensions at their respective limits. In particular, let $\mathbf{R}_i$ be the i-th column of $\mathbf{R}$. Then, for a two-link finger actuated by two tendons, the feasible torque region is a parallelogram with vertices:

$$\{(0,0), \mathbf{R}_1 f_1^+, \mathbf{R}_2 f_2^+, \mathbf{R}_1 f_1^+ + \mathbf{R}_2 f_2^+\}.$$

Figure 5 illustrates the case of $r_1 > r_2 = r_3 = r_4$ with equal maximum tensions.

It is appropriate to limit the finger operation to the condition that both joint torques have the same direction. In other words, the joints are both in either flexion or extension. This covers the behavior of fingers designed for gripping. In Fig. 5, the regions that correspond to this condition

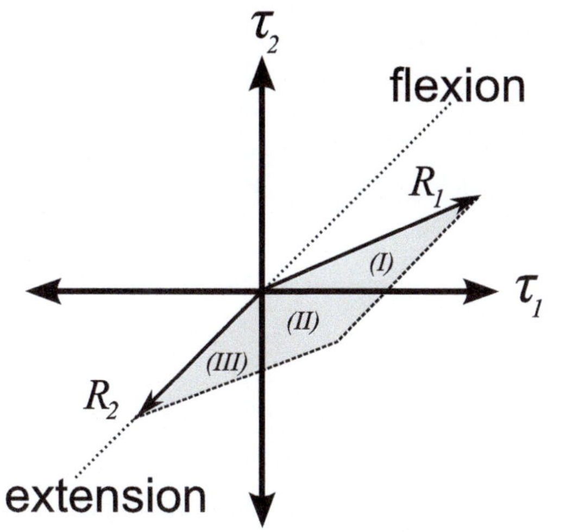

Figure 5. The space of feasible joint torques. It represents the positive span of $\boldsymbol{R}_i$, given upper bounds on the tensions. Region (I) indicates when both joints are in flexion. Region (III) indicates when both joints are in extension.

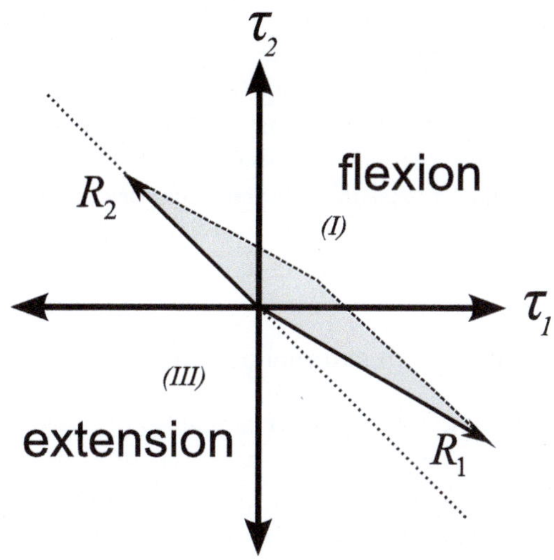

Figure 6. A different tendon arrangement can produce undesirable results. In this case, both joints cannot be extended simultaneously.

are regions *I* and *III*. Hence in flexion, $\tau_2 \leq \frac{r_1}{r_2}\tau_1$. While in extension, $\tau_2 \leq \tau_1$.

For the sake of simplicity, the joint torques in the two-joint finger can also optionally be limited to operate along just $\boldsymbol{R}_1$ or $\boldsymbol{R}_2$, that is, we pull on one tendon while maintaining a minimal tension on the other. In this case, the torques become parameterized by a single DOF. This mode of operation causes the joint torques to be both in either flexion or extension. Such a control mode is well suited for fingers designed primarily for gripping. While fingers intended for manipulation need to be fully actuated and fully controllable, gripping fingers simply need to flexibly grip objects with variable strength and variable posture. Hence, one DOF is sufficient to either specify the grip strength or to fully extend the finger. The commanded joint torques will thus drive the finger until it either comes to rest against the joint limits or wraps around an external object.

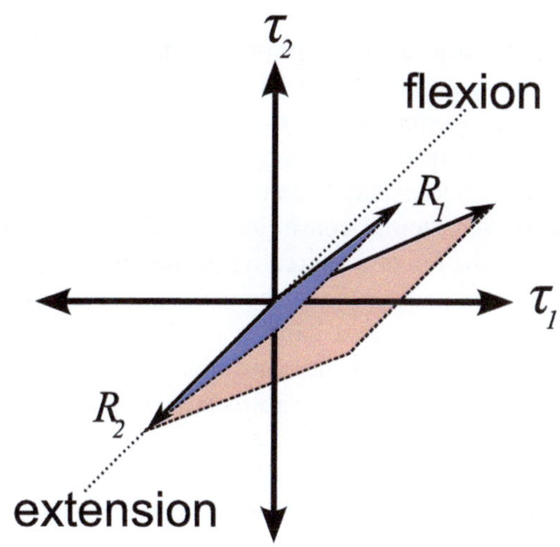

Figure 7. For limited tendon tensions, the torque space is defined by the above parallelogram. The ratio of $\frac{r_1}{r_2}$ affects the area and conditioning of the possible solution.

5 Design considerations

The tendon configuration greatly affects the feasible joint torque space, and a bad design can produce a space with undesirable characteristics. Consider the same finger in Fig. 2, but suppose we reverse the direction of wrap on the upper joint. This results in a new tendon map,

$$\mathbf{R} = \begin{bmatrix} r_1 & -r_2 \\ -r_3 & r_4 \end{bmatrix}. \qquad (9)$$

Although this arrangement is valid in many ways, it produces a potentially undesirable space of feasible joint torques – shown in Fig. 6. Assuming $r_1 > r_2 = r_3 = r_4$, as before, the feasible region excludes quadrant three, meaning that the finger can no longer command both joints simultaneously in extension, and its ability to simultaneously move both in flexion is limited.

The size of the joint radii carry several design implications as well. For a given tendon force and stroke length, larger radii increase the maximum joint torques while decreasing the joint range. Consider Fig. 5 again. To obtain good control off the diagonal, the differences in the radii must be large enough to leave $\mathbf{R}$ well conditioned. As r_1 increases with $r_2 = r_3 = r_4$, vector $\boldsymbol{R}_1$ rotates clockwise and grows in length. As $\boldsymbol{R}_1$ rotates, the area of the feasible parallelogram increases, but the ability to move along the diagonal in flexion is sacrificed. This behavior is demonstrated in Fig. 7.

It is important to note that while a more asymmetrical arrangement increases the area of the feasible torque region, it also widens the slack region. This is not too much of a concern as long as the finger is under tension control, or if the actuators are parked near a boundary of the joint space (see next section). But if the actuators are parked in the middle of their ranges, the larger slack region may be undesirable. For example, it may be possible for slack tendons to escape their guideways such that the mechanism will not return to a usable state even when the actuators re-tension the tendons.

6 Position control with external constraints

We have seen that position control of an underactuated finger can only constrain the finger to a slack region. Without joint limits or external contact constraints, this region would be infinite in extent, but with them, the slack region becomes finite. In fact, under certain conditions, the slack region can shrink to a point, that is, the underactuated finger can become fully controlled in position when acting against a constraint.

It takes a minimum of three inequalities to bound a finite region of the plane, or in general, $n+1$ inequalities to bound a finite region in n-space. This is the fundamental reason why $n+1$ tendons are required to hold an n-DOF mechanism in place without any external constraints. But once the mechanism contacts the environment or a hard joint limit, its freedom drops and n tendons may become sufficient. However, the direction of the external constraint matters: it must be oriented such that the finger can exert a positive force against it. It is easiest to understand this if the external constraint is first mapped into an equivalent joint-space constraint. Then the ability to exert forces on the constraint can be reinterpreted as the ability to exert joint torques against the joint space constraint.

The symmetric pulley case, or any case in which **R** has co-rank 1 (meaning that it is singular with a one-dimensional null space), there remains a complete one-dimensional slack space. Any single one-sided constraint can only cut this infinite line into a ray. It takes two independent constraints in opposition to reduce the slack region to a single point. We illustrate this in the next section on joint limits.

6.1 Joint limits

Let's first examine the situation for the case of joint limit constraints without additional contact. Figure 3 shows the slack region for the two-joint finger with asymmetric radii. Notice that if the tendons are pulled sufficiently, the slack region will shrink to a point somewhere on the boundary of the joint limit box. Figure 8 illustrates a configuration where the slack region has been reduced to a tiny triangle. A slight additional tug on either tendon will trap the finger at a point with q_1 at its maximum, q_1^+. It is clear that in this manner the finger can be fully constrained at any point on the southeast boundary of the joint limits, that is, any point with either q_1 at its maximum q_1^+ or q_2 at its minimum q_2^-.

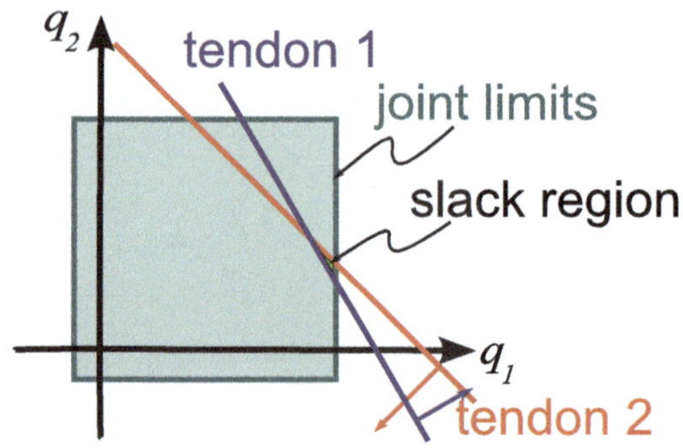

Figure 8. The asymmetric finger trapped near the maximum of q_1.

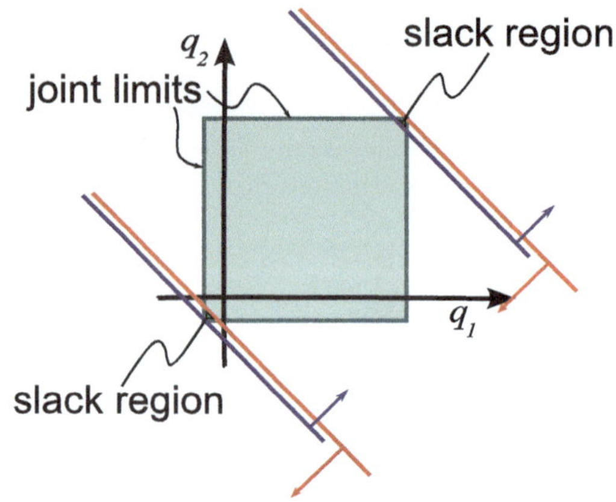

Figure 9. The symmetric finger is stable only at full flexion or full extension.

In contrast, the symmetric finger requires two additional constraints to become fully constrained. This can occur only at full flexion (q_1 and q_2 both at their maxima) or at full extension (both joints at their minima). This is illustrated in Fig. 9.

Another way to view the situation is through the feasible torque space. By reference to Fig. 5, one sees that the asymmetric finger can generate torque in either the positive τ_1 or the negative τ_2 directions and can also generate torque vectors with components that can resist movement along the $q_1 = q_1^+$ or the $q_2 = q_2^-$ joint limits. In contrast, the symmetric arrangement can only generate torques along the diagonal, $\tau_1 = \tau_2$, and thus can be stably driven only to the corners along this diagonal.

6.2 General constraints

Point contact with an external object introduces a new constraint and multiple point contacts introduce multiple constraints. These may be mapped into the joint space of the mechanism. Neglecting friction, the mechanism may reach quasi-static equilibrium in contact if and only if the inward normal to the contact constraint in joint space is within the

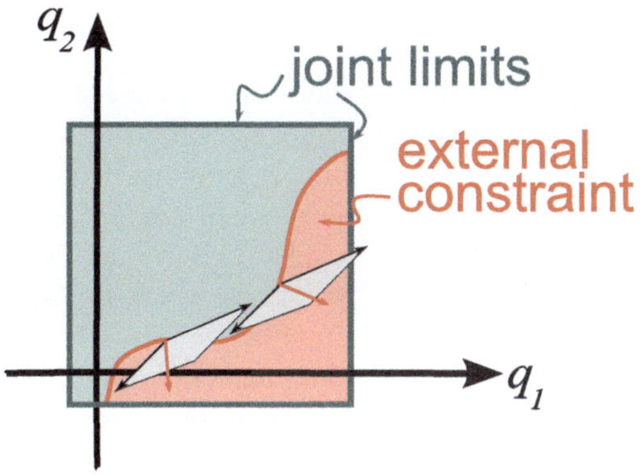

Figure 10. Equilibrium conditions for a asymmetric finger touching a general constraint. Constraint inner normal (red) is directed into the superimposed feasible torque region.

set of feasible torques. In other words, the robot must be able to exert positive force along the normal to the constraint.

For our example of the 2-DOF finger, it is convenient to visualize this by overlaying the feasible torque diagram, Fig. 5, onto the constraint curve in joint space, as shown in Fig. 10. In the figure, the feasible torque region has been superimposed at two different points along the contact constraint curve. Since the inner normal at those points are directed into the feasible torque region, a positive contact force can be applied normal to the constraint and equilibrium is possible.

We note that if friction is present, then the range of slopes for the constraint curve where stable contact is possible expands. By the same token, since friction impedes the finger from sliding along the constraint, the range of slopes along which the finger can be driven along the constraint decreases.

If the finger is close to the symmetric design, the magnitude of the torque vector that can be generated perpendicular to the main diagonal is greatly diminished. Suppose that the constraint runs along the main diagonal. Even though in principle the finger can produce a force normal to the constraint, it can at best exert a small force towards it, and this will be further diminished by any friction in the mechanism. Furthermore, the torque vectors $\boldsymbol{R}_1$ and $\boldsymbol{R}_2$ are nearly opposed, so the direction of the resultant torque is highly sensitive to small errors in the tendon forces. For the finger to be well-controlled in sliding contact along such a constraint, its design must be sufficiently asymmetric to give a well-conditioned tendon map matrix $\mathbf{R}$.

So far, we have visualized the constraints only in the joint space. One may also consider the equivalent picture at the contact point. Let $\boldsymbol{v}$ be the velocity of the contact point and let $\boldsymbol{F}$ be the force exerted on the environment by the finger at the contact point. We have the usual velocity and force relations:

$$\mathbf{J}\dot{q} = \boldsymbol{v}, \qquad \tau = \mathbf{J}^T \boldsymbol{F}, \tag{10}$$

where $\mathbf{J}$ is the Jacobian for the contact point. From these and Eq. (1),

$$\boldsymbol{F} = \mathbf{J}^{-T}\mathbf{R}\boldsymbol{f}. \tag{11}$$

Thus, in operational space, the columns of $\mathbf{J}^{-T}\mathbf{R}$ play the analogous role as the columns of $\mathbf{R}$ play in building the feasible torque region. The external constraints that can be stably contacted with a single point contact are those whose inward normal lies within the positive range space of $\mathbf{J}^{-T}\mathbf{R}$.

Two-point contact is another matter. This may occur when the fingertip is inserted into a corner feature or it may occur when the inner and outer phalanges each contact an object within a grasp. Now, the constraints give two inward normals in the joint-space diagram and the finger gives the two vectors $\boldsymbol{R}_1$ and $\boldsymbol{R}_2$, as usual. Each of these defines a feasible parallelogram, and as long as there is an overlap between them, there exists sets of tendon forces that stably direct the finger into the constraint.

7 Higher DOF fingers

Although we have illustrated all the above concepts on a 2-DOF finger, they apply equally well on any n-DOF manipulator controlled by n tendons. We discuss here some of the subtleties of the general case.

The feasible torque space, i.e. the positive range space of $\mathbf{R}$, determines what kind of constraints can be stably contacted. Suppose we consider only joint limits. If $\mathbf{R}$ has n independent columns, i.e., if it is nonsingular, then it takes only 1 additional constraint to reach an equilibrium among the total of $n+1$ forces, all nonzero. Accordingly, any face of the joint limit polytope is a candidate for stable positioning. However, we need the additional fact that all the tendon forces are positive and the net torque is directed along the inward normal of the joint limit constraint (which is to say, an outward normal of the limit polytope). Only certain faces are oriented such that this is possible. Let $\boldsymbol{e}_i$ be an elementary vector with 1 in the i-th position and zeros elsewhere. Then, the mechanism can achieve equilibrium on the positive $\boldsymbol{e}_i$ face of the joint limits if and only if there is an all positive solution $\boldsymbol{f}$ of

$$\boldsymbol{e}_i = \mathbf{R}\boldsymbol{f}, \tag{12}$$

or in other words, if $\mathbf{R}^{-1}\boldsymbol{e}_i$ has all positive elements. When this condition is satisfied, we will say that the face is "stable". The opposing face in the $-\boldsymbol{e}_i$ direction is stable if and only if $-\mathbf{R}^{-T}\boldsymbol{e}_i$ is all positive, which means that only one of two opposing faces can be stable.

Where two faces meet, that is, on a $(n-2)$-dimensional "edge" of the joint limit polytope, equilibrium is possible if at least one of the faces is stable. However, there is an additional freedom to exert forces on both faces simultaneously, hence the condition to be stable on the edge where the $\boldsymbol{e}_i$ and

$\boldsymbol{e}_j$ faces meet is that there must exist an all positive solution $(\lambda_i, \lambda_j, \boldsymbol{f})$ to

$$\boldsymbol{e}_i \lambda_i + \boldsymbol{e}_j \lambda_j = \mathbf{R}\boldsymbol{f}. \tag{13}$$

(The obvious sign adjustment is made for a face in the $-\boldsymbol{e}_i$ direction.) Accordingly, write the solution for $\boldsymbol{f}$ in terms of $\lambda = (\lambda_i, \lambda_j)^T$ as

$$\boldsymbol{f} = \left[\mathbf{R}^{-1}\boldsymbol{e}_i \;\; \mathbf{R}^{-1}\boldsymbol{e}_j\right]\lambda = W\lambda. \tag{14}$$

If either column is all positive, there is an all positive solution for $\boldsymbol{f}$. If any row of W is all negative, then there is no all-positive solution. All positive rows give a positive left-hand side no matter what positive λ is used. So the only problems are the rows with one positive and one negative entry. To this end, consider the 2×2 system

$$\boldsymbol{w} = \begin{bmatrix} a & -b \\ -c & d \end{bmatrix}\lambda, \tag{15}$$

where a,b,c,d are all positive. It is easy to show that there is an all-positive solution $(\boldsymbol{w},\lambda)$ if and only if

$$d/c > b/a. \tag{16}$$

This must be true for every 2×2 minor of W. So scanning the rows of W, one picks out the smallest d/c and the largest b/a. If these satisfy Eq. (16), then the edge is stable. Notice that it is possible for the edge to be stable even if neither of the adjoining faces is.

It is more difficult to assess lower-dimensional facets. However, let's examine the extreme case of a vertex. In particular, let's consider the vertex where all joints are at their upper limits. Hence, the question of stability comes to the existence of an all-positive solution to

$$I\lambda = \mathbf{R}\boldsymbol{f}. \tag{17}$$

If $\mathbf{R}$ has an all-positive column, then this equation has all-positive solutions. However, this is a sufficient but not necessary condition, as is clear from the discussion on edges above. In general, the question of whether an all-positive solution exists has no simple solution, but the question can always be answered using the techniques of Linear Programming. As in the $(n-2)$ dimensional edge case above, it is clearly necessary that every row of $\mathbf{R}$ have at least one positive element; otherwise the associated element of λ is always negative for nonzero positive $\boldsymbol{f}$. If we also want the opposing vertex where all joints are at their minima to be a stable point, then we need an all-positive solution to $-I\lambda = \mathbf{R}\boldsymbol{f}$, or equivalently, $\lambda = (-\mathbf{R})\boldsymbol{f}$. So the same logic that applies to $\mathbf{R}$ in the former case now applies to $-\mathbf{R}$.

Although the condition of an all-positive column is not necessary, it does offer a simple condition for the design of a basic gripper finger. For a finger designed to wrap around an external object in one direction and fully extend in the other direction, it suffices for $\mathbf{R}$ to have one all-positive and one all-negative column.

8 Conclusions

We have examined possible design and control trade-offs for underactuated tendon-driven mechanisms, in particular, n-DOF mechanisms driven by only n tendons. One would like to operate a tendon-driven mechanism without allowing slack in any tendon. One would also like to reliably drive the mechanism to certain configurations where the mechanism will be in stable equilibrium. Finally, when the mechanism is in contact with the environment, one might like to be able to move the contact point along the environmental obstacle. This article examines the trade-offs between these objectives as they depend on the tendon map matrix $\mathbf{R}$ that relates tendon tensions to joint torques. Our objective is to present the fundamental ideas, including the idea of a slack region and torque controllability. To keep the analysis simple, we ignore friction, which may occur in the finger joints and tendon transmission elements, and we assume inelastic tendons. There would be value in developing more a sophisticated analysis that includes these neglected effects.

Mechanisms can be classified according to whether $\mathbf{R}$ is singular or nonsingular. In the singular case, we assume $\mathbf{R}$ has rank $n-1$, leaving a one DOF null-space. ($\mathbf{R}$ with lower rank are not practically useful.) If the elements of this null vector are all of the same sign, then the tendons can all be in tension without generating any torque at the joints. Although this is a desirable characteristic, it comes at a price: the mechanism also has a 1-dimensional slack region that is limited only by the joint limits of the mechanism. To reach a stable equilibrium, the mechanism must meet up with at least two obstacles, either two joint limits or two point contacts with the environment (or one of each).

By choosing $\mathbf{R}$ to be nonsingular, the set of feasible torques becomes an n-dimensional parallelotope (a parallelogram in the case of $n=2$). One can drive the mechanism in any direction within the positive range space of $\mathbf{R}$. The slack region changes to a cone extending from the point where all tendons are taut. If one locks the actuators in place, the mechanism can freely move within this cone and the tendons lose tension. On the good side, the motion is limited in the sense that it cannot go beyond the vertex of the cone. This characteristic is valuable, and it gives the mechanism the capability of reaching a stable equilibrium with only a single point of contact or a single joint limit.

These concepts are examined thoroughly in the case of a 2-DOF finger. It is shown that to get a nonsingular $\mathbf{R}$ that can both flex and extend the finger, then at least one pulley radius must be different than the others. We call this an asymmetrical design. We illustrate how the asymmetric finger can be fully flexed and fully extended and also can move along two of four boundaries where one joint is at a limit. It can also scan some external obstacles. We detail how to tell which joint limits and which obstacle shapes can be scanned in quasi-static equilibrium. In contrast, a symmetric finger can only come to equilibrium when trapped at the intersection of two constraints: it cannot move in quasi-static equilibrium.

We also outline how to analyze the capabilities of mechanisms with more than two joints by an examination of the **R** matrix.

Edited by: J. L. Herder
Reviewed by: two anonymous referees

References

Abdallah, M. E., Platt, R., Wampler, C. W., and Hargrave, B.: Applied Joint-Space Torque and Stiffness Control of Tendon-Driven Fingers, in: IEEE Intl. Conf. on Humanoid Robots, Nashville, TN, 2010.

Biggers, K., Jacobsen, S., and Gerpheide, G.: Low level control of the Utah/M.I.T. dextrous hand, in: The IEEE International Conference on Robotics and Automation (ICRA), 61–66, 1986.

Dollar, A. and Howe, R.: Joint Coupling Design of Underactuated Grippers, in: Proceedings of ASME Intl. Design Engineering Technical Conferences, Philadelphia, PA, 1–9, 2006.

Hirose, S. and Umetani, Y.: The development of soft gripper for the versatile robot hand, Mech. Mach. Theory, 13, 351–359, 1978.

Lee, Y., Choi, H., Chung, W., and Youm, Y.: Stiffness Control of a Coupled Tendon-Driven Hand, IEEE Contr. Syst. Mag., 14, 10–19, 1994.

Murray, R., Li, Z., and Sastry, S.: A Mathematical Introduction to Robotic Manipulation, CRC Press, Boca Raton, FL, 1994.

Salisbury, J. and Craig, J.: Articulated hands: Force control and kinematic issues, Int. J. Robot. Res., 1, 4–17, 1982.

14

Identifying links between origami and compliant mechanisms

H. C. Greenberg, M. L. Gong, S. P. Magleby, and L. L. Howell

Department of Mechanical Engineering, Brigham Young University, Provo, UT 84602, USA

Abstract. Origami is the art of folding paper. In the context of engineering, orimimetics is the application of folding to solve problems. Kinetic origami behavior can be modeled with the pseudo-rigid-body model since the origami are compliant mechanisms. These compliant mechanisms, when having a flat initial state and motion emerging out of the fabrication plane, are classified as lamina emergent mechanisms (LEMs). To demonstrate the feasibility of identifying links between origami and compliant mechanism analysis and design methods, four flat folding paper mechanisms are presented with their corresponding kinematic and graph models. Principles from graph theory are used to abstract the mechanisms to show them as coupled, or inter-connected, mechanisms. It is anticipated that this work lays a foundation for exploring methods for LEM synthesis based on the analogy between flat-folding origami models and linkage assembly.

Keywords. Compliant mechanisms, Lamina emergent mechanisms, Origami, Orimimetics

1 Introduction

Origami is the art of folding paper where *ori-* means fold and *-kami* means paper. The scope and complexity of origami has exploded in the last twenty years creating many different schools of thought (Demaine et al., 2010). Recently in the fields of science, mathematics and engineering, origami has been used to solve complex problems such as airbag folding, shock absorption (crash box), and deployable telescopic lenses (Cromvik, 2007; Ma and You, 2010; Heller, 2003). Origami design is governed by mathematical laws, which if better understood, could be applied in engineering. Specifically, the authors propose that origami can provide inspiration for synthesis techniques for compliant mechanisms. This paper identifies key concepts that can link origami to mechanism design.

Traditionally origami has been static and representational. However, kinetic origami focuses on origami models that have some novel motion. While the term "action origami" is a more generally known term, we introduce the term kinetic origami to refer to models that exhibit a mechanical motion and to differentiate from models such as paper airplanes which are also referred to as action origami. Figure 1 shows two examples, where one is a static origami structure and the other is an action origami mechanism called a flasher hat; it is shown in its fabricated and deployed forms. The way an origami model folds also provides insight and another way to classifying origami. The class of rigidly foldable origami focuses on motion where the creases often act as joints and the faces act as links with bending stress only occurring at the creases (Tachi, 2006; Balkcom and Mason, 2008; Hull, 1994; Watanabe and Kawaguchi, 2006).

The process of folding origami has been examined using kinematic theory (Dai and Jones, 2002; Balkcom and Mason, 2008; Tachi, 2006; Buchner, 2003). This paper focuses on the motion of a finished origami and not the states in between flat and final states.

Since origami relies on the deflection of flexible materials it is a compliant mechanism. The origami mechanisms examined herein are all flat-folding in their final-folded state, and so they are part of a subgroup of compliant mechanisms called lamina emergent mechanisms (LEMs) (Jacobsen et al., 2010). Lamina emergent mechanisms are compliant mechanisms made from planar materials (lamina) with motion that emerges out of the fabrication plane.

A productive connection between origami and compliant mechanisms can be developed by drawing upon principles from graph theory to depict the origami mechanisms via simple, planar connected graphs. These graphs can show how the folds and facets interact with each other in their motion. Graphs allow for improved understanding of the interaction

Figure 1. Example of an origami structure (left) and flasher hat mechanism in its fabricated (middle) and deployed form (right).

between motion and structure of origami, and can help in understanding how to predict complex motion and develop corresponding mechanisms.

Origami's rich history and research in design optimization can provide an "orimimetic" perspective to LEM design that can lead to the generation of novel mechanisms.

Objective

Orimimetic design refers to the application of the concepts of folding to mechanism design. It can provide alternative ways to achieve a particular range of motion; compliant mechanisms have a similar aim. In addition, origami is achieved from the manipulation of a planar material which motivates its examination as a lamina emergent mechanism. The objective of this paper is to identify links between origami and compliant mechanisms to facilitate the application of origami design literature to compliant mechanism and LEM design.

2 Nomenclature

It is helpful to establish terminology to facilitate discussion. Key terms, and their use, are provided below.

2.1 Origami

Origami. Origami is defined as the art of folding paper, and in the context of engineering, it is the use of folding to solve mechanical problems (Demaine and O'Rourke, 2007).

Kirigami. Kirigami is defined as the art of folding and cutting paper (Hart, 2007). An example of kirigami is shown in Fig. 2 with the side view showing how creases are made such that the house pops out of the paper. Kinetic kirigami models are often LEMs, especially kirigami pop-up models.

Hinge creases. In traditional origami hinge creases define the boundaries between flaps. In the context of mechanisms they are creases along which lies an interface of two planar faces and it is the axis about which both facets rotate (Demaine and O'Rourke, 2007).

Construction creases. Creases used to create references in the construction of the mechanism are not directly used to create motion, are referred to as construction creases. In some cases, they coincide with hinge creases (Demaine and O'Rourke, 2007).

Structural creases. Structural creases are used to define the shape of flaps; they can be hinge creases as well. These creases are not needed in most mechanisms and they are not feasible for many materials. They may/can be substituted or eliminated in various ways (Demaine and O'Rourke, 2007).

Crease vs. fold. A fold is an action and a crease is the product of that action. Creases may be folded in one of two ways: mountain or valley.

Mountain crease. A convex crease (Hull, 2002).

Valley crease. A concave crease (Hull, 2002).

2.2 Mechanisms

Compliant mechanism. Mechanisms which transfer or transform motion, force, or energy at least in part through the deflection of flexible members are compliant mechanisms (Howell, 2001).

Lamina emergent mechanism. Lamina emergent mechanisms (LEMs) are a subset of compliant mechanisms made from planar materials (lamina) with motion that emerges out of the fabrication plane (Jacobsen et al., 2010). They are a subset of compliant mechanisms, in that they use the deflection of flexible members to achieve the desired motion. As a subset of compliant mechanisms, LEMs can provide feasible, repeatable solutions to advance the design and manufacturing of products. The advantages of LEMs include: reducing the number of parts, reducing cost, reducing weight, improving recyclability, increasing precision, and eliminating assembly (Jacobsen et al., 2010). The incorporation and use of LEMs offer many potential advantages in the design of mechanical products. These mechanisms can provide opportunities for more cost-effective, compact, easy to assemble, and modular products.

Spherical mechanism. In a spherical mechanism any point in a moving body is confined to move within a spherical surface, and all spherical surfaces of motion are concentric (Chiang, 1988).

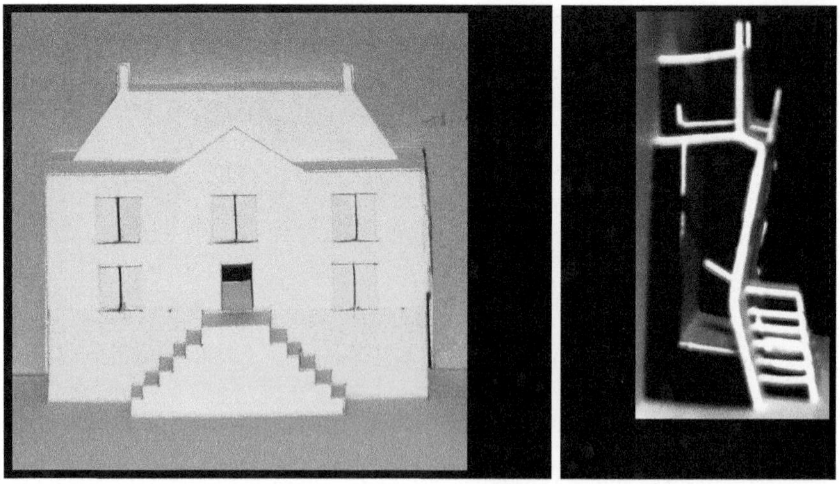

Figure 2. A kirigami house made in cardstock.

Orimimetic. Orimimetic means the ability to imitate folds. In an engineering design context, it refers to the ability to use the concept of folding to solve problems. Specifically it refers to the use of folding, either conceptually or literally, in mechanism design.

2.3 Principles from graph theory

Graph. A graph consists of points, called vertices, and connections, called edges (Marcus, 2008).

Planar graph. A graph with no edges crossing is a planar graph (Marcus, 2008).

Facet. The area enclosed by a planar graph is referred to as a facet.

Degree. The degree of a vertex is the number of edges that occur at that vertex (Marcus, 2008).

Degree sequence. The degree sequence of a graph is the degrees of each vertex listed in decreasing order (Marcus, 2008).

*Regular graph, **d**-regular graph*. A graph is considered regular if all its vertices are of the same degree. It can be referred to as a d-regular graph, where d is a non-negative integer where each vertex has degree d (Marcus, 2008).

Generically rigid. A graph is generically rigid if almost every realization of the graph is rigid. A realization of a graph is an assignment of coordinates to joints but without restrictions on edge lengths. Generic realizations avoid degeneracies such as having three points collinear, four points concyclic, or inducing parallel edges (Demaine and O'Rourke, 2007).

3 Origami background

These sections describe the relevant history and current applications of origami.

3.1 Origami history

The current origami movement began in the early 20th century, though historically origami has its beginning in several countries dating back to the Muromachi period, from 1333–1573 AD (Demaine and O'Rourke, 2007). Akira Yosihzawa and his origami works have been largely credited for the creative explosion in origami in the last century (Demaine and O'Rourke, 2007). Since the 1920's, origami design has become increasingly complex and varied. Initially origami was used to create figures and animals. As the art has progressed, the models have become increasingly complex, starting with a few to a few dozen steps to a few hundred or even a thousand steps. In the last twenty years new branches of origami have begun to be explored by more than artists; a growing number of mathematicians, educators, engineers and scientists have gathered to discuss the applications of origami in their respective fields (Demaine and O'Rourke, 2007).

Closely related to origami are other forms of paper engineering, such as kirigami, pop-up paper mechanisms, and origami architecture (Winder et al., 2009a).

3.2 Origami applications in engineering and science

Previous work has focused on studying the folding process to map the transition between initial and final states. An example is airbag folding designs which unfold smoothly from their flat starting state to their final volume (Cromvik, 2007). The Diffractive Optics Group at Lawrence Livermore Laboratory has worked on developing a telescope having a 100 m diameter lens that could collapse to fit into a space vehicle having a 4 m diameter and 10 m length. Lang used origami to identify a design that would fit and maintain the integrity of the surface when deployed into space (Heller, 2003; Wu and You, 2010; Wei and Dai, 2009). Origami has also been applied to the crash box of a car to improve energy absorbtion

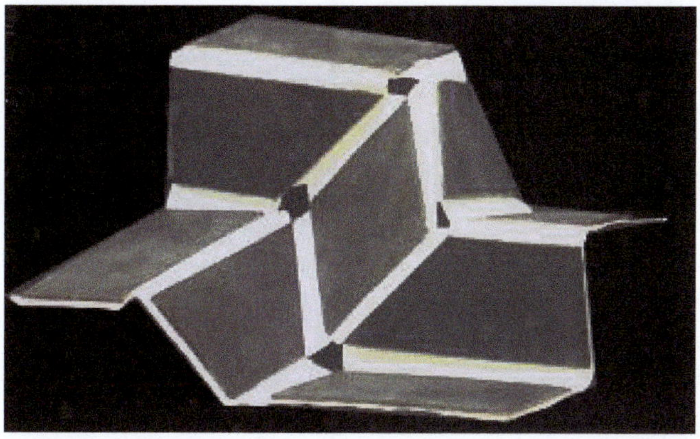

Figure 3. A common double slit kirigami representation and its corresponding PRBM both shown placed in the crease of a folded sheet of paper.

in a low speed collision (Ma and You, 2010). Miura explored methods of folding maps that could be unfolded with the simple pull of a corner (Miura, 2002). In the 1980's he invented the Miura-ori pattern, which is used as a basis for folding solar arrays (Miura and Natori, 1985). In 2005, Mahadevan published findings that this same pattern exists in leaf folding, wings, and flower petals (Mahadevan and Rica, 2005).

4 Origami mechanisms

This paper considers kinetic origami or origami mechanisms as opposed to origami structures. We will use the terms kinetic origami and origami mechanism interchangeably.

4.1 Kinematic modeling of origami mechanisms

Origami mechanisms are modeled with hinge creases as joints and facets as links (Winder et al., 2009b). Origami that is rigid-foldable is a "piecewise linear origami that is continuously transformable without the deformation of each facet" (Tachi, 2010). Rigid origami is also defined as having regions of the paper between crease lines that do not need to bend or twist in the folding process (i.e., the facet could be replaced with sheet metal and hinge creases replaced by hinges and it would still fold up) (Hull, 1994; Tachi, 2006). Figure 4 shows an example of rigid origami made by modeling the links of a square twist, a common origami unit for tessellations, converted to a polypropylene compliant mechanism.

4.2 Origami and compliant mechanisms

Origami mechanisms are compliant mechanisms; their motion is a result of the deflection of the material. As a material undergoes deformation, the resulting stored strain energy gives rise to an internal spring force. Balkcom notes that "the configuration of paper is determined by internal spring forces as well as external forces and constraints" (Balkcom and Mason, 2008). Thus, the origami can be viewed as a compliant mechanism which can be modeled using the pseudo-rigid-body model.

By focusing on origami in a fabricated state we only examine the hinge creases that contribute to the mechanism's structure and motion, not the folds that were used to construct it.

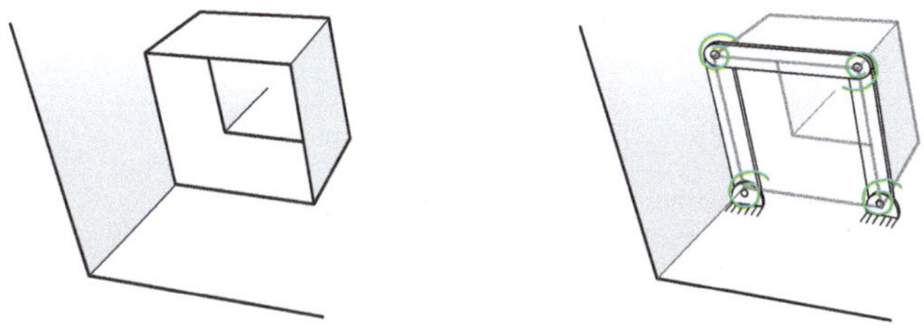

Figure 4. A rigid-foldable version of the square twist origami constructed from polypropylene sheet and paper. The polypropylene shows that it behaves rigidly, where bending is only allowed at the folds.

Pseudo-rigid-body model

The pseudo-rigid-body model (PRBM) uses rigid-body components, rigid links and springs, that have equivalent force-deflection characteristics to model the deflection of flexible members (Howell, 2001). Figure 3 shows a common kirigami double slit and its corresponding PRBM (Winder et al., 2009a).

4.3 Origami as a lamina emergent mechanism

Because origami mechanisms are made from lamina materials it follows that they are LEMs. Traditionally, origami designs are judged by their efficiency and accuracy in terms of material usage and number of folds (Lang and Hull, 2005). For a model of an animal, efficiency is often defined by the ratio of the size of the final model to the initial sheet.

A branch of origami that is of special relevance to LEMs is flat-folding origami, which is origami that folds to a flat state, and tessellations or tilings, which could be useful for LEM applications in arrays. The repetition of basic folds in tessellations leads to the development of an origami mechanism that expands and contracts thereby capturing the motion that would be required for an array-type structure.

Table 1 lists examples of origami corollaries in each of the six categories. The potential for origami insights in LEM design can be seen by mapping existing applications of origami to the six application categories for LEMs (Albrechtsen et al., 2010).

Table 1. Origami Applications by LEM Technology Category.

LEM Application Class	Origami Corollaries
Disposable mechanisms	Packaging
Novel Array mechanisms	Space sails Telescopes
Scalable mechanisms	Origami at nano-level Cellular origami
Surprising Motion mechanisms	Pop-up books
Shock-absorbing mechanisms	Crash box
Deployable mechanisms	Airbags Stents

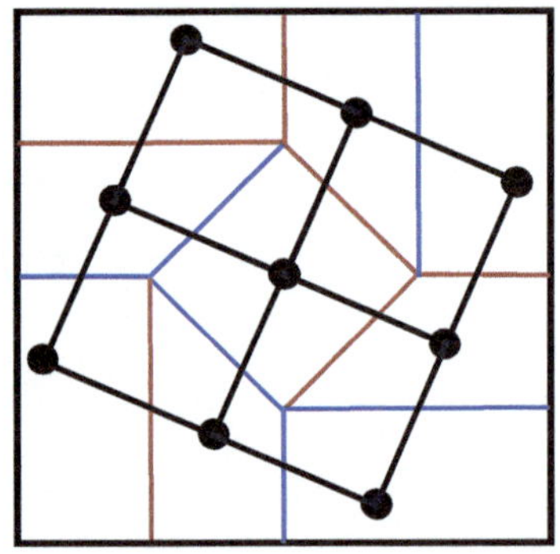

Figure 5. The crease pattern for the square twist shown in Fig. 4 is shown with its corresponding graph overlaid in black. In the crease pattern the red corresponds to mountain folds and blue to valley folds.

5 Linking origami to mechanisms with graphs

Principles from graph theory can be used as a tool to understand how origami functions as a mechanism (Dobrjanskyj and Freudenstein, 1967). Graphs can model both origami and mechanisms on an abstract level and show their similarities. The abstraction of origami mechanisms to graphs demonstrates how the origami functions as a mechanism. Origami mechanisms can be reduced to graphs which show how they behave also. Applications of this are seen in Dai's carton folding research where graph theory is applied to show an equivalent mechanism (Dai and Jones, 2002).

In a graph of a kinetic origami model folds become line segments (edges) and links become nodes (vertices). From such graphs it can be seen that origami mechanisms may be thought of as interconnected linkages, with each loop representing a linkage system. It is important to note that in graph theory the shape of an edge or the position of the vertices does not matter since graphs depict connections (Marcus, 2008).

The crease pattern can be related to the graph of the origami mechanism in some cases where all creases are hinge creases and the origami can be flattened via actuation. In this case the graph and crease pattern are considered dual graphs. Figure 5 shows the square twist with its crease pattern and corresponding graph overlaid. However, for the general case, the crease pattern may include structural and construction creases which aid in the folding of the origami but do not contribute to the motion and therefore are not considered in the origami's corresponding graph.

Any simple planar connected graph with four segments connecting four vertices that is a 2-regular graph represents a four-bar mechanism. The graphs depict the degree of interconnection for each linkage system. This is shown in how each facet relates to its neighbouring facets. Each facet in a graph represents a linkage having the same number of links as nodes. For example, a facet (enclosed region) having four vertices (links) on the boundaries separating four segments (joints) is representative of a four-bar linkage.

6 Paper mechanism examples

Four paper mechanisms are shown in Table 2 with their corresponding kinematic and graphical representations. For simplification, all four paper mechanisms are flat-folding mechanisms and are rigid-foldable.

The kinematic representations for each mechanism examine a portion of the larger mechanism and its motion. When the paper mechanisms have a flat initial state and they can be represented with kinematics, by definition they can be realized as a LEM. In addition, the graphs representing each paper mechanism show more abstractly the mechanism and how they are coupled. The graphs also indicate the type of fold, where mountain folds are shown in red and valley folds are in blue.

The four-bar double slit mechanism (Table 2, row 1) uses the PRBM to show how each crease can be modeled as a joint with a torsional spring. Its graph is a 2-regular graph. It has three valley folds, and one mountain fold as indicated by the colors of edges in the graph. Each link has only two joints that define its motion.

For the 45-degree-fold twisting mechanism, also known as the pop-up spinner card (Table 2, row 2) each center fold, along the spine of the mechanism, is a 45 degree fold (O'Rourke, 2011). The kinematic representation shows only four links along the spine. The 45-degree-fold twisting mechanism is a chain of four-bar mechanisms where those links not on the ends share two links between each neighbouring linkage system, and each inside link is constrained by three joints with one of those joints shared between each neighbouring linkage system.

The square twist mechanism (Table 2, row 3) is a series of coupled spherical mechanisms which collapse onto one another. Figure 4 shows the basic square twist, and the

Table 2. Table of Mechanisms.

Paper Mechanism	Kinematic Representation	Graph
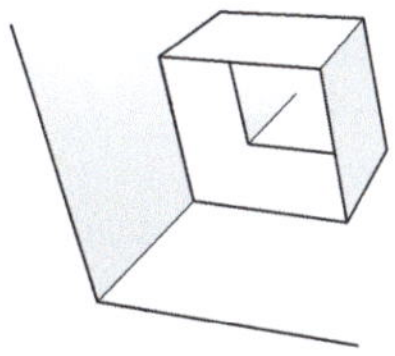 1. Four-Bar Double Slit	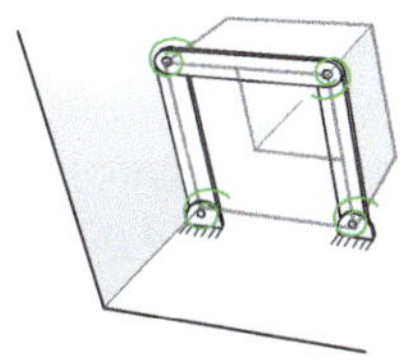	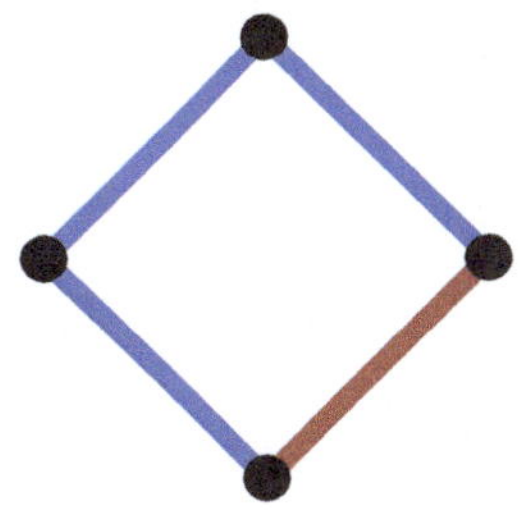
 2. 45-degree-fold Twisting Mechanism	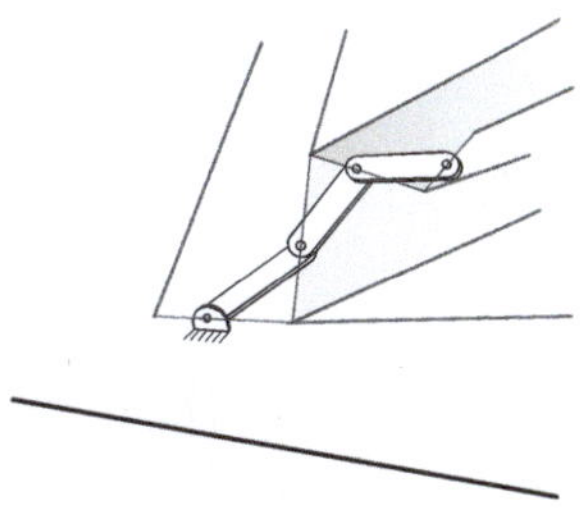	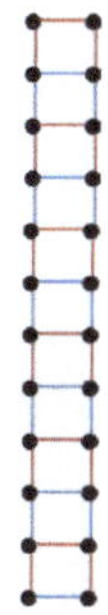
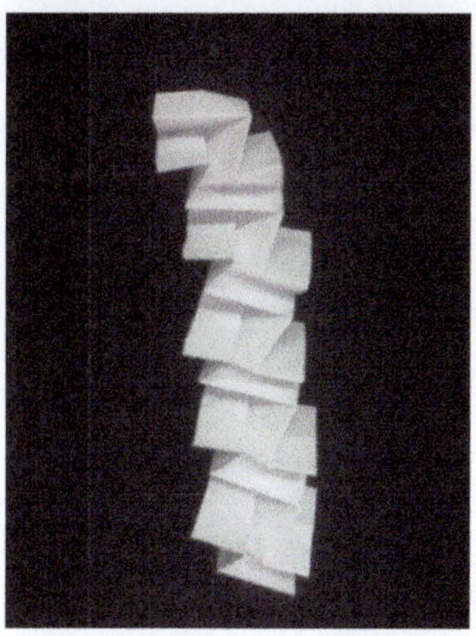 3. Square Twist	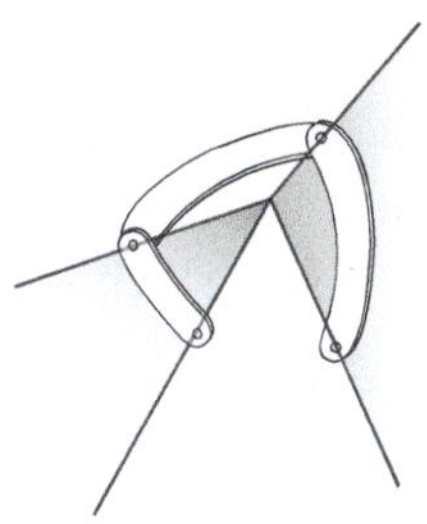	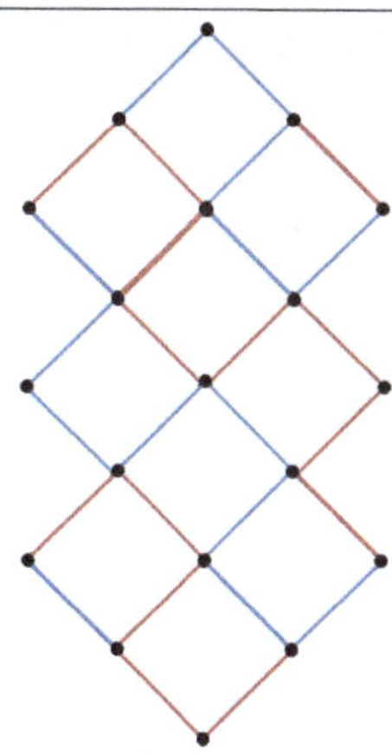
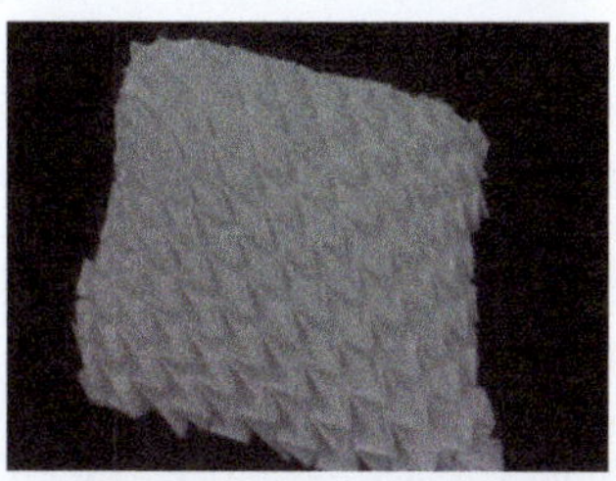 4. Water Bomb Base Tessellation	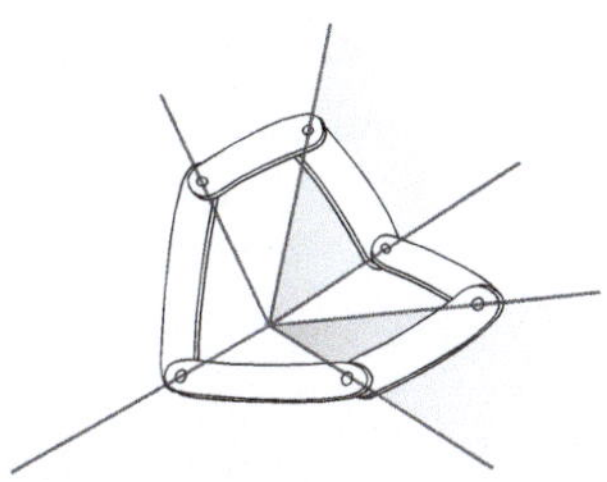	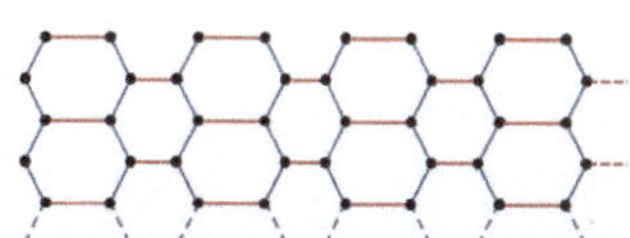

kinematic representation shows one of the four-bar linkages seen at the corner of the twisting square platform. It may be better to express the square twist as inter-connected four-bar spherical mechanisms, which may be seen in the graph representation. The outer corner links each connect to only two joints. There are two neighbouring links on each end that are connected to three joints and are shared between two linkage systems. Of the 7 nodes of degree four, 3 are shared between four linkage systems and 4 are shared between three linkage systems.

Lastly, the kinematic representation for the water bomb base tessellation is shown in Table 2, row 4. As a graph the water bomb base tessellation is a 3-regular graph. The water bomb base tessellation contains six-bar linkage systems that are interconnected. Each link is shared between three linkage systems via three joints. This is also the basic tessellation used in the origami stent design which allows for the mechanism to compact radially as well as lengthwise (Kuribayashi et al., 2006).

7 Discussion

The examples in Table 2 examine a single four-bar mechanism, four-bars connected in series, four bars in series and parallel, and a six-bar spherical mechanism. The use of graph theory helps bridge the fields of origami and mechanisms. Visually the graphs show how the origami models are interconnected spherical mechanisms. The following sections will introduce concepts for knowing whether a graph can be realized as a flat-folding mechanism.

7.1 Mobility

Given any generic graph we can determine whether it is generically rigid or flexible and subsequently realizable as a mechanism or a structure, where a mechanism would be classified as generically flexible and a structure as generically rigid. A generic graph can be realized as a mechanism in two dimensions when

$$e < 2n - 3 \tag{1}$$

or in three dimensions when

$$e < 3n - 6 \tag{2}$$

where n is the number of nodes and e is the number of edges in the graph (Demaine and O'Rourke, 2007). When

$$e = 2n - 3 \tag{3}$$

the graph is generically rigid.

7.2 Assembly

For a four-bar mechanism to be assembled it is required to meet the criteria of

$$l < s + p + q \tag{4}$$

where l refers to the length of the longest link, s to the length of the shortest link and p and q the lengths of the other links. Likewise for an origami to be folded flat it must follow certain criteria. In determining flat-foldability it is assumed models can be folded perfectly, ignoring folding error, that paper has zero thickness, and that our creases have no width (Hull, 2002). The following criteria were developed for determining local flat-foldability, where local means about a single vertex.

Theorem 1 (Kawasaki, Justin, Hull). *A single-vertex crease pattern defined by angles $\theta_1 + \theta_2 + ... + \theta_n = 360°$ is flat-foldable if and only if n [the number of creases] is even and the sum of the odd angles θ_{2i+1} is equal to the sum of the even angles θ_{2i}, or equivalently either sum is equal to* 180° (Demaine and O'Rourke, 2007).

This theorem is generally known as Kawasaki's Theorem, though also discovered independently by Justin in 1989. It's analogy in mechanisms is the condition for a changepoint mechanism where

$$s + l = p + q \tag{5}$$

Theorem 2 (Maekawa [Kasahara and Takahama], Justin). *In a flat-foldable single-vertex mountain-valley pattern defined by angles $\theta_1 + \theta_2 + ... + \theta_n = 360°$, the number of mountains and the number of valleys differ by* ±2 (Demaine and O'Rourke, 2007).

Theorem 2 is generally know as Maekawa's Theorem. It is essentially a condition for assembling a four-bar closed kinematic chain.

Theorem 3 (Kawasaki, Justin). *If an angle θ_i is a strict local minimum (i.e., $\theta_{i-1} > \theta_i < \theta_{i+1}$), then the two creases bounding angle θ_i have an opposite mountain-valley assignment in any flat-foldable mountain-valley pattern* (Demaine and O'Rourke, 2007).

This theorem is analogous to the conditions defining the classes for spherical four-bar LEMs where the shortest link has to connect to two links having opposite direction assignment.

An additional criterion is that a sheet can never penetrate a fold. To determine this it is required to have the geometry specified as well as the angles.

While criteria exist for determining the local flat-foldability (for a single vertex crease pattern), determining global flat-foldability is np-hard (non-deterministic polynomial-time hard). However we will explore the flat-foldability of the square twist to better understand these rules for assembly.

Square twist

The square twist crease pattern, shown in Fig. 6, has 2^{12} possible assignments and of those only 16 are flat-foldable (Hull, 2003). Figure 7 shows the 16 valid assignments for the square twist. It should be noted that only 6 of these 16 are unique.

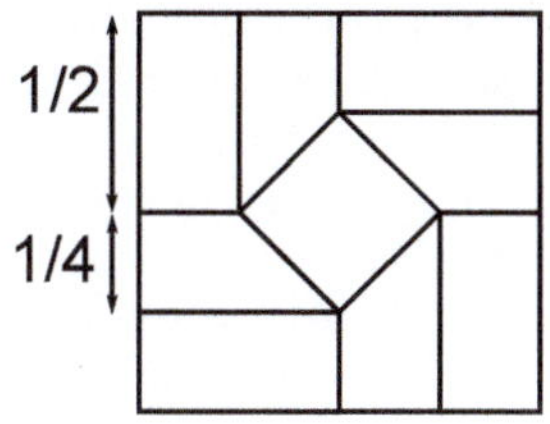

Figure 6. Crease pattern for square twist.

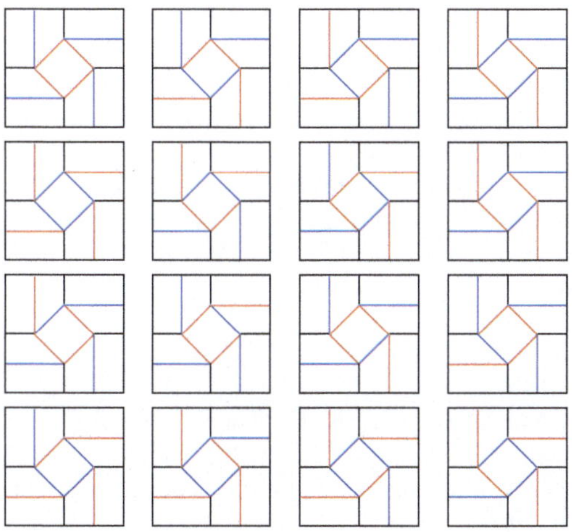

Figure 7. The 16 possible square twists given Theorem 3.

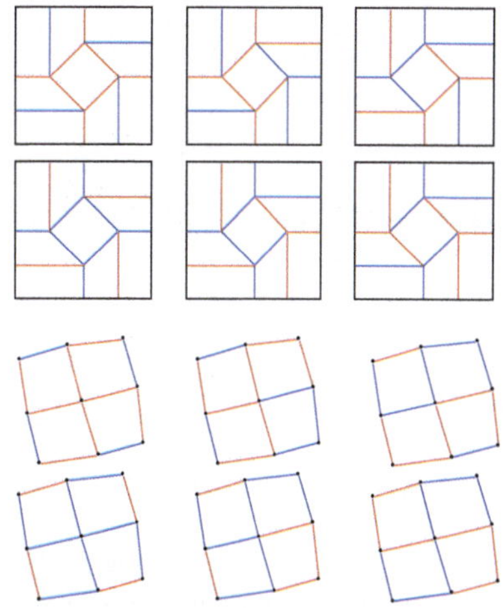

Figure 8. The 6 possible unique square twist crease patterns and their corresponding graphs given Theorem 2.

Figure 9. Folded origami models corresponding to the 6 possible square twists corresponding to those shown in Fig. 8.

Next by applying Theorem 2 we can determine the remaining crease designations which are shown in Fig. 8.

It is easier to use the graph (on the right) as opposed to the crease pattern for visualizing how Theorem 2 is applied. Note that each subgraph for one of the four four-bars has either 3 mountain and 1 valley folds or 1 mountain and 3 valley folds. Corresponding paper representations of each are shown in Fig. 9.

All six paper models in Fig. 9 are flat-folded versions of the square twist which are kinetic origami. An important distinction to make is that while an origami can be folded flat it does not mean that it is kinetic. Since all folds here are simple folds and they are non-locking folds this allows us to actuate the square twist to have the twisting motion to go from a folded state (half the size of the original square) back to the original square.

8 Conclusions

Graphs of kinetic origami and kirigami models were used to show how the fields of origami and mechanisms are related. We propose that some of the literature from the origami field can be applied to mechanisms and vice versa. This was shown by comparing the criteria for flat-foldability to some conditions required for changepoint mechanisms to be assembled. As such it is anticipated that the abstraction to graph theory will improve the ability to synthesize LEMs.

In addition, by examining the graphs of some paper mechanisms it is understood that their motion is achieved because they are a system of coupled, or interconnected, mechanisms. Graph theory includes theories for identifying whether a generic graph is rigid or flexible in n-dimensions. As such it is believed that given any generic graph we can know whether it represents a structure or mechanism. Showing the relationship between origami and mechanisms will facilitate the application of origami design principles in the design of compliant mechanisms and LEMs.

Acknowledgements. The authors would like to acknowledge Larrin Wada for his drawings. This material is based upon work supported by the National Science Foundation under Grant No. CMMI-0800606. Any opinions, findings, and conclusions or recommendations expressed in this material are those of the authors and do not necessarily reflect the views of the National Science Foundation.

Edited by: J. Herder
Reviewed by: H. Stachel and another anonymous referee

References

Albrechtsen, N., Magleby, S. P., and Howell, L. L.: Using Technology Push Product Development to Discover Enabled Opportunities and Potential Applications for Lamina Emergent Mechanisms, in: Proceedings of the ASME International Design Engineering Technical Conferences, DETC2010–28 531, ASME, 2010.

Balkcom, D. J. and Mason, M. T.: Robotic origami folding, Int. J. Robot. Res., 27, 613–627, doi:10.1177/0278364908090235, compendex, 2008.

Buchner, T.: Kinematics of 3D Folding Structures for Nanostructured Origami TM, Ph.D. thesis, Massachusetts Institute of Technology, 2003.

Chiang, C. H.: Kinematics of Spherical Mechanisms, Cambridge University Press, 1988.

Cromvik, C.: Numerical Folding of Airbags Based on Optimization and Origami, Master's thesis, Chalmers University of Technology and Göteborg University, Sweden, 2007.

Dai, J. S. and Jones, J. R.: Kinematics and mobility analysis of carton folds in packing manipulation based on the mechanism equivalent, Proceedings of the Institution of Mechanical Engineers – Part C – J. Mech. Eng. Sci., 216, 959–970, doi:10.1243/0954406021760400931, 2002.

Demaine, D. E., Dinh, G., Crimp, V. F. L., Golan, M., Hull, D. T., and Gould, V.: Independent Lens: Between the Folds, 2010.

Demaine, E. and O'Rourke, J.: Geometric Folding Algorithms, Cambridge University Press, 2007.

Dobrjanskyj, L. and Freudenstein, F.: Some applications of graph theory to the structural analysis of mechanisms, J. Eng. Ind., 89, 153–158, 1967.

Hart, G.: Modular Kirigami, San Sebastian Spain, 1–8, 2007.

Heller, A.: A Giant Leap for Space Telescopes, Science and Technology Review by Lawrence Livermore National Laboratory, 12–18, 2003.

Howell, L. L.: Compliant Mechanisms, Wiley-Interscience, 2001.

Hull, T.: On the mathematics of flat origamis, Congressus numerantium, p. 215, 1994.

Hull, T.: The Combinatorics of Flat Folds: A Survey, in: Origami3: Third International Meeting of Origami Science, Mathematics, and Education, A K Peters, Ltd., 29–38, 2002.

Hull, T.: Counting Mountain-Valley Assignments for Flat Folds, Ars Combinatoria, 67, 175–178, 2003.

Jacobsen, J. O., Winder, B. G., Howell, L. L., and Magleby, S. P.: Lamina emergent mechanisms and their basic elements, Journal of Mechanisms and Robotics, 2, 1–9, doi:10.1115/1.4000523, compendex, 2010.

Kuribayashi, K., Tsuchiya, K., You, Z., Tomus, D., Umemoto, M., Ito, T., and Sasaki, M.: Self-Deployable Origami Stent Grafts as a Biomedical Application of Ni-rich TiNi Shape Memory Alloy Foil, Mat. Sci. Eng. A-Struct., 419, 131–137, 2006.

Lang, R. J. and Hull, T. C.: Origami design secrets: mathematical methods for an ancient art, The Mathematical Intelligencer, 27, 92–95, doi:10.1007/BF02985811, 2005.

Ma, J. and You, Z.: The Origami Crash Box, in: Origami 5: the 5th International Conference on Origami in Science Mathematics and Education, 2010.

Mahadevan, L. and Rica, S.: Self-Organized Origami, Science, 307, 1740, doi:10.1126/science.1105169, 2005.

Marcus, D. A.: Graph theory: a problem oriented approach, MAA, 2008.

Miura, K.: The Application of Origami Science to Map and Atlas Design, in: Origami3: Third International Meeting of Origami Science, Mathematics, and Education, edited by: Hull, T., A K Peters, Ltd., 2002.

Miura, K. and Natori, M.: 2-D Array Experiment on Board a Space Flyer Unit, Space Power, 5, 345–356, 1985.

O'Rourke, J.: How To Fold It, Cambridge University Press, 2011.

Tachi, T.: Simulation of Rigid Origami, in: Origami 4: Fourth International Meeting of Origami Science, Mathematics, and Education, edited by: Lang, R., A K Peters, Ltd., 175–187, 2006.

Tachi, T.: Rigid-Foldable Thick Origami, in: Origami 5: the 5th International Conference on Origami in Science Mathematics and Education, 2010.

Watanabe, N. and Kawaguchi, K.-I.: The Method for Judging Rigid Foldability, in: Origami 4: Fourth International Meeting of Origami Science, Mathematics, and Education, edited by: Lang, R., 165–174, 2006.

Wei, G. and Dai, J.: Geometry and kinematic analysis of an origami-evolved mechanism based on artmimetics, in: Reconfigurable Mechanisms and Robots, 2009, ReMAR 2009. ASME/IFToMM International Conference on, 450–455, 2009.

Winder, B. G., Magleby, S. P., and Howell, L. L.: A study of joints suitable for lamina emergent mechanisms, in: 2008 ASME International Design Engineering Technical Conferences and Computers and Information in Engineering Conference, DETC 2008, 3–6 August 2008, vol. 2 of 2008 Proceedings of the ASME International Design Engineering Technical Conferences and Computers and Information in Engineering Conference, DETC 2008, ASME, New York City, NY, United states, compendex, 339–349, 2009a.

Winder, B. G., Magleby, S. P., and Howell, L. L.: Kinematic representations of pop-up paper mechanisms, Journal of Mechanisms and Robotics, 1, 1–10, doi:10.1115/1.3046128, compendex, 2009b.

Wu, W. and You, Z.: Modelling rigid origami with quaternions and dual quaternions, Proceedings of the Royal Society A: Mathematical, Physical and Engineering Science, 466, 2155–2174, doi:10.1098/rspa.2009.0625, 2010.

15

An articulated handle to improve the ergonomic performance of robotic dextrous instruments for laparoscopic surgery

B. Herman[1,2], **A. Devreker**[1,*], **F. Richer**[3], **A. Hassan Zahraee**[3], **and J. Szewczyk**[3]

[1]Université catholique de Louvain, Institute of Mechanics, Materials, and Civil Engineering, Center for Research in Energy and Mechatronics, Louvain-la-Neuve, Belgium
[2]Université catholique de Louvain, Louvain Bionics, Louvain-la-Neuve, Belgium
[3]Université Pierre et Marie Curie – CNRS, Institute for Intelligent Systems and Robotics, Paris, France
[*]now with: K. U. Leuven, Department of Mechanical Engineering, Division Production engineering, Machine design and Automation, Leuven, Belgium

Correspondence to: B. Herman (benoit.herman@uclouvain.be)

Abstract. Hand-held robotic instruments with dextrous end-effectors offer increased accessibility and gesture precision in minimally invasive laparoscopic surgery. They combine advantages of both intuitive but large, complex, and expensive telesurgery systems, and much cheaper but less user-friendly steerable mechanical instruments. However, the ergonomics of such instruments still needs to be improved in order to decrease surgeon discomfort. Based on the results of former experimental studies, a handle connected to the instrument shaft through a lockable ball joint was designed. An experimental assessment of ergonomic and gesture performance was performed on a custom-made virtual reality simulator. Results show that this solution improves ergonomics, demanding less wrist flexion and deviation and elbow elevation, while providing gesture performance similar to a robotic dextrous instrument with standard pistol-like handle configuration.

1 Introduction

Several gestures in laparoscopic surgery are rather difficult to perform with conventional non-dextrous instruments. Such a straight and elongated instrument passing through a cannula has only a reduced set of 4 degrees of freedom (DOFs). As a consequence, the orientation of the instrument end-effector is coupled to its position in the abdominal cavity, and by extension to the handle posture above the patient. These kinematic constraints often force surgeons to work in awkward and painful postures, inducing discomfort and even pain after a while (Nguyen et al., 2001).

Adding one or several distal DOFs can help to restore mobility as in open surgery, making complex gestures like suturing easier. The da Vinci surgical system (Intuitive Surgical Inc., Sunnyvale, CA) offers such a functionality, and the benefits in terms of ease of use and ergonomics are well established (Freschi et al., 2013). However, its wide diffusion and use in clinical routine are restricted by the high selling price and maintenance costs of this telesurgery robotic system, and by its size and bulk.

A couple of simpler hand-held dextrous instruments have been available since 2006, such as RealHand HD from Novare Surgical Systems, Cupertino, CA (Danitz, 2006), Radius from Tuebingen Scientific Medical GmbH, Tübingen, Germany (Schwarz et al., 2005), Roticulator from Covidien Inc., Mansfield, MA (Marczyk et al., 2013), or Autonomy Laparo-Angle from Cambridge Endo, Framingham, MA (Lee and Chamorro, 2008). Research prototypes are also under development (e.g. Awtar et al., 2012; Wang et al., 2012). The distal DOFs of these devices are actuated manually either via knobs or joysticks on the handle (*fingertip control*), or through a jointed handle that is mechanically coupled to the end-effector (*wrist control*). They are very useful especially for single-port laparoscopic procedures (MacDonald et al., 2009), where the two instruments are inserted into the same cannula and cross each other in the abdominal wall, requiring intra-abdominal bending to reach the same site in a triangulated configuration (Rettenmaier et al., 2009; Frede et al., 2007; Rosenblatt et al., 2012; Endo et al., 2011). Kolwadkar et al. (2011) demonstrated experimentally that a fingertip-controlled instrument with a miniature joystick outperforms a wrist-controlled instrument on a needle driving task in conventional laparoscopy. The same conclusion was reached by Okken et al. (2012). However, the latter study also showed that, in some cases, conventional straight instruments performed even better than both dextrous prototypes. According to the authors, this can be due to the difference in quality between commercialized standard instruments and their research prototypes, and to the absence of locking mechanisms for the jointed handle and the joystick. However, adding locking mechanisms to a miniature joystick might not be that straightforward, especially if one wishes to lock one DOF and leave the other free. For example, to perform a stitch in a plane that is not orthogonal to the longitudinal axis of the instrument, the optimal strategy is to actuate the terminal roll motion of the grasper under a constant yaw bending (Hassan Zahraee et al., 2010). Furthermore, a direct mechanical transmission that synchronizes joystick and end-effector angular configurations (*position-position mapping*) requires a smooth and continuous finger motion to actuate the intra-abdominal DOFs, and restricts their range of motion.

One solution to these limitations is to replace the direct mechanical link by actuators whose velocity can be controlled by the joystick (*position-velocity mapping*). This has several advantages over a mechanical transmission. Distal range of motion is not restricted by the joystick angular range. Using non-backdrivable transmissions between actuators and end-effector induces a self-locking behavior: if the surgeon does not tilt the joystick, the intra-abdominal DOFs remain at rest. In addition, while a high amount of force must be applied by the finger on the joystick to stitch with a mechanical transmission, the required force and torque are now provided by the actuators. Finger and hand muscular fatigue can therefore be reduced. We demonstrated on a custom-made virtual-reality (VR) simulator that the number of successful stitches was significantly higher when controlling the distal DOFs with a thumb-actuated joystick mounted on the instrument handle in a position-velocity mapping, with respect to a jointed handle with position-position mapping and a locking feature (Hassan Zahraee et al., 2010). Hand-held robotic instruments might thus combine the advantages of both solutions: a cost-effective dexterity as in mechanical dextrous instruments; with a user-friendly control of distal DOFs as offered by telesurgery systems. Three hand-held robotic instruments have so far been put on the market, all using the optimal yaw-roll kinematics: JAiMY from EndoControl Medical, La Tronche, France (Paik et al., 2010), Kymerax from Terumo Europe Advanced Surgical, Eschborn, Germany (Hackethal et al., 2012) and Dextérité Hand Held Robot from Dextérité Surgical, Annecy, France (Barrier et al., 2010). Various research prototypes were also introduced by the University of Tokyo, Japan (Yamashita et al., 2004), Toshiba Medical Systems with the Keio University School of Medicine, Tokyo, Japan (Jinno et al., 2002), the University of Darmstadt, Germany (Röse et al., 2009), Scuola Supereriore Sant'Anna, Pisa, Italy (Piccigallo et al., 2008), and the Delft University of Technology (Lassooij et al., 2012). These instruments help surgeons with their biggest challenge which is performing complex gestures in laparoscopy. However, they fail to completely address another very important problem of laparoscopists – the poor ergonomics of laparoscopic surgery.

Decoupling handle and shaft orientation with a free ball joint between them allows exploration of the entire intra-abdominal workspace without excessive wrist flexion or deviation, since the handle is not forced to rotate while swiveling the shaft. We showed on our VR simulator (Herman et al., 2011) that the ergonomic performance of such an articulated handle was significantly higher, but at the cost of a lower gesture performance (i.e. an increase in task duration and path length). This can be attributed mainly to the "floating" behavior of the free handle that might complicate reaching a precise target by inducing some loss of shaft controllability. To solve this problem of floating behavior, introducing the possibility to lock/unlock the joint between handle and shaft seems to be a promising approach. The Dextérité Hand Held Robot also features a lockable joint between the handle and the shaft. However, it is only a 1-DOF revolute joint that allows a Yaw motion, which might not be sufficient for releasing all strains in the surgeon's wrist. In this paper, we introduce the design and performance assessment of a robotic instrument equipped with two lockable joints between handle and shaft that provide 3 revolute DOFs. Our hypothesis is that it could allow a more comfortable arm and wrist configuration regardless of the shaft orientation, while restoring instrument rigidity for a precise control of fine surgical gestures. The remainder of this paper is organized as follows: Sect. 2 describes the handle mock-up and its locking modes. An experimental performance assessment using the VR simulator is detailed in Sect. 3. Results are reported in Sect. 4 and discussed in Sect. 5.

2 Articulated handle

2.1 Design overview

Figure 1 depicts the prototype of a lockable articulated handle. It is composed of a Nunchuk handle (Nintendo) with a thumb-actuated 2-DOF joystick that allows the control of the Yaw and Roll distal DOFs. Note that, for this study, these distal DOFs are implemented virtually in the simulator, whereas the actual robotized prototype (not described here) embeds DC motors to actuate them. The Nunchuk handle is connected to the instrument shaft through two lockable joints:

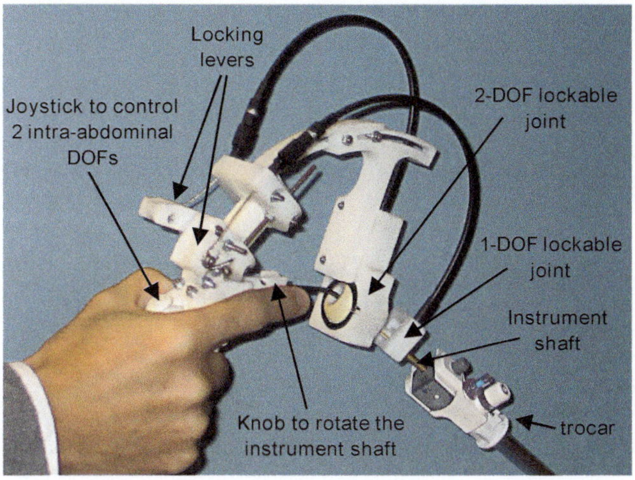

Figure 1. Prototype of articulated handle with lockable joints.

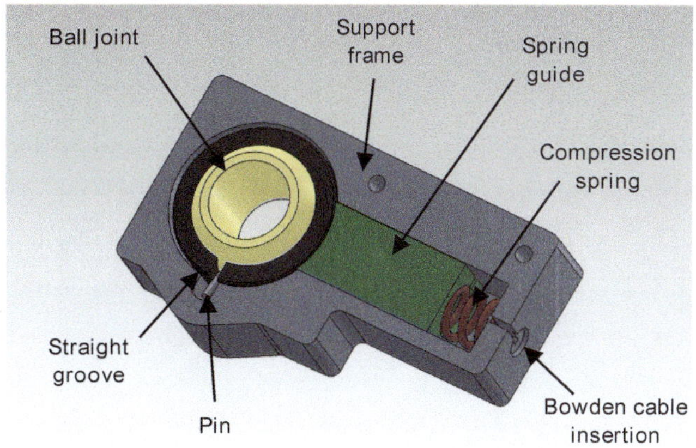

Figure 2. CAD view of the universal joint with locking compression spring.

a 2-DOF universal joint and a 1-DOF revolute joint around the shaft longitudinal axis.

As shown in Fig. 2, the universal joint is made up of a plastic ball joint with a pin partly inserted into the ball, perpendicularly to the instrument shaft. The pin can slide along a straight groove in the joint base to allow Yaw and Pitch motions of the shaft, but prevents any Roll motion (i.e. self-rotation). The universal joint allows the user to move the shaft within a cone of approximately 70° of aperture, while self-rotation is unrestricted. A star-shaped knob at the front of the handle is linked to the instrument shaft via a flexible axle in order to control its Roll motion with the index finger, as on most straight instruments.

Both joints are locked by means of a compression spring and can be released by a thumb-actuated lever that pulls a Bowden cable. The locking torque produced by the spring on the universal joint is 0.5 Nm for Pitch (i.e. left–right motion of the instrument tip) and 0.25 Nm for Yaw (i.e. up–down motion). It is obviously not sufficient to perform a task in the real surgical world, where a force of up to 50 N may be required to insert a surgical needle into muscle tissue. Nevertheless, since it is designed to operate in a VR simulator without any interaction force between the instrument tip and the environment, the only forces that the locking mechanism has to withstand are the reaction and friction forces in the trocar (neglecting the shaft's own weight). Furthermore, we determined experimentally that the locking torque is one order of magnitude higher than the friction torque inside the joint. The user can thus easily feel if the joint is locked or free during a trial run on the VR simulator, which is the purpose of this mock-up.

2.2 Locking modes

The handle can be used in one of these four modes:

1. *Standard fixed handle*, with both joints locked in their central position.

2. *Fully free handle*, with both joints released during the entire task, the shaft self-rotation being controlled by the index finger.

3. *2-DOF adjusted and locked, 1-DOF free*, in which the 2-DOF joint is locked in a convenient configuration during the entire task while the 1-DOF joint can be actuated by finger during the task.

4. *Both joints adjusted and locked* in a convenient configuration during the entire task.

The first two modes are similar to those tested previously in Herman et al. (2011), although the actuated self-rotation of the shaft through the active trocar was replaced by finger actuation using the star-shaped knob described above. For the last two modes, the handle is pre-locked in a convenient and ergonomic configuration after positioning the end-effector in the task region, before starting the task itself. According to Matern and Waller (1999), this configuration is ideal with the arm slightly abducted, retroverted, and rotated inwards at shoulder level, the elbow bent at about 90–120°, and the hand in medial configuration (i.e. without any pronation, wrist flexion or deviation). The pre-locking procedure is compatible with actual surgery, during which instrument motions remain confined in a small region of the intra-abdominal workspace for a few minutes to perform a specific action (e.g. dissection, cutting, clip placement, stitching).

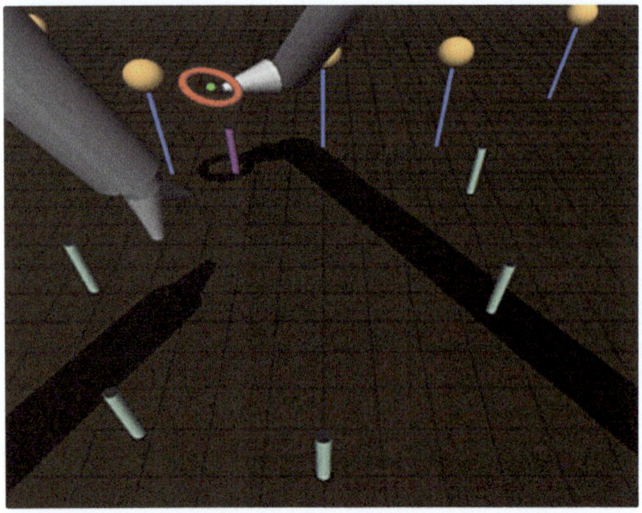

Figure 3. Screen capture of the VR simulator pick-and-place task.

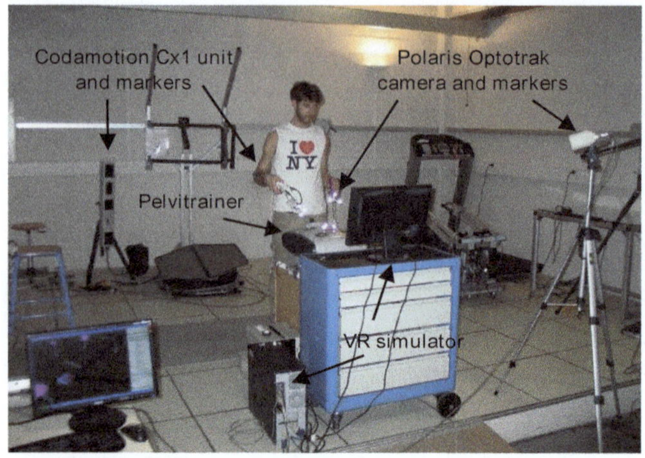

Figure 4. Overview of the experimental setup.

3 Performance assessment

The performance of these four modes was assessed through an experiment on the aforementioned VR simulator. This section describes the manipulation task, the two main comparison metrics, the experimental setup and protocol, and the statistical methods used to analyze recorded data. Since the present study derives directly from the experiment reported in Herman et al. (2011), only important information and differences with the previous study are summarized in this paper. The reader interested in further details will find them in the aforementioned reference.

3.1 Task

A single pick-and-place task (Fig. 3) was used to assess the performance of the four handle modes. It reproduces the gesture complexity of stitching but, contrary to the latter, remains feasible with the same degree of difficulty in the absence of haptic feedback. As explained above, the task was improved compared to the previous experiment so as to reduce the instrument motion required during a task, by placing the ring close to the pin at the beginning of the trial. In addition, the pin can be placed either vertically (normal to the bottom plane) or at 45° from the vertical, so as to require a ring orientation during the task.

The chronometer is triggered when the subject bursts a balloon placed on top of a pin with the instrument tip. He/she must then grasp a ring placed horizontally on the bottom of the workspace, next to the pin. The grasp must be performed on a specific portion of the ring and under a certain orientation by using the intra-abdominal DOFs. This mimics the fact that a curved needle must be grasped under specific conditions to perform a stitch. Finally, the ring must be placed on the pin after a reorientation of the end effector. The chronometer is stopped when the ring is released on the pin.

3.2 Metrics

The performance of the four handle modes was assessed using the same metrics as in Herman et al. (2011): a global gesture performance score available in most commercially-available laparoscopy simulators, and an instantaneous ergonomic score.

The global performance score (P), initially proposed by Huang et al. (2005), combines the time to complete the task (TTC), the number of errors (Err) and the motion economy (ME), defined as the ratio between actual and optimal (shortest) path length:

$$P = \text{TTC} + \text{Err} + 2 \times \text{ME}. \tag{1}$$

The weight for motion economy had been chosen previously in agreement with surgeons for the study reported in Herman et al. (2011).

The ergonomic score is based on a real-time index inspired by the Rapid Upper Limb Assessment (RULA) form (McAtamney and Corlett, 1993) that was adapted by Person et al. (2001) to laparoscopic surgery. This instantaneous score is computed from joint angles of the major upper arm (i.e. the one that holds the dextrous instrument). It ranges over an integer scale from 1 (excellent posture) to 9 (very poor ergonomics). All fine details regarding the computation of this instantaneous score are reported and depicted in Herman et al. (2011).

3.3 Experimental setup

The experimental setup is shown in Fig. 4. The mock-up of robotic instrument with an articulated handle and one standard laparoscopic grasper (not required to perform the task but used to make the situation more realistic) are inserted in a pelvitrainer through trocars that form a 10 cm equilateral triangle with a virtual laparoscope. The pelvitrainer and monitor heights are adjusted to respect the standard ergonomic prescriptions for laparoscopy. Positions and orientations of both instruments with respect to the pelvitrainer are measured using a Polaris Optotrak system (Northern Digital Inc., Waterloo, Canada) connected to the VR simulator. To compute the instantaneous ergonomic score described above, major upper limb segments are tracked using a Codamotion system (Charnwood Dynamics Ltd., Leicestershire, UK) consisting of a Hub computer unit and 3 Cx1 units, offering a tracking accuracy better than 0.3 mm.

3.4 Protocol

Six right-handed subjects with no experience in surgery and variable experience in playing music or video games took part in the experimental campaign. At the beginning of each session with a new participant, the basic principles of laparoscopic surgery were explained. Then, the purpose of the study was detailed, along with the two metrics. The experimental setup and protocol were briefly explained during the placement of the Codamotion tracking markers on the subject. After calibration of the markers, the VR simulator and the handle were introduced via a short demonstration. The participant was then asked to perform the tasks as quickly as possible and using the shortest possible path, avoiding errors (i.e. a collision between ring and pin, or an instrument out of sight). This instruction was repeated regularly during the session.

Before starting the experiment, a learning exercise similar to the task was performed so as to become familiar with the VR simulator and the instrument. The learning curve for the global performance score and the task duration were fitted after each learning trial using Matlab. The exercise was repeated until the subject reached a performance plateau. On average, 10 repetitions were required. The learning exercise was also repeated after each handle mode change before starting the 5 recorded trials, until the subject felt comfortable with the new mode and the global performance score became stable. Three to five repetitions of the learning exercise were required, depending on the subject and the handle mode.

The task was repeated five times with each handle mode, each repetition being placed at one specific position in the virtual abdomen and under a specific angle so as to cover the entire workspace. The same order of pin placement and orientation was repeated for all handle modes and participants. The experiment lasted 75 min on average, including initial introduction, learning and experimental phases, and closing discussion.

Each participant started the experiment with handle mode 1 (fixed in a central configuration), then continued with one of the six permutations of the remaining three modes, randomly selected. Since the subjects had to become accustomed to both the instrument and the VR simulator itself during the initial learning session, we decided to always start with the most basic handle mode so as to facilitate the understanding of the VR simulator. In order to check whether this not fully randomized sequence had any influence on the results, the first two subjects repeated the experiment with the first handle mode (five repetitions of the task) at the end of their session, and their performance did not seem to be different from that at beginning of the experiment, suggesting no important cross-learning effect between modes.

3.5 Statistical analysis

A total of 120 repetitions (6 subjects, 4 handle modes, 5 trials for each mode) were recorded during the experimental session. During 5 trials, the subject let the ring fall and had to redo the entire gesture, increasing task duration and motion economy significantly. These trials were therefore excluded from the performance analysis, although they were taken into account in the ergonomics analysis since the gesture was performed normally. Conversely, recording of the upper limb kinematics did not work during 4 trials and these were not taken into account for the ergonomics analysis, although the performance metrics were recorded properly and included in the statistical analysis.

Valid data were analyzed with JMP 10.0.2 software (SAS Institute Inc.). An ANOVA was performed on the global performance score P, time to task completion TTC, motion economy ME, and on the average and maximum values of the RULA-based instantaneous ergonomic score for each trial. Each model contained the following effects: handle mode, subject, and their interaction. The handle mode was defined as a fixed factor, while the subject and the two-factor interaction were defined as random. The linear model was solved using REML (Restricted Maximum Likelihood). The Tukey HSD test of multiple comparisons was used to compare modalities of significant factors.

4 Results

Results are depicted on Figs. 5–8. Figures 5 and 7 represent the box-and-whisker plot of the average RULA-based score and the global performance score, respectively. These graphs were derived from the valid data (see above), and indicate the degree of dispersion and skewness in the data. Upper and lower limits of the boxes represent the first and third quartiles, respectively, while the intermediate line is the median. Whiskers depict the lower and higher values, excluding the outliers depicted as dots. These remaining outliers were not excluded from the analysis. Table 1 summarizes the least-square means and standard errors modeled by the ANOVA for each ergonomic and performance metric.

The ANOVA performed on the average RULA-based ergonomic score turned out to be significant ($p = 0.0011$). The Tukey HSD test showed that the first handle mode (standard

Table 1. Least-squares means and standard errors modeled by ANOVA for each handle mode.

	Ergonomic scores			Performance scores			
Handle mode	Repetitions	Average RULA	Max. RULA	Repetitions	*P*	TTC	ME
Mode 1	30	3.50 (0.171)	4.43 (0.216)	29	34.0 (1.76)	29.4 (1.48)	3.26 (0.25)
Mode 2	29	3.03 (0.172)	3.88 (0.217)	27	33.2 (1.84)	28.2 (1.55)	3.55 (0.26)
Mode 3	28	3.05 (0.172)	3.88 (0.219)	30	30.2 (1.72)	25.8 (1.45)	3.15 (0.25)
Mode 4	29	2.91 (0.172)	3.57 (0.217)	29	34.1 (1.76)	29.6 (1.48)	3.24 (0.25)

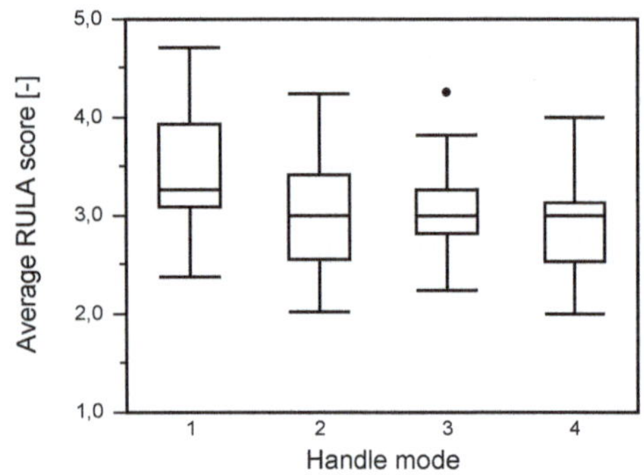

Figure 5. Box-and-whisker plot of the average RULA-based score for each handle mode. Sample size for each box is reported in Table 1. Outliers are represented as dots.

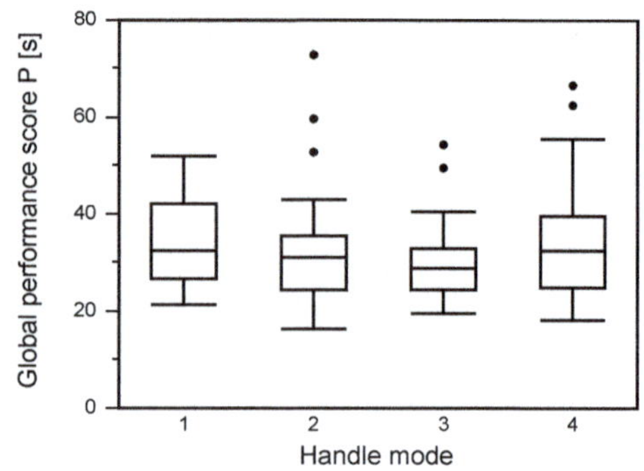

Figure 7. Box-and-whisker plot of the global performance score P for each handle mode. Sample size for each box is reported in Table 1. Outliers are represented as dots.

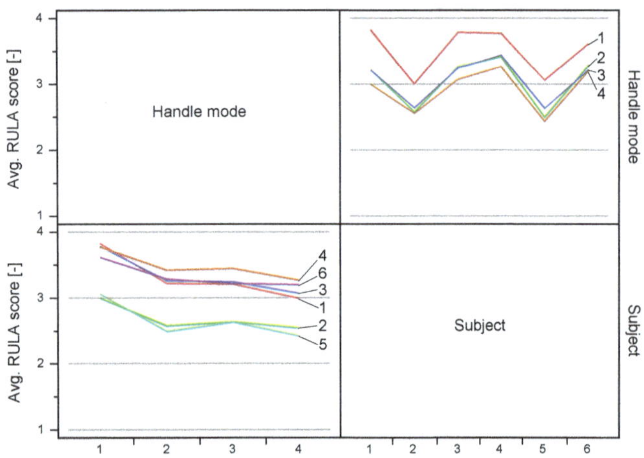

Figure 6. Interaction plot between factors "Handle mode" (fixed) and "Subject" (random) for the average RULA-based score.

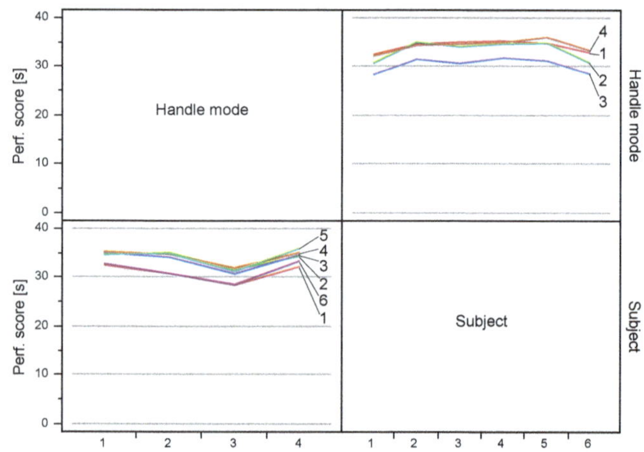

Figure 8. Interaction plot between factors "Handle mode" (fixed) and "Subject" (random) for the global performance score P.

fixed handle) was significantly less ergonomic (higher score) than all other modes ($p = 0.001$ for mode 4, $p = 0.0076$ for mode 2, and $p = 0.0114$ for mode 3), as can be seen on the upper-right plot of Fig. 6. No difference was found between the other modes ($p > 0.63$). The interaction plot on Fig. 6 suggests that there might be significant differences in terms of average RULA score between two groups of subjects: subjects 2 and 5 have a better (lower) ergonomic score than all the others. However, the plot shows that there is no interaction between factors "handle mode" and "subject": the same handle mode has poorer performance for all subjects. The ANOVA on the maximum ergonomic score is also significant ($p < 0.0001$) and the Tukey HSD results are similar, with a significantly higher value for the first mode ($p \leq 0.0015$), no significant difference between the last three modes, and no interaction.

No significant difference in global performance score was found between handle modes ($p = 0.42$). The same result was found for the task duration ($p = 0.3$) and for the motion economy ($p = 0.54$). Differences between subjects are minor, as can be seen on Fig. 8, and there is no interaction between factors.

5 Interpretation

Figure 6 and the statistical analysis show, as expected, that the ergonomics is improved significantly thanks to the articulated handle. This is in accordance with the previous study (Herman et al., 2011). It confirms the benefits of such a device for laparoscopists, enabling them to operate with a more comfortable and less tiring arm posture than with a handle fixed to the shaft of a conventional or dextrous instrument. In addition, this ergonomic improvement is obtained when the handle is free (mode 2) as in the previous study, and also when it is locked in a convenient configuration (modes 3 and 4 featured by the new design). This confirms our intuition that locking the handle partly or fully does not affect the general arm posture when gestures are concentrated momentarily in a small portion of the intra-abdominal workspace.

This increase in ergonomics is obtained with no decrease in average performance, as depicted in Fig. 8. This was expected for mode 4 where the handle is also fully locked during the task. However, it differs from what we found previously with a free handle (mode 2): in Herman et al. (2011), we reported that a free handle was less efficient than a fixed one. This difference between the two experimental results is due mainly to the fact that rolling the shaft is performed manually using a finger in this experiment, whereas it was motorized and controlled by the joystick previously. This might suggest that the DOFs decoupling proposed in this paper with manual control of the Roll shaft motion is more natural and intuitive than the previous actuated version, although this needs to be confirmed experimentally.

The significant difference found between subjects stems mainly from the fact that the simulator height could only be adjusted between two positions. Depending on his/her own height, each subject chose the most comfortable adjustment. However, arm angles could differ between subjects for the same instrument tip position inside the virtual abdomen, resulting in different average RULA-based scores. In addition, although the same instructions were given repeatedly to the subjects, some tried to maintain the most ergonomic posture possible, while others tried to work as fast as they could and paid less attention to their arm posture.

6 Conclusions

Several surgical gestures are difficult to perform using standard laparoscopic instruments. Hand-held robotic instruments with additional end-effector DOFs might be an optimal solution, combining the dexterity enhancement offered by tele-surgery robotic systems and the cost-effectiveness of purely mechanical devices. However, despite these technological improvements, surgeons still have to take uncomfortable, or even painful, postures.

This paper introduces a novel articulated handle that releases constraints between upper limb configuration and instrument tip position and orientation inside the abdominal cavity. An experimental study performed on a custom-made VR simulator tends to demonstrate that the articulated handle helps in restoring an ergonomic arm posture, without reducing gesture performance. A fully-functional instrument prototype is currently being developed for future bench-top and in vivo validation. Ongoing work also includes the implementation of force feedback on the VR simulator, so as to assess its influence on the comparison between instruments.

In addition, although the decrease in ergonomic score offered by the articulated handle is significant, it would be interesting to know to what extent it has an effect on the surgeon's comfort over a period of time longer than the duration of the experiment. A complementary study could therefore be performed using either the VR simulator or the instrument prototype under development. During this study, subjects would repeat gestures with the handle in mode 1 or 3 for at least one hour, and several physiological parameters (e.g. cardiac rythm, EMG in shoulder, arm and forearm muscles) that correlate with physical workload, comfort and ergonomics would be measured.

Finally, although our study focuses only on robotic instruments with intra-abdominal mobility, one can assume that its conclusions could be extended to standard laparoscopic instruments. Since not all surgical gestures require the high dexterity provided by the actuated distal DOFs, standard instruments (e.g. graspers, hooks, scissors) could easily be equipped with such an ergonomic handle and at a low additional cost.

Acknowledgements. The authors would like to warmly thank all their colleagues who took part in the experiments. Authors are also grateful to Catherine Rasse (Institute for Multidisciplinary Research in Quantitative Modelling and Analysis, Université catholique de Louvain) for her valuable advice on the design of this experiment and on the statistical analysis.

Edited by: J. van den Dobbelsteen
Reviewed by: two anonymous referees

References

Awtar, S., Trutna, T. T., Nielsen, J. M., Abani, R., and Geiger, J.: FlexDex^{TM}: A minimally invasive surgical tool with enhanced dexterity and intuitive control, J. Med. Devices, 4, 035003, doi:10.1115/1.4002234, 2012.

Barrier, P., Ollagnier, J., and Gautier, G.: Manipulator with decoupled movements and application to instruments for minimally invasive surgery, US Patent Application US2010/0331860A1, 2010.

Danitz, D. J.: Articulating mechanisms with joint assembly and manual handle for remote manipulation of instruments and tools, US Patent Application US2006/0201130A1, 2006.

Endo, T., Nagasawa, K., Umemura, K., Baba, T., Henmi, H., and Saito, T.: A remarkably easy knot-tying technique for single-incision laparoscopic surgery with the SILS port for gynecologic diseases, J. Minim. Invas. Gyn., 18, 500–502, doi:10.1016/j.jmig.2011.03.014, 2011.

Frede, T., Hammady, A., Klein, J., Teber, D., Inaki, N., Waseda, M., Buess, G., and Rassweiler, J.: The Radius Surgical System: A new device for complex minimally invasive procedures in urology?, Eur. Urol., 51, 1015–1022, 2007.

Freschi, C., Ferrari, V., Melfi, F., Ferrari, M., Mosca, F., and Cuschieri, A.: Technical review of the da Vinci surgical telemanipulator, Int. J. Med. Robot. Comp., 9, 396–406, doi:10.1002/rcs.1468, 2013.

Hackethal, A., Koppan, M., Eskef, K., and Tinneberg, H.-R.: Handheld articulating laparoscopic instruments driven by robotic technology. First clinical experience in gynecological surgery, Gynecol. Surg., 9, 203–206, doi:10.1007/s10397-011-0708-2, 2012.

Hassan Zahraee, A., Paik, J. K., Szewczyk, J., and Morel, G.: Towards the development of a hand-held surgical robot for laparoscopy, IEEE T. Mech., 15, 853–861, doi:10.1109/TMECH.2010.2055577, 2010.

Herman, B., Hassan Zahraee, A., Szewczyk, J., Morel, G., Bourdin, C., Vercher, J.-L., and Gayet, B.: Ergonomic and gesture performance of robotized instruments for laparoscopic surgery, in: Proceedings of the IEEE/RSJ International Conference on Intelligent Robots and Systems (IROS'11), 25–30 September 2011, San Francisco, CA, 1333–1338, doi:10.1109/IROS.2011.6094449, 2011.

Huang, J., Payandeh, S., Doris, P., and Hajshirmohammadi, I.: Fuzzy classification: Towards evaluating performance on a surgical simulator, in: Proceedings of Medicine Meets Virtual Reality (MMVR'05), 26–29 January 2005, Long Beach, CA, 194–200, 2005.

Jinno, M., Matsuhira, N., Sunaoshi, T., Hato, T., Miyagawa, T., Morikawa, Y., Furukawa, T., Ozawa, S., Kitajima, M., and Nakazawa, K.: Development of a master slave combined manipulator for laparoscopic surgery, in: Medical Image Computing and Computer-Assisted Intervention (MICCAI2002), vol. 2488 of Lect. Notes Comput. Sc., edited by: Dohi, T. and Kikinis, R., Springer, Berlin, Heidelberg, 52–59, doi:10.1007/3-540-45786-0_7, 2002.

Kolwadkar, Y. V., Brown, S. I., Abboud, R. J., and Wang, W.: Comparison of two actuation systems for laparoscopic surgical manipulators using motion analysis, Surg. Endosc., 25, 964–974, doi:10.1007/s00464-010-1300-y, 2011.

Lassooij, J., Tolou, N., Tortora, G., Caccavaro, S., Menciassi, A., and Herder, J. L.: A statically balanced and bi-stable compliant end effector combined with a laparoscopic 2DoF robotic arm, Mech. Sci., 3, 85–93, doi:10.5194/ms-3-85-2012, 2012.

Lee, W. and Chamorro, A.: Surgical instrument, US Patent US7338513B2, 2008.

MacDonald, E. R., Brownlee, E., and Ahmed, I.: New tools for a new job – Single port laparoscopic surgery equipment, Med. Equip. Insights, 2, 1–7, doi:10.4137/MEI.S9056, 2009.

Marczyk, S., Pribanic, R., and Horton Jr., K. W.: Surgical instrument with articulation and wrist rotation, US Patent Application USUS2013/0012958, 2013.

Matern, U. and Waller, P.: Instruments for minimally invasive surgery: Principles of ergonomic handles, Surg. Endosc., 13, 174–182, doi:10.1007/s004649900934, 1999.

McAtamney, L. and Corlett, E. N.: RULA: A survey method for the investigation of work-related upper limb disorders, Appl. Ergon., 24, 91–99, doi:10.1016/0003-6870(93)90080-S, 1993.

Nguyen, N. T., Ho, H. S., Smith, W. D., Philipps, C., Lewis, C., De Verab, R. M., and Berguer, R.: An ergonomic evaluation of surgeons' axial skeletal and upper extremity movements during laparoscopic and open surgery, Am. J. Surg., 182, 720–724, doi:10.1016/S0002-9610(01)00801-7, 2001.

Okken, L. M., Chmarra, M. K., Hiemstra, E., Jansen, F. W., and Dankelman, J.: Assessment of joystick and wrist control in hand-held articulated laparoscopic prototypes, Surg. Endosc., 26, 1977–1985, doi:10.1007/s00464-011-2138-7, 2012.

Paik, J. K., Morel, G., Vidal, C., and Henri, P.: Surgical Instrument, US Patent Application US2012/0083770A1, 2010.

Person, J., Hodgson, A., and Nagy, A.: Automated high-frequency posture sampling for ergonomic assessment of laparoscopic surgery, Surg. Endosc., 15, 997–1003, doi:10.1007/s004640080155, 2001.

Piccigallo, M., Focacci, F., Tonet, O., Megali, G., Quaglia, C., and Dario, P.: Hand-held robotic instrument for dextrous laparoscopic interventions, Int. J. Med. Robot. Comp., 4, 331–338, doi:10.1002/rcs.214, 2008.

Rettenmaier, M. A., Lopez, K., Graham, C. L., Brown, J. V., John, C. R., Micha, J. P., and Goldstein, B. H.: RealHand high dexterity instruments for the treatment of stage I uterine malignancy, J. Soc. Laparoend., 13, 27–31, 2009.

Röse, A., Wohlleber, C., Kassner, S., Schlaak, H., and Werthschutzky, R.: A novel piezoelectric driven laparoscopic instrument with multiple degree of freedom parallel kinematic structure, in: Proceedings of the IEEE/RSJ International Conference on Intelligent Robots and Systems (IROS 2009), 10–15 October 2009, St. Louis, MO, 2162–2167, doi:10.1109/IROS.2009.5354507, 2009.

Rosenblatt, P. L., Ferzandi, T. R., Hong, S., and Kim, D.: Laparoscopic paravaginal repair with the Autonomy Laparo-Angle device, J. Minim. Invas. Gyn., 14, S151, doi:10.1016/j.jmig.2007.08.659, 2012.

Schwarz, K. M., Schurr, M. O., and Buess, G. F.: Surgical instrument for minimally invasive surgical interventions, US Patent US6913613B2, 2005.

Wang, X., Wang, S., Li, J., Zhang, G., and Wu, Z.: Conceptual design of a novel multi-DoF manual instrument for laparoscopic surgery, Int. J. Med. Robot. Comp., 9, 75–82, doi:10.1002/rcs.1433, 2012.

Yamashita, H., Hata, N., Hashizume, M., and Dohi, T.: Handheld laparoscopic forceps manipulator using multi-slider linkage mechanisms, in: Medical Image Computing and Computer-Assisted Intervention (MICCAI2004), vol. 3217 of Lect. Notes Comput. Sc., edited by: Barillot, C., Haynor, D. R., and Hellier, P., Springer, Berlin, Heidelberg, 121–128, doi:10.1007/978-3-540-30136-3_16, 2004.

16

New multiphase choke correlations for a high flow rate Iranian oil field

M. Safar Beiranvand[1], P. Mohammadmoradi[2], B. Aminshahidy[1], B. Fazelabdolabadi[1], and S. Aghahoseini[3]

[1]Institute of Petroleum Engineering, College of Engineering, University of Tehran, Tehran, Iran
[2]Department of Chemical and Petroleum Engineering, Sharif University of Technology, Tehran, Iran
[3]Department of Petroleum Engineering, Azad University of Kharg Island, Kharg Island, Iran

Correspondence to: M. Safar Beiranvand (mahmoud.beiran@ut.ac.ir)

Abstract. The multiphase flow through wellhead restrictions of an offshore oil field in Iran is investigated and two sets of new correlations are presented for high flow rate and water cut conditions. The both correlations are developed by using 748 actual data points, corresponding to critical flow conditions of gas-liquid mixtures through wellhead chokes. The first set of correlations is a modified Gilbert equation and predicts liquid flow rates as a function of flowing wellhead pressure, gas-liquid ratio and surface wellhead choke size. To minimize error in such condition, in the second correlation, free water, sediment and emulsion (BS & W) is also considered as an effective parameter. The predicted oil flow rates by the new sets of correlations are in the excellent agreement with the measured ones. These results are found to be statistically superior to those predicted by other relevant published correlations. The both proposed correlations exhibit more accuracy (only 2.95 % and 2.0 % average error, respectively) than the existent correlations. These results should encourage the production engineer which works at such condition to utilize the proposed correlations for future practical answers when a lack of available information, time, and calculation capabilities arises.

1 Introduction

Generally almost all flowing wells utilize wellhead chokes in order to regulate the flowing rate, prevent water and gas coning and sand problems, as well as valves installed for safety purposes. As a major factor of killing the producer well, the inaccurate flow rate predictions may inevitably lead towards gas or water coning, sand entry and excessive pressures at the separator. Larger amount of produced water contributes to high operating costs and is a major environmental concern for oil production (Jin and Wojtanowicz, 2010).

The most of published multiphase flow correlations are highly empirical (Mesallati et al., 2000). Several correlations have been offered for describing critical and subcritical multi-phase flow through wellhead chokes (Al-Attar, 2009). Majority of such correlations are based on limited ranges of flow variables. According to quality and scope of data upon which they are based, the validity of those correlations is limited. The most popular correlation was developed by Gilbert (1954) but it is valid for critical flow occurring when the upstream pressure of the choke is at least 70 % higher than the downstream pressure or when the ratio of downstream pressure to upstream pressure is equal to 0.588 (Ghareeb and Shedid, 2007). The most popular correlation was developed by Gilbert (1954) but it is valid for critical flow occurring when the upstream pressure of the choke is at least 70 % higher than the downstream pressure (Ghareeb et al., 2007). However, other researchers proposed various correlations rather than Gilbert's equation in the literature.

Flow through the wellhead chokes mainly fall into two critical and subcritical conditions.

Within this context, the critical flow condition refers to the state which the flow rate reaches a maximum value that is independent of a pressure drop applied across the choke and changes in pressure downstream of the choke do not affect the flow rate. Hence, wellhead chokes are commonly

Table 1. Equation coefficient for different correlations and their accuracy (General form: $Q = A\frac{P_{\text{wh}}S^{B}(1-\frac{\text{BS\&W}}{100})^{D}}{\text{GLR}^{C}}$).

Correlation	Empirical Coefficient					
	A	*B*	*C*	*D*	Error %	Absolute Error %
This Work[1]	0.0328	2.275	0.586	0	2.95	12.1
This Work[2]	0.0382	2.151	0.5154	0.52965	2	9.8
Gilbert	0.1	1.89	0.546	0	11.26	18.37
Ros	0.574	2	0.5	0	13.45	19.2
Baxendell	0.1046	1.93	0.546	0	32.1	33
Achong	0.2618	1.88	0.65	0		

Table 2. Range of data used for the correlations.

	Flow Rate (bbl/day) (STB/Day)	GLR (SCF/STB)	Choke Size (1/64) in	Wellhead Pressure (Psi)	BS&W (Percent)
Range	3000–24 000	80–260	16–40	1400–12 000	0.1–30

operated under critical flow conditions in order to isolate the reservoir from pressure variation introduced by surface equipment. Empirical correlations are mainly used for critical flow condition. In this paper, production data from offshore field, in Iran, were used to generate two sets of new correlations for multiphase flow through wellhead surface choke.

2 Backgrounds

In 1954, Gilbert proposed his correlation relating surface production with the wellhead choke size as follows:

$$Q = A\frac{P_{\text{wh}}S^{B}}{\text{GLR}^{C}},\quad A = 0.1,\quad B = 1.89,\quad C = 0.546 \tag{1}$$

Where Q is gross liquid rate (bbl/day), GLR is gas-liquid ratio (MSCF/STB), P_{wh} is well (or tubing) head pressure (psig) and S is bean size (1/64) inch.

In 1960, Ros developed a very similar correlation to Gilbert (1954) correlation but with different correlating exponents as follows:

$$Q = A\frac{P_{\text{wh}}S^{B}}{\text{GLR}^{C}},\quad A = 0.574,\quad B = 2.0,\quad C = 0.5 \tag{2}$$

Another Gilbert form correlation was presented by Achong in 1961:

$$Q = A\frac{P_{\text{wh}}S^{B}}{\text{GLR}^{C}},\quad A = 0.2618,\quad B = 1.88,\quad C = 0.65, \tag{3}$$

There were many other correlations to predict flow rates in the critical region. Some of them were similar to the Gilbert correlation, but with different constants and exponents. One of them was proposed by Baxendell (1958):

$$Q = A\frac{P_{\text{wh}}S^{B}}{\text{GLR}^{C}},\quad A = 0.104,\quad B = 1.93,\quad C = 0.546 \tag{4}$$

These correlations are tabulated in Table 1.

3 Developments of new correlation and discussion

3.1 Levenberg-Marquardt algorithm

Levenberg-Marquardt algorithm is a method for finding the minimum of a multivariable nonlinear function which has been applied as a standard method for solving the squares minimum problem for the nonlinear functions. Levenberg-Marquardt algorithm (LMA) interpolates between Gauss-Newton algorithm (GNA) and gradient descent method. LMA algorithm is more resistant than GNA, means most times even if has been started very far from the extreme minimum, will gain the answer and of course for well-behaved functions and reasonable initial parameters, LMA is slightly slower than NGA.

Least squares problem includes finding $\boldsymbol{P}$ parameters vector for which the following cost function is minimized:

$$S(\boldsymbol{P}) = f^{T}f = \sum_{i=1}^{m}[f_i(\boldsymbol{P})]^2$$

Like as the other numerical minimization algorithms, Levenberg-Marquardt algorithm is a repeated process. For beginning the minimization, the user must present an initial guess for $\boldsymbol{P}$ parameters vector. Most cases a standard unconscious guess such as $\boldsymbol{P}^{T} = (1, 1, 000, 1)$ applies very well. In other cases, the algorithm applies just when the initial guess is close to the final answer somehow.

In each repeat step, $\boldsymbol{P}$ parameters vector is replaced with a new $p+q$ estimation. For achieving q, $f_i\,(p+q)$ functions are estimated as $f_i\,(p+q) \approx f(p) + J_q$ through their linearization, that J is Jacobin of f in p.

In a minimum of sum of S squares, we have $\nabla_q S = 0$, that upon the above linearization the following equation

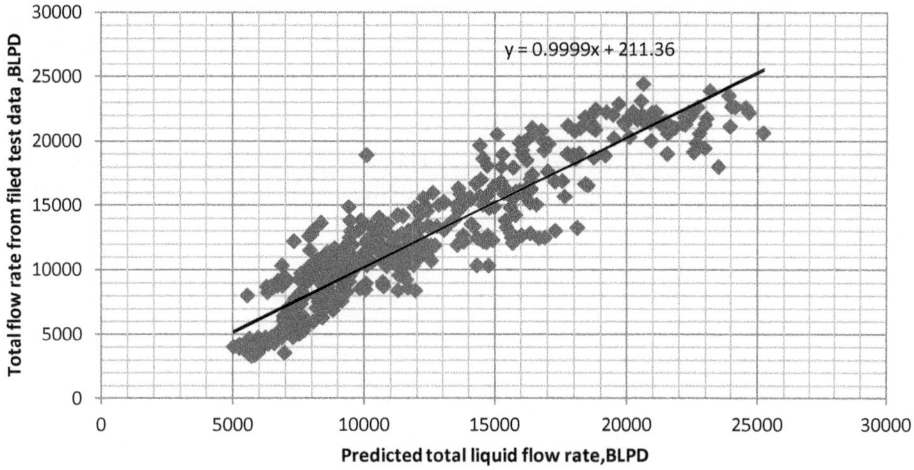

Figure 1. Predicted vs. measured flow rates for the first derived correlation.

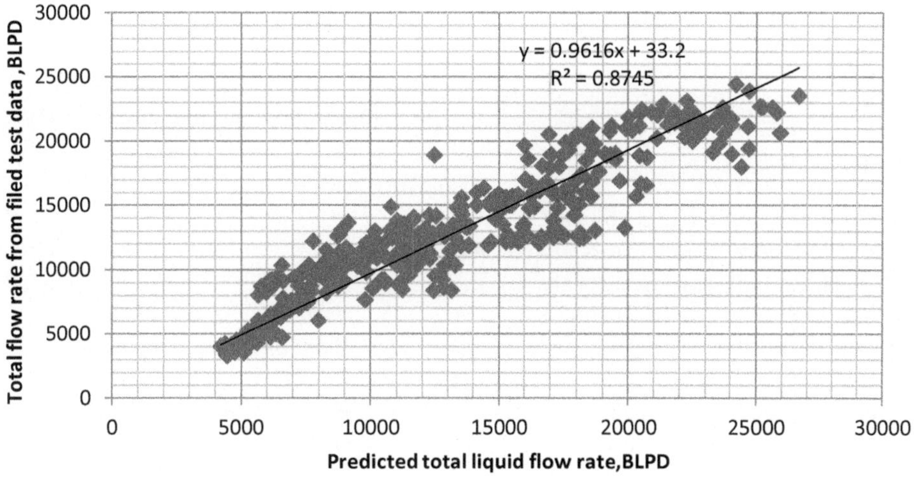

Figure 2. Predicted vs. measured flow rates for the second derived correlation.

is gained:

$$(J^T J)q = -J^T f$$

That q may be gained thereof through reversing $J^T J$. LMA key is replacement of this equation with its "attenuated version"

$$(J^T J + \lambda)q = -J^T f$$

Nonnegative attenuation coefficient λ is adjusted in each repeat. If the descent of S was fast, we give a smaller value to which that closes the algorithm to GNA, but if a repeat exhibited inadequate descent, may ascend the λ and take a step closer to gradient descent. A similar attenuation coefficient is observed in Tiknonov regularization which is useful for solving malfunctioned linear problems. If a recovered step length or descent of sum of squares for the last set of $\boldsymbol{P}$ parameters is lower than predetermined values, the repeat is finished and the last $\boldsymbol{P}$ parameter vector is deemed as the answer.

3.2 Results and discussion

748 production test data were collected for 10 wells, including the liquid flow rate, gas-liquid ratio, choke Size and wellhead pressure lied within the reign 3000–24 000 (STB/day), 80–260 (SCF/STB), 16–40 (1/64th-inch) and 1400–12 000 (Psi), respectively. Range of the data used is also tabulated in Table 2. The Levenberg-Marquardt Algorithm (LMA) is used for proposing the two sets of new correlations.

The first proposed correlation, a modified Gilbert equation based on regression analysis for the field is:

$$Q = A\frac{P_{\mathrm{wh}}S^B}{\mathrm{GLR}^C}, \quad A = 0.0328, \quad B = 2.275, \quad C = 0.586 \tag{5}$$

The second newly-developed correlation considered a parameter which was not included in the previous correlations; free water, sediment and emulsion (BS & W). This is in addition to other parameters appeared in Gilbert and other correlations, in order to minimize error in field condition, this correlation is:

$$Q = A\frac{P_{wh}S^B\left(1-\frac{BS\&W}{100}\right)^D}{GLR^C}, \quad A = 0.0382,$$
$$B = 2.1510, \quad C = 0.5154, \quad D = 0.5297 \tag{6}$$

where P_{wh} is well head pressure (Psig), GLR is gas-liquid ratio (SCF/STB), Q is gross liquid flow rate (bbl/day), S is choke size (64th of an inch) and finally, BS & W, it includes free water, sediment and emulsion and is measured as a volume percentage of the production stream. As can be seen, in the Gilbert's formula, the unit of GLR is MSCF/STB whereas in the proposed correlation the unit is SCF/STB.

Generally adopted assumptions on flow conditions are:

- Critical multiphase flow occurs at a constant critical pressure ratio,
- Both phases flow with the same velocity in the throat,
- Friction is negligible, the flow behavior is dominated by changes in kinetic energy,
- Expansion of the gas phase in the throat is polytropic.

Regarding the above assumptions, the average and absolute errors are used for comparing correlations. These errors are statistical measurements of how far estimates or forecasts are from actual values and could be applied to any two pairs of numbers, where one set is "actual" and the other is an estimate, forecast or prediction.

Two types of errors are calculated using the equations as follow:

$$e = \frac{q_{test} - q_{correl}}{q_{test}} \tag{7}$$

$$e_{abs} = \left|\frac{q_{test} - q_{correl}}{q_{test}}\right| \tag{8}$$

The first proposed correlation, a modified Gilbert equation, has exhibited an average and absolute errors of 2.95 % and 12.1 % respectively, whereas the second proposed correlation has showed an average error of roughly 2.0 %, also absolute error of 9.8 %. The average and absolute errors of other types correlations are tabulated in Table 1. Figures 1 and 2 show a good agreement between the measured and the predicted data by the new correlations. As evident from Fig. 2, a very good accuracy of predicted production rates with correlation factor (R^2) of 0.973 is obtained.

4 Conclusions

Based on 748 actual data points, the available choke correlations were reviewed and two sets of newly-established correlations were presented for an offshore field in Iran. The following conclusions are drawn:

1. The first proposed correlation, a modified Gilbert equation, has exhibited an average error of 2.95 %, whereas current correlations such as Gilbert and Ros represent average errors of 11.26 % and 13.45 %, respectively.
2. The second newly-developed correlation considered a parameter which was not included in the previous correlations; free water, sediment and emulsion (BS & W). This is in addition to other parameters appeared in Gilbert and other published correlations considered in this work, in order to minimize error in field condition. The second set of correlations exhibited an average error of 2.0 %.
3. The results strongly encourage the production engineer to utilize the proposed correlations for future practical answers in case of lack of information availability, time, and calculation capabilities arises.
4. Undoubtedly, presented correlations can be used, in order to controlling the water production as well as reach the optimum production for under consideration high flow rate Iranian oil field and similar cases.

Appendix A

Nomenclature

A	Proportionality constant
B	Bean or choke size exponent
bbl	Barrel
BS & W	Basic Sediment & Water
C	Gas-oil ratio exponent
D	Basic Sediment & Water term exponent
E	Temperature term exponent
e	Error
e_{abs}	Absolute error
F	Well head pressure exponent
GLR	Producing gas-liquid ratio at standard conditions, SCF/STB
P_{wh}	Well head pressure, psig
Q	Gross liquid flow rate, bbl/day
Q_{correl}	Calculated gross liquid flow rate from correlation, bbl/day
Q_{test}	Measured gross liquid flow rate, bbl/day
S	Choke or bean size, 1/64 inch
SCF	Standard cubic foot
STB	Stock tank barrel
T	Temperature, R
T_{SC}	Standard condition temperature, R

Edited by: A. Barari
Reviewed by: three anonymous referees

References

Achong, I.: Revised Bean Performance Formula for Lake Maracaibo Wells, internal co. report, Shell Oil Co., Houston, TX, 1961.

Al-Attar, H. H.: New Correlations for Critical and Subcritical Two-Phase Flow Through Surface Chocks in High-Rate Oil Wells, SPE 120788, 2009.

Baxendell, P. B.: Producing Wells on Casing Flow – An Analysis of Flowing Pressure Gradients, AIME, 213, 202–207, 1958.

Economides, M. J., Hill, A. D., and Ehlig-Economides, C.: Petroleum Production Systems, Prentice Hall PTR, New Jersey, p. 229, 1993.

Ghareeb, M. and Shedid, A.: A New Correlation for Calculating Wellhead Production Considering Influence of Temperature, GOR, and Water-Cut for Artificially Lifted Wells, IPTC 11101, 2007.

Gilbert, W. E.: Flowing and Gas-Lift Well Performance, Dril. And Prod. Prac. API, 143, 127–154, 1954.

Jin, L. and Wojtanowicz, A. K.: Coning Control and Recovery Improvement Using in-Situ Water Drainage/Injection in Bottom – Water-Drive Reservoir, SPE 129663, 2010.

Mesallati, A., Biznati, M., and Mansouri, N.: Multiphase-Flow choke correlations for Offshore Bouri Oil field, International Gas Union 21st World Gas Conference, 49 pp., Nice, France, 6–9 June, 2000.

Ros, N. C. J.: An Analysis of Critical Simultaneous Gas/Liquid Flow Through a Restriction and Its Application to Flow metering, Applied Science Research, 9, 374–388, doi:10.1007/BF00382215, 1961.

17

Static model for a 3-DOF underactuated finger

F. Guay and C. Gosselin

Laboratoire de robotique, Département de génie mécanique, Université Laval, Québec, Canada

Abstract. This paper introduces a static model of a three-degree-of-freedom underactuated finger. The model includes all static forces, namely actuation forces, return forces and gravity. All geometric and static parameters can be freely changed (pulley radius, member's mass, etc.). Hence, the model allows complete static simulations to be performed and it can also be used for numerical optimization.

This paper was presented at the IFToMM/ASME International Workshop on Underactuated Grasping (UG2010), 19 August 2010, Montréal, Canada.

1 Introduction

Governmental investment in prosthetic research has increased significantly over the last decade, the most notable initiative being DARPA's Revolutionizing Prosthetics program in the US (Belfiore, 2009).

One of the most sought and important criterion is to develop a hand that looks and feels natural (Belfiore, 2009). That is the person who wears and uses the prosthesis doesn't feel out of place in public. This criterion is extremely important for that the person who is to wear the prosthesis will not prematurely discard it to never wear it again. Therefore, a prosthetic hand should have 5 fingers each including 3 phalanges.

A possible approach is to develop a hand in which each finger joint is independently actuated. Even with simplifications such as modelling the finger's joints as revolute joints, this leads to a total of 15 actuators which have to be powered and controlled. The problem encountered in such an approach is the integration of such a large number of parts, the amount of time and effort for someone who is not likely technically inclined to master and use this complex assembly, the reliability and ruggedness of such an assembly and, often the problem that will be determinant, the combined weight and wearability of such a system.

Therefore, underactuation is a very attractive alternative. Underactuation is the property of a mechanism that has fewer degrees of actuation than degrees of freedom. The control of the degrees of freedom is achieved by mechanical intelligence.

Several underactuated fingers have been proposed in the literature (see for instance Birglen et al., 2008) for a literature survey). Many of the proposed fingers use tendon-based transmissions and a single degree of actuation. This principle is studied here and a finger model is proposed.

2 The finger model

The model proposed in this article considers the following torque vectors: τ_a, the actuation which makes the finger curl, τ_r, the spring action which makes the finger flex (open), τ_f, the friction forces and, an action never considered before, τ_g, the gravitational forces. τ_a and τ_r cannot be completely specified since we will impose a torque at only one joint of the finger. On the other hand, τ_f and τ_g can be fully specified for a given configuration of the finger. In order to model the behaviour of the finger, we will combine all four torque vectors. Figure 1 shows the components of these vectors and their position on the finger.

3 Underactuation

We begin by using, as a basis, the underactuation model proposed in Birglen et al. (2008); Allen-Demers et al. (2009) and adapt it to a typical finger used in a prosthesis. Here the only adaptation is to realise that our finger has 3 phalanges which are underactuated.

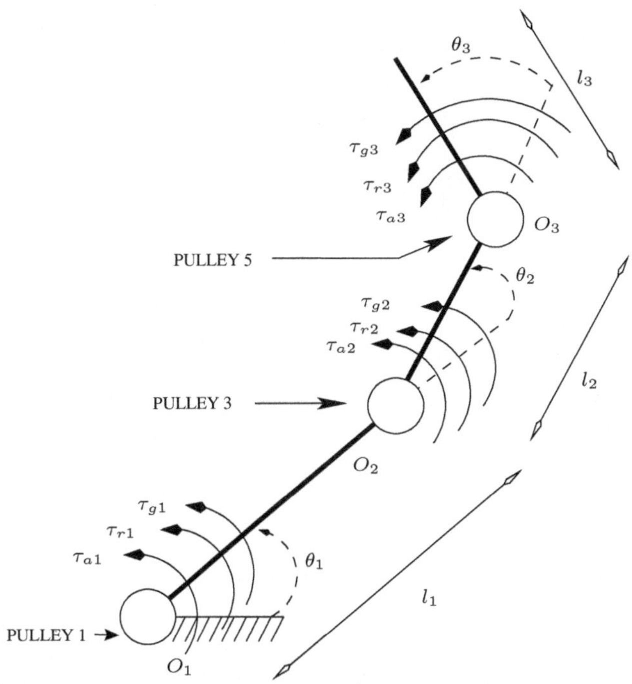

Figure 1. Finger with corresponding variables.

We equate the input and output virtual powers acting on the finger and obtain (Birglen et al., 2008):

$$t^T \omega_a = \sum_{i=1}^{3} \xi_i \circ \zeta_i \tag{1}$$

where

$$t = \begin{bmatrix} T_a \\ 0 \\ 0 \end{bmatrix}, \quad \omega_a = \begin{bmatrix} \dot{\theta}_1 \\ \dot{\theta}_2 \\ \dot{\theta}_3 \end{bmatrix},$$

$$\xi_i = \begin{bmatrix} \omega_i \\ v_i^x \\ v_i^y \end{bmatrix} \text{ and } \zeta_i = \begin{bmatrix} f_{ti} \\ f_{ni} \\ \tau_i \end{bmatrix}$$

t is the vector composed of the input torques applied on the i-th joint ($i = 1,2,3$). It is to be understood that joint 1 refers to the joint of the finger linking the latter to the palm of the hand and that joint 3 links the distal phalanx to the medial phalanx. ω_a is the vector composed of the angular velocities of the i-th joint ($i = 1,2,3$). $\dot{\theta}_a$ is the angular velocity of the actuation. ξ_i is the twist associated to the i-th point of contact and ζ_i is the corresponding wrench. They are operated with the reciprocal product, $\circ$ in the plane containing the finger. For simplicity, we assume that each phalanx is in contact with the object being grasped and that the i-th contact can be considered as a point. The twist is composed of the angular velocity, ω_i, the velocity components, v_i^x and v_i^y, of the i-th point of contact. These velocity components are in relation of the local reference frame of the i-th phalanx. The wrench is composed of the forces acting at the i-th contact point: a tangential force, f_{ti}, a normal force, f_{ni}, and a torque, τ_i.

As shown in Hunt (1978), we can define ξ_i in the following alternative form:

$$\xi_i = \sum_{k=1}^{i} \dot{\theta}_k \begin{bmatrix} 1 \\ \mathbf{E}\boldsymbol{r}_i^{O_k} \end{bmatrix} \tag{2}$$

Here, $\boldsymbol{r}_i^{O_k}$ is a vector from O_k to the i-th point of contact and $\mathbf{E}$ is the transformation matrix for a cross product in the plane:

$$\mathbf{E} = \begin{bmatrix} 0 & -1 \\ 1 & 0 \end{bmatrix} \tag{3}$$

Now, we can express the right-hand side of Eq. (1) as follows:

$$\sum_{i=1}^{3} \xi_i \circ \zeta_i = \sum_{i=1}^{3}\sum_{k=1}^{i} \dot{\theta}_k \begin{bmatrix} 1 \\ \mathbf{E}\boldsymbol{r}_i^{O_k} \end{bmatrix} \circ \zeta_i \tag{4}$$

In this article, we make the assumption that the finger is free of all external forces but the torques needed to obtain static equilibrium. Therefore, the wrench is expressed as:

$$\zeta_i = \begin{bmatrix} 0 \\ 0 \\ \tau_i \end{bmatrix}$$

And Eq. (4) becomes:

$$\begin{aligned} \sum_{i=1}^{3} \xi_i \circ \zeta_i &= \sum_{i=1}^{3}\sum_{k=1}^{i} \dot{\theta}_k \begin{bmatrix} 1 \\ \mathbf{E}\boldsymbol{r}_i^{O_k} \end{bmatrix} \circ \begin{bmatrix} 0 \\ 0 \\ \tau_i \end{bmatrix} \\ &= \tau^T \mathbf{J}\dot{\theta} \end{aligned} \tag{5}$$

where

$$\mathbf{J} = \begin{bmatrix} 1 & 0 & 0 \\ 1 & 1 & 0 \\ 1 & 1 & 1 \end{bmatrix}$$

$\mathbf{J}$ is a transformation matrix which links the global torques, τ_i (i=1,2,3) and the local torques, τ_i^* (i=1,2,3):

$$\tau^{*T} = \tau^T \mathbf{J} \tag{6}$$

τ_i^* is important since it is the amount of torque felt at the i-th joint, $i = 1,2,3$.

Combining Eqs. (5), (6) with (1), we get:

$$t^T \omega_a = \tau^T \mathbf{J}\dot{\theta} \tag{7}$$

$$= \tau^{*T} \dot{\boldsymbol{\theta}} \tag{8}$$

At this point, we define $\mathbf{T}$, the transmission matrix that relates vector ω_a to vector $\dot{\boldsymbol{\theta}}$. Explicitly,

$$\begin{bmatrix} \dot{\theta}_1 \\ \dot{\theta}_2 \\ \dot{\theta}_3 \end{bmatrix} = \begin{bmatrix} X_1 & X_2 & X_3 \\ 0 & 1 & 0 \\ 0 & 0 & 1 \end{bmatrix} \begin{bmatrix} \dot{\theta}_a \\ \dot{\theta}_2 \\ \dot{\theta}_3 \end{bmatrix}$$

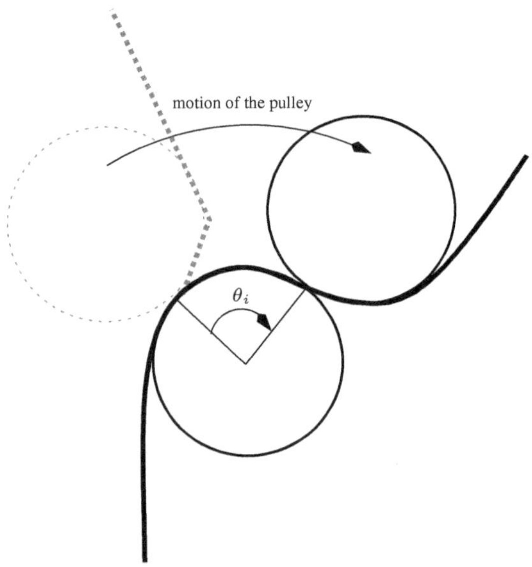

Figure 2. Isolated rotation of a joint.

or

$$\dot{\boldsymbol{\theta}} = \mathbf{T}\omega_a \quad (9)$$

Thus, combining Eqs. (8) and (9), we obtain

$$t^T \omega_a = \tau^{*T} \mathbf{T} \omega_a \quad (10)$$

which is equivalent to writing

$$\tau^* = \mathbf{T}^{-T} \mathbf{t} \quad (11)$$

4 Calculation of the transmission matrix for the actuation

In this section, we take a look at the actuation mechanism of the finger. The mechanism consists of a string attached to the distal phalanx. The string is pulled through a pulley network which produces the joint actuation. The transmission matrix is the mathematical tool that models the behaviour of the joints.

As stated in the last section, the transmission matrix is the application that transforms the actuation velocity vector, ω_a, into the time derivatives of the joints coordinates, Eq. (9). The main relation to solve is

$$\dot{\theta}_1 = X_1\dot{\theta}_a + X_2\dot{\theta}_2 + X_3\dot{\theta}_3 \quad (12)$$

First, as illustrated in Fig. 2, we notice that the isolated rotation of each joint, in joint coordinates, results in an elongation of the tendon, x_i, of the order of the radius of the joint times the rotation of the same joint:

$$x_i = r_{ia}\Delta\theta_i \quad i = 1,2,3 \quad (13)$$

In the case of tendon and pulleys, the total amount of elongation of the tendon corresponds to the rotation of actuation joint times the radius of joint 1:

$$x = r_{1a}\Delta\theta_a \quad (14)$$

And because we assume that the tendon is rigid, the total amount of elongation is the sum of the elongation for each joint rotation:

$$x = x_1 + x_2 + x_3$$

and by combining Eqs. (13) and (14), we obtain

$$r_{1a}\Delta\theta_a = r_{1a}\Delta\theta_1 + r_{3a}\Delta\theta_2 + r_{5a}\Delta\theta_3 \quad (15)$$

Now, we differentiate Eq. (15) with respect to time, which gives us

$$r_{1a}\dot{\theta}_a = r_{1a}\dot{\theta}_1 + r_{3a}\dot{\theta}_2 + r_{5a}\dot{\theta}_3$$

and reorganize in a form similar to Eq. (12):

$$r_{1a}\dot{\theta}_1 = r_{1a}\dot{\theta}_a - r_{3a}\dot{\theta}_2 - r_{5a}\dot{\theta}_3$$

or

$$\dot{\theta}_1 = \dot{\theta}_a - \frac{r_{3a}}{r_{1a}}\dot{\theta}_2 - \frac{r_{5a}}{r_{1a}}\dot{\theta}_3 \quad (16)$$

So, from Eq. (16), we can define our transmission matrix for the actuation as the following:

$$\mathbf{T}_a = \begin{bmatrix} 1 & -\frac{r_{3a}}{r_{1a}} & -\frac{r_{5a}}{r_{1a}} \\ 0 & 1 & 0 \\ 0 & 0 & 1 \end{bmatrix} \quad (17)$$

5 Calculation of the transmission matrix for the return

We now turn our attention to the return mechanism. Similarly to the actuation mechanism, the return mechanism consists of a string attached to the distal phalanx and passed through a system of pulleys. However, the string is attached to a linear spring a the base of the finger thus providing the return force.

As in the case of the actuation, we notice in Fig. 3, that the isolated rotation of each joint, in articulation coordinates, results in an elongation of the tendon, x_i, of the order of the radius of the joint times the rotation of the same joint:

$$x_i = r_{ib}\Delta\theta_i \quad i = 1,2,3 \quad (18)$$

In the case of tendon and pulleys, the total amount of elongation of the tendon corresponds to the rotation of actuation joint times the radius of joint 1:

$$x = r_{1b}\Delta\theta_a \quad (19)$$

And because we assume that the tendon is rigid, the total amount of elongation is the sum of the elongation for each joint rotation:

$$x = x_1 + x_2 + x_3$$

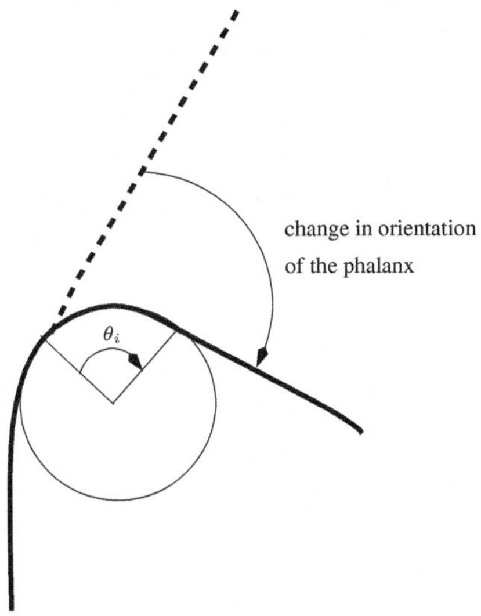

Figure 3. Isolated rotation of a joint for the return mechanism.

and by combining Eqs. (18) and (19), we obtain

$$r_{1b}\Delta\theta_a = r_{1b}\Delta\theta_1 + r_{3b}\Delta\theta_2 + r_{5b}\Delta\theta_3 \tag{20}$$

Now, we differentiate Eq. (20), which gives us

$$r_{1b}\dot{\theta}_a = r_{1b}\dot{\theta}_1 + r_{3b}\dot{\theta}_2 + r_{5b}\dot{\theta}_3$$

and reorganize in a form similar to Eq. (12):

$$r_{1b}\dot{\theta}_1 = r_{1b}\dot{\theta}_a - r_{3b}\dot{\theta}_2 - r_{5b}\dot{\theta}_3$$

or

$$\dot{\theta}_1 = \dot{\theta}_a - \frac{r_{3b}}{r_{1b}}\dot{\theta}_2 - \frac{r_{5b}}{r_{1b}}\dot{\theta}_3 \tag{21}$$

So, from Eq. (21), we can define our transmission matrix for the return as the following:

$$\mathbf{T}_r = \begin{bmatrix} 1 & -\frac{r_{3b}}{r_{1b}} & -\frac{r_{5b}}{r_{1b}} \\ 0 & 1 & 0 \\ 0 & 0 & 1 \end{bmatrix} \tag{22}$$

The transmission matrix for the return mechanism is very similar to $\mathbf{T}_a$ found for the actuation. However, the pulleys in the return mechanism do not have to be of the same dimensions used for the pulleys in the actuation mechanism. Therefore, we use here the notation "b" to designate a return pulley.

6 Torques from actuation and return

In order to find torques due to the underactuated mechanisms, we have found expressions for the transmission matrices. Now we need an expression for the vector $\boldsymbol{t}$ as defined for the use of said vector in Eq. (1). In the actuation process, its expression is straightforward as $\boldsymbol{t}_a$ is defined as

$$\boldsymbol{t}_a = \begin{bmatrix} T_a \\ 0 \\ 0 \end{bmatrix} \tag{23}$$

where the torque, T_a is the cross product of the tension put by the user on the string and the radius of the actuation pulley, r_{1a}. The amplitude is expressed as

$$T_a = F_a r_{1a} \tag{24}$$

Using Eq. (11), we now can get an analytical expression for the torque vector due to the actuation of the finger. This expression is the vector

$$\boldsymbol{\tau}_a = \begin{bmatrix} F_a r_{1a} \\ F_a r_{3a} \\ F_a r_{5a} \end{bmatrix} \tag{25}$$

We see with Eq. (25) that this design has achieved one of its objectives. The actuation torque at each joint is proportional to the tension given to the string and the actuation torques are not a function of the finger's configuration, i.e., they do not depend on θ_i, $i = 1,2,3$.

In the return process, the vector $\boldsymbol{t}_r$ is defined in the same fashion as

$$\boldsymbol{t} = \begin{bmatrix} T_r \\ 0 \\ 0 \end{bmatrix} \tag{26}$$

where the torque, T_r is the cross product of the tension put by the spring on the string and the radius of the return pulley, r_{1b}. The amplitude is expressed as

$$T_r = F_r r_{1b} \tag{27}$$

Contrary to the case of actuation, the return force, F_r, is a straightforward and arbitrary value. To express the return force, we must use Hooke's law:

$$F_r = -Ky$$

where K is the spring's stiffness and y, its elongation. Since we use preload in this design, an initial extension y_0 must be included in the value of y. The rest of the value of y is taken into account by the elongation caused by the mouvement of the finger which can be described by using Eq. (20):

$$y - y_0 = r_{1b}\theta_1 + r_{3b}\theta_2 + r_{5b}\theta_3$$

Combining these two expressions, we get a useful expression for the return force:

$$F_r = -K(y_0 + r_{1b}\theta_1 + r_{3b}\theta_2 + r_{5b}\theta_3) \tag{28}$$

We notice that the negative sign in Eq. (28) accounts for the counter-motion characteristic of the return action.

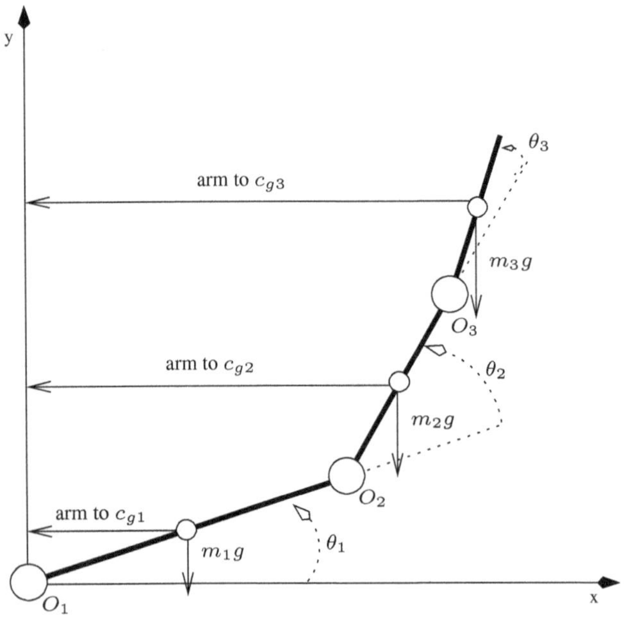

Figure 4. Gravity forces diagram for the proximal joint.

Again, using Eqs. (26), (27) and (28), we now can get an analytical expression for the torque vector due to the return action of the finger. This expression is the vector

$$\tau_r = \begin{bmatrix} -Kr_{1b}(y_0 + r_{1b}\theta_1 + r_{3b}\theta_2 + r_{5b}\theta_3) \\ -Kr_{3b}(y_0 + r_{1b}\theta_1 + r_{3b}\theta_2 + r_{5b}\theta_3) \\ -Kr_{5b}(y_0 + r_{1b}\theta_1 + r_{3b}\theta_2 + r_{5b}\theta_3) \end{bmatrix} \tag{29}$$

7 Gravity

In order to describe the effect of gravity, we must include the spatial orientation of the hand. For this purpose, we will use the Tait-Bryan angles which are a specific convention of Euler angles. These angles correspond to the XYZ convention where φ is the rotation angle around the x-axis, θ is the rotation angle around the y-axis and, ψ is the rotation angle around the z-axis.

We assume that the hand is palm up and along the Y-axis. The origin of the global coordinate system is at the wrist. This leads to:

- θ_0 expresses the angle formed by the palm and the y-axis;
- θ_1 expresses the angle formed by the proximal phalanx and the extension of the palm;
- θ_2 expresses the angle formed by the medial phalanx and the extension of the proximal phalanx;
- θ_3 expresses the angle formed by the distal phalanx and the extension of the medial phalanx;
- φ expresses the "roll" of the hand (side-to-side rotation).

Now that the hand is spatially oriented, we use the free body diagram technique to calculate the torques.

We begin to look at the distal phalanx. It is elementary to see that the resulting torque on the distal joint is the cross product of the lever arm from the joint to the centre of mass and the gravity force, the amplitude being

$$\tau_{g3} = \cos(\varphi)(m_3 g c_{g3} \cos(\theta_0 + \theta_1 + \theta_2 + \theta_3)) \tag{30}$$

where g is the gravitational acceleration and m_i is the mass of the i-th phalanx. As for the medial joint, the resulting torque is the sum of the cross products of, on one hand, the arm created by its centre of mass and the gravity force due to the mass of the medial phalanx and, on the other hand, the arm created by its centre of mass and the gravity force due to the mass of the distal phalanx. The amplitude is

$$\begin{aligned} \tau_{g2} &= \cos(\varphi)([m_2 g c_{g2} \cos(\theta_0 + \theta_1 + \theta_2)] \\ &+ [m_3 g(c_{g3} \cos(\theta_0 + \theta_1 + \theta_2 + \theta_3) \\ &+ l_2 \cos(\theta_0 + \theta_1 + \theta_2))]) \end{aligned} \tag{31}$$

Finally, Fig. 4 shows that for the proximal joint, the resulting torque is the sum of the cross products of: the arm created by its centre of mass and the gravity force due to the mass of the proximal phalanx, the arm created by its centre of mass and the gravity force due to the mass of the medial phalanx and, the arm created by its centre of mass and the gravity force due to the mass of the distal phalanx. The amplitude is

$$\begin{aligned} \tau_{g1} &= \cos(\varphi)([m_1 g c_{g1} \cos(\theta_0 + \theta_1)] \\ &+ [m_2 g(c_{g2} \cos(\theta_0 + \theta_1 + \theta_2) + l_1 \cos(\theta_0 + \theta_1))] \\ &+ [m_3 g(c_{g3} \cos(\theta_0 + \theta_1 + \theta_2 + \theta_3) + l_2 \cos(\theta_0 + \theta_1 + \theta_2) \\ &+ l_1 \cos(\theta_0 + \theta_1))]) \end{aligned} \tag{32}$$

In summary, the action of gravity can be modeled by

$$\tau_g = g\mathbf{CLm}\cos\varphi \tag{33}$$

where,

$$\mathbf{C} = \begin{bmatrix} \cos(\theta_0 + \theta_1) & \cos(\theta_0 + \theta_1 + \theta_2) & \cos(\theta_0 + \theta_1 + \theta_2 + \theta_3) \\ 0 & \cos(\theta_0 + \theta_1 + \theta_2) & \cos(\theta_0 + \theta_1 + \theta_2 + \theta_3) \\ 0 & 0 & \cos(\theta_0 + \theta_1 + \theta_2 + \theta_3) \end{bmatrix}$$

is the matrix of the finger's configuration. And,

$$\mathbf{L} = \begin{bmatrix} c_{g1} & l_1 & l_1 \\ 0 & c_{g2} & l_2 \\ 0 & 0 & c_{g3} \end{bmatrix}$$

is the matrix of geometric parameters. And,

$$\boldsymbol{m} = \begin{bmatrix} m_1 \\ m_2 \\ m_3 \end{bmatrix}$$

is the vector of the phalanx masses. The orientation of the palm is simply determined by the sign of the constant g. As a convention, g is negative for palm up, positive for palm down. We can also eliminate the effect of gravity by setting the gravitational acceleration to 0.

Table 1. Prototype physical parameters.

	i	r_{ia}	r_{ib}	l_i	m_i	c_{gi}
Proximal	1	6	2.95	45	0.87	21.65
Medial	2	2.75	2.75	25	0.5	9.19
Distal	3	1.5	2	n/a	0.45	10.85

8 The Model

We now have all the pieces of the puzzle. After the development above, it is natural to conclude that the vector representing the total torque at each joint of the finger is the sum of the expressions given in Eqs. (25), (29) and (33). Allen-Demers et al. (2009) shows a similar result.

The torque vector composed of the torque at each joint is expressed as

$$\tau^* = \tau_a + \tau_r + \tau_g \quad (34)$$

If the reader refers to the beginning of this section, there is one component missing from Eq. (34), namely the torque due to friction. Eq. (34) can be used to simulate the overall behaviour of the finger since friction is negligeable for our purposes. But manufacturing affects the amount of friction in the systems, so the torque due to friction must be obtained empirically. So a corrected model can be written as

$$\tau^* = \tau_a + \tau_r + \tau_g + \tau_f \quad (35)$$

9 Numerical simulations

Equation (34) allows the simulation of the behaviour of the finger under arbitrary physical parameters. One can study the behaviour of the finger, more precisely the closing sequence, according to the stiffness of the return spring, to the imposed preload, to the orientation of the finger (i.e. palm up or down). Another effect of a physical parameter that needs to be studied is the combination of pulleys of different radii and how it affects the closing sequence and ultimately, the grasping force that can be provided.

The numerical software, Matlab, is used to perform numerical simulations. Equation (34) is programmed and the physical parameters of the prototype of the finger are inserted, as shown in Table 1. The routine sets the configuration of the finger as being flat, i.e., all angles to be zero. An arbitrary and progressively increasing cable tension is applied and once one of the joints' torque becomes positive, the variable θ_i is progressively incremented until the joint angle is 90° from its initial position.

Figure 5 shows the result of this numerical simulation. We first notice that initially, the torques are negative for all three joints. There is no movement since the joints are designed such that they only move in a positive range of angles. Physically, the range in which the torques are negative corresponds to the situation where the actuation does not overcome the return forces, in this case provided by the return spring and gravity.

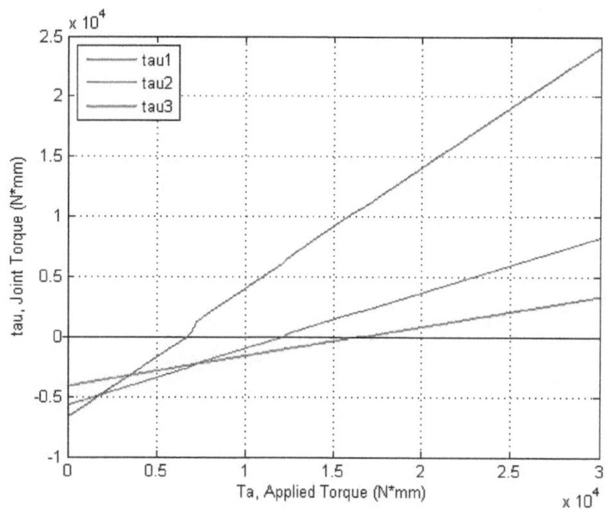

Figure 5. Closing sequence for the finger prototype.

We also see that the torque for each joint becomes positive at three distinct points. Furthermore, the torques become positive in numerical order such that joint 1 moves before joint 2 which moves before joint 3. This shows that the design attained another of its objectives which is its closing sequence. We have here a closing sequence where the proximal joint moves fully through its range of motion then the medial joint moves through its range of motion and then the distal joint moves through its range of motion, all seperately. This simulation is consistent with the physical motion of the prototype whose paramters were used for the simulation.

Figure 6 shows the simulation with a small preload, smaller than 10. We see the same general features that were observed in Fig. 5. However, the negative tension zone is much smaller. This is no surprise since a smaller preload leads to a smaller return force and so a smaller actuation force is needed to overcome the former forces.

Also, we observe that the joint torques still become positive in numerical order. However, the crossing points are very close to each other which suggests a closing sequence that will not follow that crossing order. Multiple joints will move at the same time and so the closing sequence will be uncontrolled.

Finally, Fig. 7 shows a simulation where there is no return. This simulation shows that the closing sequence does not correspond to the design goal. Since there are no return forces and so only gravity forces oppose the actuation, we can see that the distal joint will close first since it is lighter. The proximal joint will be last since it is opposed to the finger's total mass.

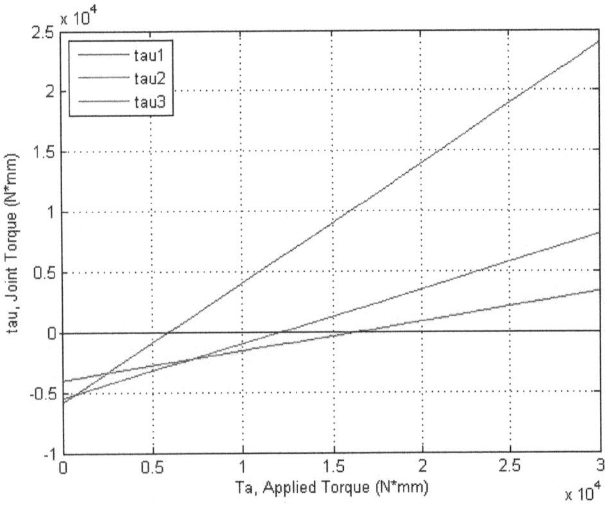

Figure 6. Closing sequence with small preload.

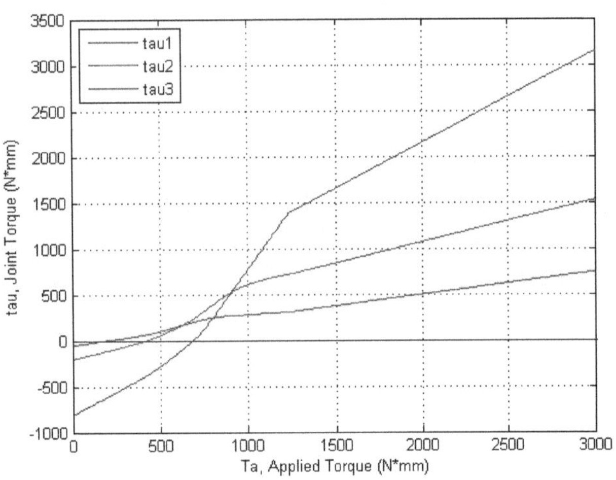

Figure 7. Closing sequence with no return.

10 Conclusions

The model presented in this paper is simple and flexible. It expands the work in Birglen et al. (2008); Allen-Demers et al. (2009) by adapting the equations to the specific case of human-like fingers. It also completes said work by adding the effect of physical forces such as gravity and that, in any spatial orientation of the hand. Furthermore, each force can be easily isolated. Each joint can be also isolated. Hence, the influence of the type of force on a specific joint can be simulated to study its effect along the full range of motion. Each physical parameter is also easily isolated. The effect of combinations of pulley ratios can be easily studied.

This leads to a better understanding and characterisation of the intelligence of the mechanism or of the effect of the return force. Such characterisations can help with the optimisation of the design.

Even though this model was focusing on the internal torques acting on the individual joint, it can be easily extended to external forces. A stability characterisation is needed to see the adaptability of the design.

Acknowledgements. The authors would like to acknowledge the financial support of the Natural Sciences and Engineering Research Council of Canada (NSERC) and of the Canada Research Chair program.

Edited by: J. L. Herder
Reviewed by: two anonymous referees

References

Allen-Demers, L. and Gosselin, C.: Kinematic Design of an Ejection-Free Underactuated Anthropomorphic Finger, Proceedings of the 2009 IEEE International Conference on Robotics and Automation, 2009.

Belfiore, M.: The Department of Mad Scientists; How DARPA Is Remaking Our World, from the Internet to Artificial Limbs, Smithsonian Books, New York, NY, 2009.

Birglen, L., Laliberte, T., and Gosselin, C.: Underactuated Robotic Hands, Springer, Berlin, Germany, 2008.

Hunt, K.: Kinematic Geometry of Mechanisms, Oxford University Press, Oxford, England, 1978.

18

Grasp planning for a reconfigurable parallel robot with an underactuated arm structure

M. Riedel, M. Nefzi, and B. Corves

Department of Mechanism Theory and Dynamics of Machines (IGM),
RWTH Aachen University, Aachen, Germany

Abstract. In this paper, a novel approach of grasp planning is applied to find out the appropriate grasp points for a reconfigurable parallel robot called PARAGRIP (Parallel Gripping). This new handling system is able to manipulate objects in the six-dimensional Cartesian space by several robotic arms using only six actuated joints. After grasping, the contact elements at the end of the underactuated arm mechanisms are connected to the object which forms a closed loop mechanism similar to the architecture of parallel manipulators. As the mounting and grasp points of the arms can easily be changed, the manipulator can be reconfigured to match the user's preferences and needs. This paper raises the question, how and where these grasp points are to be placed on the object to perform well for a certain manipulation task.

This paper was presented at the IFToMM/ASME International Workshop on Underactuated Grasping (UG2010), 19 August 2010, Montréal, Canada.

1 Introduction

In recent years, manufacturing industry is governed by fundamental changes with regard to the conditions, like progressive globalization and rapid technological development as well as changes in the resources situation (Möller, 2008; Müller et al., 2010). Classical demands on handling systems are currently undergoing change. In the past higher load capacity, greater precision and higher speeds were the main demands in the present situation priorities are increasingly shifting towards customized production and flexible solutions to component dependent problems. Currently available handling systems are not completely efficient to fulfil the increased demands (Nyhuis, 2008) and one of the major problems is the automated handling of large components in small production series such as aerospace systems, shipbuilding, wind turbine construction or manufacture of solar panels etc.

A single robot with one large customized Gripper is not usually capable of moving the spacious component without subjecting it to any forces.

The concept of handling with co-operating robots represents a more versatile approach as the component can be gripped and supported at different points by several robots depending on the shape of the object (Feldmann et al., 2007). One disadvantage of having many industrial robots handling one part is the high overall number of actuators, e.g. 18 actuators for 3 robots to perform a 6 DOF object motion. The additional drives lead to a high redundant system which results in complex control architecture and high costs.

A novel reconfigurable handling system is presented in (Nefzi et al., 2006; Müller et al., 2010) and shown in Fig. 1. The system "PARAGRIP" provides the same versatility and flexibility as co-operating robots but at significant lower procurement and operating costs. It is based on a concept with a different actuator configuration featuring simpler and less number of drives. Due to this drive architecture the separate arms are underactuated and there is need that these arms should work together in combination. The contact element which connects the object with the arm is mounted on a free movable, spring centered wrist joint. This joint needs to be aligned passively when the contact element touches the surface of the object.

This paper deals with the grasping behavior and the optimization of grasp positions for the underactuated arms.

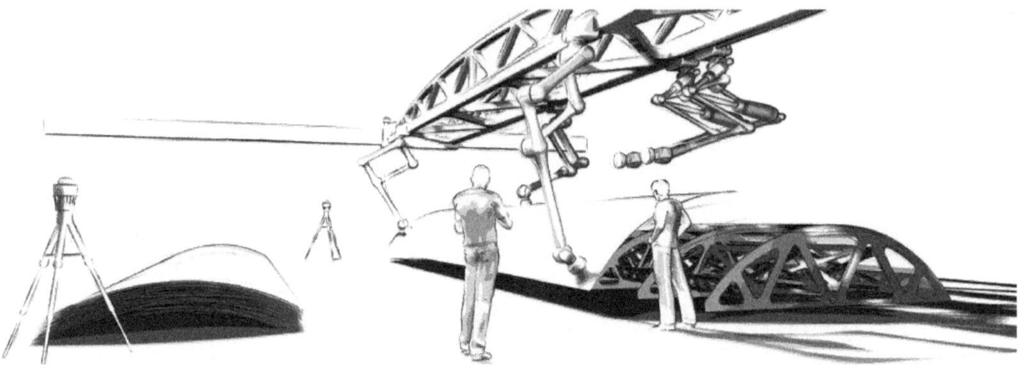

Figure 1. Design concept of the adaptable and reconfigurable handling system PARAGRIP.

Firstly we present the novel handling idea and the structure of the robot arms. Then we proceed towards the grasping process of the underactuated arms. Finally, we will present an approach for finding the optimal grasp points for a given object motion.

2 Handling concept

The aim of the new handling concept is to increase the flexibility and the automation in the assembly processes of large components in small series (Müller et al., 2010). This can be achieved by using a modular and adaptable assembly system which enables reconfiguration of the kinematic layout depending on the task. The main idea of this handling principle is based upon autonomous gripping and moving objects by using multiple arms within a parallel kinematic structure which is shown in Fig. 2. Whenever the grasping is done, the parallel kinematic structure is regenerated and integrates the object within the robot structure as a moveable platform. In contrast to mechanical handling systems that mimic human hands (Bicchi, 2000; Yoshikawa and Nagei, 1991) the object is not clamped or clutched to make a manipulation possible. The contact element at the end of each arm provides a temporary stable connection which can transfer forces and moments in all directions. The resulting closed-loop mechanism is formed by the arms and the object.

A complete manipulation sequence is schematically shown in Fig. 3. Here the object becomes part of the robot when the contact areas of the grippers are located on the object by the separate arms. Adhesion forces are generated by electro-magnets, vacuum cups, etc to ensure that the kinematic chains remain closed. This newly generated kinematic structure is similar to the structure of parallel robots. In contrast to cooperating robots, the configuration of the architecture in each object pose is determined by only six drives. After the motion the object is released and the connection between the arms is cut.

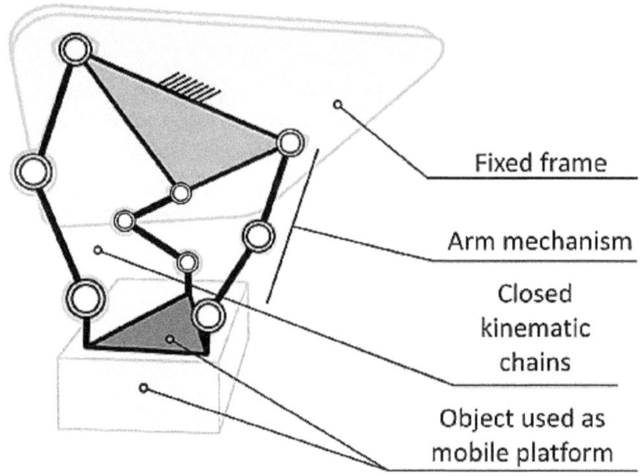

Figure 2. New handling principle.

3 Robot architecture

To perform a 6 DOF object motion with the prescribed handling concept a mobility number of six for each arm is necessary independently from the number of arms that are in contact with the object. A favorable configuration consists of three arms with six main actuators and three additional smaller actuators for grasping. During manipulation phase the additional drives can be disconnected, driven passively or support actively the object motion. As shown in Fig. 4 each arm mechanism is a combination of a five-bar-linkage with a parallelogram arrangement, a revolute joint around the vertical axis and a spherical wrist joint, similar to the structure presented in (Ebert-Uphoff I. and Gosselin, 1998).

Figure 5 illustrates the architecture of the wrist. It consists of two nested links (W1, W2) and one link (W3), which supports a contact element. Link 2 is the distal part of the arm mechanism. The axes of the three revolute joints (R, S and T) intersect at one center of rotation.

There are two wrist configurations: in the first configuration all joints are passive and in the second one the joint R is

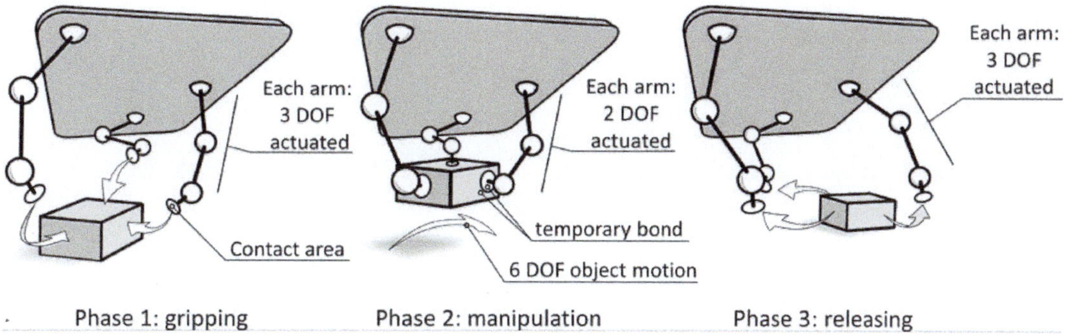

Figure 3. Movement sequence when handling an object.

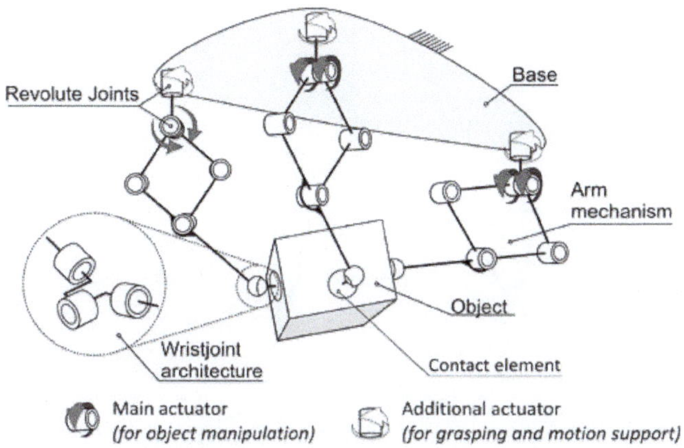

Figure 4. Robot structure.

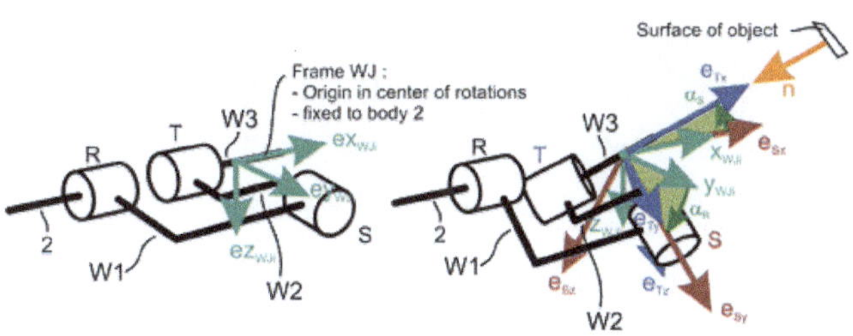

Figure 5. Wrist joint structure with frames and denotations. Left: center joint position. Right: general joint position.

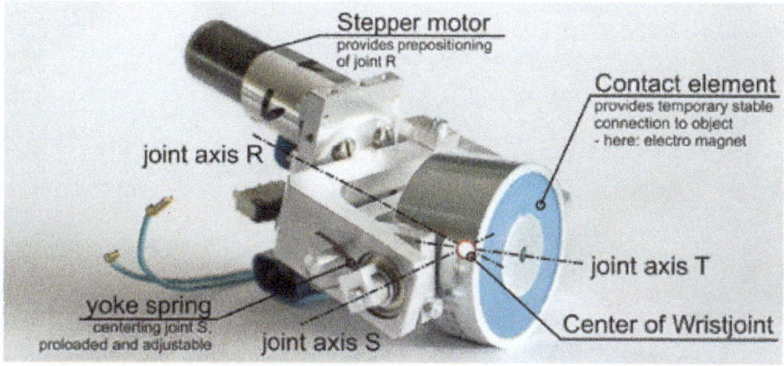

Figure 6. Wrist joint prototype.

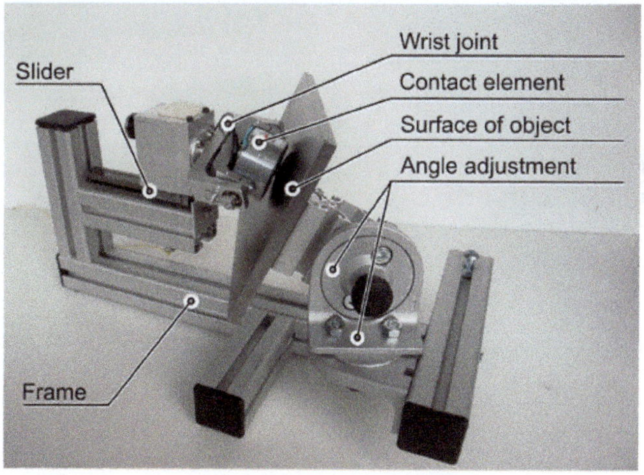

Figure 7. Wrist joint test bench.

driven by a low-cost stepper motor for grasping. The wrist is underactuated, which means all passive joints should be spring-centered to avoid collisions and grasp failure caused by undefined link positions. To reduce the necessary spring torque the common center of gravity of both links W2 and W3 together, is located at the center of rotation. The centering torque of joint S can be adjusted by tensioning two rotary springs.

The main drawback of this architecture is a singular joint configuration in center position, when axis R and T are collinear, see left part of Fig. 5. In this situation a rotation of link W3 about an axis perpendicular to R and S is not possible. This aspect has a major effect on grasping behavior and needs to be carefully considered in motion- and grasp planning. Nevertheless, this wrist architecture fits best our requirements, which provides a wide range of motion. Rotation angle of ±90° in each axis is possible, in comparison to ball- or universal joints with ±15° and ±45° respectively. Furthermore, the chosen wrist joint has a short distance between the center of rotation and the surface of the contact element which enhances the passive grasp behavior. A picture of the above discussed wrist prototype including the electromagnet is shown in Fig. 6.

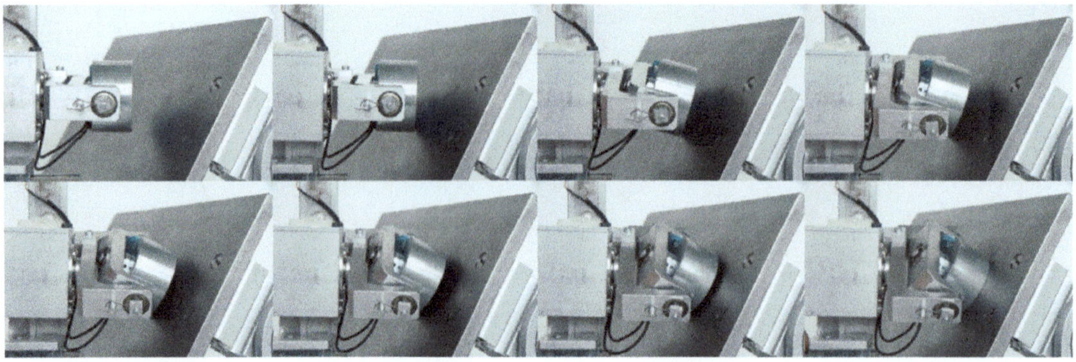

Figure 8. Wrist joint passive alignment.

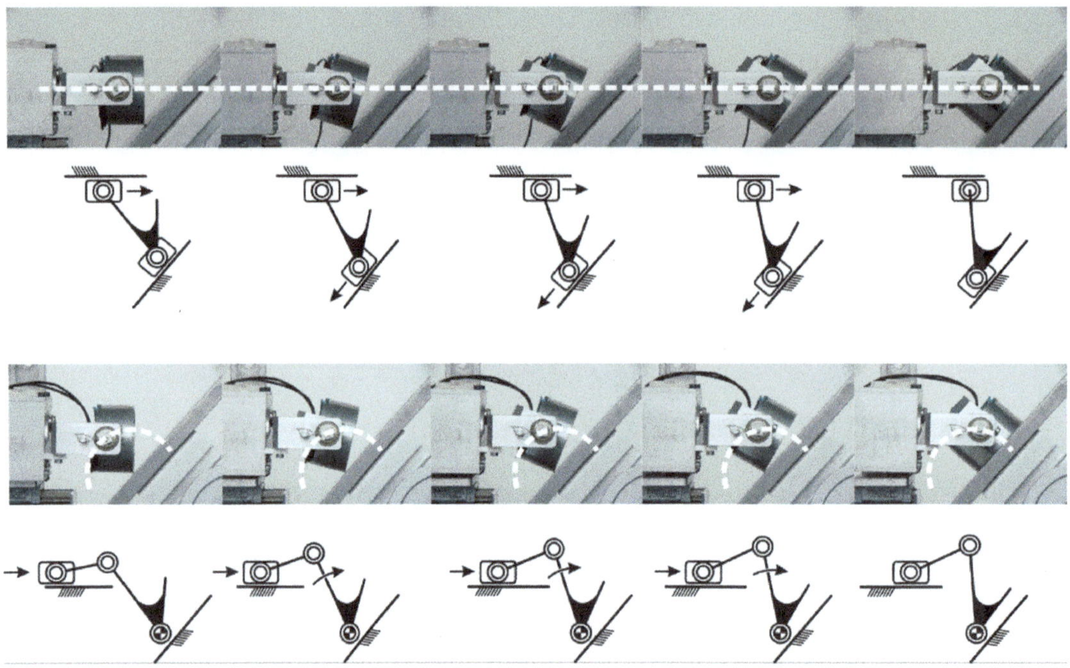

Figure 9. Comparison of linear and circular grasp motion.

4 Grasp process

One of the main advantages of this novel handling concept is its cost-effective build-up due to the reduced number of actuators. As explained in the previous sections the 3 DOF wrist joint can be designed which free of drives. Hence, during grasping phase this underactuated structure needs to be driven by the external contact forces to ensure a correct alignment of contact element to the surface of the object which is similar to movement of the phalanx at a grasp process with an underactuated multi-finger gripper investigated in (Laliberté et al., 2002; Krut, 2005). This passive alignment depends here on several factors like the joint angles, friction conditions, grasp motion etc.

To calculate the necessary joint angles α_R and α_S for grasping, the orientations of the systems of coordinates S and T need to be determined, see Fig. 5. System S is attached to link W1, T to W3. Unit vector $\boldsymbol{e}_{Sy}$ is collinear to joint axis S, vector $\boldsymbol{e}_{Tx}$ is in the direction of joint axis T. Vector $\boldsymbol{n}$ is normal to the surface. The vectors can be computed as follows:

$$\boldsymbol{e}_{Sx} = \boldsymbol{e}_{W_{Jx}}, \quad \boldsymbol{e}_{Sy} = \frac{n \times \boldsymbol{e}_{W_{Jx}}}{\| n \times \boldsymbol{e}_{W_{Jx}} \|}, \quad \boldsymbol{e}_{Sz} = \boldsymbol{e}_{Sx} \times \boldsymbol{e}_{Sy}$$
$$\boldsymbol{e}_{Tx} = -\boldsymbol{n}, \quad \boldsymbol{e}_{Ty} = \boldsymbol{e}_{Sy}, \quad \boldsymbol{e}_{Tz} = \boldsymbol{e}_{Tx} \times \boldsymbol{e}_{Ty} \tag{1}$$

The joint angle α_r is the angle between the vectors $\boldsymbol{e}_{WJy}$ and $\boldsymbol{e}_{Sy}$ in direction of vector $\boldsymbol{e}_{WJx}$, α_s between $\boldsymbol{e}_{Sx}$ and $\boldsymbol{e}_{Tx}$ in direction of $\boldsymbol{e}_{Sy}$.

To investigate the grasp behavior a test bench has been developed which is shown in Fig. 7. The wrist joint prototype is located on a slider and can be moved towards a test surface. This surface is adjustable in every combination of the angles

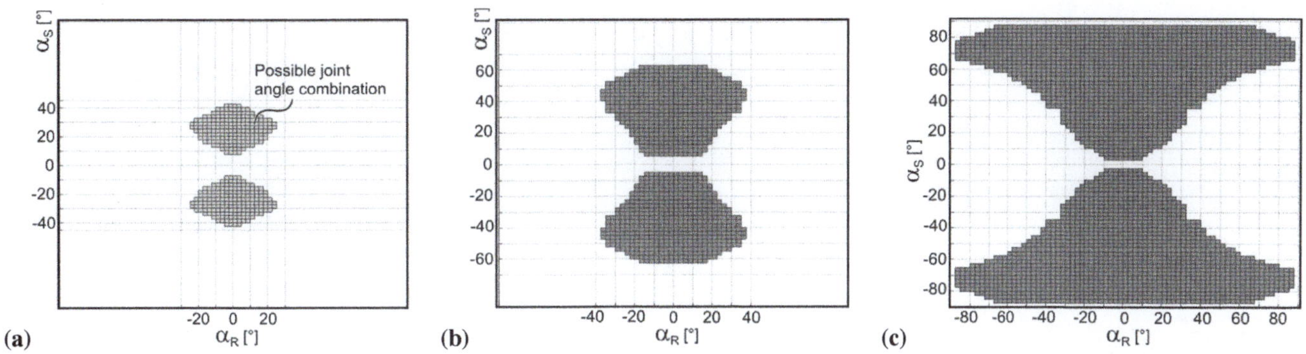

Figure 10. Successful passive alignment of the contact element for **(a)** linear motion ($\mu = 0.2$), circular motion ($\mu = 0.2$) **(b)** and circular motion and contact modification ($\mu = 0.6$) **(c)** while grasping.

α_R and α_S. The test bench can be configured in many variations to test different grasping conditions by changing grasp paths, friction conditions etc. Figure 8 shows a sequence of a grasping procedure, where the contact element aligns passively to the test surface by rotating about the both angles α_R and α_S.

A change in the configuration of the test bench enables the guidance of the joint on different paths. Figure 9a shows a linear grasp motion, when the center of the wrist joint is guided on a straight line. Here the contact point needs to slide on the surface which is similar to a double-slider-mechanism. At circular motion the contact element rolls on the fixed contact point by moving the center of the wrist joint on a radius like a reversed slider-crank-mechanism as shown in Fig. 9b.

The test results show that there is a significant difference in the passive alignment of the contact element for these grasping paths. It can be shown, that independent from friction conditions the circular grasp motion is always more successful. Successful grasping means that the contact element is correctly attached to the surface of the object without any misalignment.

An additional enhancement can be achieved by modifying the contact conditions using supplementary friction elements like rubber rings at the edge of the contact element.

Figure 10 shows the area of a successful passive alignment in the space of the wrist joint angles α_R and α_S for friction coefficient $\mu = 0.2$ respectively $\mu = 0.6$ at contact point and different grasp motions.

To understand the grasp behavior Fig. 11 depicts the contact situation for an arbitrary joint angle α_S and the center position of α_R when the edge of the contact element first hits the object.

When hitting the object a contact force F_K arises from Point K in the direction of the center of the wrist joint C_{WJ}. If this force is inside the friction cone, normal and frictional forces (F_N and F_R) can be transferred from the surface to the contact element. Otherwise the maximum possible frictional force (F_{R1}) is lower than the required frictional force ($F_{R1} + F_{R2}$) in the contact point in such a way as no static

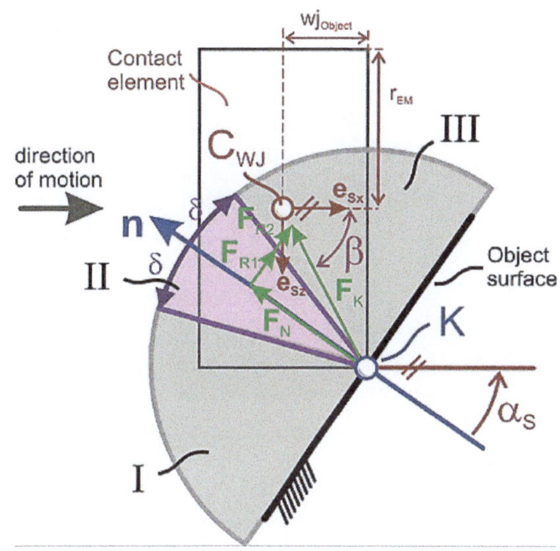

Figure 11. Force condition at contact point while grasping.

equilibrium can be found. Consequently the contact point has to slip away. As grasping with a linear motion requires a sliding of the contact point, the center of the wrist joint C_{WJ} needs to be located in the area III to succeed grasping. A force direction in area II causes jamming and the object will be displaced unintended. Here grasping is not successful. For circular motion both areas II and III are feasible. If the contact force shows in direction of area I the contact element will slip away in the wrong direction in both motion cases.

The maximum possible joint angle $\alpha_{S\ max}$ for a successful grasping is computed for $\alpha_R = 0°$ in the following way:

$$\delta - \arctan(\mu), \quad \beta = \arctan\left(\frac{r_{FM}}{w\,j_{offset}}\right) \tag{2}$$

Linear grasp motion: $\alpha_{S\ max} = \beta - \delta = 43.1°$ for μ=0.2
Circular grasp motion: $\alpha_{S\ max} = \beta + \delta = 65.7°$ for μ=0.2
$\alpha_{S\ max} = 85.4°$ for μ=0.6

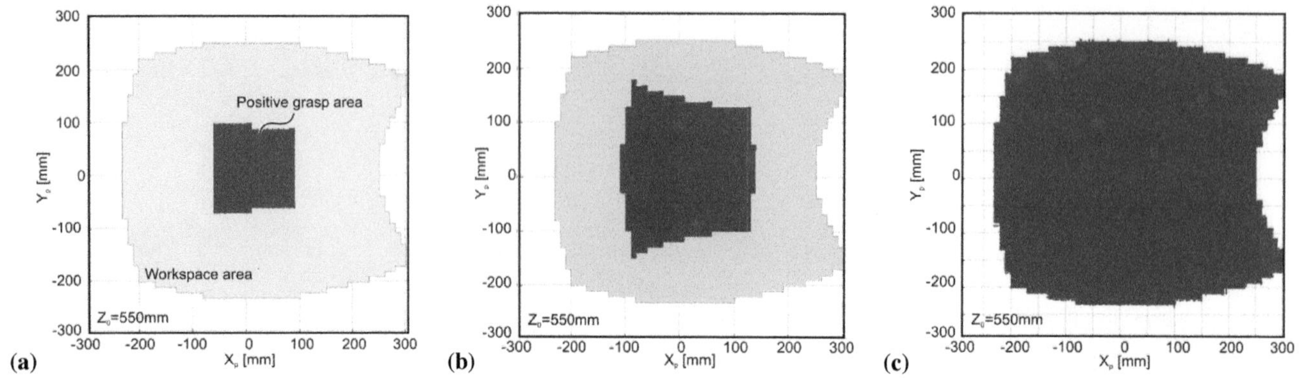

Figure 12. Successful grasp- and workspace area for linear motion (**a**), circular motion (**b**) and circular motion with contact element modifications (**c**).

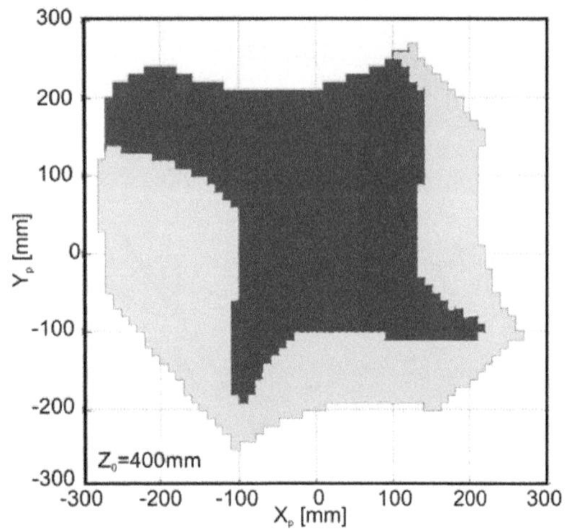

Figure 13. Grasping behavior for object orientation $\phi = 45°$ and wrist joint configuration I.

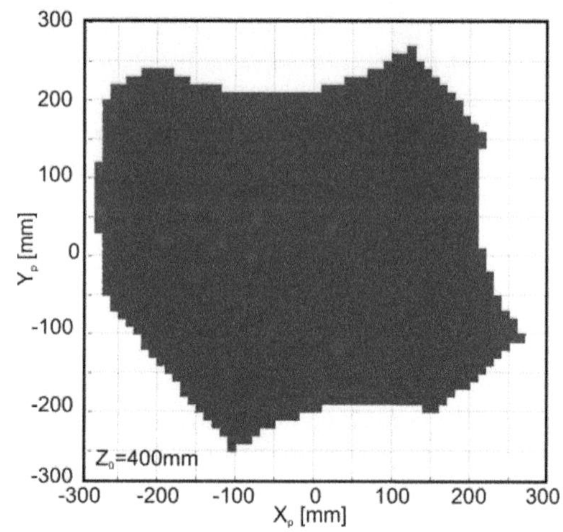

Figure 14. Enhanced grasping with object orientation $\phi = 45°$ and wrist joint configuration II.

These angle limits can also be seen well in the experimental results. The equations also show that independent from friction conditions a circular grasp motion is always better. Above all higher friction at contact points affects the passive alignment at the circular motion in a positive way and at the linear motion in a negative way.

Investigation of positive grasp behavior in Cartesian space is more interesting than in joint space. Figure 12 shows the workspace for object manipulation in a single layer (light grey part) and the area (dark grey) where a cube with an edge length of 150 mm can be grasped. Analogous to the case of Fig. 10, Fig. 12a shows the result for linear motion, Fig. 12b for circular motion and Fig. 12c for a circular motion with a modified contact element. In the last case the higher friction causes a very good grasp behavior so that the object can be picked up all over the workspace. Investigations with different object orientations on the ground show that even with a circular grasp motion and an enhanced contact element some object poses at the edges of the workspace area cannot be grasped because of an improper combination of the joint angles α_R and α_S, see Fig. 13. This drawback can be avoided by using the before mentioned wrist joint configuration II with an actuator in joint axis R. This actuator is a very small, low cost stepper drive, which controls actively the position of the angle α_R before or while grasping. During object manipulation the stepper is driven passively by the close loop kinematic chains of the complete robot architecture.

Figure 14 shows that a prepositioning of joint angle α_R enables to pick up the object in any pose on this workspace layer.

To sum up the tests show that a positive passive alignment of the contact element can be achieved for any object pose on ground level by the three modifications depicted in Fig. 15:

- Guiding the wrist joint center on a circle
- Enhancing the frictional conditions at contact point by modifying the contact element
- Prepositioning of joint angle α_R by a low cost stepper

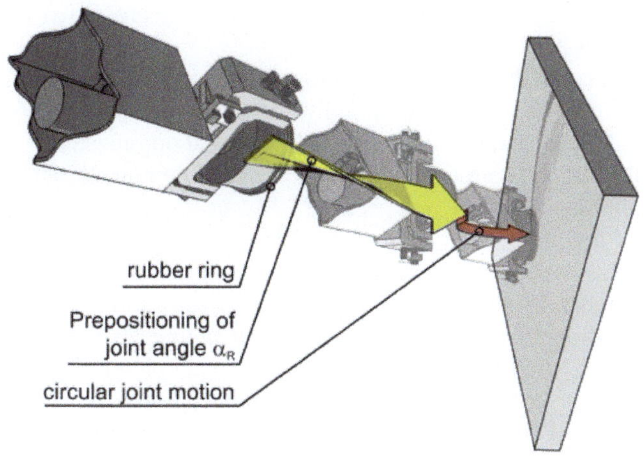

Figure 15. Enhancements for successful grasping

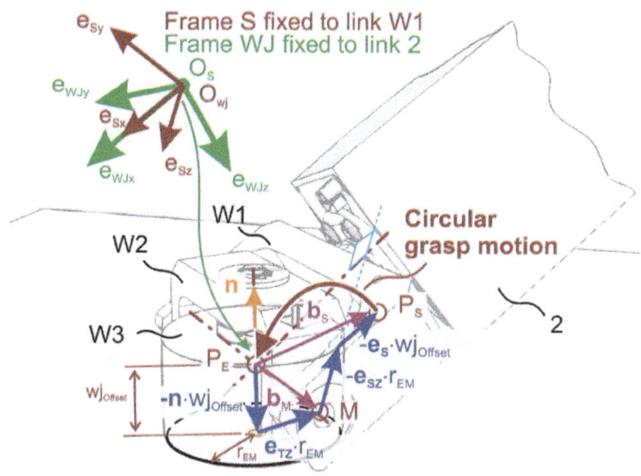

Figure 16. Computation of the circular grasp motion.

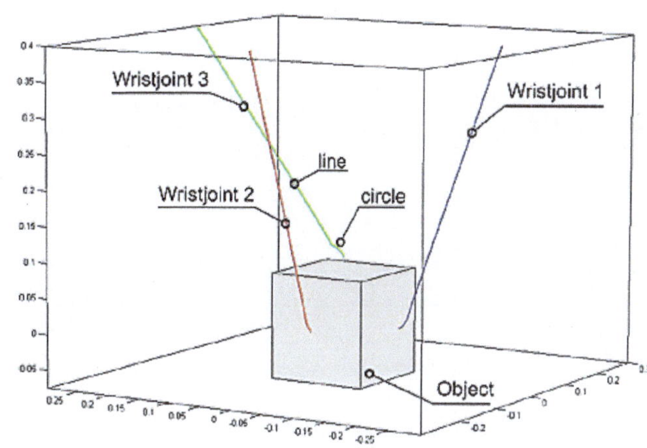

Figure 17. Visualization of the computed grasp paths.

The friction conditions can be enhanced by using a ring of rubber on the edge of the contact element. The necessary joint angle α_R for grasping can be computed with the help of Eq. (1) and is used for prepositioning.

Finally the exact circular path in 3-D space needs to be determined to achieve optimal grasp behavior. This can be done with the vectors starting from origin O_{wj} which are shown in Fig. 16.

$$\boldsymbol{b}_S = -\boldsymbol{e}_n \cdot wj_{\text{offset}} + \boldsymbol{e}_{TZ} \cdot r_{EM} - \boldsymbol{e}_{SZ} \cdot r_{EM} - \boldsymbol{e}_{SX} \cdot wj_{\text{offset}}$$
$$\boldsymbol{b}_M = -\boldsymbol{e}_n \cdot wj_{\text{offset}} + \boldsymbol{e}_{TZ} \cdot r_{EM} \quad (3)$$

where wj_{Object} is the distance between the joint center and the contact surface and r_{EM} is the radius of the contact element. The computed complete grasp paths of the wrist joint centers for all three arms are shown in Fig. 17. The first arbitrary part, here is a straight motion, and the second part is the circle which may have different curvatures at transition. This requires to be considered in trajectory planning to avoid vibrations.

5 Grasp point optimization

As the handling concept is based on the integration of the object into the kinematic structure, some robot performances can be directly affected by choosing adequate positions of the connecting points while grasping. The influence of a bad choice of these grasp points becomes very clear when taking a look at the following two examples. If the grasp points lie at or close to an imaginary line the complete robot configuration becomes uncertain in every object pose. In this singular situation the robot structure cannot withstand a wrench on the object and has a possibility to collapse. In a different case the grasp points are positioned well for force transmission but may reduce the mobility of the object, as for example large rotations could cause collisions.

That means by choosing the connection points some kinematic dimensions can be designed depending on the task to enhance workspace and dexterity, force- and velocity transmission along with accuracy and stiffness performance. As the selection of an appropriate grasp point combination is very challenging to a normal user an automatic optimization strategy is required.

A feasible optimization sequence which supports the user when planning the robot motion is shown in Fig. 18.

Among the information about the manipulation task, the optimization aim and the restrictions especially the possible grasp positions on the object are of importance. Therefore these positions could be generated automatically by scanning CAD data or be can be inputted manually by the user. The optimization algorithms can handle both, discrete and continuous descriptions of possible grasp areas, like a list of coordinates or a set of boundary for given surfaces. After computing an appropriate configuration the object motion can be performed.

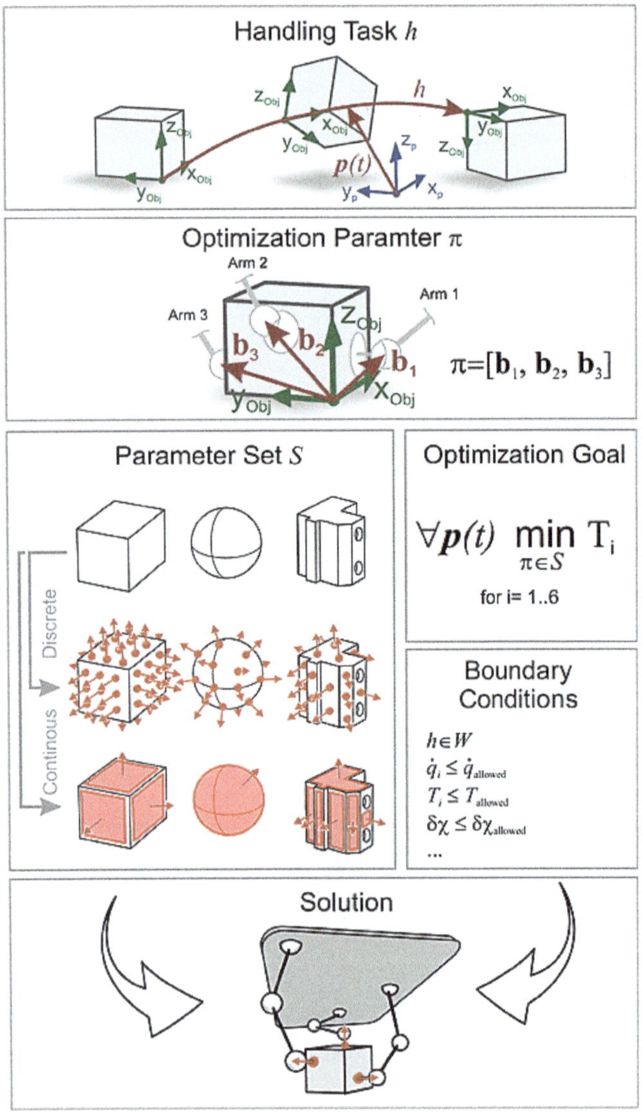

Figure 18. Grasp point optimization for e.g. minimizing drive torque.

The finding of adequate kinematic lengths on the object is similar to a dimensional synthesis of parallel manipulators. In (Angeles and Park, 2008; Nefzi et al., 2008) some optimization strategies are discussed which deal with enhancing kinetostatic performances indices all over the workspace to obtain design solutions that perform well for any trajectory.

These computations, which are used during the design process of the robot, could take several hours or even days to find one or more than one set of solutions depending on the complexity of the problem. This type of time consuming optimization for grasp planning is not acceptable as the user needs to wait after defining the handling tasks in the robot interface. The same applies to industrial applications with assembly belts featuring optical object detection via cameras, where a time critical planning of robot motion required. Depending on the information about possible grasp areas two approaches has been investigated.

The first one deals with a given set of discrete grasp points and tries to find the best solution by browsing through these points. The second one uses the gradient-based optimization algorithm fgoalattain to search inside a set of combinations of continuous surfaces.

Both algorithms are implemented in MATLAB and integrated in a graphical user interface to control robot motion. Figure 18 shows schematically the optimization process, starting from the definition of the handling task h in Workspace W and the parametervector $\boldsymbol{\pi}$, going on with the specification of the parameter set S, the optimization goal and the boundary conditions, to run the optimization algorithm and find a good solution for coupling points on the object.

The optimization ensures that the chosen pointes can be grasped successfully with an underactuated wrist joint architecture by considering the combinations of the angles α_R and α_S, as shown in Fig. 10. Furthermore collisions are observed along with the permissible range of joint motions. These restrictions guarantee that the workspace criterion is not violated. An optimization goal could be for example to find the grasp points which features the minimum actuator torque T_i for all actuators along the path. As all kinematic performances influence each other, there is a necessity to define additional boundary conditions .These limits define n maximum tolerable actuator velocities $q_{i\_allowed}$ and torques $T_{i\_allowed}$, necessary object pose accuracy $\delta\chi_{allowed}$ or required stiffness $f_{allowed}$ of the robot.

To reduce computation time in contrast to general dimensional synthesis of parallel robots three simple steps are included in the grasp point optimization.

As in most cases, the handling task is explicitly given and an optimization for each object pose in the complete workspace is not target-aimed and time consuming. Hence, all performance values are only computed for the object poses, velocities and accelerations along the given path. The roughness of the path discretization determines significantly the computation time. The main time benefit arises from a pre-selection of possible grasp points or areas. Here single points or areas are checked regarding the orientations of wrist joint vector $\boldsymbol{e}_{WJx}$ and surface vector $\boldsymbol{n}$ (see Fig. 5) without determining the robot performance. The reduced number of points is listed in a new set of feasible grasp combinations. An additional option to make the grasp planning even faster can be chosen by the user when the determination of adequate point is very time critical. Here the algorithm doesn't search for the best point combination, but tries to find a solution which just satisfies the given limits, as quickly as possible.

Table 1. Results of an optimization example.

Task definition				
Path	*cubic spline*	P1	P2	P3
	$[x,y,z]^T$ in [m]	$[-0{,}15;0{,}1;0]^T$	$[-0{,}05;0{,}15;0{,}1]^T$	$[0{,}1;0{,}15;0]^T$
	$[\varphi,\phi,\psi]^T$ in [deg]	$[0{,}0{,}0]^T$	$[0{,}45{,}0]^T$	$[0{,}90{,}0]^T$
Trajectory	law of motion	*5th order polynomial*		
Motion	duration	*1*	s	
	length	*0,48*	m	
	max. transl. velocity	*0,91*	m/s	
	max. transl. acceleration	*6,52*	m/s²	
Object	Shape	*Cube*		
	Size	*150x150x150*	mm x mm x mm	
	Mass	*1*	kg	
Optimization goal				
	minimize drive torque along path			
Boundary conditions				
Restrictions	Grasp criterion	*true*		
	Workspace criterion	*true*		
Limits	Maximal drive torque	*15*	Nm	
	Maximal drive velocity	*1000*	rpm	
	Maximal object error δX	*3*	mm	

Results

Optimization method	Case	Grasp point vectors b_i [x;y;z] in respect to frame F_OBJ in [m] b_1	b_2	b_3	Optimization Time in [s] @2,66GHz	Grasp criterion	Workspace criterion	max T in [Nm]	max q in [rpm]	max δX in [mm]
non	A	-0,075 -0,060 0,030	0,030 0,075 0,060	0,000 0,030 0,075	0	true	true	**31,0**	784,4	2,2
	B	-0,075 0,060 0,010	0,060 0,075 0,060	0,000 0,030 0,075	0	true	true	11,9	690,1	**6,8**
	C	-0,075 0,060 -0,060	0,075 0,060 0,060	0,075 0,000 0,000	0	true	**false**	-	-	-
	D	-0,075 0,060 -0,060	-0,075 0,060 0,060	0,075 0,000 0,000	0	**false**	**false**	-	-	-
discrete	E	-0,075 0,060 -0,060	0,060 0,075 0,060	0,000 0,000 0,075	734,59	true	true	8,6	647,1	1,2
	F	-0,075 0,060 -0,060	0,060 0,075 0,060	0,000 0,000 0,075	0,51	true	true	8,6	647,1	1,2
	G	-0,075 -0,060 -0,060	-0,060 0,075 -0,060	0,000 0,000 0,075	0,10	true	true	13,3	647,1	2,6
continuous	H	-0,075 -0,049 -0,060	0,043 0,075 -0,024	0,060 -0,035 0,075	4757	true	true	7,5	575,6	1,0
	I	-0,075 -0,049 -0,060	0,043 0,075 -0,024	0,060 -0,035 0,075	84,50	true	true	7,5	575,6	1,0
	J	-0,075 -0,060 -0,060	-0,060 0,075 -0,060	0,060 0,029 0,075	0,75	true	true	8,2	567,3	1,3

Table 1 shows the results of different optimization strategies for a given task. The first four solutions (A to D) are chosen arbitrarily and the next three (E to G) are optimized from a set of discrete start parameters and the last three solutions (H to J) were optimized by the basis of gradient on continuous areas.

Although the selection of user-defined grasp points takes no computation time, the choice may not satisfy all the conditions. Grasping the object in the configuration (A) causes a high maximum actuator torque during motion. Otherwise this motion is accomplished, but with a bad performance. The same applies to point combination (B), where the low accuracy of the object pose is not tolerable. However in the cases (C and D) the motion is not possible at all. In C the object can only be picked up, but the grasp points would cause collision between the robot links and the object during motion. Furthermore the object in (D) cannot even be grasped in the initial pose. Automatic grasp planning helps the user to prevent from choosing bad points.

For cases (E to G) a limited set of only 54 grasp points on the cube (9 per surface) is given, which can be combined theoretically in 148824(= 54!/51!) ways by the three arms. In (E) all those combinations are evaluated. This method finds the best solution in the set, but with a duration of 735 seconds which is very time consuming. In (F) the pre-selecting process minimizes the number of points to be analyzed, which must lead to the same solution as in (E) but with a significant computing time reduction to only 0.068%. Aborting the grasp planning process when the first solution is found (G) enhances the computation speed additionally to 19.6% of the total time in contrast to (F). The solution found here is usually not the best one in the set, but still satisfies the requirements as defined in the boundary conditions.

Analogous to (E to G) the last three cases also shows one optimization without pre-selection (H), one with pre-selection (I) and one with pre- and "first-best" selection (J). In these cases the parametervector π is optimized over continuous areas by using the gradient-based MATLAB function *fgoalattain*. In contrast to a discrete search these algorithms can also find solutions in between the set, which leads to better performances, however the computations takes more time. To prevent optimization from finding solutions only in local minima, each area is checked from several different start points.

In the end the user has to decide which optimization method is best for each individual case, by trading off for performance goal and computation durance.

6 Conclusions

In this paper, a novel approach to grasp planning was applied to the search of appropriate grasp points for a reconfigurable robot called PARAGRIP (Parallel Gripping). This new handling system is based on a parallel kinematic structure, which is generated temporarily after grasping for manipulation. As the object is integrated in the resulting closed loop mechanism, the location of the connection points are of importance for the motion performance. It was shown that these points could be carefully selected in an automatic optimization process. This grasp planning is not only focused on generating of a collision free motion with good performance but also considers the special grasp process with an underactuated arm structure.

Acknowledgements. As parts of this work are within the scope of the excellence cluster "Integrative production technology for high-wage countries (EXC 128)", the authors thank the German Research Foundation for its support.

This project is part of the sub-category "ICD D-3.2: Flexible Assembly Systems for Selfoptimising Automation" is realized in collaboration with the Laboratory for Machine Tools and Production Engineering at RWTH Aachen University.

Edited by: J. L. Herder
Reviewed by: two anonymous referees

References

Angeles, J. and Park, F. C.: Performance Evaluation and Design Criteria, Springer Handbook of Robotics, 229–244, 2008.

Bicchi, A.: On the Closure Properties of Robotic Grasping, Int. J. Robot. Res., 14, 319–334, 1995.

Bicchi, A.: Hands for dexterous manipulation and powerful grasping: A difficult road toward simplicity, IEEE. Trans. Robot. Automat., 16, 652–662, 2000.

Birglen, L., Laliberté, T., and Gosselin, C. M.: Optimal Design of Underactuated Fingers, Underactuated Robotic Hands, Springer Berlin, ISSN 1610-7438, Vol. 40, 117–138, 2008.

Butterfass, J., Grebenstein, M., Liu, H., and Hirzinger, G.: DLR II: Next generation of a dexterous robot hand, Proc. IEEE Int. Conf. on Robotics and Automation, 109–120, 2001.

Ebert-Uphoff, I. and Gosselin, C. M.: Kinematic Study of a new Type of special parallel Platform Mechanism, DECT98/MECH-5962, Proceedings of ASME Design Engineering Technical Conference, 1998.

Feldmann, K., Ziegler, C., and Michl, M.: Bewegungssteuerung für kooperierende Industrieroboter in der Montageautomatisierung, wt werkstattstechnik online, 97, 9, p. 713, 2007.

Hao, F. and Merlet, J.-P.: Multi-Criteria optimal design of parallel manipulators based on interval analysis, Mech. Mach. Theory, 40, 157–171, 2005.

Kragten, G. A. and Herder, J. L.: The ability of underactuated hands to grasp and hold objects, Mech. Mach. Theory, 45, 3, 408–425, 2010.

Krut, S.: A Force-Isotropic Underactuated Finger, Proceedings of the 2005 IEEE International Conference on Robotics and Automation, Barcelona, Spain, 2005.

Laliberté, T., Birglen, L., and Gosselin, C. M.: Underactuation in robotic grasping hands, Journal of Machine Intelligence and Robotic Control, 4, 3, 77–87, 2002.

Lee, J.-J. and Tsai, L.-W.: Structural Synthesis of Multi-Fingered Hands, J. Mech. Design, 124, 272–276, 2002.

Möller, N.: Bestimmung der Wirtschaftlichkeit wandlungsfähiger Produktionssysteme, Forschungsbericht IWB, Band 212, 2008.

Müller, R., Riedel, M., Vette, M., Corves, B., Esser, M., and Huesing, M.: Reconfigurable Self-Optimising Handling System, Precision Assembly Technologies and Systems, edited by: Ratchev, S., 5th International Precision Assembly Seminar, IPAS 2010, 1. Edn., Springer Berlin, ISBN 978-3-642-11597-4, 255–262, 2010.

Nefzi, M., Riedel, M., and Corves, B.: Development and design of a multi-fingered gripper for dexterous manipulation, Mechatronic Systems, 4(1), doi:10.3182/20060912-3-DE-2911.00026, 2006.

Nefzi, M., Riedel, M., and Corves, B.: Towards an Automated and Optimal Design of Parallel Manipulators, edited by: Arreguin, J. M. R., Automation and Robotics, I-Tech, Wien, ISBN 978-3-902613-41-7, 143–156, 2008.

Nyhuis, P.: Wandlungsfähige Produktionssysteme, Heute die Industrie von morgen gestalten, Garbsen, PZH, Produktionstechn. Zentrum, 2008.

Riedel, M., Nefzi, M., Hüsing, M., and Corves, B.: An adjustable gripper as a reconfigurable robot with a parallel structure, Proceedings of the Second International Workshop on Fundamental Issues and Future Research Directions for Parallel Mechanisms and Manipulators, 2008.

Yoshikawa, T. and Nagai, K.: Manipulating and Grasping forces in Manipulation by Multifingered Robot Hands, Int. J. Robotics and Automation, 7, 66–67, 1991.

19

A statically balanced and bi-stable compliant end effector combined with a laparoscopic 2DoF robotic arm

J. Lassooij[1], N. Tolou[1], G. Tortora[2], S. Caccavaro[2], A. Menciassi[2], and J. L. Herder[1]

[1]Delft University of Technology, Faculty of Mechanical, Maritime and Materials Engineering, Department of Biomechanical Engineering, Mekelweg 2, 2628 CD Delft, The Netherlands

[2]The BioRobotics Institute, Scuola Superiore Sant'Anna, Piazza Martiri della Libertà 33, 56127 Pisa, Italy

Correspondence to: N. Tolou (n.tolou@tudelft.nl)

Abstract. This article presents the design of a newly developed 2DoF robotic arm with a novel statically balanced and bi-stable compliant grasper as the end effector for laparoscopic surgery application. The arm is based on internal motors actuating 2 rotational DoFs: pitch and roll. The positive stiffness of the monolithic grasper has been compensated using pre-curved straight guided beams that are preloaded collinear with the direction of actuation of the grasper. The result is a fully compliant statically balanced laparoscopic grasper. The grasper has been successfully adapted to a robotic arm. The maximum force and stiffness compensations were measured to be 94 % and 97 % (i.e. near zero stiffness) respectively. Furthermore, the feasibility of adjusting for bi-stable behavior has been shown. This research can be a preliminary step towards the design of a statically balanced fully compliant robotic arm for laparoscopic surgery and similar areas.

1 Introduction

In recent years, surgical technologies have been advancing with many devices proposed to reduce the invasiveness of surgery (Breedveld, 1999). The main benefit of minimally invasive surgery (MIS) derives from the small incisions needed to perform complex procedures, thus reducing patient trauma and leading to shortened recovery times and improved cosmetics compared to open procedures (Forgione, 2009). Conventional MIS is performed by means of long slender instruments. Nowadays, laparoscopy represents the standard technique in surgery, but it requires high skills for two main reasons: (1) lost hand-eye-coordination and reduced freedom of movement, and (2) missing direct manual contact to the operation area. Therefore, many sophisticated procedures still cannot be performed minimally invasively. Robotic surgery and compliant mechanisms (CM) play important roles in overcoming the drawbacks of conventional MIS. In current robotic platforms the tools, are not manipulated directly by the surgeon anymore, but are held by specialized robot arms and remotely commanded by the surgeon who comfortably sits at an input console, such as in the Intuitive Surgical's da Vinci system (Alterovitz, 2009) and in the RAVEN robotic platform for telesurgery (Lum et al., 2009). An alternative approach is to introduce small robots with embedded actuators inside the body, as demonstrated by several researchers (Rentschler et al., 2006; Hu et al., 2009; Fowler et al., 2010; Lehman et al., 2008; Oleynikov et al., 2005).

Compliant mechanisms can help in overcoming the second drawback, i.e. missing direct manual contact to the operation area. Because the surgeon has to use long slender instruments instead of manually feeling the tissue, the sense of force feedback is lost. This is mainly due to the large amount of friction in laparoscopic graspers, and is the main reason of increased errors made by surgeons (Joice, 1998). Compliant mechanisms do not use rigid body joints to gain part or all of their motion. Instead, they use the relative flexibility of their members (Howell, 2001). Therefore, these mechanisms offer several advantages, including single-piece production, absence of Coulomb friction, no need for lubrication, and compactness, compared to conventional rigid body mechanisms. However, the energy storage in the compliant mechanism can be a big drawback, because the input-output relationship is affected; part of the input energy is not transferred

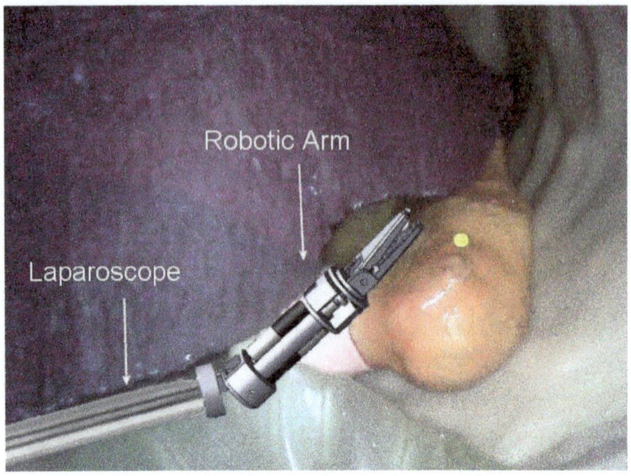

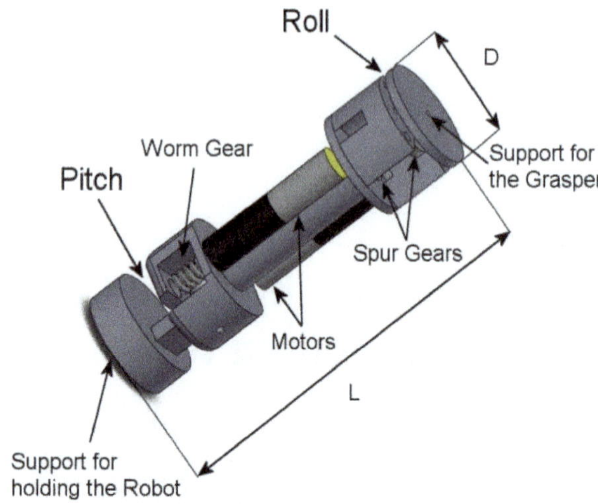

Figure 1. **(a)** View of the robot in a possible working scenario in a phantom with silicone organs (provided by EndoCAS Center, Pisa, Italy); the grasper will be substituted by the compliant grasper of Fig. 2; **(b)** 3-D sketch of the robot.

to the output, but used for the deformation of the compliant segments of the mechanism (Tolou et al., 2010a). Especially in the laparoscopic application, this can lead to low force feedback in spite of cancelling the friction, and can fatigue surgeons' muscles due to the constant effort required to operate the device. These drawbacks can be prevented by including a pre-stressed structure. Therefore, the total amount of potential energy in the system will be altered, which may lead to two situations. In the first situation, the potential energy in the system will be constant over a finite range of motion, resulting in static equilibrium over this range of motion. Therefore, no force is required to operate the system or to hold it at any position within the balanced range, and force feedback is restored. This kind of mechanism is called a statically balanced compliant mechanism (SBCM) (Herder and van den Berg, 2000). In the second situation, the potential energy in the system has two minimums, thus making the system bi-stable (e.g. the grasper will be stable in the open and in the closed positions). This enables the grasper to provide the necessary force to hold objects (e.g. tissue or a needle) for longer periods of time. Conventionally, the SBCM used spring based bi-stable mechanisms (Herder and van den Berg, 2000), leaf springs (De Lange, 2008) or bi-stable straight guided beams with compliant (Stapel and Herder, 2004; Hoetmer et al., 2009) or a partially compliant structure (Tolou and Herder, 2009) as the static balancer. However, all aforementioned works rely on preloading in a direction perpendicular to instrument actuation, and therefore protrudes outside the grasper and is difficult to be tuned for fabrication errors and improved performance. Collinear preloading has only been introduced in a few studies for other applications, such as: human skeleton (Chen and Zhang, 2011), MEMS (Tolou et al., 2010b) and precision stages (Dunning et al., 2011).

This article presents a laparoscopic 2DoFs robotic arm with a statically balanced fully compliant laparoscopic grasper (namely SBCG) as the end effector. The design is a step toward the development of fully compliant robotic arms. The 2DoF robotic arm was designed with internal motors actuating the DoFs. The grasper was designed as a monolithic structure, where the positive stiffness was cancelled using pre-curved straight guided beams as static balancers. The concept allows the static balancers to be integrated inside the compliant grasper, to be tuned for fabrication errors, and to be adjusted for different force displacement behaviors: bistable, zero stiffness with near zero actuation force, constant positive force and constant negative force. The performance of the novel SBCG and the laparoscopic robot will be investigated experimentally.

The remainder of this paper is structured as follows: in Sect. 2 the concepts of the robotic arm and SBCG are presented and followed by Dimensional Design, Fabrication and Measurement in Sect. 3. Results are presented and discussed in Sect. 4. Finally, conclusions are made in Sect. 5.

2 Conceptual design

2.1 Robotic arm

The robotic arm is conceived to be used as an extension of traditional laparoscopic tools, endoscopes, and existing robotic arms. This results in two main design criteria: increasing the DoFs, thus the dexterity of the system, and making the operation in narrow space easier. An example of the intended application and a 3-D sketch of the robotic arm are shown in Fig. 1. The robotic arm is jointly attached to a traditional laparoscope to increase the dexterity. In addition, the robotic arm could magnetically or mechanically be coupled to the abdominal wall. The proposed design for the

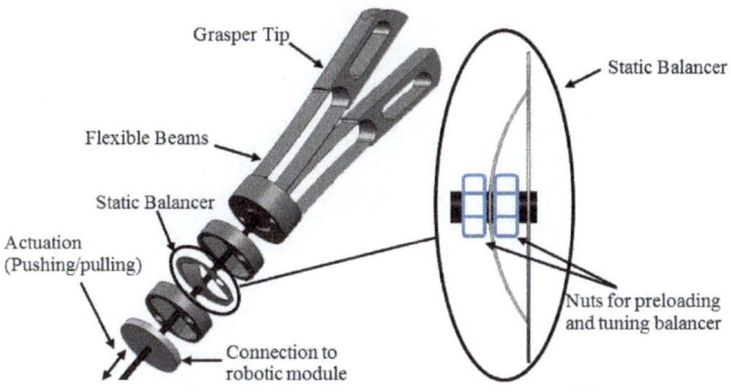

Figure 2. 3-D sketch of the SBCG at relaxed configuration.

robot arm is serial. This implies that the robotic arm can be easily integrated with traditional tools, thanks to its cylindrical shape. The robotic arm provides two additional DoFs as well as the actuation of the SBCG. The proposed design is a 2DoFs cylindrical link embedding two rotational motors and respective mechanisms, for pitch and roll DoFs.

Regarding the actuation of the compliant grasper, a cable actuation system is proposed. One end of the actuating wire is attached to the central beam of the compliant grasper, thus when the cable is pulled the grasper is closed, as explained in Sect. 3.1.2. The other end of the wire is attached to a winding rod. Cable actuation exhibits several advantages: the actuator can be placed remotely, i.e. outside the part of the instrument that is inserted into the patient, thus not influencing the dimensions of the robotic arm. An external spring is used to let the cable spring back to its natural position. The spring also ensures that the cable is always tensioned.

2.2 Statically balanced compliant grasper

The design of the compliant grasper has been inspired from the work of (Herder and van den Berg, 2000). The initial (i.e. relaxed) configuration of the compliant grasper is half open, to reduce the stresses in both the compliant grasper and the static balancer (see Fig. 2). The opening and closing of the compliant grasper corresponds to actuation by pushing or pulling the middle beam as shown in Fig. 2. In Fig. 3a, the force-displacement curve of the grasper is linear with a positive stiffness and passes through the origin; the origin of the graph refers to the relaxed configuration of the grasper. The closed position of the compliant grasper is marked A. B is the equilibrium position where the compliant grasper is half open and relaxed. C indicates the open position of the compliant grasper.

To statically balance the compliant grasper, negative stiffness should be added to compensate the positive stiffness of the device (Herder, 1998). Consequently, near zero stiffness with a near zero actuation force may be achieved. Bi-stable compliant mechanisms have been an interesting mechanism with nonlinear stiffness behavior (Jensen et al., 2001; Hartono, 2001; Qiu et al., 2004; Li and Zhou, 2005; Sönmez,

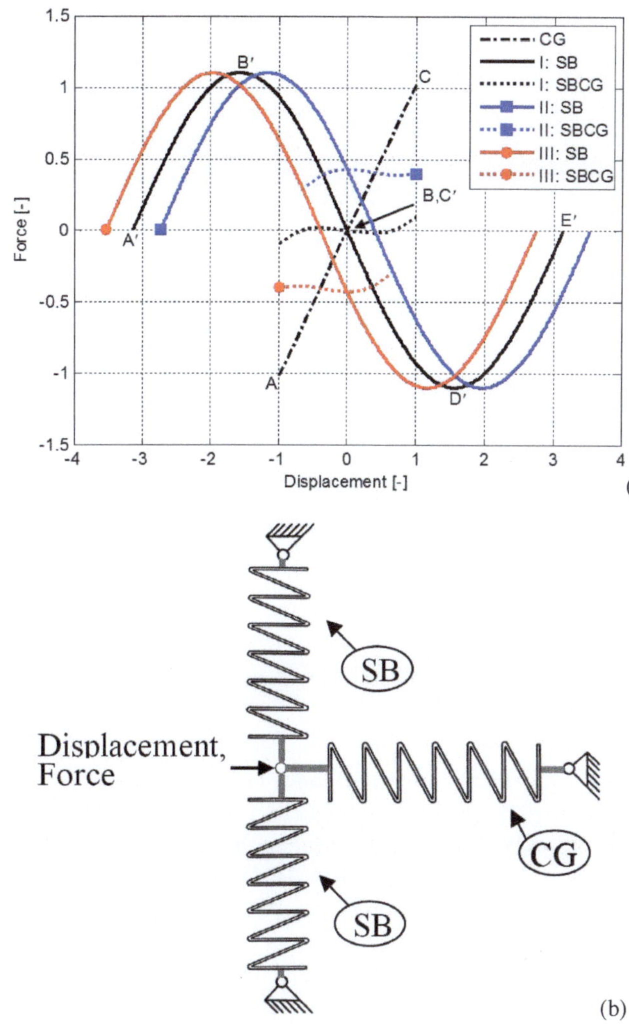

Figure 3. **(a)** The force-displacement curve and **(b)** the corresponding configuration of a typical bi-stable mechanism (static balancer, SB) and the compliant grasper (CG) as a pin-pin bi-stable linear compression spring mechanism and a linear extension spring respectively. The shown configuration in **(b)** corresponds with position B of **(a)**. **(a)** also shows the different cases (I–III) of zero stiffness with zero (black lines), constant positive (blue lines ■) and constant negative actuation force (red lines •). Force-displacement behaviors accomplished by positioning the center of the static balancer (SB) relative to the compliant grasper (CG). Positioning the unstable equilibrium (C′) can be performed by nuts.

2008; Todd et al., 2010). It has been shown that these mechanisms have the largest ratios of both compensated force and statically balanced stroke relative to the size (Dunning et al., 2011). A typical force-displacement curve of such a mechanism is shown in Fig. 3a. For clarity, in this figure the bi-stable mechanism and the grasper considered as pin-pin linear springs: a bi-stable compression spring mechanism and an extension spring respectively (Fig. 3b). In Fig. 3a, five points are identified in the force-displacement curve of the bi-stable mechanism: first stable equilibrium position (A′),

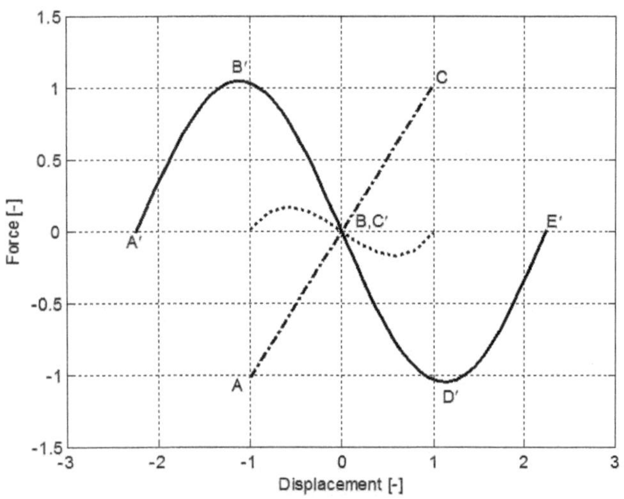

Figure 4. Conceptual dimensionless force-displacement behavior of statically balanced compliant grasper (SBCG) for the case of bi-stable behavior. Force-displacement behaviors accomplished by using stiffer balancer; stiffness of the static balancers may be adjusted by varying thicknesses and initial shape.

initial buckling (first bifurcation point) (B′), unstable equilibrium point (C′), snap through (second bifurcation point) (D′), and the second stable equilibrium point (E′). The negative stiffness of bi-stable mechanism (B′-D′) is tuned to have the same stiffness as the compliant grasper (A-C) but with the opposite sign. The cooperative action of force-displacement behaviors of the grasper and bi-stable mechanism can result into cases of zero stiffness with (I) zero actuation force, (II) constant positive force and (III) constant negative force as shown in Fig. 3a or (IV) bi-stable behavior (Fig. 4). Case I (black dotted line) occurs when the unstable equilibrium position (C′) of the bi-stable beam (black solid line) falls on the equilibrium position of the grasper (B; black dash dotted line). Case II (blue dotted line ■) occurs when the unstable equilibrium position (C′) of the bi-stable beam (blue solid line ■) falls after the equilibrium position of the grasper (B; black dash dotted line). Case III (red dotted line •) occurs when the unstable equilibrium position (C′) of the bi-stable beam (red solid line •) falls before the equilibrium position of the grasper (B; black dash dotted line). Case IV occurs when the negative stiffness of the static balancer (B′-D′) is greater than the positive stiffness of the grasper (A-C; i.e. over compensation). More on this subject can be found in Tolou et al. (2010b) and Pluimers et al. (2012).

In this research, straight guided beams with initially pre-curved shape were employed as the bi-stable mechanism. However, in practice the buckling behavior of such beams are not symmetric (opposite to Fig. 3a) because of the fixed-guidance at the end tips. The proposed concept presents a straightforward method to preload the structure along the axis, unlike the conventional ones that are preloaded orthogonal to the direction of instrument actuation (Hoetmer et al., 2009); therefore, the final design is more compact. The method presented here is also capable to be tuned for relevant force displacement behaviors I–IV. This can be accomplished by using the nuts shown in Fig. 2. To study the functionality of this concept for different balancers, a modular design was made and illustrated in Fig. 2.

3 Dimensional design and fabrication/measurement

3.1 Robotic module

3.1.1 Medical requirements

Medical professionals estimate that a force of 5–10 N and a minimum speed of $360^\circ s^{-1}$ are needed during surgical tasks in traditional robotic surgery to perform all possible tasks. As a result, a typical manipulator requires a total power of up to 4 W. These requirements are less strict when specific manipulation tasks are performed (Parittotokkaporn et al., 2010).

3.1.2 Actuation system and control

The robotic arm provides two additional DoFs as well as the actuation of the compliant grasper. According to medical requirements, a suitable diameter for the robotic modules should be slightly less than 12 mm to be inserted through a 12 mm internal diameter laparoscopic trocar. Two SBL04 micro motors (Namiki, Akita, Japan) were selected as actuators for the 2 DoFs because of their small size (4 mm in diameter and 17.4 mm in length including the gearbox) and large torque (up to 5.7×10^{-3} Nm). One motor drives an off-the-shelf 3 mm diameter worm gear (Didel, Belmont, Switzerland), which is used to transmit motion between the axis of the motor and the orthogonal pitch joint axis. The other motor is used for the roll mechanisms. It drives two modulus 0.3 spur gears with unitary transmission ratio. The robotic arm has a total length of 41.8 mm excluding the grasper.

A Faulhaber motor has been used to pull a steel cable inside a 1 mm sheath. When activated the motor closes the grasper by pulling the cable. When the motor is idle, the passive opening of the compliant grasper is due to a compression spring, placed between the compliant grasper and the tip of the robotic arm, as shown in Fig. 7b. If the compliant grasper is not perfectly balanced and has a small residual positive stiffness or force, the spring may be neglected.

A control strategy based on the preemptive priority pseudokernel approach was selected to drive multiple brushless DC motors in real time using one microcontroller instead of utilizing one chip dedicated to each motor (Susilo et al., 2009). The Faulhaber motor is driven thanks to its commercial driver.

3.1.3 Manufacturing

This model was produced using rapid prototyping techniques (VisiJet XT 200 and Invision Si2, Inition, ThingLab, London, UK). An acrylic material composed of urethane acrylate polymer (35–45 %) and triethylene glycol dimethacrylate ester (45–55 %) was used. At this stage, both devices were tested separately. Integration between working prototypes remains relevant and will be done once they are manufactured in final form.

3.2 Statically balanced compliant grasper

3.2.1 Scaling

To evaluate the proposed concept (see Sect. 2.2), a prototype was made. Similar to the robotic arm, the diameter of the compliant grasper must be slightly smaller than 12 mm. For ease of prototype fabrication and verification, the model was scaled up by 3.5 times. The model maintained the stiffness and stresses of the small scale compliant grasper during scale-up. The stiffness (k), in a flexible part is dependent on the length (l), thickness (b), and height (h) of the model, as shown in the force-moment relations based on the deflection of a cantilever beam with end load (Gere and Timoshenko, 1999) (where "I" is the area moment of inertia):

$$\delta = \frac{Fl^3}{3EI} \rightarrow k = \frac{F}{\delta} = \frac{3EI}{l^3}$$

$$\text{with}: \; I = \frac{bh^3}{12} \rightarrow k = \frac{Ebh^3}{4l^3} = \frac{Eb}{4}\left(\frac{h}{l}\right)^3 \quad (1)$$

As can be seen in Eq. (1), the stiffness remains the same if the length and height are scaled with the same ratio. Therefore, all the geometrical values of both the compliant grasper and the static balancer were scaled up, except the thickness. The stresses were kept the same by scaling up the actuation displacement with the same ratio as the geometrical dimensions. Because the stiffness remained constant, the force increased with the same ratio as the actuation displacement. Since the force was scaled up with the same factor as the geometrical dimensions, the stresses maintained the same. With these new values, the opening displacement of the 3.5-times scale-up compliant grasper could be calculated.

3.2.2 Finite element modeling

To provide a basis for concept evaluation, finite element modeling (FEM) was performed using the commercial FEM package ANSYS™ 11.0, and the forces-displacements behaviors were analyzed (ANSYS Inc. Version 11.0 Manual). Because of large deflections, a non-linear static analysis has been performed. PLANE82 and BEAM3 elements were used to mesh the grasper and the static balancer, respectively. The first element type consist of 2-D eight node elements with two translational degrees of freedom per node. It provides more accurate results for mixed (quadrilateral-triangular) automatic meshes and can tolerate irregular shapes without significant loss of computational accuracy. The uni-axial BEAM3 elements require the least amount of computation time, while the actual out-of-plane properties can also be provided using the real constant capability of ANSYS™. The BEAM3 element has three degrees of freedom at each node with tension, compression, and bending capabilities. The material is assumed to be isotropic and to follow linear elastic stress-strain behavior.

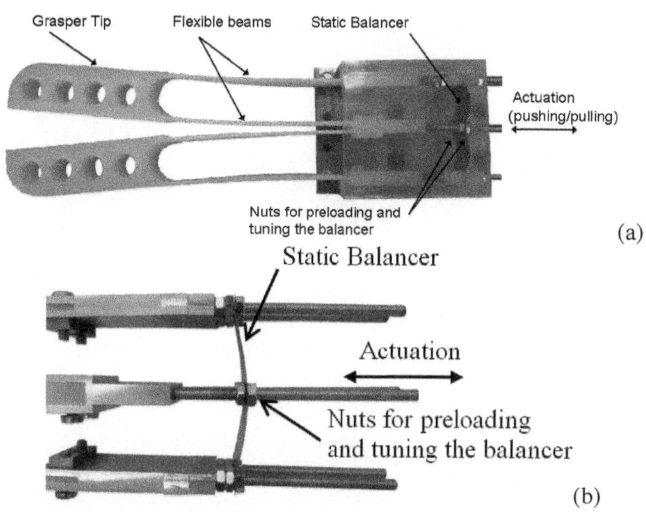

Figure 5. **(a)** Prototype of scaled up SBCG with **(b)** a detail of the tuning equilibrium positions using the nuts.

3.2.3 Manufacturing

The first prototype of the SBCG (see Fig. 5) was made using a 3-D printer (Solido SD300 Pro). The printer uses PVC sheets (0.168 mm thick, $E = 3.37\times10^9$) to build up the model. The balancers were cut out of a high quality stainless steel (ST 301) sheet, instead of using 3-D printer and PVC material, to avoid material hysteresis. The thickness was chosen to be between 0.15 ± 0.05 mm. The location of the center of the balancer was made adjustable with respect to the grasper using two nuts that are attached to a screw thread, as shown in Fig. 5. This allowed for preloading and aligning the center of unstable equilibrium for the static balancer (point C′ in Fig. 3) with stable equilibrium of the compliant grasper (point B in Fig. 3). The screw thread was also used for actuating the SBCG. The sides of the balancer were attached to the grasper via two other screw threads and nuts.

3.2.4 Measurement

Prototype verification was conducted by measuring the force-displacement curves of the compliant grasper and the SBCG. The prototype was mounted on the setup shown in Fig. 6. The compliant grasper was attached to the force

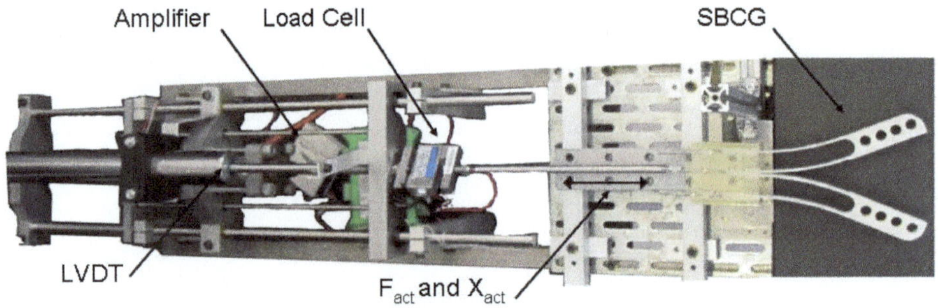

Figure 6. The top view of test set-up: the stiffness characteristic and force displacement of SBCG was determined by measuring actuation force (Fact; using a load cell) and displacement (Xact; using LVDT) during opening and closing of the SBCG.

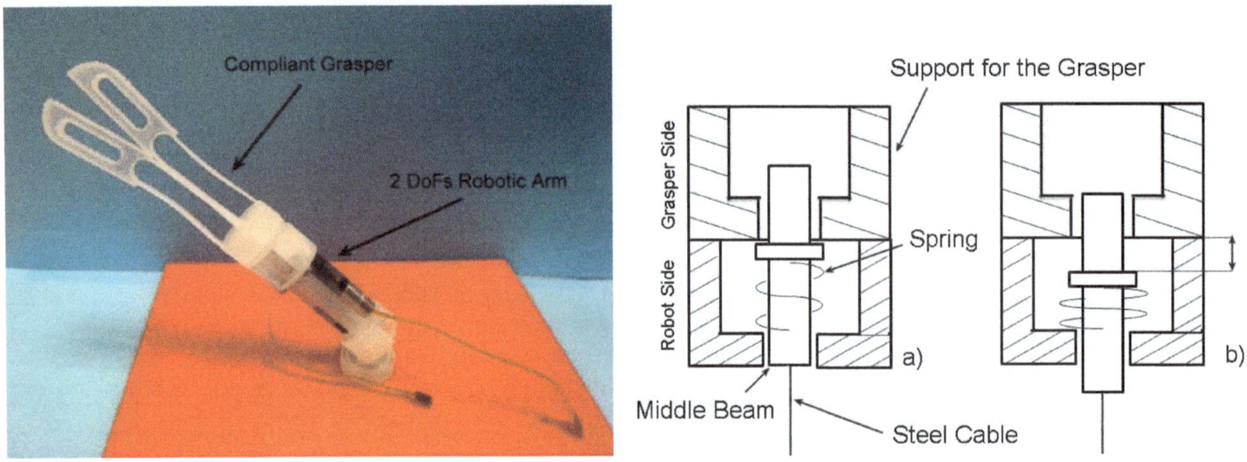

Figure 7. (a) Fabricated robot and grasper mock-up prototypes; **(b)** Mechanisms for coupling the grasper to the robotic arm, a) initial configuration; b) actuated configuration.

sensor by a pulling rod. For each experiment, the compliant grasper was first brought from its relaxed to its closed position. Then the compliant grasper was brought to its fully opened position and back to its closed position. During this movement, the pushing/pulling force (FETE RIS components – B3G-C3-50kg-6B, resolution = 0.06 N, range = [0,50] kg) and displacement (Positek – P101.200CL100, resolution: 0.045 mm, range = [0,200] mm, namely LVDT) were measured. An amplifier (Scaime CPJ 25) and DaQ-mx Data Acquisition system (NI USB6008) were used to read the data.

3.3 System integration

The SBCG has been mechanically integrated on the robotic arm. For this purpose, the proximal end of the grasper has been designed to fit the roll mechanisms of the robotic module. The result of the integration of the mock-up prototypes is shown in Fig. 7. The mechanism for coupling the SBCG to the robot is detailed in Fig. 7b. A double silicone coating should be added once the robot is in its final version to meet the sterilization requirements.

Although the single devices have been demonstrated separately, the whole system will be tested in the future to assess the overall performance in terms of dexterity and capability of generating proper forces for each DoF (pitch, roll, grasping) in a real working scenario.

4 Results and discussion

The limited size of the robot strictly affects the torque and speed with respect to bigger devices. The evaluation of these parameters is very important for the characterization of the whole system. Torque, speed and precision results have been adapted to the current robot design on the basis of the work described in Tortora et al. (2011) and are summarized below. The measured pitch torque is up to 14.6 mNm with a reduction 1:79, whereas the rotation torque is up to 4.35 mNm. Maximum speeds of $90^\circ\,s^{-1}$ for the pitch and of $190^\circ\,s^{-1}$ for the roll were obtained. The pitch ranges from 0° to 180°, depends only on the supported torque on the pulling cable and can be considered unlimited. The values are affected by the manufacturing of the first prototype of the robot, which has been obtained using rapid prototyping. Specifically, the

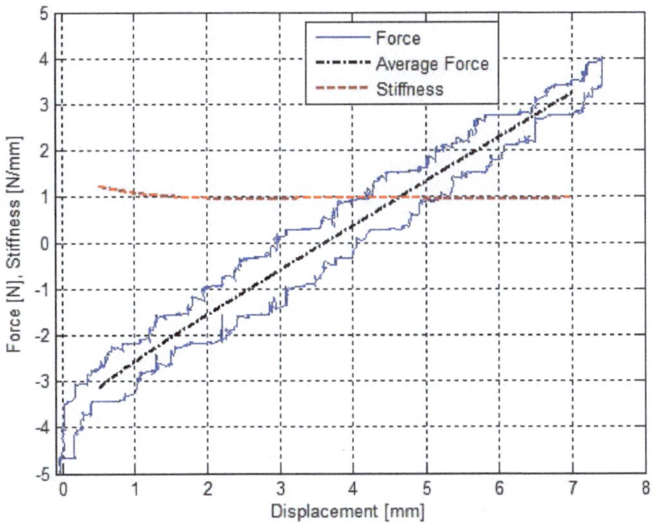

Figure 8. Force-displacement and stiffness curves of the compliant grasper from the measurement.

acrylic material has low-performance mechanical properties at the moment. Further experiments with the current design will be carried out once the robot is manufactured in a more suitable material (e.g. aluminium or titanium) and the SBCG is integrated. In this case, higher torques and speed are expected.

In Figs. 8 and 9, the force-displacement (solid line), the average force-displacement (dashed-dotted line), and stiffness-displacement (dashed line) curves of the compliant grasper and SBCG are shown, respectively. For each measurement the force-displacement of both pushing and pulling were recorded because of hysteresis, and the average was made for ease of design. The average force has been calculated by fitting a polynomial to the curves at the intervals of [0.5 7] mm, [0.3 4.7] mm and [0.2 5.45] mm and of the orders 5 N, 5 N and 7 N for Figs. 8 and 9a and b, respectively. The stiffness has been calculated by differentiating the average force. As shown in the force-displacement curve of the compliant grasper depicted in Fig. 8, the grasper has a positive constant stiffness of $1\,\mathrm{N\,mm^{-1}}$. In Fig. 9, the results of the combined compliant grasper and static balancer can be seen. Figure 9a shows the compliant grasper with the 0.1 mm static balancer. The result is a zero stiffness, near zero with constant positive force SBCG for a range of about 5 mm. This means the negative stiffness of the static balancer is equal to the positive stiffness of the compliant grasper, but the unstable equilibrium position of the static balancing mechanism falls slightly after the equilibrium position of the compliant grasper (Case II). When the displacement becomes larger than 5 mm, the force suddenly increases as the static balancer has reached the end of its working range in which it can balance the grasper (between B′ and D′ of Fig. 3). After this point, the balancer no longer has a negative stiffness, so

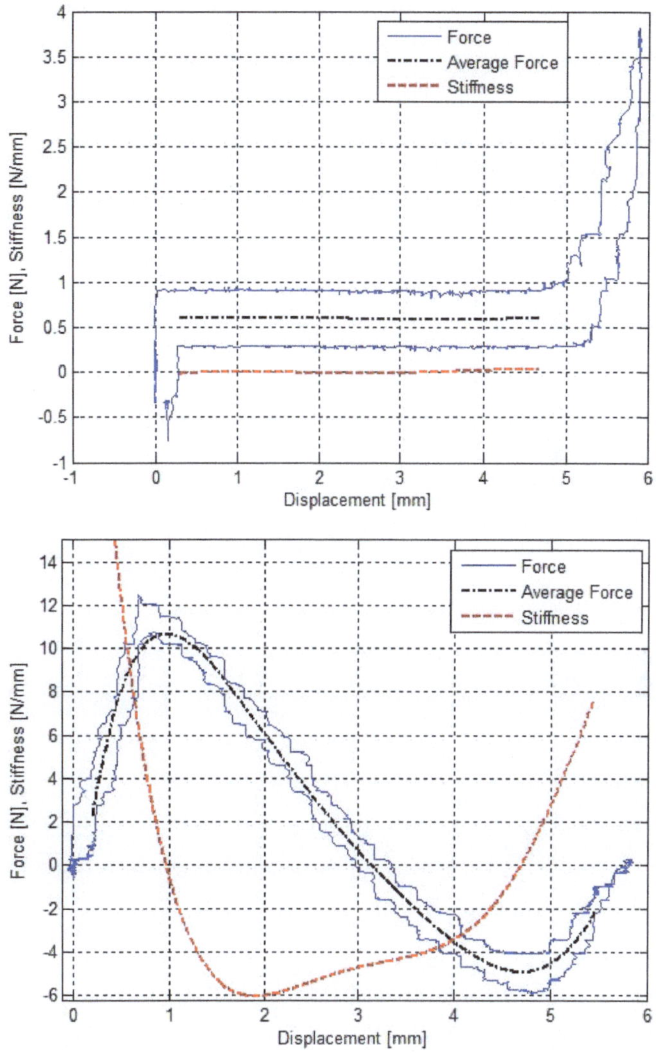

Figure 9. Force-displacement and stiffness curves of the SBCG for **(a)** statically balanced and **(b)** bi-stable behaviors from the measurements; **(a)** the compliant grasper was balanced with the 0.1 mm thickness static balancer and **(b)** the compliant grasper was balanced with the 0.15 mm thicknes static balancers.

the total force increases instead of remaining constant. Again the graph shows hysteresis of about 0.6 N. Due to the limited adaptability of the prototype, the compliant grasper could not be perfectly balanced. However, zero stiffness and near zero actuation force has been achieved.

Figure 9b shows the compliant grasper combined with the 0.15 mm static balancer. This graph shows a bi-stable behavior of the SBCG in a range of 5.8 mm due to over compensation. In all the figures hysteresis of about 1.5 N can be seen. The hysteresis could occur due to slight damping in the measurement setup (friction introduced by bearings, sliding joints and positioning sensor) and the material. The difference in the hysteresis might be explained by the magnitude of the force in the direction of actuation. It can be seen that the larger the force in this direction, the larger the hysteresis.

It should be noted that for the larger scale, both the static balancer and the compliant grasper worked successfully. However, in this scale, the static balancer has a thickness of 0.1 mm, thus the smaller model requires a static balancer of the thickness of approximately 0.03 mm; it remains to be seen if this is feasible. On the other hand, further investigation is needed on the minimum dimensional limitations of the grasper to support the applied forces during the surgery (i.e. robustness). For the future work, beside aforementioned issues, more research needs to be done to find suitable materials and production methods for compliant mechanisms of this scale.

5 Conclusions

In this article, a new 2DoF robotic arm with a fully compliant statically balanced grasper as the end-effector has been presented. The robotic arm provides pitch and roll degrees of freedom. The grasper as the end-effector has been designed fully compliant and is statically balanced using straight-guided beams that are preloaded collinear to the direction of grasper actuation. The prototype shows the adaptively of the grasper to the arm. The measured actuation forces of the grasper has been successfully compensated up to 94 %, and a near zero stiffness has been obtained with up to 97 % reduction. This allows the robotic arm to get force feedback from the end effector. Furthermore, bi-stable behavior is also achieved due to adjustability of the concept to some different stiffness behaviors: zero stiffness with zero (case I), constant positive (case II) and constant negative (case III) actuation forces, and the bi-stable (case IV) behavior. This design may be extended to a fully compliant statically balanced 2DoF robotic arm integrated with the current grasper.

Acknowledgements. This research is part of VIDI Innovational Research Incentives Scheme grant for the project "Statically balanced compliant mechanisms", NWO-STW 7583 and in part supported by the European Commission in the framework of the ARAKNES FP7 European Project EU/IST-2007-224565. Further thanks to R. Luttjeboer, his help was crucial in the manufacturing of the scaled up prototype.

Edited by: C. Kim
Reviewed by: two anonymous referees

References

Alterovitz, R.: Surgical Robotics, Robotics & Automation Magazine, 16, 16–17, 2009.

Breedveld, P., Stassen, H. G., Meijer, D. W., and Stassen, L. P. S.: Theoretical background and conceptual solution for depth perception and eye-hand coordination problems in laparoscopic surgery, Min. Invas. Ther. and Allied Technol., 8, 227–234, 1999.

Chen, G. and Zhang, S.: Fully-compliant statically-balanced mechanisms without prestressing assembly: concepts and case studies, Mech. Sci., 2, 169–174, doi:10.5194/ms-2-169-2011, 2011.

De Lange, D. J. B. A., Langelaar, M., and Herder, J. L.: Towards the design of a statically balanced compliant laparoscopic grasper using topology, Proceedings ASME DETC 26th Biennial, Mechanisms and Robotics Conference, 10–13 September, Baltimore, Maryland, paper number DETC2008/MECH-49794, 2008.

Dunning, A. G., Tolou, N., and Herder, J. L.: Review Article: Inventory of platforms towards the design of a statically balanced six degrees of freedom compliant precision stage, Mech. Sci., 2, 157–168, doi:10.5194/ms-2-157-2011, 2011.

Forgione, A.: In vivo microrobots for natural orifice transluminal surgery, Current status and future perspectives, Surg. Oncol., 18, 121–129, 2009.

Fowler, D., Hu, T., Nadkarni, T., Allen, P., and Hogle, N.: Initial trial of a stereoscopic, insertable, remotely controlled camera for minimal access surgery, Surgical Endoscopy, 24, 9–15, 2010.

Gere, J. M. and Timoshenko, S. P.: Mechanics of Materials, Fourth SI edition, Stanley Thomas Publishers, Cheltenham, 1999.

Hartono, W.: On the Post-buckling Behavior of Elastic Fixed-End Column with Central Brace, ZAMM_Z. Angew. Math. Mech., 81, 605–611, 2001.

Herder, J. L.: Design of spring force compensation systems, Mech. Mach. Theory, 33, 151–161, 1998.

Herder, J. L. and van den Berg, F. P. A.: Statically balanced compliant mechanisms (SBCM's), an example and prospects, Proceedings ASME DETC 26th Biennial, Mechanisms and Robotics Conference, 10–13 September, Baltimore, Maryland, paper number DETC2000/MECH-14144, 2000.

Hoetmer, K., Kim, C., and Herder, J. L.: An extended building block approach for the design of statically balanced compliant mechanisms, 2009 ASME International Design Engineering Technical Conferences, 30 August–2 September 2009, San Diego, California, Paper number DETC2009-87451, 2009.

Howell, L. L.: Compliant Mechanisms, New York, Wiley, 2001.

Hu, T., Allen, P. K., Hogle, N. J., and Fowler, D. L.: Insertable Surgical Imaging Device with Pan, Tilt, Zoom, and Lighting, Int. J. Robot. Res., 28, 1373–1386, 2009.

Jensen, B. D., Parkinson, M. B., Kurabayashi, K., Howell, L. L., and Baker, M. S.: Design Optimization Of A Fully-Compliant Bi-stable Micro-Mechanism, Proceedings of 2001 ASME International Mechanical Engineering Congress and Exposition, 11–16 November, New York, NY, 2001.

Joice, P., Hanna, G. B., and Cuschieri, A.: Errors enacted during endoscopic surgery; a human reliability analysis, Appl. Ergon., 29, 409–414, 1998.

Lehman, A. C., Berg, K. A., Dumpert, J., Wood, A. W., Visty, Q., Rentschler, M. E., Platt, S. R., Farritor, S. M., and Oleynikov, D.: Surgery with cooperative robots, Comput. Aided Surg., 13, 95–105, 2008.

Li, S. R. and Zhou, Y. H.: Post-buckling of a hinged-fixed beam under uniformly distributed follower forces, Mech. Res. Commun., 32, 359–367, 2005.

Lum, M., Friedman, D., Rosen, J., Sankaranarayanan, G., King, H., Fodero, K., Leuschke, R., Sinanan, M., and Hannaford, B.: The RAVEN – design and validation of a telesurgery system, Int. J. Robot. Res., 28, 1183–1197, 2009.

Oleynikov, D., Rentschler, M., Hadzialic, A., Dumpert, J., Platt, S. R., and Farritor, S.: Miniature robots can assist in Laparoscopic cholecystectomy, Surg. Endosc., 19, 473–476, 2005.

Parittotokkaporn, T., Degenaar, P., Davies, B. L., and Baena, R.: Force vs. Displacement during Tool Insertion: Techniques and Modeling Approaches, Proc. Hamlyn Symposium, 2010.

Pluimers, P. J., Tolou, N., Jensen, B. D., Howell, L. L., and Herder, J. L.: A Compliant On/Off Connection Mechanism for Preloading Statically Balanced Compliant Mechanisms, DETC2012-71509, ASME IDETC 2012, Chicago, USA, 2012.

Qiu, J., Lang, J. H., and Slocum, A. H.: Microelectromechanical Systems, A curved-beam bi-stable mechanism, 13, 137–146, 2004.

Rentschler, M. E., Dumpert, J., Platt, S. R., Lagnernma, K., Oleynikov, D., and Farritor, S. M.: Modeling, Analysis, and Experimental Study of In Vivo Wheeled Robotic Mobility, IEEE T. Robot., 22, 308–321, 2006.

Stapel, A. and Herder, J. L.: Feasibility study of a fully compliant statically balanced laparoscopic grasper, Proceedings ASME Design Engineering Technical Conferences, 28 September–2 October, Salt Lake City, Utah, DETC2004-57242, 2004.

Sönmez, T. C. C.: A Compliant Bistable Mechanism Design Incorporating Elastical Buckling Beam Theory and Pseudo-Rigid-Body Model, J. Mech. Design, 130, 042304-1/13, 2008.

Susilo, E., Valdastri, P., Menciassi, A., and Dario, P.: A miniaturized wireless control platform for robotic capsular endoscopy using advanced pseudokernel approach, Sensor. Actuat. A-Phys., 156, 49–58, 2009.

Todd, B., Jensen, B. D., Schultz, S. M., and Hawkins, A. R.: Design and Testing of a Thin-Flexure Bistable Mechanism Suitable for Stamping From Metal Sheets, J. Mech. Design, 132, 071011-1/7, 2010.

Tolou, N. and Herder, J. L.: Concept and Modeling Of A Statically Balanced Compliant Laparoscopic Grasper, in: Proceeding of ASME 2009 International Design Engineering Technical Conferences and Computers and Information in Engineering Conference, 30 August–2 September, San Diego, CA, USA, Paper no. DETC2009-86694, 163–170, doi:10.1115/DETC2009-86694, 2009.

Tolou, N., Gallego, J. A., and Herder, J. L.: Statically Balanced Compliant Micro Mechanisms (SBMEMS): A Breakthrough in Precision Engineering, Mikroniek, 50, 20–25, 2010a.

Tolou, N., Henneken, V. A., and Herder, J. L.: Statically Balanced Compliant Micro Mechanisms (SB-Mems): Concepts And Simulation, in: Proceeding of ASME 2010 International Design Engineering Technical Conferences and Computers and Information in Engineering Conference, 15–18 August 2010, Montreal, Quebec, Canada, Paper no. DETC2010-28406, 447–454, doi:10.1115/DETC2010-28406, 2010b.

Tortora, G., Dimitracopoulos, A., Valdastri, P., Menciassi, A., and Dario. P.: Design of Miniature Modular in vivo Robots for Dedicated Tasks in Minimally Invasive Surgery, AIM 2011, Budapest, Hungary, 3–7 July 2011.

20

Design and performance assessment of an underactuated hand for industrial applications

C. Meijneke, G. A. Kragten, and M. Wisse

Faculty of Mechanical, Maritime and Materials Engineering, Dept. of BioMechanical Engineering, Delft University of Technology, Mekelweg 2, 2628 CD Delft, The Netherlands

Abstract. The Delft Hand 2 (DH-2) is an underactuated robot hand meant for industrial applications, having six degrees of freedom (DoF), one actuator (DoA) and no sensors. It was designed to provide a cheap and robust hand to grasp a large range of objects without damaging them. The goal of this paper is to assess the design and performance of the DH-2, demonstrating how the design was optimized for its intended application area and how the hand was simplified to make it commercially attractive. Performance tests show that the DH-2 has a payload of 2 kg for an object range of 60 to 120 mm, it can close or open within 0.5 s, and it only uses open-loop control by means of the input voltage of the motor. The results demonstrate that the industrial need of a simple, cheap and effective robotic hand can be achieved with the principle of underactuation and the use of conventional components.

This paper was presented at the IFToMM/ASME International Workshop on Underactuated Grasping (UG2010), 19 August 2010, Montréal, Canada.

1 Introduction

The principle of underactuation (Birglen et al., 2008) can make a robotic hand adaptive to the shape of objects, enabling it to grasp a diversity of objects. The demand of grasping various objects can be found in, for example, service robots for grasping in household environments, or sorting systems in industry. Although a lot of robot hands were made using this principle, only very few are commercially available (i.e., the Barrett hand (Townsend, 2000) and the SARAH (Laliberté and Gosselin, 2001)). This small number might be caused by unfamiliarity with underactuation, or because the solutions are still too complex or too expensive. To get the industry interested, there is a need of cheap, simple and robust solutions.

In 2008, an underactuated hand – called Delft Hand 1 (Meijneke and Wilbers, 2009) – was built at Delft University of Technology. The objectives were to learn about the performance of underactuated grasping, and to get industry interested. Although the majority of feedback from the industry was positive, issues like closing speed of the hand, the robustness of the design and the grasp reliability needed to be improved.

To address these issues, a second prototype – called Delft Hand 2 (DH-2) – was developed. Figure 1 shows the design of this hand featuring three similar fingers consisting of two phalanges, which are all driven by one actuator without any sensory feedback.

The objectives of this paper are firstly to assess the critical design choices of the DH-2 leading to a simple, cheap and robust hand. Secondly, the performance of the DH-2 is assessed. The critical performance factors, according to industrial partners, are the (1) payload, (2) the closing and opening time of the hand, (3) and the range of object sizes which can be grasped by the hand. Simple and robust means that the number of actuators and sensors is minimized and external disturbances can be resisted.

The structure of this paper is as follows. In Sect. 2 the design requirements are defined, and the conceptual and dimensional design choices are assessed based on performance calculations. Section 3 describes the experiments to measure the payload for a range of object sizes, the relation between the motor voltage (input) and the contact forces on objects, and the closing or opening time of the hand. The results of these experiments are given in Sect. 4, followed by Discussion and Conclusion.

Figure 1. Picture of the DH-2 prototype.

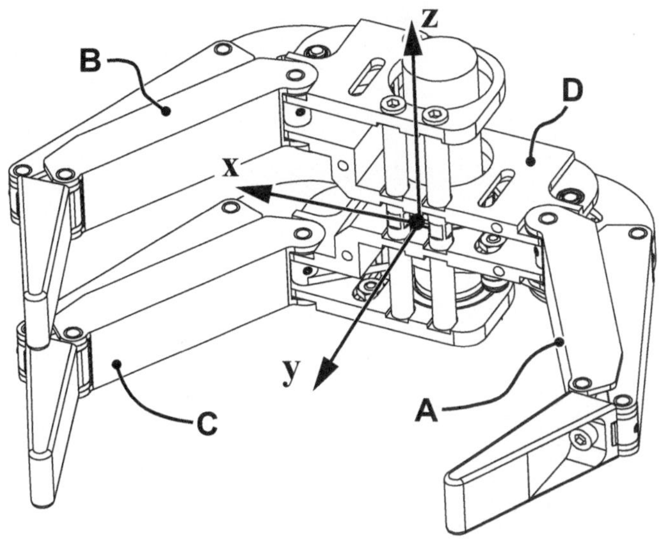

Figure 2. Design drawing of DH-2, with one finger A, two opposite fingers B and C and a palm D.

2 Design

This section shows the main requirements for the design of the DH-2 prototype and the design choices made to satisfy these requirements. The model used to optimize the performance will be discussed and some predictions of the performance are made.

2.1 Requirements

The following functional requirements were defined based on interviews with industrial partners involved in areas like distribution centers and fruit and vegetable sorting: (1) a minimum payload of 2 kg, (2) a maximal closing time of 0.5 s, and (3) a product range of 60 mm to 120 mm diameter. The weight of most products were much smaller than 2 kg, but accelerations in pick and place tasks increase the required payload. The cycle time to do profitable pick and place tasks determined the time to fully open and close the grasper, which had to be less then 0.5 s. The graspable range refers to the diameter (D_{obj}) of a (long) cylindrical object, because the majority of the target products approximates this shape. We consider an object to be graspable when the robot hand can successfully fix the object in a *power grasp* posture (Napier, 1956).

2.2 Mechanical design

In DH-2 the motor torque is distributed between the fingers and phalanges by means of differential mechanisms. To ensure robustness, a belt transmission and a gear differential were chosen in the DH-2 to distribute the motor torque to the three fingers. Each finger consists of two phalanges driven by a four-bar linkage mechanism as inspired by the MARS and SARAH (Laliberté and Gosselin, 2001). To prevent entrapment of objects between the linkages, the parts of the fingers where designed to form a closed casing (Fig. 2). When the closing and opening sequence of the phalanges are critical, linear springs can be mounted in the finger casings.

To obtain a simple and cheap design, we minimized the amount of mechanical and electronic parts. It was therefore decided not to use any force or position sensor. In addition. the fingers can be both opened and closed by only one Maxon RE13 3.5W motor (with a GP13 67:1 gearbox). When a voltage of 24 V is applied to the motor this results in a constant actuation torque of 1 Nm applied at the base of finger A and 0.5 Nm on each of the opposite fingers B and C (Fig. 2) as defined in (Birglen et al., 2008, p. 35). Hence, only two cables for the power supply to the motor were required. This approach also provided the possibility to fit the full mechanism in a palm of 30 mm thickness and 75 mm width.

The absence of sensory feedback implies open-loop control of the hand. It was chosen to control the contact forces of the hand simply by the input voltage to the motor. The relation between this input voltage and contact forces is experimentally assessed in the next chapter.

2.3 Dimensional design

The main dimensions of the linkage mechanism of the fingers and the width of the palm were determined by performance calculations according to the model proposed in Kragten and Herder (2010). In this model the equilibrium configuration and the contact forces of the fingers were calculated as a function of the objects size and object position. Friction in the mechanism of the fingers and between the fingers and the object was neglected in this model. A contact point between a phalanx and the object can be geometrically described by a loop closure vector equation, as shown in Fig. 3. For the

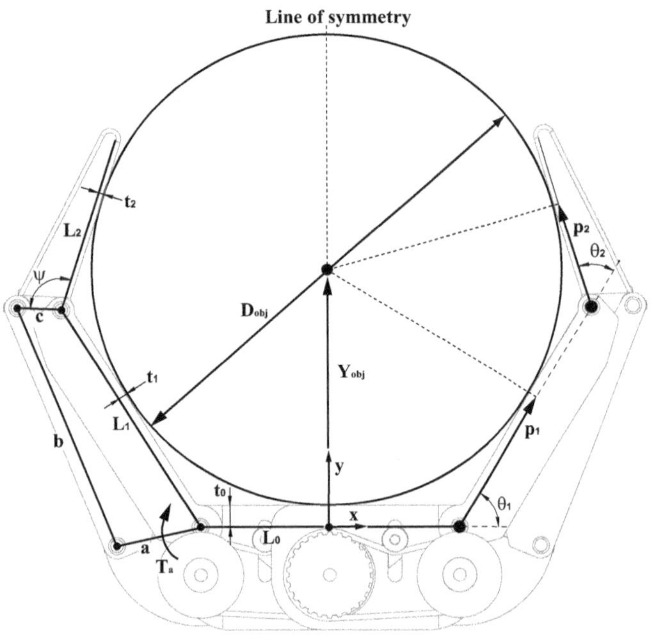

Figure 3. Schematic representation of a symmetric *power grasp* of the DH-2 on a cylindrical object of radius r_{obj} and position (X_{obj}, Y_{obj}). On the left finger, the length of the phalanges L_i, the dimensions of the linkage mechanism a,b,c, the angle ψ between the distal phalanx L_2 and link c, the thickness of the phalanges t_i, and the actuation torque T_a are shown. On the right finger the contact point position on the phalanges p_i and the rotation of the phalanges θ_i are shown.

proximal and distal phalanx, these equations were as follows:

$$\binom{L_0}{0} + \mathbf{R}_{\theta_1}\binom{p_1}{t_1} = \binom{X_{obj}}{Y_{obj}} + \mathbf{R}_{\theta_1}\binom{0}{-r_{obj}} \tag{1}$$

$$\binom{L_0}{0} + \mathbf{R}_{\theta_1}\binom{L_1}{0} + \mathbf{R}_{\theta_1}\mathbf{R}_{\theta_2}\binom{p_2}{t_2} = \binom{X_{obj}}{Y_{obj}} + \mathbf{R}_{\theta_1}\mathbf{R}_{\theta_2}\binom{0}{-r_{obj}} \tag{2}$$

where index 1 and 2 denote the proximal and distal phalanx, respectively; p_i is the contact point position on the phalanx; t_i is the thickness of the phalanx; θ_i is the rotation of the phalanx; $\mathbf{R}_{\theta_i}$ is the rotation matrix; and X_{obj} and Y_{obj} is the position of the object with radius r_{obj}.

The magnitude of the normal contact forces was calculated according to (Birglen et al., 2008, p. 65):

$$F_1 = -\frac{L_1(-p_2 + h\cos\theta_2)}{p_1 p_2(h + L_1)} \cdot T_a \tag{3}$$

$$F_2 = \frac{h}{p_2(h + L_1)} \cdot T_a \tag{4}$$

where

$$h = c(\cos\theta_2 - \psi - \sin\theta_2 - \psi\cot\beta); \tag{5}$$

with

$$\cot\beta = \frac{c\sin(\theta_2 - \psi)\sqrt{4a^2b^2 - N^2} + M(L_1 + c\cos(\theta_2 - \psi))}{-(L_1 + c\cos(\theta_2 - \psi))\sqrt{4a^2b^2 - N^2} + Mc\sin(\theta_2 - \psi)} \tag{6}$$

Table 1. Design parameter used for DH-2.

Parameter	L_0	L_1	L_2	a	b	c	ψ
Value [mm]	30	60	40	20	60	10	105°

with

$$M = -L_1(L_1 + 2c\cos(\theta_2 - \psi)) + a^2 - b^2 - c^2 \tag{7}$$

$$N = L_1(L_1 + 2c\cos(\theta_2 - \psi)) - a^2 - b^2 + c^2 \tag{8}$$

Simultaneously solving Eqs. (1) and (2) for each finger at a specific object position and object size resulted in the equilibrium configuration of the fingers (p_i, θ_i). Substituting this configuration in Eqs. (3) and (4) yielded the contact forces F_i. In the case of a negative proximal contact force, a new configuration was calculated where this phalanx had no contact with the object (i.e., Eq. (1) was omitted and $F_1 = 0$, Eq. (3)). A grasp equilibrium of the fingers with the object existed when the resultant of the contact forces is zero or when the fingers compress the object against the palm. The maximum allowable weight or force disturbance on the object was equal to the resultant of the contact forces pointing in the direction of the palm at the particular object position where this resultant force was maximal.

An iterative search was done to find the dimensions of the design parameters that would result in a hand that on the one hand could grasp the required range of object sizes and on the other hand was able to resist large disturbance forces. Based on this search the dimensions were chosen according to Table 1, and the in-plane thickness t_i of the phalanges were chosen as small as possible. The mechanical limits of the fingers were chosen $\theta_1 = 45° - 90°$ and $\theta_2 = 0° - 90°$ (Fig. 3), which allowed the hand to grasp all objects in the required range within the required closing and opening time.

In addition, if the actuation torque $T_a = 1$ Nm would be applied to the fingers, the maximum allowable force disturbance on the object in the direction perpendicular to the palm (y-direction) would be at least 32 N and 5 N for the object with a diameter of 60 and 120 mm, respectively. Since friction in the finger mechanism or between the object and fingers was not taken into account for these performance predictions, higher allowable forces can be expected for the real prototype.

3 Materials and methods

This section describes the experiments to assess the performance of the DH-2 with the maximum input voltage of $U_{in} = 24$ V applied to the motor. Furthermore, we will investigate the relation between U_{in} and the contact forces.

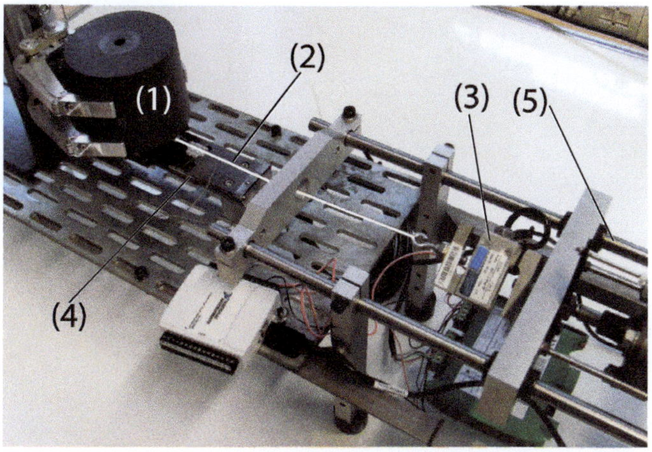

Figure 4. Setup for y-payload test where (1) is the test cylinder, (2) a cable (highlighted), (3) the load sensor, (4) the linear guidance for the object, and (5) the frame of the tensile tester.

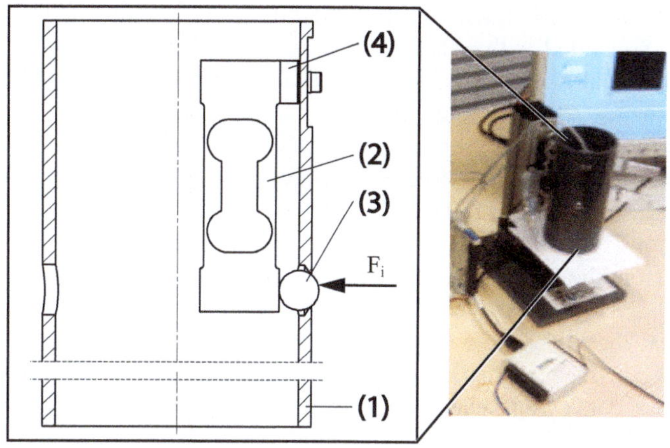

Figure 5. Overview of the test setup containing, (1) a PVC cylinder available in diameter $D_{obj} = 63, 75, 110$ mm, (2) the load cell, (3) a steel ball, and (4) the sensor mount.

3.1 Payload

The maximum payload of the hand depends on the size and shape of the object and the orientation of the hand with respect to gravity. In this study, the payload was determined for cylindrical objects with a diameter of 63, 75, and 110 mm, while the hand was oriented with respectively the y-axis and z-axis parallel to gravity (see Fig. 2).

The maximal payload parallel to the y-axis was measured by slowly pulling the grasped objects out of the hand in this direction and measuring the required force with the setup in Fig. 4. The objects were mounted on a horizontal linear guide (THK, RSR15WZM, friction force less than 0.5 N), such that the objects could freely move along the y-axis. The objects were initially placed against the palm of the hand, while $U_{in} = 24$ V was supplied to the motor of the hand (i.e., the hand and object were in the grasp equilibrium configuration). Then the object was pulled out of the hand by a manually driven tensile tester at a speed of approximately 3 mm/s, while the required pull force was measured with a load cell (Zemic Europe, B3G-C3-50kg-6B, capacity 50 kg), and the object position was measured using a LVDT (Schaevitz Sensors, 2000 LCIT). Each experiment was replicated 5 times. At least 1 min waiting time were kept between the experiments to have comparable motor temperatures at each experiment. To calculate the mean force as a function of the object displacement relative to the palm, the position where the pull force was 5 N was chosen as the zero position of the object. Then the measured pull force was re-sampled at a displacement interval of 1 mm by linear interpolation between the nearest measured positions.

For each object size, the maximum of the mean force was determined and regarded as the maximum payload F_{ymax}.

The maximal payload at an orientation parallel to the z-axis depends on the friction between the fingers and the object:

$$F_{zmax} = \sum F_i \cdot \mu_i \tag{9}$$

where μ_i is the friction coefficient at the corresponding contact surface and $\sum F_i$ the sum of the contact forces when the object is in grasp equilibrium and the hand is actuated at $U_{in} = 24$ V. The measurement of the contact forces applied to the three objects is described in the next section. Since the contact friction between the object and the fingers depends on their material properties, the minimum contact friction needed to hold objects of these sizes was calculated, assuming the required payload $F_{zmax} = 20$ N.

3.2 Input voltage – contact force relation

The magnitude of the contact forces of the fingers on cylindrical objects of available diameter $D_{obj} = 63,75,110$ mm was measured as a function of the motor input voltage (U_{in}). The measurements were executed using the setup displayed schematically in Fig. 5. A load cell (HBM PW4MC3 3 kg) was mounted inside the cylinder. A steel ball was placed in a countersunk hole on the inside of the cylinder and was held in place by the sensor. A small part of the ball protruded through the hole of the cylinder to act as a contact point where the force could be measured. The cylinder was thus placed in the hand in such orientation that the contact force between the objects and the proximal phalanx (F_1), distal phalanx (F_2) and palm (F_{palm}), respectively, could be measured on finger A, assuming that this force is equal to the sum of the corresponding contact forces on the opposite fingers B and C (see Fig. 2). The cylinder was vertically supported by a platform on which the aforementioned orientations were plotted for better repeatability of the cylinder orientation.

The load cell signal was amplified to a ±10 V signal and A/D converted with a data acquisition unit (NI DAQ USB-6008) resulting in a measurement resolution of 0.05 N. During the test three parameters were varied: (1) The input voltage U_{in} of the motor, varying from 10 V limited by static friction of the mechanism to 24 V limited by the nominal voltage in steps of 2 V, (2) The available object diameters D_{obj} of 63 mm, 75 mm and 110 mm, and (3) the contact point varying from the palm F_{palm}, the proximal phalanx F_1, and distal phalanx F_2. The full experiment was replicated 3 times. To check the linearity of the measured forces F_i as a function of U_{in}, a linear regression analysis was applied to the experimental data using a least squares method. The data was fitted to the following line:

$$F_i(U_{\text{in}}) = a\cdot(U_{\text{in}} - 10\ \text{V}) + b, \quad U_{\text{in}} \geq 10\ \text{V} \tag{10}$$

where a is the slope and b the 10 V intercept. The contact forces at the maximal input voltage $F_i(U_{\text{in}} = 24\ \text{V})$ was used to determine the maximal z-payload F_{zmax} as mentioned in the previous section.

3.3 Closing and opening time

The time to fully close and open the hand was measured. A high-speed camera (Casio, EX-F1, 300fps) was used to record a full opening and closing cycle at an input voltage $U_{\text{in}} = 24$ V. The opening and closing time was calculated as follows:

$$\Delta t_c = t_{\text{closed}} - t_{\text{open}} \tag{11}$$

$$\Delta t_o = t_{\text{open}} - t_{\text{closed}} \tag{12}$$

where Δt_c and Δt_o are the closing and opening time respectively, t_{closed} the time when grasper was fully closed, and t_{open} the time when the grasper was fully opened. This experiment was replicated 3 times and the mean closing and opening time was calculated.

4 Experimental results

Figure 6 shows the force-displacement characteristics to measure the maximal payload in y-direction for objects of diameter $D_{\text{obj}} = 63, 75$ and 110 mm respectively. Each individual measurement as well as the mean of the five repetitions are shown. The points $A = 131.7$ N, $B = 102.0$ N and $C = 26.3$ N correspond to the mean maximal force F_{ymax}.

Figure 7 shows the contact forces measured at the proximal phalanx, distal phalanx and palm for each object size. The raw data (*), the linear regression line, its formula (F_{reg}) along with the individual R^2 values are shown in the respective subplot. The sum of the contact forces at $U_{\text{in}} = 24$ V for object sizes $D_{\text{obj}} = 63, 75$ and 110 mm are 67.6, 63.6 and 63.6 N respectively, was obtained by the summation of two times the proximal phalanx force (F_1), two time the distal phalanx force (F_2), and the palm force (F_{palm}). According to Eq. (9), the minimal required friction coefficients to achieve a payload of 20 N are thus 0.29, 0.30 and 0.36, assuming that the contact friction is equal for all contact points. For cylinder diameter 110 mm the force on the proximal is zero because with these design parameters there is no 5 point grasp possible.

Table 2 shows the measured opening and closing times and their average.

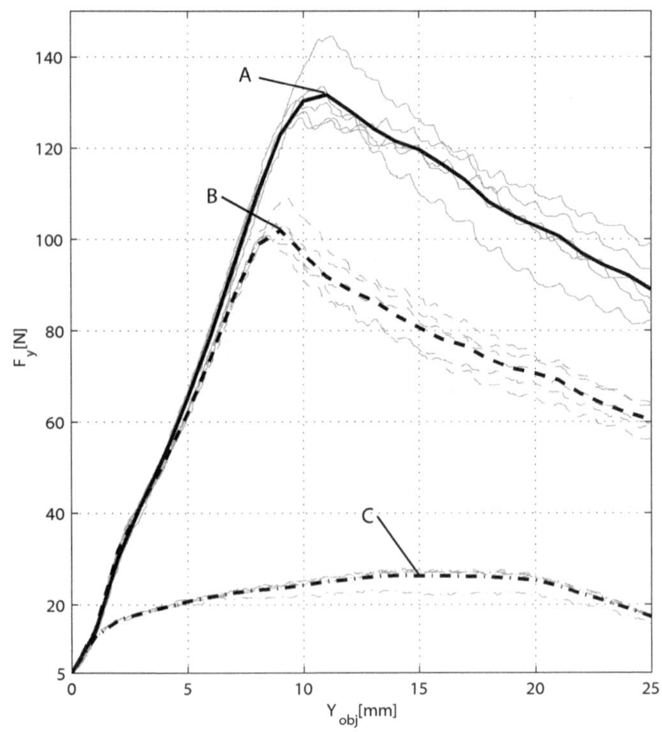

Figure 6. Results for the y-payload measurements where the solid line (—), dashed line (−−) and dash-dotted line (−.−) are the mean forces (F_y) on cylinders of 63, 75, 110 mm respectively, as a function of the y displacement of the object (Y_{obj}). In gray the raw data is presented with using the same line type convention. A, B and C denote the maximum of the mean force of the respective cylinder diameter.

Table 2. Closing and opening time of the hand in [s].

Repeat	1	2	3	Mean
Closing time	0.421	0.432	0.425	0.426
Opening time	0.450	0.432	0.430	0.437

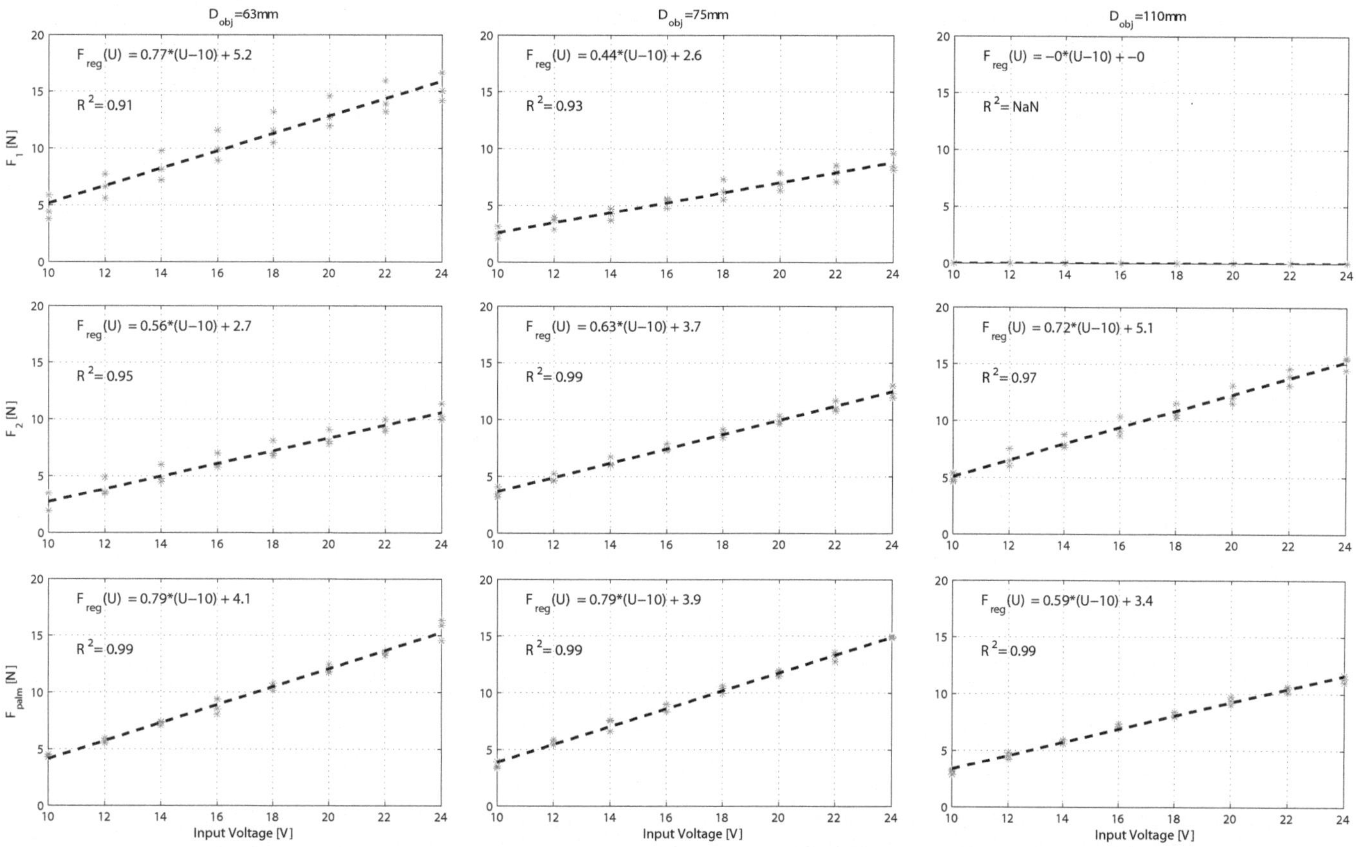

Figure 7. Results of the contact force measurements where, the stars (∗) are the raw data points , and the dashed line (−−) the linear regressed test results , as function of the input voltage [V]. From left to right: cylinder diameter 63, 75 and 110 mm. From top to bottom: Contact forces on the proximal phalanx, distal phalanx, and palm.

Table 3. Comparison of the test results with the requirements where F_{ymax} is the maximal force in y-direction, $\mu_{\min}$ the minimal required friction coefficient for 2 kg payload in z-direction, D_{obj} the object diameter, and Δt_i the opening and closing time.

Criteria	Unit	Measured	Required	
			min	max
F_{ymax} ⌀63	[N]	131.7	20	–
F_{ymax} ⌀75	[N]	102.0	20	–
F_{ymax} ⌀110	[N]	26.3	20	–
$\mu_{\min}$ ⌀63	[–]	0.29	–	–
$\mu_{\min}$ ⌀75	[–]	0.30	–	–
$\mu_{\min}$ ⌀110	[–]	0.36	–	–
D_{obj}	[mm]	63 to 110	60	120
Δt_c	[s]	0.426	–	0.500
Δt_o	[s]	0.437	–	0.500

5 Discussion

The requirements to the DH-2 were a simple and robust design that was able to grasp objects of a payload of 2 kg and of a size between ⌀60 to ⌀120 mm within 0.5 s closing or opening time. In this paper, the performance was assessed. In Table 3 all measured values are compared with there corresponding requirements. Though the maximal payload perpendicular to the palm (parallel to the y-axis) is strongly decreasing for increasing object size, the minimal disturbance force requirement is satisfied for the measured object sizes. The payload parallel to the palm (in z-direction) is less dependent on the object size but greatly depends on the friction between the object and the fingers. This friction coefficient has to be greater then 0.36 for the largest object. This means that when the payload reaches critical values, friction coating or synthetic phalanges is necessary, but for lower payloads blank metal phalanges suffices. The friction enhancements can also be unevenly distributed over the contact points if this would be more convenient. The tested object sizes do not cover the full required range of objects sizes. However, the benchmark test in Kragten et al. (2010) showed that the DH-2 can grasp objects between ⌀55 to at least ⌀120 mm in a power grasp configuration. The closing and opening time satisfy the required 0.5 s.

It was chosen to design the hand without any force or position sensors, to keep the hand cheap and less vulnerable. The motor input voltage was chosen to control the magnitude of the contact forces. The relation between the input voltage and the contact forces was measured. Applying the linear regression to the contact force measurement resulted in a regression coefficient $R^2 > 0.9$ indicating a strong linear relationship. In addition, we observed that the sum of the contact forces on the object hardly changed for different object sizes. This means that the input voltage of the motor seems suited for open-loop control of the grasp forces of the hand.

Comparing the design of the DH-2 with the DH-1, we observed that the choice of gears and belt transmissions instead of a cable-pulley transmission greatly improved the reliability of the design. The main dimensions of the fingers of the DH-2 also changed based on performance calculations. The DH-2 can grasp larger object sizes compared to the DH-1, while the total length of the fingers remained constant.

6 Conclusions

This paper revealed the design choices to develop a cheap, simple and robust underactuated robotic hand for industrial purposes and showed the resulting prototype. Tests to assess the payload, closing time and graspable range showed that: (1) A minimal payload of 2 kg can be grasped, providing that the contact friction is at least 0.36, (2) the hand can be closed or opened within 0.5 s, and (3) objects sizes from ⌀55 to ⌀120 mm can be held in a power grasp configuration. Hence, the DH-2 satisfies the main requirements provided by industrial partners.

It was shown that the magnitude of the contact forces can be controlled by open-loop control on the input voltage on the motor, making it possible to leave out force sensors and electronics needed for closed-loop control.

The key factors that lead to a commercially attractive, functional and reliable underactuated hand were (1) the choice of a simple mechanism consisting of conventional linkages, gears and belt transmissions; (2) the use of a minimal amount of motors and sensors; and (3) a dimensional design based on performance predictions by a grasp model.

Acknowledgements. We gratefully acknowledge the help of Tim Krijger with the contact force measurements and Gerwin Smit for the use of the tensile tester for the payload measurement.

Edited by: J. L. Herder
Reviewed by: two anonymous referees

References

Birglen, L., Laliberté, T., and Gosselin, C.: Underactuated robotic hands, Vol. 40, Springer Tracts in Advanced Robotics, Springer Verlag, Berlin, Heidelberg, 2008.

Kragten, G. and Herder, J.: Ability to hold grasped objects by underactuated hands: performance prediction and experiments, Mech. Mach. Theory, 45, 408–425, 2010.

Kragten, G., Meijneke, C., and Herder, J.: A Proposal of Benchmark Tests for Underactuated or Compliant Hands, Mech. Sci., in press, 2010.

Laliberté, T. and Gosselin, C.: Underactuation in space robotic hands, in: Proceedings of the 6th International Symposium on Artificial Intelligence and Robotics & Automation in Space, 2001.

Meijneke, C. and Wilbers, F.: Slimme Robothanden, Mikroniek, 5–8, 2009 (in Dutch).

Napier, J.: The prehensile movements of the human hand, J. Bone Joint Surg., British Volume, 38, 902–913, 1956.

Townsend, W.: The BarrettHand grasper- programmably flexible part handling and assembly, Ind. Robot, 27, 181–188, 2000.

21

ROBOTRAN: a powerful symbolic generator of multibody models

N. Docquier[1,*], A. Poncelet[1], and P. Fisette[1]

[1]Université catholique de Louvain, Center for Research in Mechatronics, B 1348 Louvain-la-Neuve, Belgium
[*]Postdoctoral Researcher of the Fonds de la Recherche Scientifique – FNRS

Correspondence to: N. Docquier (nicolas.docquier@uclouvain.be)

Abstract. The computational efficiency of symbolic generation was at the root of the emergence of symbolic multibody programs in the eighties. At present, it remains an attractive feature of it since the exponential increase in modern computer performances naturally provides the opportunity to investigate larger systems and more sophisticated models for which real-time computation is a real asset.

Nowadays, in the context of mechatronic multibody systems, another interesting feature of the symbolic approach appears when dealing with enlarged multibody models, i.e. including electrical actuators, hydraulic devices, pneumatic suspensions, etc. and requiring specific analyses like control and optimization. Indeed, since symbolic multibody programs clearly distinguish the modeling phase from the analysis process, extracting the symbolic model, as well as some precious ingredients like analytical sensitivities, in order to export it towards any suitable environment (for control or optimization purposes) is quite straightforward. Symbolic multibody model portability is thus very attractive for the analysis of mechatronic applications.

In this context, the main features and recent developments of the ROBOTRAN software developed at the Université catholique de Louvain (Belgium) are reviewed in this paper and illustrated via three multibody applications which highlight its capabilities for dealing with very large systems and coping with multiphysics issues.

1 Introduction

Before the appearance of efficient computer architectures for scientific numerical computations, only analytical methods were available for modeling systems. The analyses often used rather restrictive hypotheses (truncated and/or linearized models, for instance). The emergence of powerful processors and reliable and user-friendly languages and software led the scientific community to develop numerical programs able to cover a wide range of applications in a given field (e.g. structural dynamics, multibody dynamics, fluid mechanics, electronic circuits, …). Regarding multibody dynamics, numerous so-called multibody programs were developed all over the world as from the seventies (Schiehlen, 1990), each of them being described as a general purpose code although, in reality, faced with the huge variety of applications, they all impose some restrictions on the modeling and analysis processes.

1.1 System modeling versus system analysis

From our point of view, it is useful to distinguish modeling from analysis. The *modeling* phase is the analytical, numerical or symbolical process which, once and for all, sets up the equations of motion describing a multibody system (MBS in short) for a given set of system parameters and generalized coordinates. For instance, the direct dynamic model is illustrated in Fig. 1 as a block whose input are the system parameters δ (joints location, body masses, and inertia, etc.) and the generalized positions q and velocities $\dot{q}$ and whose outputs are the generalized accelerations $\ddot{q}$ and the Lagrange multipliers λ. The *analysis* phase denotes any numerical process which uses a *model* (or part of it) to generate the expected

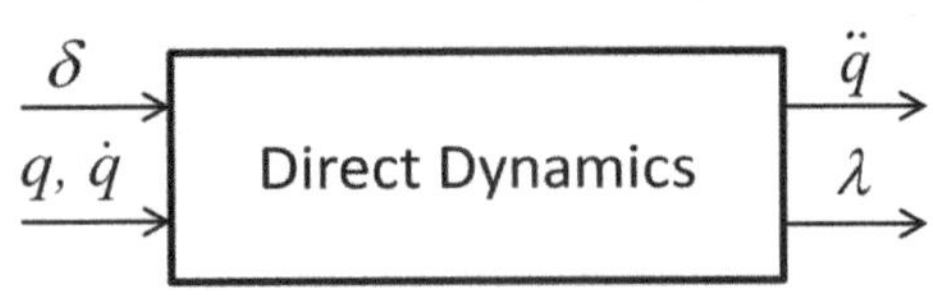

Figure 1. Direct Dynamic Model: main input/output.

results. Equilibrium solution, modal analysis, time integration, optimization are examples of analysis processes which, in case of complex MBS, must be performed numerically via specific algorithms. All of these numerical processes require a repeated evaluation of the model, often more than a million times in a time integration or optimization process. This means that, whatever the chosen formalism and the available computer resources, the computation of the model must be as efficient as possible.

1.2 Numerical versus symbolic implementation

The classical way of implementing these two phases, modeling and analysis, in a computer program is to do so in a purely numerical manner. Although this is mandatory for the analysis phase, it is not the only possibility for the modeling phase.

Before thinking about generating it automatically using a computer, the most natural way to model a system such as that of Fig. 1, is to do it *manually* – using a pen or some general purpose program – on the basis of a given formalism (Newton/Euler equations, Virtual Work Principle, etc.). While appearing to be archaic in the computer era, and not always reliable, this method is still used for simple systems , simplified models and, hopefully, for training undergraduate students dealing with Newtonian mechanics. Although it is subject to human error, manual generation often yields optimized mathematical equations, in terms of arithmetical and trigonometrical expressions: this remark is fundamental in the context of symbolic generation.

Numerical generation means that, each time a computation is required by some analysis process, the model is rebuilt by numerous calls of subroutines, each involved in some specific parts of the equations or specific vector/tensor operations (library subroutines). This multitude of subroutines is the consequence of the universal nature of multibody programs, and the multiple calls are partially responsible for the heaviness of the numerical generation, compared to manual generation. Experiments in multibody dynamics show that for most applications, many data values, e.g. components of geometrical vectors, inertia parameters, joint velocities, etc., are simply equal to zero. All these zero quantities are treated in a numerical process in the same way as the non-zero ones: a manual generation of the model would eliminate all the unnecessary arithmetical operations beforehand, yielding an optimal form of the equations of motion.

Symbolic generation of multibody models tries to take advantage of both the numerical and manual techniques. So-called symbolic multibody programs manipulate only arithmetical operators (+, ., ., /) and strings of alphanumeric characters (*mi*, *qdi*, *dij*, ...) to generate – in a set of files – the analytical form of the equations using a desired syntax (e.g. C, Java, Fortran, MATLAB, etc.). For a given multibody application, this symbolic generation is performed only once, as in manual generation. From the multibody modeling point of view, these symbolic generators exhibit the same level of generality as their numerical competitors in the sense that they can handle systems with any topology and containing several degrees of freedom. However, they allow drastic simplifications, from the most trivial (addition/multiplication by zero) to the most complex ones (simplification of long trigonometric expressions). Comparison tests (Samin and Fisette, 2003) showed that, all other things being equal, a symbolic multibody model performs a time simulation between five and ten times faster than a purely numerical one. This is not at all negligible. This speed enhancement motivated the development of symbolic multibody codes in the eighties since computers were at that time quite inefficient in simulating even medium-sized systems (around 10 d.o.f.). The computational superiority of the symbolic approach still remains attractive since the exponential increase in modern computer performances naturally provides the opportunity to investigate larger systems or more sophisticated models (Samin and Fisette, 2003).

More recently, another interesting aspect of the symbolic approach has emerged in the framework of the analysis of hybrid mechatronic systems involving enlarged multibody models (i.e. including electrical actuators, hydraulic devices, etc.) and requiring enlarged analysis (control, optimization, etc.). Indeed, since symbolic multibody programs clearly separate the modeling part from the analysis process, it is quite straightforward to extract the symbolic model in order to export it towards another environment (e.g. for control or optimization purposes). In other words, symbolic multibody models are portable, and this is very attractive for the analysis of multiphysics applications (Samin et al., 2007).

1.3 State-of-the-art

An interesting state-of-the-art of multibody symbolic software is proposed in (Kurz et al., 2010). This paper presents the new features of the research software Neweul-M^2 (involving Maple and MuPad algebra symbolic engines), successor of the Neweul program developed at the end of the seventies and which is one of the first symbolic computer implementation.

The common denominator that emerges when analyzing and comparing multibody symbolic software is their intrinsic versatility in terms of:

- Model type and underlying formalisms (inverse or direct models, kinematic, dynamic equations, sensitivity matrices, etc.).
- Generated languages (Fortran, C, Java, Matlab, etc.).
- Coupling with other engineering environments (Matlab, Simulink, Comsol, etc.).

In addition to this, they all take advantage of modern languages (e.g. Java) and graphical interfaces (GUI, CAD) for data introduction and results presentation. This aspect, which is naturally stressed for visibility or commercial purposes, is far from being trivial in our discipline. Indeed, the large variety of mechanical devices in the three-dimensional space (e.g. joints with multiple d.of. possibly constrained) and of force laws (e.g. issuing from friction, pneumatic pressure or even look-up table, etc.) which is specific to MBS, definitively requires a high-level reflexion to make a general purpose program both flexible, user-friendly and efficient. With respect to this, a particular attention has been paid to ROBOTRAN to preserve a good equilibrium between the user leeway and the automated processes (see Sect. 4).

Naturally guided by specific educational objectives, research topics and/or industrial projects and collaboration opportunities, symbolic programs have their own specificities and have evolved in different directions.

In short, Maplesim (successor of DynaFlexPro) focuses on graphical way of modeling, the linear graph theory being at the root of the approach (Shi and McPhee, 2000). As simulation engine, it uses Maplesoft, a world leader in mathematical and analytical software. Maplesim is able to deal with rigid and flexible MBS and, thanks to the graph approach, is intrinsically well-suited to deal with multiphysics applications.

The research software Neweul-M^2 (Kurz et al., 2010) is based on the Newton-Euler equations and on the virtual work principle to generate the differential [differential/algebraic] equations of motion of open-loop [closed-loop] systems, via the state-of-the-art computer algebra systems Maple or Mupad.

MOBILE (Kecskeméthy, 1993) was developed by A. Kecskeméthy in the nineties to model multibody mechatronic systems using an original object-oriented approach. In particular, the underlying formulation is able to provide an analytical closed-form solution for most of 3-D loop constraints, on the basis of the so-called kinematic pair approach (Kecskeméthy et al., 1997).

Fast simulation is at the root of the SD/FAST program, providing the equations as C or Fortran source code, which can be compiled and linked into any computers environment to perform real time simulation on standard computers.

Carsim (and all its vehicle companion programs) is the successor of Autosim (developed by M. Sayers) and clearly focuses on vehicle dynamics (race cars, passenger cars, trucks, etc.). It is based on a Lisp symbolic multibody program to generate the vehicle equations of motion in symbolic form for real-time simulation purpose.

MotionGenesis, the successor of the symbolic program Autolev (developed by Th. Kane), is scientist-oriented, the user being clearly involved in the model generation. The developers also emphasize the high performances of the program in terms of code compactness and time simulation.

As regards ROBOTRAN, the purpose of the present paper is to highlight the recent developments and features of the program, not only in terms of modeling features and computational performances but also by referring to the flexibility of the approach for educational, research and industrial purposes.

In Sect. 2, the ROBOTRAN underlying formalisms are briefly reviewed; more details can be found in (Samin and Fisette, 2003). Section 3 presents the symbolic capabilities which highly takes advantage of the recursive nature of the formalisms to generate the equations in compact form. In particular, the recent developments of the symbolic engine allow ROBOTRAN to generate symbolically the dynamic model and the sensitivity equations of very large MBS with closed-loop constraints. In Sect. 4, the ROBOTRAN user environment is shortly described as the latter is really part of multibody program appeals, considering the large variety of possible systems to analyses. Some recent applications will be presented in Sect. 5 before concluding.

2 ROBOTRAN formalisms

Before presenting the symbolic engine and the advanced features of the program, it is necessary to establish the notations and to summarize the underlying formalisms. This is the purpose of the present section. More details can be found in Samin and Fisette (2003).

2.1 Dynamics of tree-like multibody systems

Initially developed for Robotic applications (Maes et al., 1990)[1], the formalisms underlying ROBOTRAN are based on the use of *relative joint coordinates*. As usual in this case, equations are firstly established for a tree-like structure (i.e. with no explicit constraints). Constrained systems (i.e. containing loops of bodies or user constraints) are modeled by first restoring a tree-like structure whose dedicated formalisms are thus necessary for any kind of MBS.

2.1.1 Direct dynamics

To predict the motion of a MBS, the *direct dynamics* of MBS (see Fig. 1) is required to compute the generalized accelerations $\ddot{q}$ (joint accelerations) for a given configuration $(q, \dot{q})$ of the MBS to which forces and torques are applied.

[1]ROBOTRAN stands for "ROBOt TRANslator".

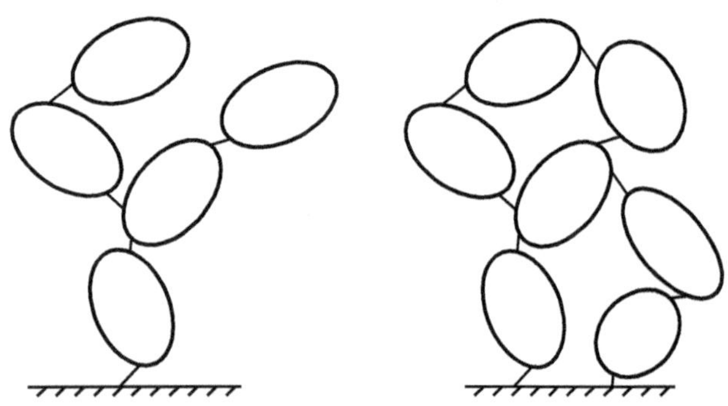

Figure 2. Tree-like versus Closed MBS.

In a synthetic form, the so-called *semi-explicit* direct dynamics model reads:

$$M(q,\delta)\ddot{q} + c(q,\dot{q},\delta,fr,tr,g) = \phi(q,\dot{q}) \quad (1)$$

or, in an *explicit* form:

$$\ddot{q} = f(q,\dot{q},fr,tr,\phi,\delta) \quad (2)$$

to be solved with respect to $\ddot{q}$. In case of the semi-explicit form (1), this can be performed via the Cholesky decomposition of mass matrix M. In the previous equations:

- M $[n \cdot n]$ is the generalized mass matrix of the system (which is symmetric and positive definite),
- c $[n \cdot 1]$ is the non linear dynamic vector which contains the gyroscopic, centrifugal effects as well as the contribution of gravity g, external resultant forces fr and torques tr,
- q $[n \cdot 1]$ denotes the relative generalized coordinates,
- δ $[10n \cdot 1]$ gathers together the dynamic parameters of the system (body masses, centers of mass location and the six components of body tensors inertia
- ϕ $[n \cdot 1]$ represents the generalized joint forces (torques): their explicit computation is typical in relative coordinate formulations, the reason being strongly related to robotic applications and inverse dynamics issues.

To obtain the above models, the symbolic engine of ROBOTRAN implements several formalisms[2].

The semi-explicit form (1) is obtained via the so-called Newton-Euler recursive algorithm[3] with mass matrix extraction (Samin and Fisette, 2003).

The explicit direct dynamics model (2) is symbolically generated

- either via the so-called *Order-N* formulation, inspired from (Schwertassek and Rulka, 1989), which requires three recursive steps to get the acceleration with an $O(N)$ complexity only
- or via the previous Newton-Euler recursive algorithm including a fully symbolic Cholesky decomposition of the mass matrix (Postiau et al., 2001).

In terms of equations complexity (i.e. the number of arithmetical operations versus the number of d.o.f.) to produce the generalized accelerations $\ddot{q}$, a detailed comparison between formalisms in relative coordinates has indicated the obvious superiority of the recursive formulations with respect the the non-recursive one, and a certain competition between the Order-N and the Newton/Euler recursive algorithms (Samin and Fisette, 2003). The latter is surprisingly more efficient – although having an $O(N)^2 + O(N)^3$ complexity – for most of practical applications dealing with rigid bodies.

2.1.2 Inverse dynamics

The *inverse dynamics* of a multibody system is the computation of the generalized joint forces (torques) ϕ to be applied to the joints for a given configuration $(q,\dot{q},\ddot{q})$ of the system to which external forces and torques are applied:

$$\phi = f(q,\dot{q},\ddot{q},\delta,fr,tr,g) \quad (3)$$

in which the dimension of ϕ and q are equal for a fully actuated system.

Inverse dynamic models (3) are typically used in robotics to control the actuator torques ϕ when following a desired trajectory $(q(t),\dot{q}(t),\ddot{q}(t))$, in biomechanics to predict the net torque in the human body joints when walking or jumping, etc.

Equations (3) are generated in ROBOTRAN via both non-recursive (Virtual principle) and recursive (Newton/Euler) formalisms; the latter, being implicit with respect to the accelerations $\ddot{q}$, has an $O(N)$ complexity.

2.1.3 Reaction dynamics

The *reaction model* is a particular inverse model which is also of practical use in robotics. It consists in computing the components, in the inertial frame, of the vector reaction force F^{r} and torque T^{r} at a reference point of the bed-plate of the robot. Defining the following $[6 \cdot 1]$ column vector,

$$\phi^{\mathrm{r}} = \begin{pmatrix} F^{\mathrm{r}} \\ T^{\mathrm{r}} \end{pmatrix} \quad (4)$$

the inverse reaction model reads:

$$\phi^{\mathrm{r}} = \phi^{r}(q,\dot{q},\ddot{q},\delta,fr,tr,g) \quad (5)$$

To compute it, ROBOTRAN automatically inserts 6 locked degrees of freedom (3 translations followed by 3 rotations)

[2]Although only the recursive ones are available from the web interface.

[3]For which lots of declination exist in the literature.

between the inertial base and the first body of the MBS. Then, using the Newton/Euler recursive algorithm, the dynamic equations of (only) these joints are symbolically generated to obtain (5).

The ROBOTRAN reaction model has been successfully used to identify dynamic parameters of robots, by establishing the relationship between [a set of] exciting trajectories and the reaction forces and torques: the main advantage with respect to an inverse dynamics approach relates to the absence of the joint friction forces which cancel each other out in (5) by the action/reaction principle (Chenut et al., 2002).

2.2 Dynamics of constrained multibody systems

2.2.1 Direct dynamics

Dynamics and constraints equations

In reality, most multibody applications contain loops of bodies (parallel robot, robotical orthesis, railway bogie, etc., as shown in Fig. 3) which impose the generalized joint coordinates q to satisfy algebraic constraints at any time, denoted $h_{\text{loop}}(q) = 0$. Constraints can result from other physical phenomena (e.g.: geometrically constrained motion, rolling without slipping condition, etc.): those will be referred to user constraints and denoted $h_{\text{user}}(q) = 0$. Gathering these m constraints together, we can write:

$$h(q) = \begin{pmatrix} h_{\text{loop}}(q) \\ h_{\text{user}}(q) \end{pmatrix} = 0 \quad [m \cdot 1]$$

In order to fully describe the system, these constraints and their first and second time derivatives must be added to the equations of motion[4], in which constraint forces are introduced via the well-established Lagrange multipliers technique:

$$M(q)\ddot{q} + c(q, \dot{q}, fr, tr, g) = \phi(q, \dot{q}) + J^t \lambda \tag{6}$$

$$h(q) = 0 \tag{7}$$

$$\dot{h}(q, \dot{q}) = J(q)\dot{q} = 0 \tag{8}$$

$$\ddot{h}(q, \dot{q}, \ddot{q}) = J(q)\ddot{q} + \dot{J}\dot{q}(q, \dot{q}) = 0 \tag{9}$$

where:

- $J = \frac{\partial h}{\partial q^t}$ denotes the constraint Jacobian matrix (dimension: $[m \cdot n]$),
- $\dot{J}\dot{q}(q, \dot{q})$ $[m \cdot 1]$ is the quadratic term (expression in $\dot{q}^i \dot{q}^j$) of the constraints at acceleration level (dimension: $[m \cdot 1]$),
- λ represents the Lagrange multipliers associated with the constraints (dimension: $[m \cdot 1]$).

[4] In which, for legibility reasons, we will no longer indicate the dependence with respect to the dynamic parameters δ.

Coordinate partitioning reduction

Various methods can be used to solve system (6–9). Among these, one can opt for a full reduction of the system to a purely differential form, which can be obtained by means of the *Coordinate Partitioning technique* (Wehage and Haug, 1982). The Jacobian matrix J is assumed to have full rank m. In this case the constraints $h(q) = 0$ are independent and m generalized coordinates can be locally expressed as functions of the $(n - m)$ others. In this way, it becomes possible to reduce the original DAE[5] system (6–9) to a set of $(n - m)$ differential equations (ODE)[6] in those $(n - m)$ independent coordinates. This reduced set will represent the equations of motion of the constrained multibody system, where $(n - m)$ also corresponds to its number of degrees of freedom.

Let us summarize the steps required to obtain these equations of motion. After reordering the vector of generalized coordinates q (and the columns of the constraint Jacobian J), we can perform the following partition:

$$q = \begin{pmatrix} u \\ v \end{pmatrix} ; J = \begin{pmatrix} J_u & J_v \end{pmatrix} \tag{10}$$

where u denotes the subset of $(n{-}m)$ *independent* coordinates and v denotes the subset of *dependent* coordinates. When correctly choosing the subset v, the m by m matrix J_v will be regular. By "correctly", we mean that, to establish this partitioning for a given application, we can rely:

- on an intuitive reasoning, based on the system configuration (e.g. for a planar slider-crank mechanism, the crank rotation can be chosen as the independent variable u whatever the crank position),
- on the LU factorization of the *full* Jacobian matrix $J(q)$, with column permutation on the basis of the largest pivot. The resulting left $[m \cdot m]$ square block will be the "best" candidate.

Once the coordinate partitioning is established, the reduction method simply uses matrix permutations and operations to produce the final system. Let us first partition the generalized mass matrix M and the vector c according to the coordinate partitioning (10):

$$\begin{pmatrix} M_{uu} & M_{uv} \\ M_{vu} & M_{vv} \end{pmatrix} \begin{pmatrix} \ddot{u} \\ \ddot{v} \end{pmatrix} + \begin{pmatrix} c_u \\ c_v \end{pmatrix} = \begin{pmatrix} \phi_u \\ \phi_v \end{pmatrix} + \begin{pmatrix} J_u{}^t \\ J_v{}^t \end{pmatrix} \lambda \tag{11}$$

where $J_v{}^t$ refers to the transpose of matrix J_v. Since this matrix is regular, eliminating the unknowns λ using the lower part of system (11) produces:

$$\begin{pmatrix} M_{uu} & M_{uv} \end{pmatrix} \begin{pmatrix} \ddot{u} \\ \ddot{v} \end{pmatrix} + B_{vu}{}^t \begin{pmatrix} M_{vu} & M_{vv} \end{pmatrix} \begin{pmatrix} \ddot{u} \\ \ddot{v} \end{pmatrix} \tag{12}$$
$$+ c_u + B_{vu}{}^t c_v = \phi_u + B_{vu}^t \phi_v$$

[5] For "differential algebraic equations".

[6] For "ordinary differential equations".

Figure 3. Closed-loop MBS (ROBOTRAN applications).

where we define the so-called coupling matrix: $B_{vu} \triangleq -(J_v)^{-1} J_u$. The algebraic constraints have to be solved in order to eliminate the dependent variables v. While analytical solutions can exist for specific cases, the algebraic constraints (7) are generally nonlinear and require an iterative procedure to be solved: the Newton-Raphson algorithm – with possible relaxation – can be used for successive estimations of v:

$$v^{k+1} = v^k - (J_v)^{-1} h|_{v=v^k} \tag{13}$$

where the right hand side is evaluated for $v = v^k$ and the values of u corresponding to the instantaneous system configuration.

Using the first (Eq. 8) and second derivatives (Eq. 9) of the constraints, the generalized velocities and accelerations $\dot{v}$ and $\ddot{v}$ are respectively given by:

$$\dot{v} = B_{vu}\dot{u} \tag{14}$$

$$\ddot{v} = B_{vu}\ddot{u} + b \quad \text{with} \quad b \triangleq -J_v^{-1}(\dot{J}\dot{q}) \tag{15}$$

and can also be eliminated from the differential Eq. (13). This produces the final *reduced* system:

$$\left(M_{uu} + M_{uv}B_{vu} + B_{vu}{}^t M_{vu} + B_{vu}{}^t M_{vv}B_{vu}\right)\ddot{u}$$
$$+\left(M_{uv} + B_{vu}{}^t M_{vv}\right)b + (c_u + B_{vu}{}^t c_v) - (\phi_u + B^t_{vu}\phi_v) = 0$$

which can be concisely written as:

$$\mathcal{M}(u)\ddot{u} + \mathcal{F}(\dot{u}, u) = 0 \tag{16}$$

The set of ordinary differential Eq. (16) constitutes the equations of motion of the constrained MBS described in terms of the $n - m$ independent generalized coordinates u.

- $\ddot{v}$ can be computed directly from system (15).
- As regards the Lagrange multipliers λ, the lower part of Eq. (11) can be used to recover them:

$$\lambda = (J_v{}^t)^{-1}\{M_{vu}\ddot{u} + M_{vv}\ddot{v} + c_v - \phi_v\} \tag{17}$$

2.2.2 Inverse dynamics

As for tree-like MBS, one can be interested in computing the value of the joint torques of a closed MBS for a given trajectory $(q(t), \dot{q}(t), \ddot{q}(t))$. This is the case of parallel manipulators for instance. The coordinate partitioning can also be used to reduce the inverse dynamic model (3) subject to kinematic constraints (7).

Let us denote ψ the left-hand-side of Eq. (6). The latter then reads:

$$\psi = \phi + J^t\lambda \tag{18}$$

Applying the previous coordinate partitioning $(q = u, v)$ to (18) and recalling the definition of the coupling matrix B_{vu}, we obtain:

$$\psi_u = \phi_u + B^t_{vu}(\psi_v - \phi_v) \tag{19}$$

To compute the inverse dynamics in a general case, let us first split the joint generalized force ϕ into an active component ϕ^{a} (corresponding to actuators) and a passive component ϕ^{p} (e.g. friction, spring-type law, etc.):

$$\phi = \phi^{\mathrm{a}} + \phi^{\mathrm{p}} \tag{20}$$

Assuming that the actuators are located on each independent joints u (which also assumes that there are as many actuators as degrees of freedom $n - m$),

$$\phi^{\mathrm{a}} = \begin{pmatrix} \phi^{\mathrm{a}}_u \\ 0 \end{pmatrix}, \tag{21}$$

the inverse dynamics (19) becomes:

$$\phi^{\mathrm{a}}_u = \psi_u - \phi^{\mathrm{p}}_u + B^t_{vu}(\phi^{\mathrm{p}}_v - \psi_v) \tag{22}$$

In practical situations, actuators are not necessary located on the independent joints u, because the $\{u, v\}$ partitioning results from a numerical requirement (matrix conditioning) and not from physical considerations. However, nothing prevents us from considering two distinct partitioning inside a unique inverse dynamic model:

- the $q = \{u, v\}$ partitioning to assemble the MBS and to solve the constraints (Eqs. 13, 14 and 15),
- a second coordinate partitioning, $q = \{q_{\mathrm{a}}, q_{\mathrm{p}}\}$, based on the actuated and non actuated joints.

Using this second partitioning, the reduced inverse dynamic model (22) simply becomes:

$$\phi^{a}_{q_a} = \psi_{q_a} - \phi^{p}_{q_a} + B^{t}_{q_p q_a}(\phi^{p}_{q_p} - \psi_{q_p}) \tag{23}$$

which requires that the constraints Jacobian sub-matrix J_{q_p} be regular.

When dealing with *overactuated* MBS for which the number of actuators (ϕ^a) is larger than the number of d.o.f., there are an infinite number of solutions for the inverse dynamics which becomes an underdetermined system of the form:

$$A(q)\phi^{a} = b(q,\dot{q},\ddot{q}) \tag{24}$$

where A is a m by n rectangular matrix (with $m < n$). Additional criteria are thus needed to solve the system (24). One can use an optimization process to satisfy some specific criteria as in (Raison et al., 2010) to deal with human muscle overactuation, or the Moore-Penrose pseudo-inverse solution:

$$\phi^{a} = A^{+}\, b \tag{25}$$

where, assuming rank$(A) = m$, $A^{+} \stackrel{\Delta}{=} A^{t}(A A^{t})^{-1}$.

Equation (25) gives the solution of minimum Euclidean norm $||\phi||_2$. This approach has been successfully used for instance in (Ganovski et al., 2004) to minimize actuator torques of overactuated parallel robots following trajectories with singular configurations.

In case of *underactuated* MBS (i.e. less actuators than d.o.f.), instead of using the previous inverse models, the direct dynamics form (16) can be used by splitting the independent coordinates u into free variables u_f and actuated variables u_a whose value is constrained according to a prescribed motion or trajectory. Equation (16) then becomes:

$$\begin{pmatrix} \mathcal{M}_{u_f u_f} & \mathcal{M}_{u_f u_a} \\ \mathcal{M}_{u_a u_f} & \mathcal{M}_{u_a u_a} \end{pmatrix} \begin{pmatrix} \ddot{u}_f \\ \ddot{u}_a \end{pmatrix} + \begin{pmatrix} \mathcal{F}_{u_f} \\ \mathcal{F}_{u_a} \end{pmatrix} = \begin{pmatrix} 0 \\ \lambda \end{pmatrix} \tag{26}$$

This system can be seen as an hybrid direct/inverse model where the upper part (related to the free motion) refers to direct dynamics (unknown $\ddot{u}_f$) and the lower part refers to the collocated inverse dynamics whose unknowns (λ) correspond to the actuated joint forces ϕ^a:

$$\phi^{a} = \begin{pmatrix} \mathcal{M}_{u_a u_f} & \mathcal{M}_{u_a u_a} \end{pmatrix} \begin{pmatrix} \ddot{u}_f \\ \ddot{u}_a \end{pmatrix} + \mathcal{F}_{u_a} \tag{27}$$

in which $\ddot{u}_f$ must be computed in parallel via a time simulation for instance.

3 Symbolic engine

There do exist commercial general purpose symbolic computation packages like Maple and Mathematica. Why do we not use these to generate multibody equations ? There are two main reasons for this.

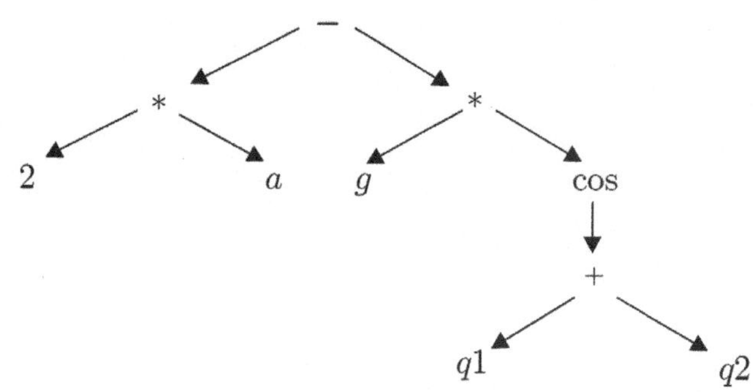

Figure 4. Tree representation of a mathematical expression.

The first and most important one relates to the amount of computer memory required to generate medium-sized and large models (up to 300 d.o.f. in our case) symbolically.

The second reason concerns the possible simplifications of the symbolic expressions. Although the simplification capabilities of commercial packages are extremely powerful, the condensation of multibody equations relies on specific rules which can be applied more easily by developing a dedicated symbolic program; this was the main motivation to develop the ROBOTRAN program. Its capabilities in terms of symbolic manipulation are briefly summarized in the following sub-sections, dealing with:

- expression simplification
- memory allocation
- advanced features
  - fully symbolic generation of constrained MBS
  - recursive differentiation of multibody models.

3.1 Expression simplification

A mathematical expression in multibody equations uses simple arithmetical operators $+, -, \cdot, /, (), =$ and functions: mainly sin() and cos(), occasionally: sqrt(), atan(), ...

Let us for example consider the following expression

$$2 \cdot a - g \cdot \cos(q1 + q2) \tag{28}$$

which, when applying mathematical priority rules, is equivalent to

$$(2 \cdot a) - (g \cdot (\cos(q1 + q2))) \tag{29}$$

This expression can be represented by the tree shown in Fig. 4. For instance, the second multiplication is an expression whose nature is $\cdot$ and points towards two arguments: the expressions g and cos(). In a tree representation, the leaves represent alphanumeric symbols, e.g. 2, a, g,... in example (28). They are also considered as expressions (or "leaf expressions") but they have no sub-expression to which they point: they simply contain the string they represent.

The tree representation and the dynamic increase in size of the expressions during the elaboration of the equations leads us to represent an expression by a C-structure, *dynamically* generated and handled via *pointers*. At the root of the program, there is the C-structure *expression* which contains – at least – the following fields:

- the *nature* of the expression (+, –,string, etc.),
- the address(es) (i.e. pointers) of the *argument*(s) to which the expression points: these arguments are either strings – for a leaf expression – or other expressions which have been created via a previous dynamic memory allocation. To ensure the basic level of symbolic simplification (i.e. expression $c+a+b-a$ replaced by $c+b$), two ordering rules are used.

The first one re-organizes a given expression on the basis of a so-called *multibody priority (e.g. a mass symbol has a higher priority than a force symbol but has a lesser priority than a coefficient), as illustrated in the following example:*

$$4 < m^i < d^{ij} < F^j_{ext} < \dot{q}^i < b \cdot c < (a-d) \tag{30}$$

In case of equal multibody priority, a (sub-)expression is then re-organized according to the lexicographical sequence of the ASCII table of characters, as for instance:

$$K^{ii} < m^i; \;\; 2 < 3; \;\; d^{ii} + d^{ij} < d^{ii} + d^{jk}; \;\; \text{etc.} \tag{31}$$

When dealing with a given symbolic expression (whatever its length), the symbolic engine (layer 1) of ROBOTRAN recursively uses the above priority rules to ensures that the final form of any expression will be purged of consecutive equal terms – or sub-expressions – with opposite signs.

3.2 Trigonometric simplification

A revolute joint i, with a generalized joint coordinate q^i, of a multibody system induces an elementary rotation matrix which contains the trigonometric functions $\cos(q^i)$ and $\sin(q^i)$. The evaluation of rotation matrices between any pair of bodies i and j, such that j is a descendant of i in the MBS structure, is obtained by multiplying the elementary rotation matrices associated with each revolute joint k belonging to the kinematic chain $\{i, i+1, i+2, ..., j-1, j\}$ for instance:

$$R^{j,i} = R^{j,j-1} R^{j-1,j-2} \dots R^{i+1,i} \tag{32}$$

Thus, an optimized trigonometric engine is required to condensate trigonometric expressions like

$$K212 \cdot qpp8 \cdot (C2 \cdot C2 \cdot C3 \cdot S4 + S2 \cdot S24 \\ -S2 \cdot S34 \cdot S5 + S2 \cdot S2 \cdot C3 \cdot S4) \tag{33}$$

in which Cj, Sk and Sjk represent $\cos(q^j)$, $\sin(q^k)$ and $\sin(q^j+q^k)$ respectively.

The ROBOTRAN trigonometric simplification process, which is performed on line (i.e. it is not a post-process), has two main levels. The first and lowest one systematically detects and performs the fundamental trigonometric simplifications according to well-known formulae. For instance, let us consider a symbolic expression a:

- if $a = C1 \cdot C1 + S1 \cdot S1$, the process returns 1,
- if $a = 2 \cdot C1 \cdot S1$, trigo(a) the process creates and returns the auxiliary variable $S11$, which stands for $\sin(2q^1)$,
- if $a = C1 \cdot C2 - S1 \cdot S2$, the process creates and returns the auxiliary variable $C12$, which stands for $\cos(q^1+q^2)$,
- etc.

while the previous level is able to detect and simplify elementary trigonometric formulae, it is not able to deal with expressions like

$$C2 \cdot C4 \cdot C56 \cdot C56 \cdot S8 + C2 \cdot C4 \cdot S56 \cdot S56 \cdot S8 \\ +C2 \cdot S4 \cdot S56 \cdot C8 + S2 \cdot C4 \cdot S56 \cdot C8 \\ -S2 \cdot S4 \cdot C56 \cdot C56 \cdot S8 - S2 \cdot S4 \cdot S56 \cdot S56 \cdot S8 \tag{34}$$

for which judicious groupings and factorings must be performed in order to make a maximum of trigonometric formulae appear, which can then be simplified. This is the purpose of the second level of the process. The following examples illustrate the power of the method. The trigonometric expressions are generated by a direct model (of a railway bogie) based on the virtual power principle (Cj, Sk and Sjk represent $\cos(q^j)$, $\sin(q^k)$ and $\sin(q^j+q^k)$ respectively):

- $C2 \cdot C4 \cdot C56 \cdot C56 \cdot S8 + C2 \cdot C4 \cdot S56 \cdot S56 \cdot S8$
  $+ C2 \cdot S4 \cdot S56 \cdot C8 + S2 \cdot C4 \cdot S56 \cdot C8$
  $- S2 \cdot S4 \cdot C56 \cdot C56 \cdot S8 - S2 \cdot S4 \cdot S56 \cdot S56 \cdot S8$
  becomes: $C24 \cdot S8 + S24 \cdot S56 \cdot C8$
- $C2 \cdot S2 \cdot C4 \cdot C56 \cdot C56 \cdot S8 + C2 \cdot S2 \cdot C4 \cdot S56 \cdot S56 \cdot S8$
  $+ C2 \cdot S2 \cdot S4 \cdot S56 \cdot C8 + S2 \cdot S2 \cdot C4 \cdot S56 \cdot C8$
  $- S2 \cdot S2 \cdot S4 \cdot C56 \cdot C56 \cdot S8 - S2 \cdot S2 \cdot S4 \cdot S56 \cdot S56 \cdot S8$
  becomes: $S2 \cdot (C24 \cdot S8 + S24 \cdot S56 \cdot C8)$

3.3 Recursive scheme condensation

In multibody dynamics, a *recursive scheme* denotes any formalism (kinematic, dynamic, direct, inverse, etc.) in relative coordinates, written as one or more algorithmic loops covering the MBS from the base body to the terminal bodies. For instance, the so-called Recursive Newton Euler formalism represents a recursive scheme consisting of two algorithmic loops: one for the forward kinematics (for $i = 1 : N^{body}$), the second for the backward dynamics (for $i = N^{body} : 1$). The so-called Order-N formalism (Schwertassek and Rulka, 1989) is also a recursive scheme which performs three recursions: forward kinematics, backward dynamics and forward kinetics to directly obtain the explicit direct dynamics (2).

In such formalisms, the relation which expresses – for instance – the angular velocity ω^3 of a given body 3 with respect to its parent body 2, ω^2, is written in vector form

$$\omega^3 = \omega^2 + \Omega^{23} \tag{35}$$

where Ω^{23} stands for the *relative* angular velocity between body 2 and 3.

A recursive ROBOTRAN implementation of the vector Eq. (35) is given hereafter for a specific MBS:

$$\begin{aligned} \text{OM13} &= qd(3) + \text{OM12} \\ \text{OM23} &= \text{OM22} \cdot C3 + \text{OM32} \cdot S3 \\ \text{OM33} &= -\text{OM22} \cdot S3 + \text{OM32} \cdot C3 \end{aligned} \tag{36}$$

where OMij denotes the i-th component of the j-th body angular velocity in body i fixed frame.

The two above equations clearly highlight the *recursive* nature of the analytic Eq. (35) on the one hand and of the corresponding symbolic expressions (36) on the other hand.

Via this very simple example, one can easily extrapolate the reasoning to a full formalism in which such a recursivity between adjacent bodies can apply to position, velocity, acceleration, forces, torques, etc. to end up in the final model (Eqs. 1 and 2 for instance). The ROBOTRAN implementation is based on this technique.

Although recursive formulations intrinsically have a compact form (when compared with *in extenso* formulations which do not exploit the above-mentioned recursivity (Samin and Fisette, 2003), they paradoxically perform superfluous evaluations: depending on the type and succession of joints of the application, some components of the vector components (in Eq. 36 for instance) are superfluous for the final scalar form. Whereas a general purpose multibody program which generates the model *numerically* is not able to detect these superfluous equations, a symbolic multibody program can do so. Thus, in addition to removing useless terms in equations (see the previous sections), *entire* equations can be detected symbolically as being superfluous (up to 30 % for direct dynamics!). The ROBOTRAN recursive condensation process is based on a linked-list (of equations) and C-pointers, and removes those equations before "engraving" the final result. The process is illustrated in Fig. 5 via an academic example which computes a given result R from data A, B, C, D via a recursive approach: symbolic equations $B1 = \ldots$, $B3 = \ldots$, $D1 = \ldots$ are clearly useless for the result and can be completely disregarded and removed from the list before printing the equations.

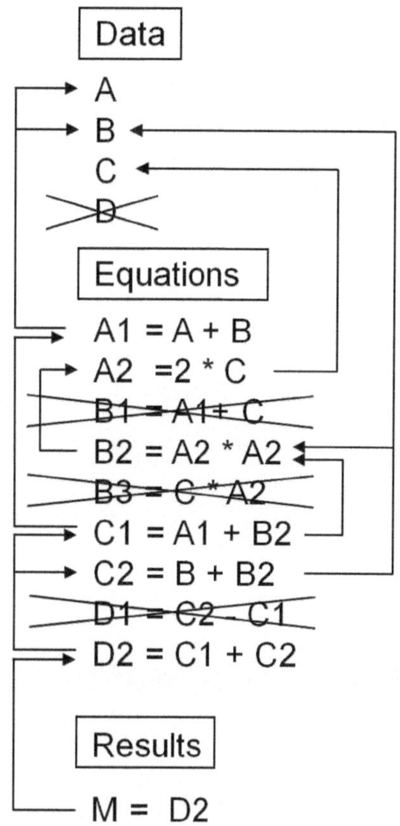

Figure 5. Recursive scheme condensation.

3.4 Memory allocation

As mentioned above, the amount of computer memory required by symbolic programs exponentially increases when manipulating large expressions or system equations: the generation can simply fail or require a very long computer time. Symbolic manipulation requires dynamic memory allocation to create and store new expressions (C-structures in our case). The symbolic process briefly described above, which recursively re-organizes any new expressions in accordance with the ordering rules, can generate in the memory thousands of auxiliary expressions which are not necessary in the final tree representation of a given (complex) expression. Experiments showed that, whatever the multibody formalism used, these short-lived expressions induced by symbolic manipulations represent more than 90 % of the whole set of expressions generated! This explains why symbolic programs can lead to an explosive use of computer memory. To solve this delicate problem, one may try to cut long equations into segments and evaluate and print them separately; then, after dealing with each segment, we would clear out all the generated expressions from the memory. By experience, this solution is mediocre because it strongly degrades the symbolic engine capabilities in terms of simplification.

To solve this problem, the following three elements were introduced into ROBOTRAN.

1. When generating a symbolic equation, in order to keep track of each new symbolic expression (complex or not, short-lived or not) created by the generation process, the address (pointer) of this expression is systematically

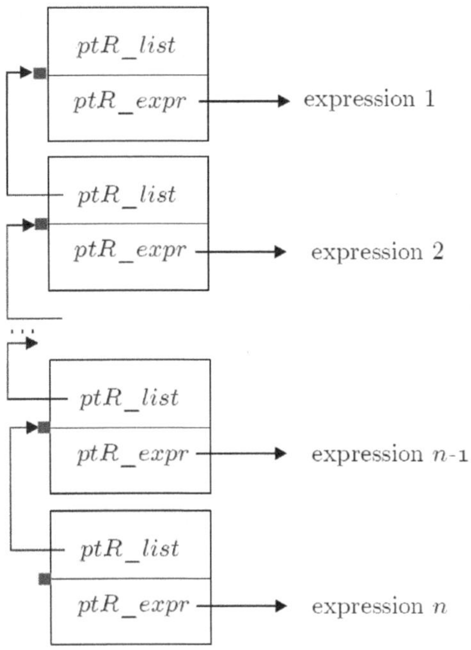

Figure 6. Dynamic linked list of expressions.

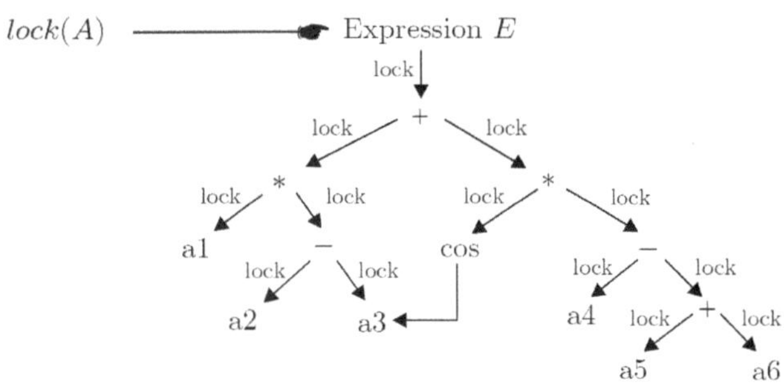

Figure 7. Locking an expression in memory.

stored in a *dynamic linked-list* which grows as and when it is needed, as shown in Fig. 6.

2. In order to protect any expression (e.g. A, $a2$, cos(), '+', etc.), we add to the corresponding C-structure, a boolean field "lock". The value lock = TRUE tells ROBOTRAN that the corresponding expression cannot be removed from memory, whereas lock = FALSE states that it can. FALSE is the default value.

3. We introduce the procedure expr_lock(E) to lock a given expression E, i.e. by setting to TRUE the "lock" field of each node belonging to the *final* tree of expression E. The procedure is thus intrinsically recursive as illustrated in Fig. 7.

By thus locking a given equation E in its *final, simplified* form, we do not lock the intermediate "short-lived" expressions (more than 90 %!) created during the evaluation of E,

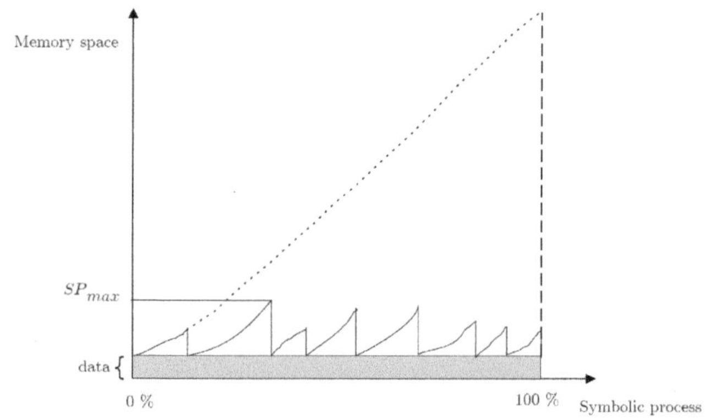

Figure 8. Memory allocation for a ROBOTRAN generation process.

those which do not contribute to its final tree: for these expressions, "lock" keeps its default value FALSE. Thus, to free the memory in an optimal way, once the evaluation of E is finished, and its tree has been purged of superfluous operations, ROBOTRAN:

1. recursively protects the final form of expression E (lock(E)),

2. frees the memory by removing every expression x which has not been locked by covering the list of Fig. 6 from tail to head. Each element of the list is also removed to end up with an empty list ... ready for the evaluation of a subsequent equation.

Thanks to this methodology, the memory space required by and during a ROBOTRAN process is illustrated in Fig. 8 and has thus a "toothed" shape, rather than a monotonic growing one as in classical symbolic packages. The maximum memory space (SP_{max}) is reached by the greediest equation.

This freeing process is of the utmost importance in multibody dynamics since it allows us to eliminate the most critical bottleneck (i.e. the memory space) of the generation of large multibody models, up to 300 d.o.f. in our case.

3.5 Advanced features

3.5.1 Fully symbolic generation of constrained MBS

Up to now, the symbolic capabilities of the ROBOTRAN program were only exploited for generating the dynamics and kinematics of tree-like MBS: this means that for closed loop systems (like vehicle suspensions, parallel robots, etc.), only the main ingredients of Eqs. (6), (7), (8) and (9) were generated symbolically but *separately*. The subsequent coordinate partitioning reduction (16) being performed numerically. While being far more efficient than a pure numerical model (Fisette, 1994), we recently noticed that there was still a lot of superfluous operations in the numerical processes underlying the reduction (empty mass and Jacobian

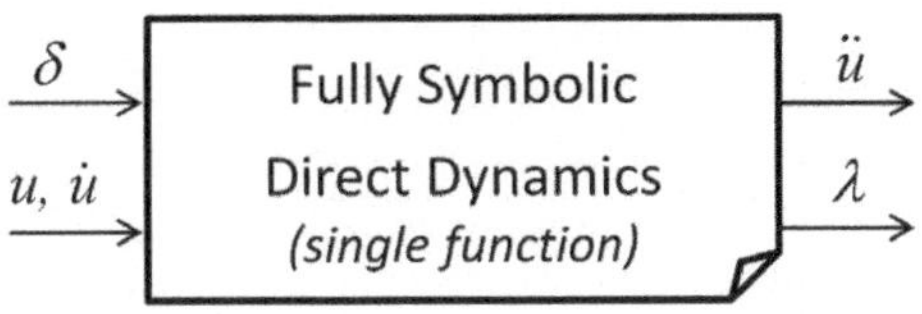

Figure 9. Closed-loop MBS: fully symbolic generation of the direct dynamics.

sub-matrices etc.) The latter could be avoided if the symbolic generation mentioned above would be applied *globally* to the model, i.e. from the input $(u, \dot{u})$ to the output $(\ddot{u})$ in case of direct dynamics of constrained systems (Fisette et al., 2002), (Poncelet et al., 2010).

Let us rewrite the semi-explicit form (16) in an *explicit* way (exactly as the form 2) for tree-like MBS) in terms of the independent accelerations $\ddot{u}$:

$$\ddot{u} = f(u, \dot{u}, fr, tr, \phi, \delta) \tag{37}$$

Thanks to the ROBOTRAN symbolic engine capabilities, in particular in terms of memory requirement (see Sect. 3.4), it is now possible to generate the independent accelerations $\ddot{u}$ according to (Eq. 37) in a fully symbolic manner and in the form of a *unique* function (C, Matlab, etc.) which successfully computes, as a unique global recursive scheme:

1. the constraints and their solving at position, velocity and acceleration levels,
2. the external forces and torques (interfaced with possible external user constitutive equations),
3. the dynamics of the restored tree-like MBS,
4. the reduction and the resolution of system (16) with respect to the generalized accelerations $\ddot{u}$.

Regarding the constraints solution at position level, despite robust and appealing formulations (e.g. the kinematic transformer technique, Kecskeméthy et al., 1997), it is not always possible to solve them analytically via a closed-form kinematic solution, because of their inherent nonlinearities. In case of complex 3-D closed-loop structures, as depicted in Fig. 3, we must often resort to a numerical iterative process to converge towards an accurate solution. As far as we are concerned, the Newton-Raphson algorithm (with possible relaxation) has been chosen to solve the constraints $h(q) = 0$ (Eq. 7) via the necessary iterations on the dependent coordinates v (see Eq. 13):

$$\Delta v^k = (J_v)^{-1}\, h|_{v=v^k} \tag{38}$$

Being a numerical iterative process, it would be illusory to implement it symbolically. However, it we examine the RHS of the previous iterative formula, it mainly contains kinematic ingredients, namely the constraint Jacobian sub-matrix

```
[udd, λ] = direct_dynamics(u, ud, δ)
/* Recursive Forward Kinematics */
ωj = ωi+ ...
...
/* Loop closure */
→ while (norm(h) > ε)
     /* Recursive Contraints Kinematics */
     Jjk = Jik+...
     ...
     Δv = ...
  end
/* Recursive Backward Dynamics*/
Fj = Fi+ ...
Lj = Li+ ...
/* DAE=> ODE Reduction */
...
udd = ...
λ = ...
return
```

Figure 10. Recursive symbolic computation of the constraints: local iterative process.

J_v and the constraints themselves $h(u, v)$, which are computed within the recursive generation (see Sect. 3.3). Thus, since the Newton-Raphson algorithm simply amounts to repeat the evaluation of the RHS of Eq. (38) until convergence, it is rather straightforward to insert – in the symbolic output file – suitable statements upstream (e.g.: "while $\|h(q)\| > \epsilon$") and downstream ("$\Delta v^k = ...$, end") the RHS computation of (38)). Figure 10 illustrates this "trick" which has been systematized in ROBOTRAN when generating the explicit direct model (37) whose efficiency in terms of CPU time is rather impressive.

The only drawback of this fully symbolic approach, in comparison with the semi-symbolic one of Sect. 2.2.1, lies in the fact that the $\{u, v\}$ partitioning must be symbolically hard-coded in the block of Fig. 9. This requires a pre-process to fix the partitioning via a numerical technique (e.g. a LU factorization of the Jacobian matrix $J(q)$ with full pivoting) which is able to find a robust $\{u, v\}$ partition and variable permutation.

3.5.2 Recursive differentiation of multibody models

Various numerical analyses require the derivatives of multibody (kinematic or dynamic) equations with respect to a given parameter or variable or a set of those. This is for instance the case for model linearization, control design, deterministic optimization and sensitivity analysis. In the present section, we will focus on the latter topic to illustrate the symbolic differentiation process recently implemented in ROBOTRAN (Poncelet et al., 2010).

Sensitivity analysis of MBS

In the multibody dynamic context, a typical objective function $\psi(p)$ for sensitivity analysis purpose (see Eberhard, 1996; Ding et al., 2007) can be written as:

$$\psi(p) = G^1(t^1, u^1, \dot{u}^1, p) + \int_{t^0}^{t^1} F(t, u, \dot{u}, \ddot{u}, p)\,\mathrm{d}t \tag{39}$$

in which:

- p denotes the parameter;
- t^0 and t^1 are the initial and final simulation time;
- G^1 refers to the final state (e.g. the configuration at time t^1 of a given body of the system);
- F depends on the dynamic behavior of the system in the time interval $[t^0, t^1]$, such as the root mean square (rms) acceleration of the car driver, the mean power dissipated at a wheel/ground contact, etc.

For the general objective function (39), sensitivity analysis consists in computing $\frac{\mathrm{d}\psi}{\mathrm{d}p}$, that is:

$$\frac{\mathrm{d}\psi}{\mathrm{d}p} = \frac{\partial G^1}{\partial u^1}\left.\frac{\mathrm{d}u}{\mathrm{d}p}\right|_{t^1} + \frac{\partial G^1}{\partial \dot{u}^1}\left.\frac{\mathrm{d}\dot{u}}{\mathrm{d}p}\right|_{t^1} + \frac{\partial G^1}{\partial p} \tag{40}$$
$$+ \int_{t^0}^{t^1} \left(\frac{\partial F}{\partial u}\frac{\mathrm{d}u}{\mathrm{d}p} + \frac{\partial F}{\partial \dot{u}}\frac{\mathrm{d}\dot{u}}{\mathrm{d}p} + \frac{\partial F}{\partial \ddot{u}}\frac{\mathrm{d}\ddot{u}}{\mathrm{d}p} + \frac{\partial F}{\partial p} \right)\mathrm{d}t$$

Let us first point out that in our case the only variables are the *independent* coordinates u (and $\dot{u}$). Indeed, the remaining variables v (and $\dot{v}$) have already been eliminated from the model during the reduction process (from the DAE (6–9) to the ODE (16) or (37)), v and $\dot{v}$ being expressed in terms of u and $\dot{u}$ according to the constraints solution.

Within expression (41), the unknown sensitivity matrices $\frac{\mathrm{d}u}{\mathrm{d}p}(t)$, $\frac{\mathrm{d}\dot{u}}{\mathrm{d}p}(t)$ and $\frac{\mathrm{d}\ddot{u}}{\mathrm{d}p}(t)$ can be computed via the so-called *direct method* (Eberhard, 1996) which consists in solving the differential equations for sensitivity matrices *simultaneously* with the equations of motion as explained here below.

Semi-explicit approach

Considering the semi-explicit form of the dynamic Eq. (16), the above sensitivity matrices can be calculated via the following equations in which, for sake of simplicity, we have defined $\Gamma(u,\dot{u},\ddot{u},\delta) \triangleq M_{\text{red}}(u,\delta)\ddot{u} + F_{\text{red}}(u,\dot{u},f,\delta,p)$:

$$M_{\text{red}}\frac{\mathrm{d}\ddot{u}}{\mathrm{d}p} + \frac{\partial \Gamma}{\partial \dot{u}}\frac{\mathrm{d}\dot{u}}{\mathrm{d}p} + \frac{\partial \Gamma}{\partial u}\frac{\mathrm{d}u}{\mathrm{d}p} + \frac{\partial \Gamma}{\partial p} = 0 \tag{41}$$

This equation can be time integrated simultaneously with the equations of motion (16). The main reason of the "symbolic versus numerical" benefits (factor 8 to 10) comes more from the *recursive* nature of the dynamic equations which is at the root of the symbolic elimination of useless equations in ROBOTRAN (see Sect. 3.3), than from the symbolic simplification of the expressions themselves. In particular, this recursivity is fully exploited when computing the explicit form (37) as stated in Sect. 3.5.1.

In view of Eq. (41), the computation of $\frac{\mathrm{d}\ddot{u}}{\mathrm{d}p}$, $\frac{\mathrm{d}\dot{u}}{\mathrm{d}p}$ and $\frac{\mathrm{d}u}{\mathrm{d}p}$ is required, the first one being the direct result of the model differentiation, the next two one being obtained via time integration.

Explicit approach

In view of the complexity of the semi-explicit form (41), which requires the partial derivative of the Γ term, it is clear that it would be far more advantageous to compute the derivatives $\frac{\mathrm{d}\ddot{u}}{\mathrm{d}p}$ by *directly* differentiating the explicit form (37):

$$\ddot{u} = f(u,\dot{u},p) \Rightarrow \frac{\mathrm{d}\ddot{u}}{\mathrm{d}p} = \frac{\mathrm{d}f}{\mathrm{d}p}(u,\dot{u},\frac{\mathrm{d}\dot{u}}{\mathrm{d}p},\frac{\mathrm{d}u}{\mathrm{d}p},p) \tag{42}$$

Thanks to the – recent – availability of the explicit model (37) in ROBOTRAN under the form of a fully symbolic recursive scheme (see Sect. 3.5.1), the computation of $\frac{\mathrm{d}\ddot{u}}{\mathrm{d}p}$ will be greatly facilitated, for tree-like as for constrained MBS: in ROBOTRAN, a direct symbolic differentiation can be achieved straightforwardly according to (42).

However, as for the model generation, the main complexity comes once again from the constraints at position level $h(q) = 0$ (assumed to be previously solved with respect to v). Without entering into details, the derivative of the dependent coordinates $\frac{\mathrm{d}v}{\mathrm{d}p}$ (and later on, the derivative of the velocities $\frac{\mathrm{d}\dot{v}}{\mathrm{d}p}$) are needed *explicitly* to compute the sensitivity model (42).

Starting from the differentiation of the implicit form of the constraints (which must be satisfied for any value of p),

$$\frac{\mathrm{d}h}{\mathrm{d}p} = \frac{\partial h}{\partial p} + \frac{\partial h}{\partial u}\frac{\mathrm{d}u}{\mathrm{d}p} + \frac{\partial h}{\partial v}\frac{\mathrm{d}v}{\mathrm{d}p} = 0, \tag{43}$$

we can isolate $\frac{\mathrm{d}v}{\mathrm{d}p}$,

$$\frac{\mathrm{d}v}{\mathrm{d}p} = -\left(\frac{\partial h}{\partial v}\right)^{-1}\left[\frac{\partial h}{\partial p} + \frac{\partial h}{\partial u}\frac{\mathrm{d}u}{\mathrm{d}p}\right] = -J_v^{-1}\frac{\partial h}{\partial p} + B_{vu}\frac{\mathrm{d}u}{\mathrm{d}p} \tag{44}$$

Symbolic implementation

Since the symbolic engine of ROBOTRAN (see next section) blindly differentiates any expression on the basis of recursive chain rules, any expression of the model derivative (e.g. $\frac{\mathrm{d}f}{\mathrm{d}p}$ must be the derivative of an "existing" expression in the model itself (e.g. f). For the particular case of (44),we can observe that it exactly corresponds to the derivative of the following relation (which is, incidentally, rather similar to the Newton-Raphson iteration 13):

$$v = -J_v^{-1} h(q) \tag{45}$$

Indeed, remembering that the constraints are satisfied (i.e.: $h(q) = 0$), by differentiating (45), we obtain:

$$\frac{dv}{dp} = -J_v^{-1} \frac{dh}{dp} = -J_v^{-1} \left[\frac{\partial h}{\partial p} + \frac{\partial h}{\partial u} \frac{du}{dp} \right] \tag{46}$$

which is the desired result (44) associated with the implicit constraints derivation. Sensitivity analysis of large constrained MBS (with more than 100 d.o.f.) led us to develop a specific procedure in ROBOTRAN to symbolically differentiate a given recursive scheme with respect to a given (set of) parameter(s) p. In the context of differentiation, the equations produced by a recursive multibody formalism (see Fig. 5) can be advantageously considered as interwoven functions $(f(g(h(...(p)))))$. However, if the corresponding differentiation rules are blindly applied to such a recursive scheme, we have observed that they produce a very large non-optimized symbolic output (e.g. $\frac{df}{\partial p}$ in Eq. 42) even for medium-sized multibody models: the interest of the recursive computation is thus completely lost.

Therefore, we take advantage of the condensation procedure described in Sect. 3.3 to solve this problem. When evaluating a given recursive scheme, we assume – a priori – that each equation depends, explicitly or not, on the set of system parameters or variables (e.g. $p_1, p_2, \ldots p_k$) with respect to which the differentiation must be performed. For instance, in the following equation:

$$\text{AUXJ} = \text{AUXI} + 2 \cdot P$$

AUXJ explicitly depends on the variable P via the second term. A priori, it may also depend implicitly on P via the first term AUXI.

We thus systematically create and evaluate a new *recursive* variable, for instance AUXJ_P, for the *total* derivative of the current equation with respect to P

$$\text{AUXJ\_P} = \text{AUXI\_P} + 2$$

even if, in the end, it appears that this new auxiliary variable is 0 or simply useless. If it is useless, the elimination process (of Fig. 5) will detect it and remove the corresponding equation from the list, before printing. Such a technique gives rise to a *compact recursive computation* of the derivatives.

To illustrate this, let us consider in (47) and (48), the symbolic evaluation of one element $J_{(3,1)}$ of the Jacobian matrix $\frac{\partial x}{\partial q^t}$ of a position vector $x(q)$ associated with a kinematic chain composed of nine joints.

In (47), the classical differentiation rule applied to $x(q)$ is far more consuming in terms of single operations (it contains 220 multiplications, additions and subtractions) than when obtained via the proposed recursive total differentiation (involving only 62 single operations).

$$\begin{aligned} J(3,1) = {} & q8 \cdot (C1 \cdot C7 \cdot (-C3 \cdot S4 \cdot C5 + S3 \cdot S5) - S1 \cdot (S2 \cdot (S3 \cdot S4 \cdot \\ & (-C5 \cdot C7 + S5 \cdot C6 \cdot S7) + S7 \cdot (-C3 \cdot C5 \cdot C6 + S3 \cdot C4 \cdot S6)) \\ & + C7 \cdot (C2 \cdot C4 \cdot C5 - S2 \cdot C3 \cdot S5)) + S7 \cdot (C1 \cdot (C3 \cdot C4 \cdot S6 \\ & + C6 \cdot (C3 \cdot S4 \cdot S5 + S3 \cdot C5)) - S1 \cdot C2 \cdot (-C4 \cdot S5 \cdot C6 + S4 \cdot S6))) \\ & + D13 \cdot S1 \cdot S2 + D14 \cdot (C1 \cdot S3 + S1 \cdot S2 \cdot C3) + D15 \cdot (C1 \cdot S3 + S1 \\ & \cdot S2 \cdot C3) + D16 \cdot (C1 \cdot (C3 \cdot S4 \cdot S5 + S3 \cdot C5) + S1 \cdot (C2 \cdot C4 \cdot S5 \\ & - S2 \cdot (-C3 \cdot C5 + S3 \cdot S4 \cdot S5))) + D17 \cdot (C1 \cdot C6 \cdot (C3 \cdot S4 \cdot S5 \\ & + S3 \cdot C5) + S1 \cdot (-S2 \cdot S3 \cdot (C4 \cdot S6 + S4 \cdot S5 \cdot C6) + C6 \cdot (C2 \cdot C4 \\ & \cdot S5 + S2 \cdot C3 \cdot C5)) + S6 \cdot (C1 \cdot C3 \cdot C4 - S1 \cdot C2 \cdot S4)) + D18 \cdot \\ & (C1 \cdot S7 \cdot (C3 \cdot S4 \cdot C5 - S3 \cdot S5) - S1 \cdot (S2 \cdot (S3 \cdot S4 \cdot (C5 \cdot S7 \\ & + S5 \cdot C6 \cdot C7) + C7 \cdot (-C3 \cdot C5 \cdot C6 + S3 \cdot C4 \cdot S6)) + S7 \cdot (-C2 \\ & \cdot C4 \cdot C5 + S2 \cdot C3 \cdot S5)) + C7 \cdot (C1 \cdot (C3 \cdot C4 \cdot S6 + C6 \cdot (C3 \cdot S4 \\ & \cdot S5 + S3 \cdot C5)) - S1 \cdot C2 \cdot (-C4 \cdot S5 \cdot C6 + S4 \cdot S6))) + D19 \cdot (C1 \\ & \cdot S7 \cdot (C3 \cdot S4 \cdot C5 - S3 \cdot S5) - S1 \cdot (S2 \cdot (S3 \cdot S4 \cdot (C5 \cdot S7 + S5 \\ & \cdot C6 \cdot C7) + C7 \cdot (-C3 \cdot C5 \cdot C6 + S3 \cdot C4 \cdot S6)) + S7 \cdot (-C2 \cdot C4 \\ & \cdot C5 + S2 \cdot C3 \cdot S5)) + C7 \cdot (C1 \cdot (C3 \cdot C4 \cdot S6 + C6 \cdot (C3 \cdot S4 \cdot S5 \\ & + S3 \cdot C5)) - S1 \cdot C2 \cdot (-C4 \cdot S5 \cdot C6 + S4 \cdot S6))); \end{aligned} \tag{47}$$

$$\begin{aligned} & \text{RO22} = S1 \cdot S2; \text{RO32} = -C1 \cdot S2; \text{RO82} = -S1 \cdot C2; \text{RO92} = C1 \cdot C2; \\ & \text{RO23} = \text{RO22} \cdot C3 + C1 \cdot S3; \text{RO33} = \text{RO32} \cdot C3 + S1 \cdot S3; \\ & \text{RO53} = -\text{RO22} \cdot S3 + C1 \cdot C3; \text{RO63} = -\text{RO32} \cdot S3 + S1 \cdot C3; \\ & \text{RO54} = \text{RO53} \cdot C4 + \text{RO82} \cdot S4; \text{RO64} = \text{RO63} \cdot C4 + \text{RO92} \cdot S4; \\ & \text{RO84} = -\text{RO53} \cdot S4 + \text{RO82} \cdot C4; \text{RO94} = -\text{RO63} \cdot S4 + \text{RO92} \cdot C4; \\ & \text{RO25} = \text{RO23} \cdot C5 - \text{RO84} \cdot S5; \text{RO35} = \text{RO33} \cdot C5 - \text{RO94} \cdot S5; \\ & \text{RO85} = \text{RO23} \cdot S5 + \text{RO84} \cdot C5; \text{RO26} = \text{RO25} \cdot C6 + \text{RO54} \cdot S6; \\ & \text{RO36} = \text{RO35} \cdot C6 + \text{RO64} \cdot S6;\ \text{RO27} = \text{RO26} \cdot C7 - \text{RO85} \cdot S7; \\ & \text{RO87} = \text{RO26} \cdot S7 + \text{RO85} \cdot C7; \text{RL23} = \text{RO22} \cdot D13;\ \text{RL24} = \text{RO23} \cdot D14; \\ & \text{JT34\_1} = \text{RL23} + \text{RL24}; \text{RL25} = \text{RO23} \cdot D15; \text{JT35\_1} = \text{JT34\_1} + \text{RL25}; \\ & \text{RL26} = \text{RO25} \cdot D16; \text{JT36\_1} = \text{JT35\_1} + \text{RL26}; \text{RL27} = \text{RO26} \cdot D17; \\ & \text{RL28} = \text{RO27} \cdot D18 + \text{RO87} \cdot q(8); \text{JT37\_1} = \text{JT36\_1} + \text{RL27}; \\ & \text{JT38\_1} = \text{JT37\_1} + \text{RL28}; \text{RL29} = \text{RO27} \cdot D19; J(3,1) = \text{JT38\_1} + \text{RL29}; \end{aligned} \tag{48}$$

For larger models and in particular for the explicit direct dynamics (37) of constrained multibody systems, the advantage of the recursive differentiation is amazing. In fact, the *explosive* increase in size of the classical differentiation technique (based on partial differentiation of interwoven functions) is quite understandable since it amounts to destroying the recursivity of the original scheme, leading to an in extenso formulation. Although the proposed recursive differentiation process is very consuming in terms of both memory storage and symbolic CPU time – because thousands of "potential" total derivatives are computed, these drawbacks are negligible in ROBOTRAN since the storage requirement is drastically controlled during the symbolic process as explained in Sect. 3.4.

4 ROBOTRAN computer framework

From the practical point of view, modeling a physical system using the multibody approach in ROBOTRAN involves several steps that can be summarized as follows:

– *drawing the multibody system*, which consists in defining the system topology (body structure, connecting

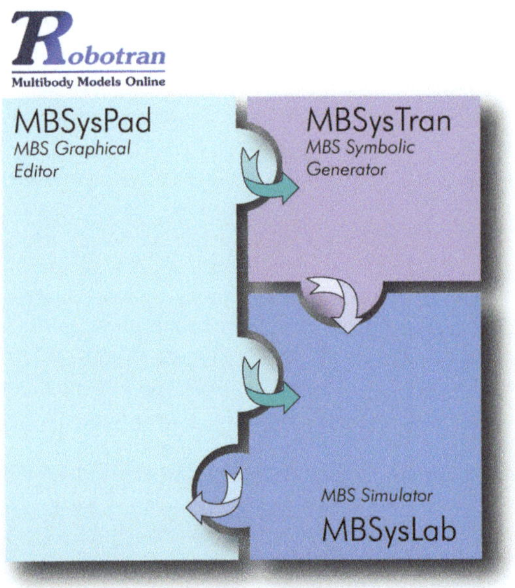

Figure 11. Illustration of the programs composing the ROBOTRAN framework.

joints, external/internal force, loop constraints, etc.) and the numerical data,

- *writing the multibody equations*, which relies on the symbolic engine for the equation of motion (using the Newton/Euler recursive formulation or the Virtual Power Principle) and requires a user intervention for specific constitutive law,
- *simulating the multibody model*, which requires the use of numerical methods so as to exploit the symbolic equations (for instance, numerical integration algorithm for a direct dynamic problem),
- *analyzing the multibody simulation results*, which needs efficient tools for presenting numerical results clearly or producing 3-D animation of the complete system.

In order to go through those various stages, the ROBOTRAN software is composed of several computer programs that are strongly related one to each other as illustrated in Fig. 11 and explained in the following subsections.

4.1 Drawing the multibody system

The first step of any multibody modeling process consists in identifying the involved bodies and the joints which connect them. This often requires a "pre-process" engineering work performed independently of any software. The main originality of ROBOTRAN is to introduce the system data and topology as it would be drawn on a sheet of paper using simple "potato" shapes like in Fig. 2. This way of thinking has guided the design of the graphical editor *MBsysPad*. It can be described as a "topology oriented" graphical editor (instead of a "3-D CAD oriented" editor which comes with most of commercial multibody programs): it relies on a 2-D graphical representation which highlights the MBS topology of the system rather than its 3-D representation. For instance, Fig. 12 illustrates the MBsysPad 2-D diagram of a 5-point suspension quarter car model. This 2-D sketch is composed of several components.

- *Bodies* are represented by various 2-D shapes that can be chosen so as to ensure the readability of the diagram. Specific points of a body are introduced using *anchor points* (arrows in the 2-D diagram).
- ROBOTRAN defines six simple *Joints* with 1 d.o.f.: 3 rotational joints about x-, y- or z-axis and 3 translational joints along x-, y- or z-axis. Combining several single joints permits to model any kind of complex joints. For instance, in Fig. 12, the 4 arms are connected by a *R1–R3* joint sequence which represents a universal joint. In such a way, the relative d.o.f. between two bodies appear directly and explicitly on the 2-D diagram.
- *Links* define point-to-point forces between two bodies (represented by a spring on the 2-D sketch).
- *Cuts* imposes a constraint between two bodies in order to deal with system with kinematic loops. For instance, in Fig. 12, the ball joints between the wheel carrier and the arms are modeled with a *ball cut* that ensures that the two connected points always coincide by imposing 3 algebraic constraints.
- *External force* or torque can be imposed on the system ("F" symbol in the diagram).

This 2-D representation gives a straightforward access to the element properties that can be modified via an edition panel such as the one appearing in Fig. 12 (on the right) for editing the body properties.

This approach, specific to ROBOTRAN, strengthens the software ergonomics, focusing on the work on the model itself, rather than on "cosmetic" features, by making the tree-like structure of the MBS, loop closure constraints and internal or external forces appearing clearly and explicitly on a single view of the system. Nevertheless, a 3-D representation can also be built in parallel so as to obtain a global view of the model which may be useful for instance for producing 3-D animation of the simulation results (see Fig. 13 which illustrates the 3-D representation of the 5-point suspension).

4.2 Writing the multibody equations

The process of writing the equations can be divided into two steps. The first one refers to the symbolic generation of the equations of motion of the MBS. The second one consists

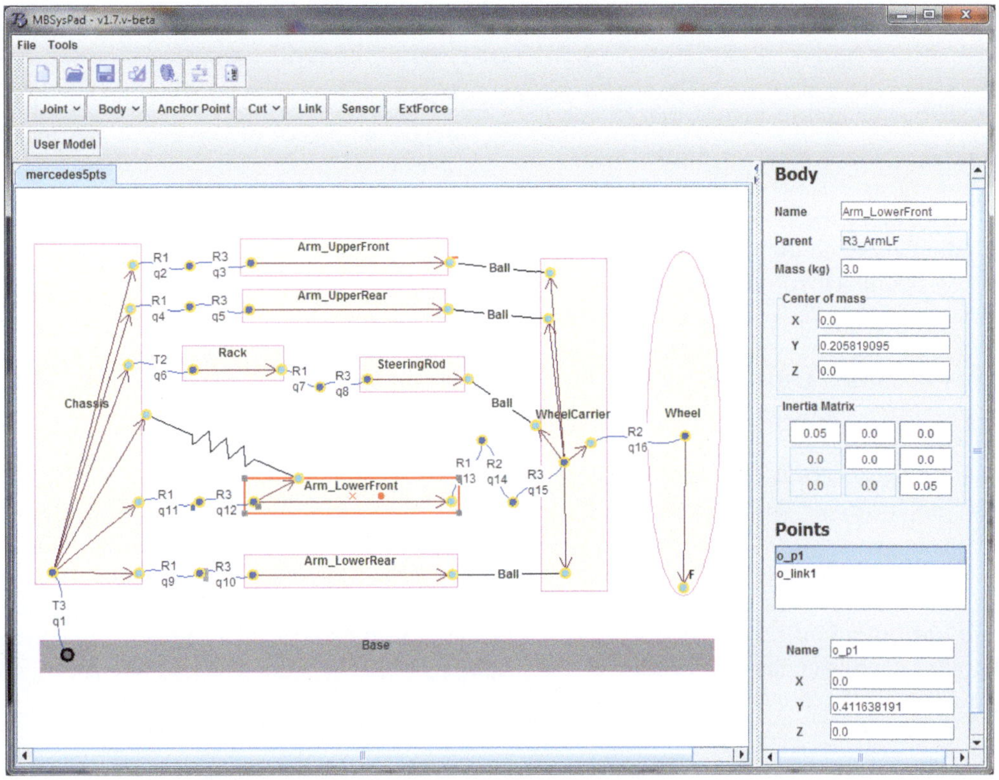

Figure 12. Representation of a 5 points suspension model in MBsysPad 2-D diagram.

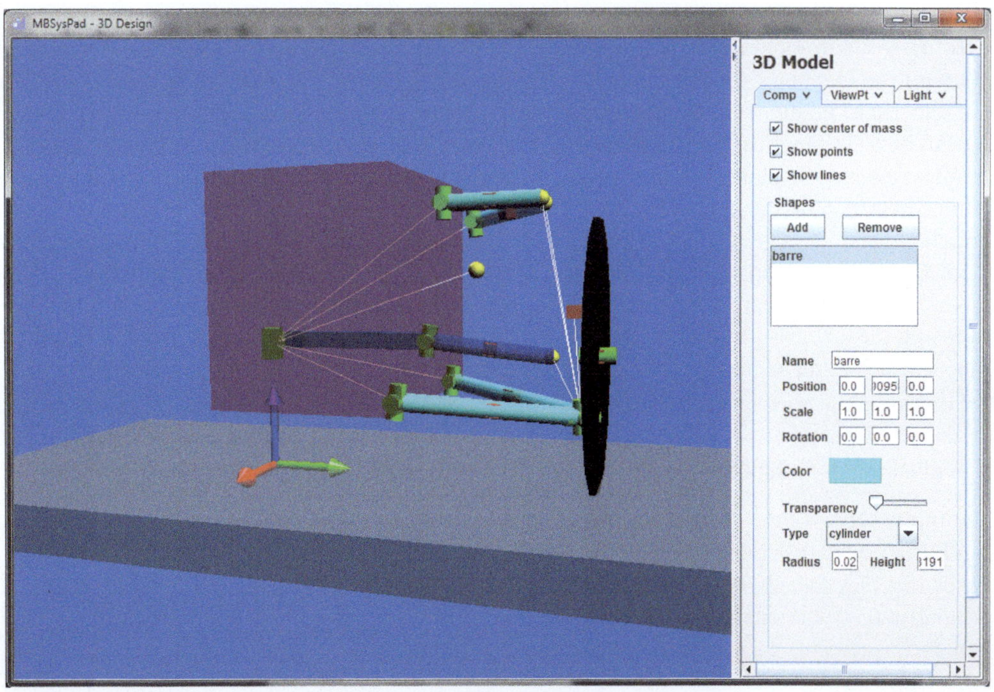

Figure 13. Representation of a 5 points suspension model in MBsysPad 3-D view.

in writing constitutive laws which are specific to the application, using symbolic ingredients produced by the previous step.

4.2.1 Symbolic equations

The symbolic equations are generated automatically on the basis of the symbolic formalism described in Sect. 3. Practically, this symbolic generation process (*MBsysTran* in Fig. 11) is performed online via the ROBOTRAN web server and does not require any additional software on the client computer. The symbolic equations consist in a set of C-code or M-code functions for calculating the various matrices and vectors of the equations of motion (i.e., mass matrix, dynamic vector, constraint vector, Jacobian matrix, etc. (see Eqs. 1, 5, 18 or Eqs. 6–9)).

The symbolic engine also generates helpful functions for specific features such as link forces or external forces. For example, for the link forces (i.e. point-to-point internal forces), a symbolic function calculates all the kinematics of the link (distance and distance time variation between the connected points) and performs all the operations required for projecting the force into the joint coordinate space. The user can thus skip this tedious work and concentrate on the tasks specific to his project.

4.2.2 User equations

For introducing force constitutive laws, ROBOTRAN relies on an "open approach" in which the user has the freedom and the responsibility of writing its own *user equations*. For instance, for the 5-point suspension in Fig. 12, the spring-damper element is model by a link force (i.e., point-to-point internal force) for which all the kinematics is calculated by the symbolic process. It remains to implement the suspension constitutive law which can be a simple linear spring-damper equation or a more complex law for which the user can benefit of all the functionalities provided by the language chosen for generating the symbolic files.

This flexible and powerful method applies for writing force constitutive laws (either internal link forces or external forces) but also for imposing the trajectory of a joint, adding specific user constraints or considering additional state equations for mechatronic MBS.

4.3 Simulating and analyzing the multibody system

In order to analyze the multibody model, all equation files (symbolic equations and user equations) must be assembled in a unique program. For this purpose, ROBOTRAN is distributed with a set of Matlab and C functions which constitute the *MBsysLab* environment.

4.3.1 MBsysLab modules for Matlab

The symbolic and user functions written in Matlab language can be used with various modules:

- the *coordinate partitioning module* checks or determines the choice of dependent and independent variables (see Sect. 2.2.1),
- the *equilibrium module* finds the equilibrium position of a given system,
- the *direct dynamics module* performs a time integration of the equations of motion,
- the *inverse dynamics module* calculates the joint forces for a given trajectory,
- the *modal analysis module* determines the eigen modes of a linearized multibody model.

This approach is very efficient for the model prototyping since it allows to benefit of the Matlab language flexibility and to call functions provided by other Matlab toolboxes. A module can for instance be called by an optimization algorithm or a user function can call any specific Matlab function.

4.3.2 MBsysLab for Simulink

MBsysLab also contains modules for building C-code S-Function in Simulink. In this case, symbolic and user files are written in C language and compiled into a binary file that is embedded in a unique Simulink block. It is thus very straightforward to incorporate the multibody model in a classical Simulink block diagram. This is a very powerful way of dealing with control or robotics applications for example. Furthermore, since all the code is compiled, this approach leads to very high performances in terms of calculation time making possible real time simulations or optimization of complex systems in a user-friendly environment.

Finally, it must be noticed that those modules are not specific to Simulink and can be combined with a own written integration algorithm so as to obtain a simulation tool completely independent of Matlab/Simulink. Additionally, the generated code can be transfered to an onboard card for hardware-in-the-loop control.

5 Illustrative applications

Three applications are shortly described in this section namely, (i) the performance of a modern car equipped with the Kinetic^{TM}-type hydraulic suspension, (ii) the analysis of a truck-mounted attenuator and (iii) the modeling of a grand piano action, to respectively highlight the capabilities (i) to simulate multiphysics systems with real-time performances, (ii) to deal with large multibody systems and (iii) to build complex systems in an open-type environment.

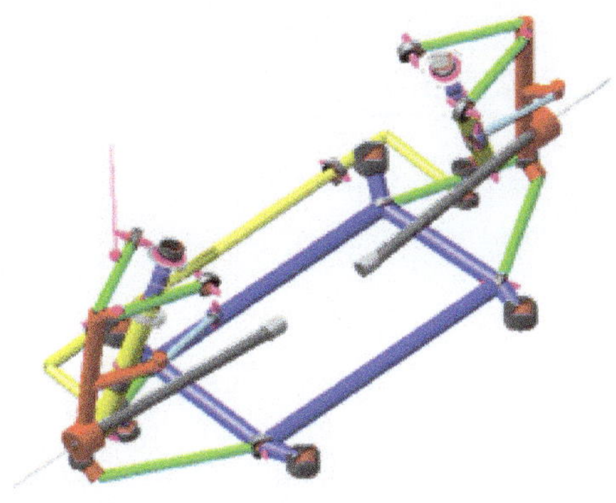

Figure 14. CAD representation of the front wheel-axle unit of the modeled Audi A6.

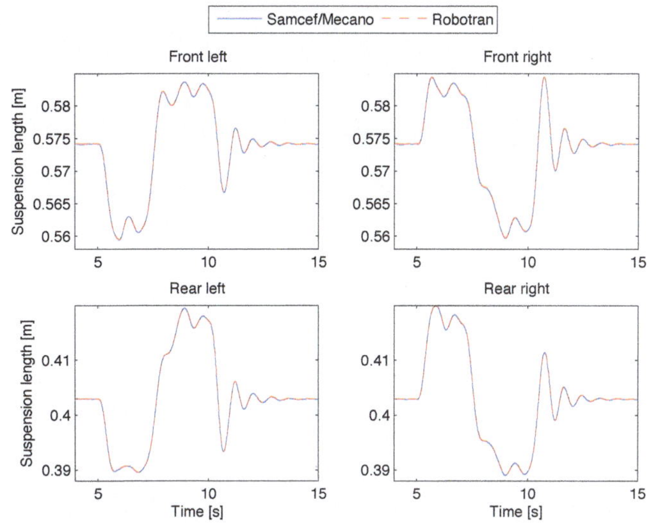

Figure 15. Suspension crushing during a line change manoeuvre.

5.1 A modern car equipped with a Kinetic™ H2 suspension

This first application example deals with the modeling of a full modern car (Audi A6) equipped with multi-link suspensions at front and at rear. Figure 14 shows the front wheel-axle unit consisting of a part of the chassis, the suspension rods, the dampers, the anti-roll bar, the tie rod and the rubber bushings at the various connection. From the multibody modeling point of view, this vehicle is a rather complex model since it involves:

- more than 80 generalized coordinates,
- multi-link suspensions (rear and front), which induce 16 three-dimensional kinematic loops,
- a longitudinallateral wheelground model with saturation effect.

The symbolic equations of this model have been generated in C-language and compiled into a Simulink S-Function so as to perform time-efficient simulation and to easily couple the mechanical model to a hydraulic model in a second step. A line change manoeuvre has been simulated by imposing the motion of the direction rack while the vehicle is running at 10 m s^{-1}. The result have been compared to the ones obtained with the multibody software SAMCEF/MECANO. The later relies on a finite-element numerical approach which is a complementary solution to the one proposed by ROBOTRAN, opening the way to the coupling of MBS with structural analysis for instance. As illustrated in Fig. 15 which shows the crushing of the four suspensions during the line change manoeuvre, the two approaches (symbolic generation and relative coordinates for ROBOTRAN, numerical generation and nodal coordinates for SAMCEF/MECANO) give similar results which reinforce the confidence in our model.

The next step has consisted in modeling the *Kinetic H2* suspension system developed by Tenneco Automotive. This innovative device consists in replacing the classical hydraulic dampers by double acting cylinders connected together by two distinct hydraulic circuits. This system breaks the tradeoff between a high roll stiffness and a low warp stiffness, as demonstrated by Docquier et al. (2010).

From a practical point of view, the hydraulic model, which involves 22 state equations, has been implemented in C-language. As illustrated in Fig. 16, two options have been retained for coupling it to the MBS. Firstly, the *H2* model has been implemented as a separated Simulink S-Function and coupled a posteriori to the multibody model of the car. Secondly, the hydraulic state equations have been combined to the mechanical state equations and compiled in a unique S-Function, giving one monolithic set of equation to the Simulink time integrator. As shown in Fig. 17, the roll angle of the car during the line change manoeuvre is smaller when the car is equipped with the Kinetic H2 system. Furthermore, the two approaches for coupling the hydraulic and mechanical models result in exactly the same results. However, as illustrated in Table 1, when using the ode45 time integrator of Simulink based on the Dormand-Prince algorithm, the strongly coupled permits larger time steps, resulting in a smaller simulation time. When using the ode15s integrator, the number of time steps is smaller in the case of the strong coupling but the simulation time is equivalent while the integrator does not converge in the case of the weak coupling (2 block diagram).

5.2 A truck-mounted attenuator

This project has consisted in testing and modelling Truck Mounted Attenuators (TMA) made of recycled materials, proposed by the ArGEnCo Laboratory of the Université de Liège (ULg, Belgium). Such a device is used on motorway in order to protect people on working site from inattentive drivers. The principle consists in assembling several cubic

Figure 16. Illustration of the two approaches for coupling the hydraulic and mechanical models. Left: weak coupling using two separated blocks. Right: strong coupling using one monolithic block.

Table 1. Calculation performance for the line change manoeuvre simulation of the model of the Audi A6 equipped with the Kinetic H2 suspension. The simulation was performed on a computer running with Windows 7 and equipped with an Intel Core 2 Duo CPU (2.53GHz), 4 GB RAM.

	ode45 integrator		ode15s integrator	
	weak coupling (2 blocks)	strong coupling (1 block)	weak coupling (2 blocks)	strong coupling (1 block)
Simulation time	66 s	38 s	failed	40 s
Number of time steps	13 754	7946		5767

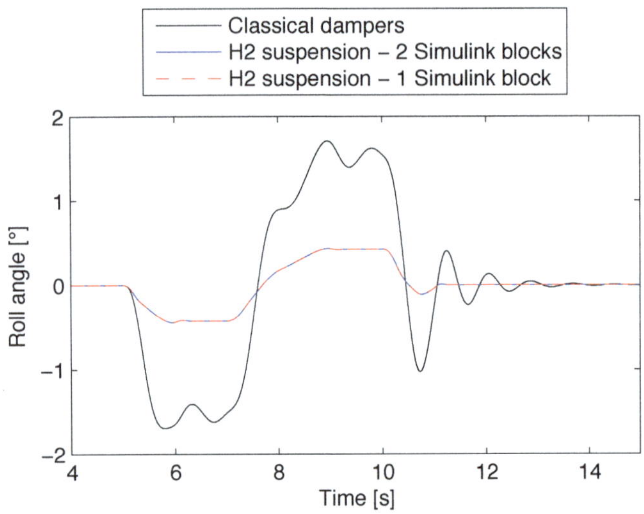

Figure 17. Impact of the Kinetic H2 suspension on the roll angle during the line change manoeuvre.

boxes made of deployed steel containing the recycled materials compacted to an initial preload. Thanks to experimental tests performed at the ULg laboratory, an analytic formulation was established in order to describe the various phases of the material constitutive law: the elastic loading phase, the plastic loading phase and the elastic unloading phase. Then, a multibody model of the TMA has been implemented in ROBOTRAN using a lumped mass approach: each block is splitted into several bodies linked together by prismatic joints. The constitutive law is applied via external forces acting on each bodies, following the action/reaction principle.

Figure 18. MBS model of a truck-mounted attenuator (TMA).

By comparing models involving 1, 3 and 10 bodies for one block, we observe discrepancies during the collision (oscillations) but a good match in terms of maximum deceleration of the impacting mass (600 kg), as illustrated in Fig. 19 (more details can be found in Abedrabbo et al., 2011). Also, comparing those results in terms of the Acceleration Severity Index (ASI), which is the reference for crash calculations, shows differences that are lower than 0.4 % between the different models, and 7.4 % with experiments which is really acceptable.

Afterward, a realistic TMA was designed for absorbing the shock between a truck and a car running at 90 km h^{-1} or a bus running at 70 km h^{-1}. In order to analyse the TMA performance during the collision, the MBS model of the TMA described above has been used to simulate a longitudinal impact between a car (or a bus) and this TMA carried by a truck, leading to a model containing more than 300 degrees of freedom. The truck, the car and the bus, being more rigid than the attenuator modules, were modeled as rigid MBS, with articulated suspension and wheels. Two deformation laws of the front part of each impacting vehicle (i.e., the car or the bus)

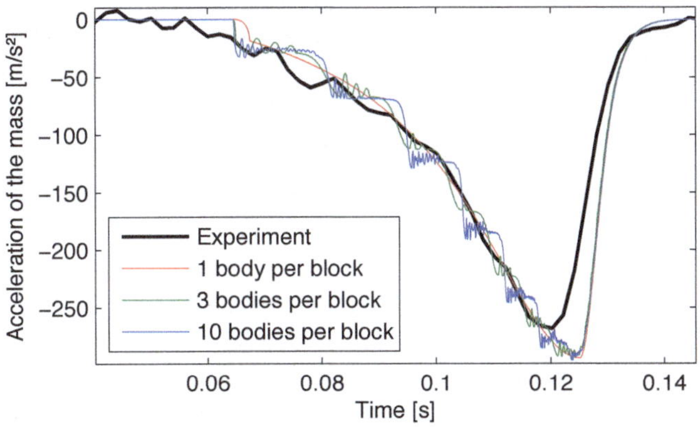

Figure 19. Influence of the number of bodies used to discretize a TMA block and comparison with experimental results.

Figure 20. Simulation of the collision of a bus or a car into the TMA.

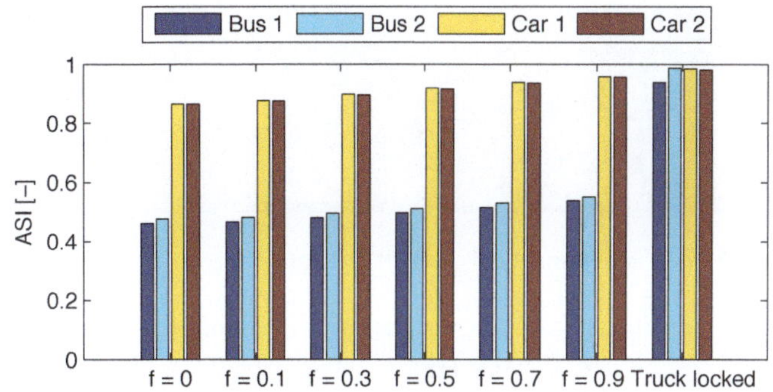

Figure 21. Impact of the friction coefficient f between the ground and the wheels of the truck. The Bus 2 and Car 2 cases correspond to the simulation of a vehicle with a front deformation law stiffer than the Bus 1 and Car 1 cases.

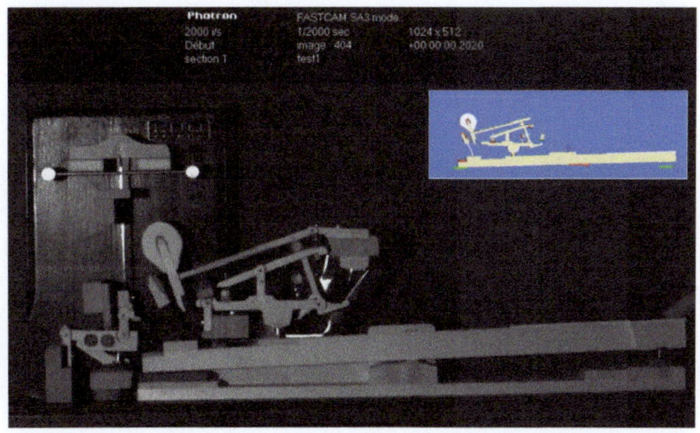

Figure 23. Model versus experiment: full system in action.

have been considered, the second one being stiffer than the first one.

Several parameters have been investigated such as:

- the influence of the own mass of the TMA blocks on the ASI measured at the driver head level,
- the influence of the deformation law of the front of the impacting vehicle,
- the influence of the road adherence conditions which impact the safety of workers located in front of the truck.

As an illustrative result, Figs. 20 and 21 illustrate a typical ROBOTRAN simulation and the ASI measured during the crash for various values of the coefficient of friction f between the ground and the wheels of the truck carrying the TMA. The impacting vehicle is supposed to slide perfectly on the ground. The extreme case where the truck is completely locked on the ground is also considered. It clearly appears that the ASI is higher for the car than for the bus, due to its smaller mass. The impact of the friction coefficient is smaller but the diagram reveals that the truck should have the possibility to move with respect to the ground.

5.3 A grand piano action

The goal of the underlying project – presently in progress in close collaboration with the MIM, the Musical Instruments Museum of Brussels – has a double nature: organological and didactic. Indeed, using a multibody approach, we have carried out a virtual demonstrator of a grand piano action mechanism (see Fig. 22) in order to understand, demystify and parameterize the "from key-to-hammer" transmission, and especially the double escapement principle patented by the French Sébastien Erard in 1821. Double escapement actions allow notes to be repeated more easily than in single piano actions.

As one might imagine, a particular attention has been paid to the modeling of the intermittent contacts (they are twelve in number in the mechanism!) for both the geometrical (shapes) and dynamic (constitutive laws) points of view.

To ensure a reasonable match between the multibody model and the real mechanism, the force law parameters have been tuned via experimental validations using a high-speed camera (2000 black and white frame/sec): they have been carried out for the whole system in action (Fig. 23) or for some specific sub-systems, like a single oscillating hammer

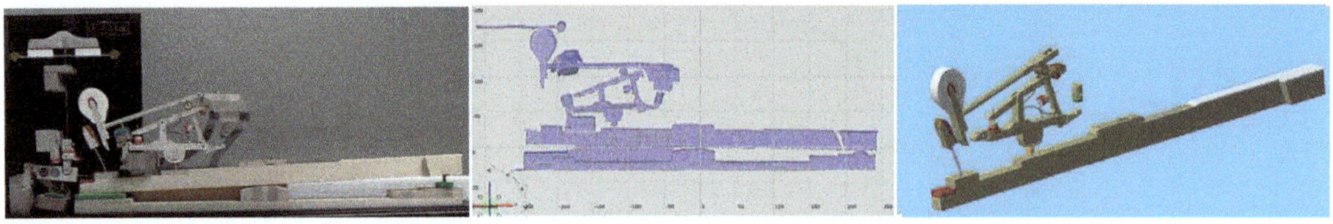

Figure 22. Real action mechanism (Left) – Scan of the action (Center) – Multibody model (Right).

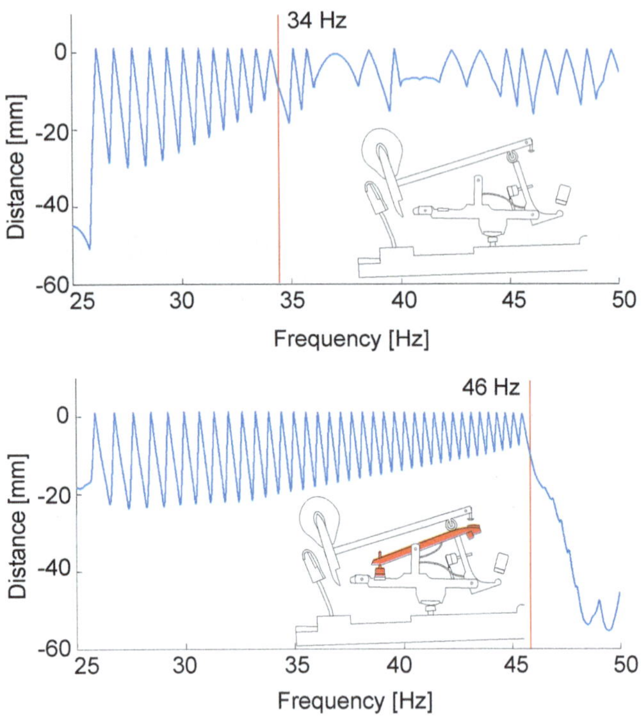

Figure 24. Single (top) versus double escapement (bottom) comparison.

(not shown). After identification, a satisfactory correspondence was obtained, considering the objectives of the project. Let us briefly show and comment on a significant result related to the comparison between single and double escapement configurations. The analyzed piano action mechanism is a modern one (Fig. 22) with the double escapement system, which allows the pianist to repeat a given note with a shorter stroke (for the key and the finger) and thus at a higher speed. By virtually "removing" some parts in the model, the double escapement can be virtually deactivated. To compare the dynamic capabilities of the two mechanisms, a staccato-type sinusoidal input with increasing frequency is applied on the key front. In Fig. 24, we plot the vertical distance [m] between the top of the hammer and the string, versus the blow frequency [Hz]. This result clearly shows that single escapement is unable to correctly repeat the motion above 34 Hz, while the original system, thanks to the double escapement principle, can reach 46 Hz (more details in Bokiau et al., 2012). This kind of result is greatly appreciated by our collaborators at the MIM, because such a comparison appears to be quite instructive and also difficult to observe experimentally.

6 Conclusions

In this paper, the main assets and recent developments of ROBOTRAN software are reviewed. The symbolic generation capabilities of the software have been greatly enhanced to deal with large models (in terms of d.o.f.) whose symbolic generation requires less than one second via a web server. The fully symbolic generation of constrained MBS is a new feature of the program that notably improves the CPU time performances but, above all, allows us to provide a completely free-standing symbolic function (in C, Matlab, etc.) ready for use for various scientific computer environments. More recently, the recursive symbolic differentiation of MBS direct dynamics with possible nonlinear constraints has been implemented and has revealed its superiority with respect to standard symbolic engines in terms of symbolic simplification and equation conciseness. A novel user interface based on an intuitive 2-D representation of the MBS – the 3-D representation being automatically constructed in the background – and the open-type architecture of the program are presented. Finally, three illustrative applications are briefly discussed to highlight the capabilities of ROBOTRAN in building complex and large models and dealing with multiphysics applications.

For the next future, the main developments will mainly concern the periphery of the ROBOTRAN symbolic generator because we wish to keep the latter as open as possible for the user, and to interface the symbolic models with specific external software. These couplings refer to multiphysics modeling (in the continuity of our present research work), the geometrical and dynamic contact between 3-D profiled bodies with a real-time approach, the coupling with FEM models (flexible bodies), CFD software (multibody-fluid interaction) and DEM software (discrete element modelling for granular materials). As regards flexible bodies, our past investigations for beams (Fisette et al., 1997) and plates (El Ouatouati et al., 1999) have shown that in case of small deformations, the floating-frame approach in relative coordinates was a suitable option for symbolic generation, leading to very efficient models: the revival of those models is part of our future objectives for the symbolic core of ROBOTRAN.

Acknowledgements. This research work has been supported by the Belgian National Fund for Scientific Research (F.R.S.-FNRS) which is acknowledged.

Edited by: O. Brüls

References

Abedrabbo, G., Poncelet, A., Sepulveda, P., Cescotto, S., and Fisette, P.: Multibody Simulation of a Crash Test Attenuator Made of Recycled Materials, in: Proceedings of the 17th International Symposium on Plasticity & Its Current Applications, Mexico, 2011.

Bokiau, B., Poncelet, A., Fisette, P., and Docquier, N.: Multibody Model of a Grand Piano Action Mechanism Aimed at Understanding and Demystifying the Escapement Principle, in: Proceedings of the 2nd Joint International Conference on Multibody System Dynamics, Stuttgrart, Germany, 2012.

Chenut, X., Fisette, P., and Samin, J.-C.: Recursive Formalism with a Minimal Dynamic Parameterization for the Identification and Simulation of Multibody Systems. Application to the Human Body, Multibody Syst. Dyn., 8, 117–140, doi:10.1023/A:1019555013391, 2002.

Ding, J.-Y., Pan, Z.-K., and Chen, L.-Q.: Second order adjoint sensitivity analysis of multibody systems described by differential–algebraic equations, Multibody Syst. Dyn., 18, 599–617, doi:10.1007/s11044-007-9080-4, 2007.

Docquier, N., Poncelet, A., Delannoy, M., and Fisette, P.: Multiphysics Modeling of Multibody Systems: Application to Car Semi-Active Suspensions, Vehicle Syst. Dyn., 48, 1439–1460, 2010.

Eberhard, P.: Analysis and optimization of complex multibody systems using advanced sensitivity analysis methods, ZAMM Zeitschrift fur Angewandte Mathematik und Mechanik, 76, 40–43, 1996.

El Ouatouati, A., Fisette, P., and Johnson, D.: A Fully Symbolic Model of Multibody Systems Containing Flexible Plates, Nonlinear Dynamics, 18, 357–382, 1999.

Fisette, P.: Génération symbolique des équations du mouvement de systèmes multicorps et application dans le domaine ferroviaire, Ph.D. thesis, Université catholique de Louvain, 1994.

Fisette, P., Johnson, D., and Samin, J.: A Fully Symbolic Generation of the Equations of Motion of Multibody Systems containing Flexible Beams, Comput. Method. Appl. M., 142, 132–152, 1997.

Fisette, P., Postiau, T., Sass, L., and Samin, J.: Fully symbolic generation of complex multibody models, Mech. Struct. Mach., 30, 31–82, doi:10.1081/SME-120001477, 2002.

Ganovski, L., Fisette, P., and Samin, J.-C.: Piecewise Overactuation of Parallel Mechanisms Following Singular Trajectories: Modeling, Simulation and Control, Multibody Syst. Dyn., 12, 317–343, doi:10.1007/s11044-004-2532-1, 2004.

Kecskeméthy, A.: Mobile – An object-oriented tool-set for the efficient modeling of mechatronic systems, in: Proceedings of the Second Conference on Mechatronics and Robotics, 27–29 September, 27–29, 1993.

Kecskeméthy, A., Krupp, T., and Hiller, M.: Symbolic Processing of Multiloop Mechanism Dynamics Using Closed-Form Kinematics Solutions, Multibody Syst. Dyn., 1, 23–45, doi:10.1023/A:1009743909765, 1997.

Kurz, T., Eberhard, P., Henninger, C., and Schiehlen, W.: From Neweul to Neweul-M2: symbolical equations of motion for multibody system analysis and synthesis, Multibody Syst. Dyn., 24, 25–41, doi:10.1007/s11044-010-9187-x, 2010.

Maes, P., Samin, J.-C., and Pierre, W.: Multibody Systems Handbook, chap. Robotran, 246–264, Springer-Verlag, 1990.

Poncelet, A., Collard, J.-F., and Fisette, P.: Parameter Sensitivity Analysis: Symbolic Computation for Large Multibody Systems, in: Proceedings of the First Joint International Conference on Multibody System Dynamics, Lappeenranta, Finland, 2010.

Postiau, T., Sass, L., Fisette, P., and Samin, J.-C.: High-Performance Multibody Models of Road Vehicles: Fully Symbolic Implementation and Parallel Computation, Vehicle Syst. Dyn., 35, 57–83, 2001.

Raison, M., Aubin, C., Detrembleur, C., Fisette, P., Mahaudens, P., and Samin, J.: Quantification of global intervertebral torques during gait: comparison between two subjects with different scoliosis severities, Studies in Health Technology and Informatics, 158, 107–111, 2010.

Samin, J. and Fisette, P.: Symbolic modeling of multibody systems, Springer, 2003.

Samin, J.-C., Brüls, O., Collard, J.-F., Sass, L., and Fisette, P.: Multiphysics modeling and optimization of mechatronic multibody systems, Multibody Syst. Dyn., 18, 345–373, doi:10.1007/s11044-007-9076-0, 2007.

Schiehlen, W.: Multibody Systems Handbook, Springer-Verlag, Berlin, Germany, 1990.

Schwertassek, R. and Rulka, W.: Real-Time Integration Methods for Mechanical System Simulation NATO ASI Series, Series F, vol. 69, chap. Aspects of Efficient and Reliable Multibody Systems Simulation, Springer-Verlag, Berlin, Germany, 1989.

Shi, P. and McPhee, J.: Dynamics of Flexible Multibody Systems using Virtual Work and Linear graph Theory, Multibody Syst. Dyn., 4, 355–381, 2000.

Wehage, R. A. and Haug, E. J.: Generalized Coordinate Partitioning for Dimension Reduction in Analysis of Constrained Dynamic Systems, J. Mech. Design, 104, 247–255, doi:10.1115/1.3256318, 1982.

22

Active gesture-changeable underactuated finger for humanoid robot hand based on multiple tendons

D. Che and W. Zhang

Key Laboratory for Advanced Materials Processing Technology, Ministry of Education,
Department of Mechanical Engineering, Tsinghua University, Beijing 100084, China

Abstract. The concept called gesture-changeable under-actuated (GCUA) function is utilized to improve the dexterities of traditional under-actuated hands and reduce the control difficulties of dexterous hands. Based on GCUA function, a novel mechanical finger by multiple tendons: GCUA-T finger, is designed. The finger uses tendon mechanisms to achieve GCUA function which includes traditional underactuated (UA) grasping motion and special pre-bending (PB, or pre-shaping) motion before UA grasping. Operation principles and force analyses of the fingers are given, and the effect of GCUA function on the movements of a hand is discussed. The finger can satisfy the requirements of grasping and operating with low dependence on control system and low cost on manufacturing expenses, which develops a new way between dexterous hand and traditional under-actuated hand.

This paper was presented at the IFToMM/ASME International Workshop on Underactuated Grasping (UG2010), 19 August 2010, Montréal, Canada.

1 Introduction

In recent years, breathtaking developments in robotics have been witnessed all over the world. The field of robotic hands including dexterous hands and under-actuated hands are emphasized and developed.

Over the past 30 years, researchers have made plentiful achievements on the study of dexterous hand. For instance, Stanford/JPL dexterous hand was designed and analyzed by J. Salisbury et al. (1982), which has three 3-DOF fingers actuated by 12 DC motors, each joint of this hand can be flexed and extended independently by one actuator. Gifu II hand, designed by H. Kawasaki et al. (1986), has 5 fingers whose all joints are actuated by servomotors, which can perform dexterous object manipulation like the human hand; Utah/MIT dexterous hand was designed by S. Jacobsen et al. (1992), which has four 4-DOF fingers with 32 independent tendons and 32 pneumatic cylinders (2004), the hand can be used as a high flexible tool for the study of machine dexterity. Dexterous hand can do almost all the movements and gestures of human hand. In fact, almost each DOF of a dexterous hand need an actuator to drive, which makes the hand high depend on control and high cost on manufacturing and using expenses.

On the contrary, under-actuated hands use fewer motors to drive more DOFs, and the under-actuated hands have a very amazing feature: self-adaptation in grasping, which let the hands easy to control. Many studies have done in the field of under-actuated hands: L. Birglen et al. (Birglen, 2003, 2004 and 2005), designed many kinds of under-actuated grippers and gave force analyses on them. A. Dollar et al. (Dollar, 2006 and 2007), gave a SDM robust robotic grasper which uses a single actuator to drive 8 DOFs; L. Tan et al. (2009), designed a multi-fingered hand using hydraulic actuation with fluidic actuators, the hand has 14 DOFs which can bend when hydraulic pressure is applied by a water pump.

This paper introduced the concept called gesture-changeable under-actuated (GCUA) function. Based on GCUA function, a new robot finger was designed for humanoid robot hands based on multiple tendons.

2 GCUA function

Traditional under-actuated (UA) function can drive many DOFs of a robotic hand with motors whose number is less than that of the hand's DOFs. There are two main kinds of under-actuated hands with traditional UA function:

1. When a hand grasp an object, it cannot change its initial gesture which is fixed (almost keep straight) until its phalange touches some objects. An example of this kind of under-actuated hand is TH-3 hand which designed by Tsinghua University (Zhang, 2009). TH-3 hand is based on gear-rack transmission, which cannot bend its phalanges without resistances of objects, this makes it feel hard to grasp gliding objects stably.
2. When a hand grasp objects, all or part of its joints must bend in a fixed order which cannot be changed when the mechanism has been designed. For instance, a multi-function mechanical hand with shape adaptive which was designed by G. Guo, X. Qian, and A. Gruver (1993), is based on link and gear transmission. The hand can grasp different objects with shape adaptive function. However, with gear transmission, the rotating angles of middle joint and distal joint are proportional, which limits its dexterity.

Traditional UA function can make robotic hand low depend on sensing and control, however, it cannot ensure the hands enough flexibility to grasp or operate. A concept called gesture-changeable under-actuated (GCUA) function was proposed (Zhang, 2009) to improve the dexterities of traditional under-actuated hands and reduce the control difficulties of dexterous hands. The GCUA function includes two main motions: under-actuated (UA) motion and pre-bending (PB, or pre-shaping) motion, which are shown in Fig. 1.

UA motion preserves advantages of traditional under-actuated hand, which makes the hand low depend on senor and control system. With UA motion, the hand can grasp different objects self-adaptively.

The hand can change its initial gesture with PB motion, which can achieve different grasping postures. After it has got a satisfied posture, the hand can grasp different objects with UA motion. PB motion is extremely important when high grasping dexterity and stability are required.

3 Design of GCUA-T fingers by tendons

3-joint GCUA-T finger uses tendon mechanism to achieve GCUA function, which is shown in Fig. 2.

Proximal joint is fixed in base, which is between the base and 1st middle phalange; middle joint is fixed in 1st middle phalange, which is between 1st and 2nd middle phalanges; distal joint is fixed in 2nd middle phalange, which is between 2nd middle phalange and distal phalange; UA transmission and PB transmission are both fixed in base; 1st return spring

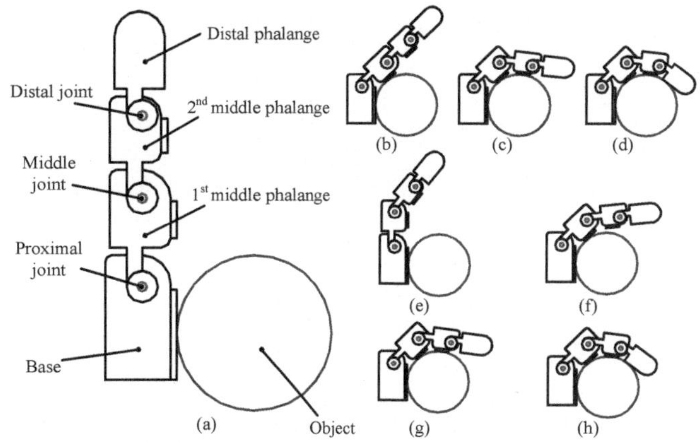

Figure 1. UA motion and PB motion of 3-joint GCUA finger. **(a, b, c, d)** UA motion; **(a, e)** PB motion; **(f, g, h)** UA motion with the changed initial gesture; **(a)** Initial gesture (keep straight); **(b)** Proximal joint rotates until 1st middle phalange touches the object; **(c)** Middle joint rotates until 2nd middle phalange touches the object; **(d)** Distal joint rotates until distal phalange touches the object; **(e)** Middle joint rotates to change the finger's initial gesture.

combines 2nd middle phalange and distal phalange with its two ends, similarly 2nd return spring combines 1st middle phalange and 2nd middle phalange with its two ends.

Assembly methods of UA tendon, RS tendon, and PB tendon are shown in Fig. 3.

UA tendon combines UA transmission and distal phalange with its two ends, which wraps around distal joint, middle joint and proximal joint; RS tendon uses its two ends to combine 1st middle phalange and UA transmission, which wraps around proximal joint and cross through the base; PB tendon combines 2nd middle phalange and PB transmission with its two ends, which wraps around middle joint and proximal joint; The wrapping directions of UA tendon and PB tendon are the same, which are different from that of RS tendon, so that UA tendon and PB tendon will drive the finger to be flexed, on the contrary, RS tendon will drive the finger to be extended. PB tendon can also be assembled in another method, which is shown in Fig. 3d.

4 The work process of GCUA-T finger

With UA tendon and RS tendon, UA motion will proceed in this way: 1st motor rotates forward, UA transmission pulls UA tendon and RS tendon to move down. Since the wrapping directions of these two tendons are different, UA tendon will tension, simultaneously RS tendon will relax; In this case, UA tendon drives distal phalange to rotate around distal joint. With the constraint of 1st return spring and 2nd return spring, 1st middle phalange, 2nd middle phalange and distal phalange rotate around proximal joint together as a rigid body. When 1st middle phalange is blocked by objects, 2nd middle phalange and distal phalange will rotate around middle

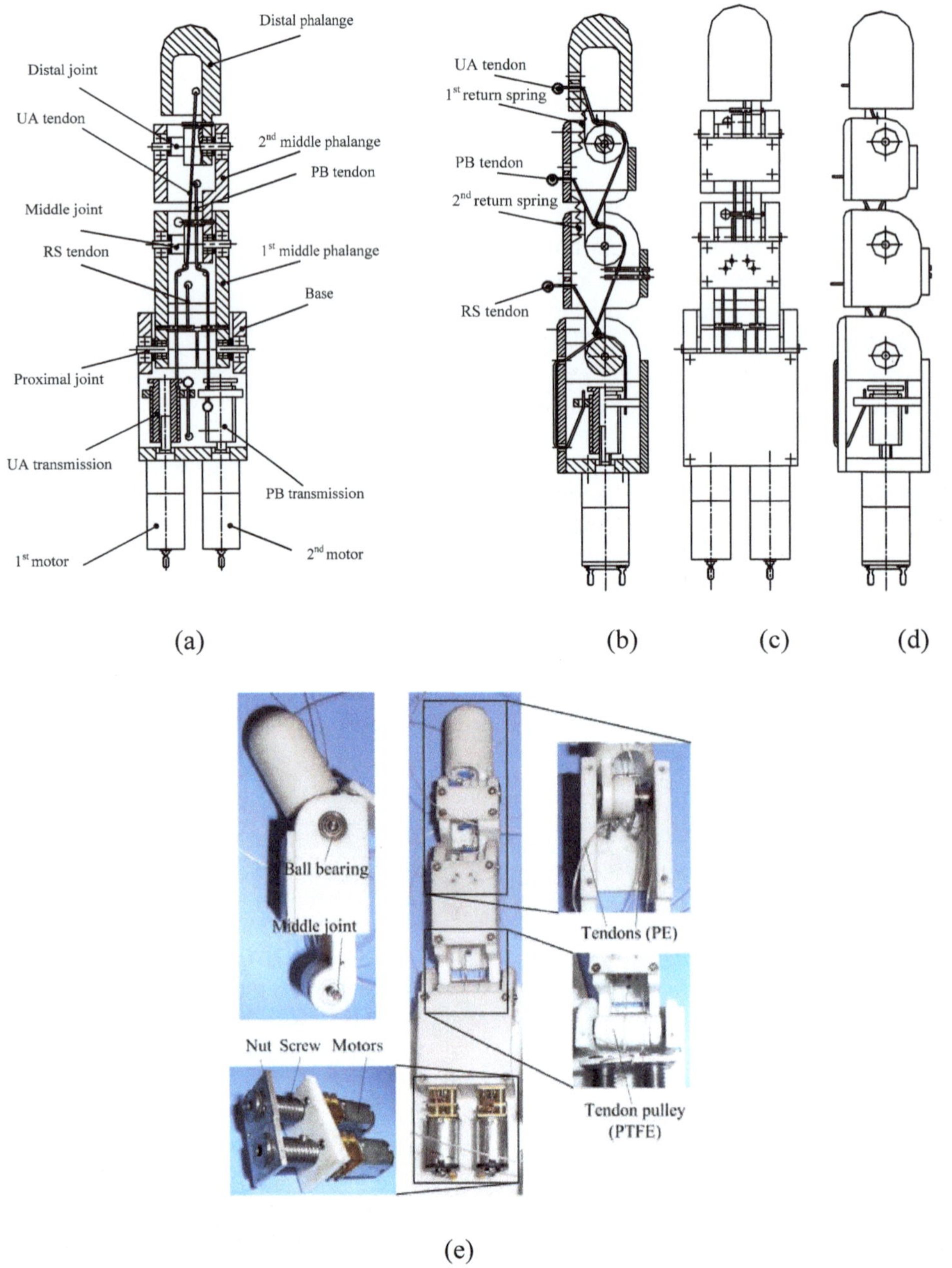

Figure 2. Structure of 3-joint GCUA-T finger. (**a**) Front cutaway view; (**b**) Side cutaway view; (**c**) Front view; (**d**) Side view; (**e**) 3-joint GCUA-T finger.

joint together, simultaneously 2nd return spring will extend; Similarly, when 2nd middle phalange is blocked by objects, distal phalange will rotate around distal joint, and 1st return spring will extend. In this way, the finger can grasp objects self-adaptively.

When 1st motor rotates back, UA transmission will pull these two tendons up, so that UA tendon will relax, and RS tendon will tension. RS tendon will drive the finger to extend and leave the object. When UA transmission goes back to the initial status, the finger will be back to the initial gesture too.

If PB tendon is assembled in the method which is shown in Fig. 3b, PB motion will proceed like this: When 2nd motor rotates forward, PB transmission pulls PB tendon to change the finger's initial gesture, since RS tendon prevents 1st middle phalange from bending and 1st return spring makes the top two phalanges become a rigid body, 2nd middle phalange and distal phalange will rotate around middle joint together, so that the finger can change its initial gesture and then grasp things with UA motion. In this process of PB motion, if either of middle phalange and distal phalange is blocked by

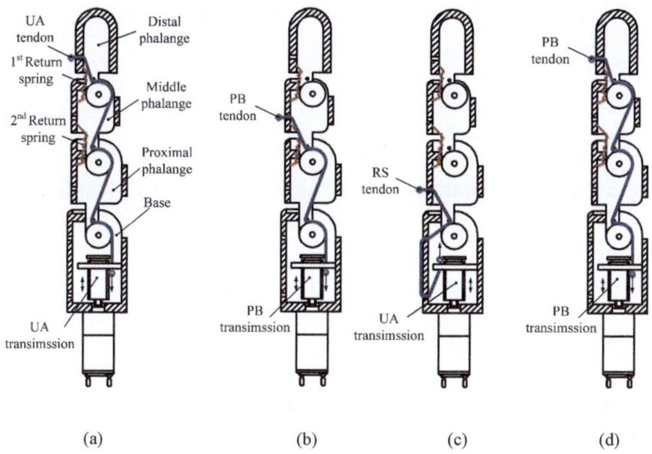

Figure 3. Assembly method of finger's tendons. **(a)** UA tendon; **(b)** PB tendon; **(c)** RS tendon; **(d)** another assembly method of PB tendon.

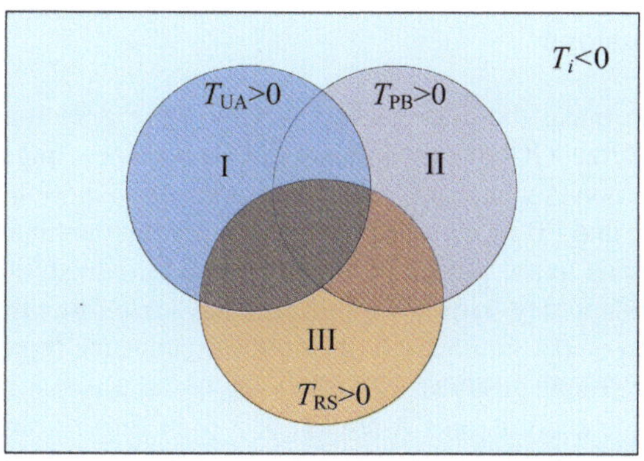

Figure 4. Operation areas of GCUA-T 3-joint finger.

objects, the finger will stop bending. However, when PB tendon is assembled in another method which is shown in Fig. 3b, PB motion will change. In this case, if 2nd middle phalange is blocked by objects, distal phalange can keep rotating around distal joint until it touches objects.

When 2nd motor rotates back, PB transmission will pull PB tendon up, and PB tendon will relax, after that, return springs will drive the top two phalanges to go back to their initial status.

Indeed, tendons can only restrict the movement of the finger in a single-direction, so that if one tendon relaxes, it will not have effect on other tendons, in this way, UA motion and PB motion can proceed independently. Assume T_{UA}, T_{PB}, and T_{RS} as the driving torques of UA tendon, PB tendon and RS tendon with regard to proximal joint, and suppose that when $T_i > 0$, the tendon will tension, on the contrary, when $T_i < 0$, the tendon will relax, the operation areas can be given, which is shown in Fig. 4.

In area I, since $T_{UA} > 0$, $T_{RS} < 0$, and $T_{PB} < 0$, both RS tendon and PB tendon relax, UA tendon tensions, so that UA

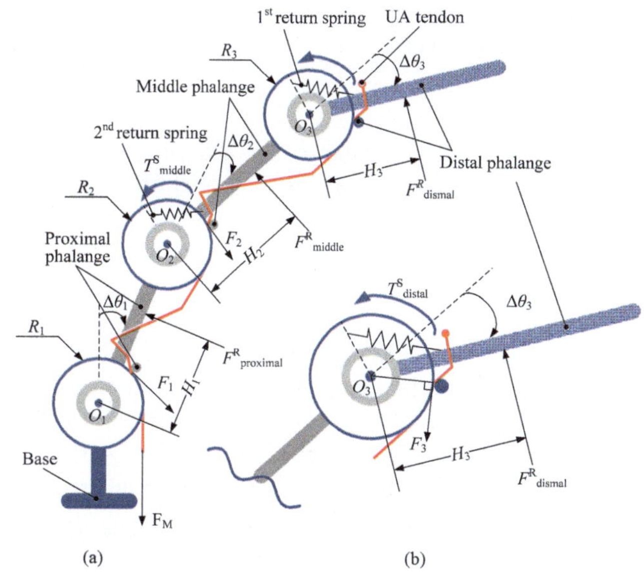

Figure 5. Force analysis of GCUA-T finger when UA motion proceeds. **(a)** Force analysis on the whole finger; **(b)** Force analysis on distal phalange.

motion proceeds and the finger is flexed; In area II, since $T_{UA} < 0$, $T_{RS} > 0$, and $T_{PB} > 0$, both PB tendon and RS tendon tension, UA tendon relaxes, so that PB motion proceeds and the finger changes its initial gesture; In area III, since $T_{UA} < 0$, $T_{RS} > 0$, and $T_{PB} < 0$, both UA tendon and PB tendon relax, RS tendon tensions, so that UA motion proceeds and the finger is extended.

5 Force analysis of GCUA-T finger

Force analysis on GCUA-T finger is given when UA motion proceeds, which is shown in Fig. 5. When 1st motor rotates forward, UA transmission pulls UA tendon to bend the finger's phalanges. Firstly, only consider force analysis on distal phalange. When distal phalange touches objects, the objects will stop distal phalange from rotating forward with F^R_{distal}. Let the torque of F^R_{distal} with regard to distal joint O_3 be T^R_{distal}, which can be expressed as:

$$T^R_{distal} = F^R_{distal} H_3 \tag{1}$$

where H_3 is the force arm of F^R_{distal} with regard to distal joint O_3.

Considering the physical property of UA tendon, the tendon's tension has the same magnitude in any point of the whole tendon, i.e.

$$F_3 = F_M \tag{2}$$

UA tendon drives distal phalange to bend with F_3 which is resulting force of UA tendon's tension F_M. In this case, the driving torque T^{M}_{distal} caused by UA tendon with regard to distal joint O_3, which can be given by

$$T^{M}_{distal} = F_M R_3 \tag{3}$$

Let the angle between distal phalange and 2nd middle phalange be $\Delta\theta_3$, the torque T^{S}_{distal} caused by 1st return spring with regard to distal joint is obtained as follows,

$$T^{S}_{distal} = -K_1 \Delta\theta_3 \tag{4}$$

where K_1 is the stiffness of 1st return spring.

Force analysis on other phalange is similar with that on distal phalange, when 1st middle phalange and 2nd middle phalange are blocked by objects, the objects will prevent them from bending with $F^{R}_{proximal}$ and F^{R}_{middle}, the force arm of $F^{R}_{proximal}$ with regards to proximal joint O_1 is H_1, similarly the force arm of F^{R}_{middle} with regards to middle joint O_2 is H_2, the following equation can be obtained,

$$\boldsymbol{T}^{R} = J_P \boldsymbol{F}^{R} \tag{5}$$

where $\boldsymbol{T}^{R}$ is the external resisting torque vector, $\boldsymbol{F}^{R}$ is the external resisting force vector, i.e.

$$\boldsymbol{F}^{R} = \begin{bmatrix} -F^{R}_{proximal} \\ -F^{R}_{middle} \\ -F^{R}_{dismal} \end{bmatrix} \tag{6}$$

$$J_P = \begin{bmatrix} H_1 & 0 & 0 \\ 0 & H_2 & 0 \\ 0 & 0 & H_3 \end{bmatrix} \tag{7}$$

When UA transmission keeps still, if external force drive 1st middle phalange to bend, there will be a resisting torque T^{R}_{RS} caused by RS tendon. Let the angle between the base and 1st middle phalange be $\Delta\theta_1$, the angle between 1st middle phalange and 2nd middle phalange be $\Delta\theta_2$, the following equation can be obtained,

$$\boldsymbol{T}^{S} = \begin{bmatrix} -T^{R}_{RS} \\ -K_2\Delta\theta_2 \\ -K_1\Delta\theta_3 \end{bmatrix} \tag{8}$$

where $\boldsymbol{T}^{S}$ is the internal resisting torque vector, K_2 is the stiffness of 2nd return spring.

Considering the principle of tendon mechanism, the relationship among $\Delta\theta_1$, $\Delta\theta_2$, and $\Delta\theta_3$ can be given,

$$\Delta\theta_1 R_1 + \Delta\theta_2 R_2 + \Delta\theta_3 R_3 = \Delta L \tag{9}$$

where R_1, R_2 are the radii of proximal joint and middle joint respectively, and ΔL is the displacement of UA tendon's distal end from initial status to current status.

According to the driving torque caused by UA tendon, the following equation can be given,

$$\boldsymbol{T}^{M} = F_M V \tag{10}$$

where $\boldsymbol{T}^{M}$ is the driving torque vector caused by UA tendon.

$$V = \begin{bmatrix} R_1 \\ R_2 \\ R_3 \end{bmatrix} \tag{11}$$

Considering force balance and torque balance of the whole finger, if the finger keeps still, the following equation must be satisfied,

$$T^{M} + T^{R} + T^{S} = 0 \tag{12}$$

Let $\mathbf{S} = T^{M} + T^{R} + T^{S}$, the grasping stability can be determined by matrix $\mathbf{S}$. When the elements of matrix $\mathbf{S}$ are all more than zero, the finger can grasp objects stably with UA motion. If some elements of matrix $\mathbf{S}$ are less than zero, the finger cannot grasp objects stably only with UA motion. In this case, PB motion will be proceeded to improve the grasping stability of the finger.

6 Conclusions

In order to make the under-actuated hand more dexterous and human-like, GCUA function is utilized into a new finger through tendon mechanisms, which consists of two motions: UA motion and PB motion. With GCUA function, the finger can change its initial gesture with PB motion and grasp objects self-adaptively with UA motion, at the same time, the cooperation of UA motion and PB motion enables the finger to do some simple gestures.

With force analysis on UA tendon of 3-joint finger, some stability requirements are given.

GCUA-T fingers based on the GCUA function develop a new way between dexterous hand and traditional under-actuated hand.

Acknowledgements. This paper was funded by the National Natural Science Foundation of China (No. 50905093), Hi-Tech R&D Program (2007AA04Z258) and the foundation of National Innovation Experiment Program for University Students (No. 091000311).

Edited by: J. L. Herder
Reviewed by: two anonymous referees

References

Birglen, L. and Gosselin, C. M.: On the Force Capability of Underactuated Fingers, Proceedings of the 2003 IEEE International Conference on Robotics and Automation, Taipei, Taiwan, September, 1139–1145, 2003.

Birglen, L. and Gosselin, C. M.: Kinetostatic Analysis of Underactuated Fingers, IEEE T. Robotic. Autom., 20(2), 211–221, 2004.

Birglen, L. and Gosselin, C. M.: Fuzzy Enhanced Control of an Underactuated Finger Using Tactile and Position Sensors, Proceedings of the 2005 IEEE International Conference on Robotics and Automation, Barcelona, Spain, April, 2320–2325, 2005.

Dollar, A. M. and Howe, R. D.: Designing Robust Robotic Graspers for Unstructured Environments, Proceedings of the Robotics: Science and Systems Conference, Workshop on Manipulation for Human Environments, Philadelphia, PA, August 2006.

Dollar, A. M. and Howe, R. D.: The SDM Hand as a Prosthetic Terminal Device: a Feasibility Study, Proceedings of the 2007 IEEE 10th International Conference on Rehabilitation Robotics, Noordwijk, The Netherlands, June, 978–983, 2007.

Guo, G., Gruver, W. A., and Qian, X.: A New Design for a Dexterous Robotic Hand Mechanism, IEEE Contr. Syst. Mag., 12(4), 35–38, 1992.

Guo, G., Qian, X., and Gruver, A.: Multi-function Mechanical Hand with Shape Adaptation, US Patent No. 5378033, 1993.

Jacobsen, S. C., Iversen, E. K., Knutti, D. F., Johnson, R. T., and Biggers, K. B.: Design of the UTAH/M.I.T. Dextrous Hand, Proceedings of the IEEE International Conference on Robotics and Automation, San Francisco, USA, April, 1520–1532, 1986.

Kawasaki, H., Komatsu, T., Uchiyama, K., and Kurimoto, T.: Dexterous Anthropomorphic Robot Hand with Distributed Tactile Sensor: Gifu Hand II, Proceedings of the IEEE International Conference on Systems, Man, and Cybernetics, Tokyo, Japan, October, 782–787, 1999.

Salisbury, J. K. and Craig, J. J.: Articulated Hands: Force Control and Kinematic Issues, The International Journal of Robotics Research, 1(4), 4–17, 1982.

Tan, L., Xie, S., Lin, I., and Lin, T.: Development of a Multifingered Robotic Hand, Proceedings of the 2009 IEEE International Conference on Information and Automation, Zhuhai/Macau, China, June, 1541–1545, 2009.

Zhang, W., Che, D., Chen, Q., and Du, D.: A Dexterous and Self-adaptive Humanoid Robot Hand: Gesture-changeable Under-actuated Hand. The Second Inter Conf on Intelligent Robotics and Applications(ICIRA), 15–19 Dec 2009, Singapore, 515–525, 2009.

Zhang, W., Che, D., Liu, H., Ma, X., Chen, Q., Du, D., and Sun, Z.: Super Under-Actuated Multi-Fingered Mechanical Hand with Modular Self-Adaptive Gear-Rack Mechanism, Ind. Robot, 36(3), 255–262, 2009.

23

Basic principles and aims of model order reduction in compliant mechanisms

M. Rösner and R. Lammering

Institute for Mechanics, Helmut-Schmidt-University/University of the Federal Armed Forces Hamburg, Holstenhofweg 85, 22043 Hamburg, Germany

Abstract. Model order reduction appears to be beneficial for the synthesis and simulation of compliant mechanisms due to computational costs. Model order reduction is an established method in many technical fields for the approximation of large-scale linear time-invariant dynamical systems described by ordinary differential equations. Based on system theory, underlying representations of the dynamical system are introduced from which the general reduced order model is derived by projection. During the last years, numerous new procedures were published and investigated appropriate to simulation, optimization and control. Singular value decomposition, condensation-based and Krylov subspace methods representing three order reduction methods are reviewed and their advantages and disadvantages are outlined in this paper. The convenience of applying model order reduction in compliant mechanisms is quoted. Moreover, the requested attributes for order reduction as a future research direction meeting the characteristics of compliant mechanisms are commented.

1 Introduction

A new approach to develop a feed unit of small machine tools for small workpieces is based on the application of compliant mechanisms. Currently, non-intuitive design and optimization techniques are in progress as well as controlling, measuring and calibration strategies. To describe the mechanical behavior of a feed unit in an accurate way, very large and sparse finite element models arise. This leads to numerical simulations which require an unacceptable amount of time and memory space and motivates the introduction and application of model order reduction (MOR) methods. In this paper, MOR approaches are reviewed taking a first step towards its implemention in compliant mechanisms. To the authors' knowledge, applications and investigations of MOR in CMs are not existent.

In the present case, the feed unit consists of two major parts: (a) the compliant mechanism (CM) is a mechanical structure consisting of flexure hinges and rigid regions, including piezo-electric actuators and (b) the measure and control system supplying appropriate input signals for the mechanical structure via the actuators. Figure 1 illustrates a prototype of a CM appropriate to novel machine tools (the actuators are not shown), as described by Wulfsberg et al. (2010), embedded in the mechatronic system.

The utilization of CMs assuring the trajectory of the tool is constituted by their performance. CMs distinguish from previous standard machine tools being potentially more accurate, better scalable, cleaner, low-maintenance, less noisy and cheaper in manufacturing which makes them particularly suitable for small-scale applications.

2 System theoretical background

Assuming a linear time invariant system (LTI), the performance of the compliant mechanism can be specified by a system of second order ordinary differential equations (ODE) derived from linear finite element discretization, see e.g. Koutsovasilis and Beitelschmidt (2008). This system describes the reaction of the mechanical structure in answer to inputs from the control system and can be written in the time domain by means of the state space form in vectorial representation as:

$$\begin{aligned} \mathbf{M}_e\ddot{\boldsymbol{q}}(t)+\mathbf{D}_e\dot{\boldsymbol{q}}(t)+\mathbf{K}_e\boldsymbol{q}(t) &= \mathbf{B}_e\boldsymbol{u}(t), \\ \boldsymbol{y}(t) &= \mathbf{C}_e\boldsymbol{q}(t), \end{aligned} \tag{1}$$

where $\mathbf{M}_e \in \mathbb{R}^{n\times n}$ mass matrix, $\mathbf{D}_e \in \mathbb{R}^{n\times n}$ damping matrix, $\mathbf{K}_e \in \mathbb{R}^{n\times n}$ stiffness matrix, $\mathbf{B}_e \in \mathbb{R}^{n\times p}$ input matrix with $\boldsymbol{u}(t) \in \mathbb{R}^p$ input vector forming an external force vector, $\mathbf{C}_e \in \mathbb{R}^{m\times n}$

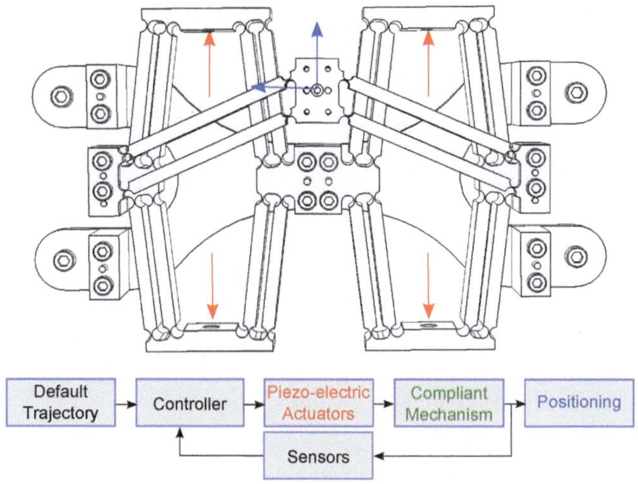

Figure 1. Prototype of a compliant mechanism used in a feed unit for novel machine tools as component of the mechatronic system.

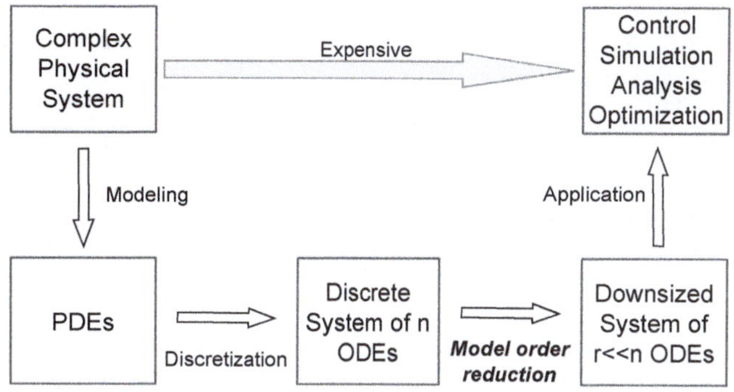

Figure 2. Classification of model order reduction in the simulation process of a dynamical system.

output matrix selecting the $y(t) \in \mathbb{R}^m$ output vector of interest of the $q(t) \in \mathbb{R}^n$ internal state vector corresponding to the structural characteristics of the CM. Moreover, n is the order of the system, also referred to as state space dimension. In case $m = p = 1$ and therefore $\mathbf{B}_e$ and $\mathbf{C}_e$ switch over to vectors as well as $u(t)$ and $y(t)$ to scalars, the systems is called single input – single output (SISO) otherwise multiple input – multiple output (MIMO).

To accurately reproduce the structural behavior, in most instances, the set of ODEs stated in Eq. (1) reaches a numerousness of degrees of freedom (DOF), stated by n in the present case. These large scale systems entail prohibitive computational time and memory space for calculation making them impractical for simulation, optimization, analysis and control. Therefore, MOR techniques are implemented to gain a low dimensional approximation of the high order system. MOR can be arranged in the simulation process, as shown in Fig. 2.

Depending on the reduction method, different representations of the system are required, which are illustrated below.

Several approximation methods rely on first order systems. The transformation of Eq. (1) to a first order descriptor system is realized by applying the state vector $x(t) = [q^T | \dot{q}^T]^T$ leading to the state-space dynamical system

$$\begin{aligned} \mathbf{E}\dot{x}(t) &= \mathbf{A}x(t) + \mathbf{B}u(t), \\ y(t) &= \mathbf{C}x(t), \end{aligned} \tag{2}$$

where $\mathbf{A} \in \mathbb{R}^{n_s \times n_s}$ system matrix, $\mathbf{B} \in \mathbb{R}^{n_s \times p}$ input matrix, $\mathbf{C} \in \mathbb{R}^{m \times n_s}$ output matrix, $\mathbf{E} \in \mathbb{R}^{n_s \times n_s}$ descriptor matrix and $n_s = 2n$. Since some reduction methods are not based on the representations written in Eqs. (1) and (2), respectively, but on the transfer function matrix $\mathbf{H}(s)$. Its derivation is described in the following.

Using the Laplace transformation, the system specified in Eq. (1) is converted from time to frequency domain resulting in an algebraic system of equations given by

$$\begin{aligned} s^2\mathbf{M}_e Q(s) + s\mathbf{D}_e Q(s) + \mathbf{K}_e Q(s) &= \mathbf{B}_e U(s), \\ Y(s) &= \mathbf{C}_e Q(s), \end{aligned} \tag{3}$$

where s is the complex-valued scalar parameter and $Q(s)$, $U(s)$, $Y(s)$ are the Laplace transformed of $q(t)$, $u(t)$, $y(t)$ and provided that homogeneous initial conditions $q(t=0) = \dot{q}(t=0) = 0$ are existent.

The transfer function matrix $\mathbf{H}(s)$ of the system can be specified by combining the two equations in Eq. (3), as

$$Y(s) = \underbrace{\mathbf{C}_e[s^2\mathbf{M}_e + s\mathbf{D}_e + \mathbf{K}_e]^{-1}\mathbf{B}_e}_{\mathbf{H}(s)} U(s), \tag{4}$$

alternatively for the descriptor system in Eq. (2)

$$\mathbf{H}(s) = \mathbf{C}(s\mathbf{E} - \mathbf{A})^{-1}\mathbf{B}, \tag{5}$$

where the transfer function matrix $\mathbf{H}(s)$ has a dimension of $(m \times p)$ in the MIMO case and, accordingly, is a scalar in the SISO case. The transfer function $\mathbf{H}(s)$ describes the correlation between input $U(s)$ and output $Y(s)$, disregarding the internal states of the system.

Due to the fact that most MOR procedures are accomplished by means of projection, the approach is explained by a system represented in Eq. (2). The objective is to approximate the state vector $x(t) \in \mathbb{R}^{n_s}$ to a low-dimensional subspace. This is achieved by the substitution

$$x(t) = \mathbf{V}x_r(t) + \epsilon(t), \tag{6}$$

where $\mathbf{V} \in \mathbb{R}^{n_s \times q}$ is the transformation matrix, $x_r(t) \in \mathbb{R}^q$ the reduced state vector, $\epsilon(t)$ the residual and $q \ll n_s$. Inserting the projection Eq. (6) into the linear system of ODEs in Eq. (2) results in a low-dimensional approximation

$$\begin{aligned} \mathbf{E}\mathbf{V}\dot{x}_r(t) &= \mathbf{A}\mathbf{V}x_r(t) + \mathbf{B}\mathbf{V}u(t) + \epsilon(t), \\ y(t) &= \mathbf{C}\mathbf{V}x_r(t). \end{aligned} \tag{7}$$

Table 1. Characteristics of condensation-based methods.

Advantages	Disadvantages
– stability is preserved	– not automatable
– invariant to system representation	– unnecessarily large reduced order models
– physical information is preserved	– not expandable to nonlinear systems
– independent from the number of the systems inputs and outputs	– no presetting of a minimal dimension of the reduced system possible
– error estimation possible	– computationally intense
– implemented in software (Matlab, ANSYS)	– inappropriate for large scale systems

In general, this system is overdetermined having q unknowns but n_s equations. To solve this problem and find a unique solution, Eq. (7) is pre-multiplied by a second transformation matrix $\mathbf{W} \in \mathbb{R}^{n_s \times q}$ such that $\mathbf{W}^T \epsilon(t) = 0$, the residual is equal to zero and

$$\underbrace{\mathbf{W}^T\mathbf{E}\mathbf{V}}_{\mathbf{E}_r}\dot{x}_r(t) = \underbrace{\mathbf{W}^T\mathbf{A}\mathbf{V}}_{\mathbf{A}_r}x_r(t) + \underbrace{\mathbf{W}^T\mathbf{B}}_{\mathbf{B}_r}u(t), \qquad y(t) = \underbrace{\mathbf{C}\mathbf{V}}_{\mathbf{C}_r}x_r(t). \tag{8}$$

The system in Eq. (8) is called general *reduced order model* (ROM) by projection. The number of inputs $u(t) \in \mathbb{R}^p$ and outputs $y(t) \in \mathbb{R}^m$ remains the same though the order of the system, expressed by the dimension of the reduced system matrices $\mathbf{E}_r \in \mathbb{R}^{q\times q}, \mathbf{A}_r \in \mathbb{R}^{q\times q}, \mathbf{B}_r \in \mathbb{R}^{q\times p}$, $\mathbf{C}_r \in \mathbb{R}^{q\times m}$ and the reduced state vector $x_r(t) \in \mathbb{R}^q$ decreases from n_s to q.

For most of the methods, the aim of model order reduction is to provide the projection matrices $\mathbf{W}$ and $\mathbf{V}$ such that their calculation is computationally efficient as well as automatable and the systems characteristics are preserved. The existence of a predefined error bound is desired in addition to the applicability in large-scale systems with an order up to a few hundred thousand. Details are specified e.g. in Eid (2009) and Rudnyi and Korvink (2006).

3 Reduction methods based on condensation

The reduction of a given system using condensation methods is the most commonly applied reduction method. Applying the time independent transformation matrices $\mathbf{W}^T$ and $\mathbf{V}$ to the first equation in Eq. (1) gives the reduced system

$$\mathbf{M}_r\ddot{x}_r(t) + \mathbf{D}_r\dot{x}_r(t) + \mathbf{K}_r x_r(t) = \mathbf{B}_r x_r(t), \tag{9}$$

with $\mathbf{M}_r = \mathbf{W}^T\mathbf{M}_e\mathbf{V}$, $\mathbf{D}_r = \mathbf{W}^T\mathbf{D}_e\mathbf{V}$, $\mathbf{K}_r = \mathbf{W}^T\mathbf{K}_e\mathbf{V}$ and $\mathbf{B}_r = \mathbf{W}^T\mathbf{B}_e$ where the transformation matrices have to ensure the conservation of the potential and kinetic energy of the system. The reduced damping matrix $\mathbf{D}_r$ may be modelled as linear combination of the mass and stiffness matrices (Rayleigh damping) with $\mathbf{D}_r = \alpha\mathbf{M}_r + \beta\mathbf{K}_r$.

The construction of the transformation matrix $\mathbf{V}$ is carried out either by static, modal or mixed condensation which are explained briefly below, see Siedl (2008), Gasch and Knothe (1989), Bennini (2005) and Gugel (2009). In Table 1, the main advantages and disadvantages are listed for MOR using condensation-based reduction methods.

3.1 Static condensation

Static condensation is based on the partition of the systems degrees of freedom into dependent (MDOF – master degree of freedom, index m) and independent (SDOF – slave degree of freedom, index s) ones in a physical expedient way. This procedure is also the basis for sub-structure modeling and the generation of super elements. For the static case, $\ddot{q}(t) = \dot{q}(t) = 0$, an external force vector is introduced by $\mathbf{B}_e u(t) = F(t)$. The aforementioned partition into MDOF and SDOF applied to Eq. (1) results in

$$\begin{bmatrix}\mathbf{K}_{mm} & \mathbf{K}_{ms}\\ \mathbf{K}_{sm} & \mathbf{K}_{ss}\end{bmatrix}\begin{bmatrix}q_m(t)\\ q_s(t)\end{bmatrix} = \begin{bmatrix}F_m(t)\\ F_s(t)\end{bmatrix}. \tag{10}$$

The second row in Eq. (10) provides the so called static correlation between the MDOF and SDOF given by

$$q_s(t) = -\mathbf{K}_{ss}^{-1}\mathbf{K}_{sm}q_m(t) + \mathbf{K}_{ss}^{-1}F_s(t). \tag{11}$$

One obtains the reduced system by substituting $q_s(t)$ in the first row of Eq. (10) by Eq. (11). Rearranging ensues

$$\mathbf{K}_{red}q_m(t) = F_{red}, \tag{12}$$

where $\mathbf{K}_{red} = [\mathbf{K}_{mm} - \mathbf{K}_{ms}\mathbf{K}_{ss}^{-1}\mathbf{K}_{sm}^{-1}]$ and $F_{red} = F_m(t) - \mathbf{K}_{ms}\mathbf{K}_{ss}^{-1}F_s(t)$. Therewith, the transformation matrices in Eq. (8) have the following structure

$$\mathbf{V} = \begin{bmatrix}\mathbf{I}\\ -\mathbf{K}_{ss}^{-1}\mathbf{K}_{sm}\end{bmatrix}, \mathbf{W} = \begin{bmatrix}\mathbf{I}\\ -\mathbf{K}_{ms}\mathbf{K}_{ss}^{-1}\end{bmatrix}, \tag{13}$$

where $\mathbf{I}$ is the identity matrix. The identification of $\mathbf{K}_{red}$ in this way is exact. In case of approximately reducing the mass matrix $\mathbf{M}_e$ and damping matrix $\mathbf{D}_e$ also with the transformation matrices in Eq. (13) from the static approach, the procedure is called Guyan reduction, as stated by Stelzmann et al. (2002) and Lienemann (2006).

3.2 Modal condensation

Using modal condensation, only a selected number of eigenmodes is taken into account. The involved reduction is based

Table 2. Characteristics of Krylov subspace methods.

Advantages	Disadvantages
– iterative method	– no support in Matlab
– easy to implement	– no global error bound
– suitable for very large scale systems	– order depends on number of inputs/outputs
– computational efficient	– result may depend on systems representation
– stationary exact	– no guarantee for preserving stability
– robust calculation	– expansion point required

on the limitation of the explorative frequency range. Resolving the eigenvalue problem, neglecting damping and external forces, results in the modal matrix $\mathbf{\Phi}$ assembled by the eigenvectors φ_i

$$\mathbf{\Phi} = \begin{bmatrix} \varphi_1, & \varphi_2, & \cdots, & \varphi_l \end{bmatrix} \tag{14}$$

where $\mathbf{\Phi} \in \mathbb{R}^{l \times l}$ is getting downsized by choosing a minor frequency range resulting in the reduced modal matrix $\mathbf{\Phi}_\mathrm{r} \in \mathbb{R}^{l \times k}$. Using an orthogonal projection $\mathbf{W} = \mathbf{V}$ as mentioned by Rickelt-Rolf (2009), the transformation matrices are chosen such that

$$\mathbf{W} = \mathbf{V} = \mathbf{\Phi}_\mathrm{r}. \tag{15}$$

3.3 Mixed condensation

Mixed condensation, known as reduction by Craig and Bampton (1968), combines static and modal reduction methods. In a first step, the system in Eq. (9) is similarly re-sorted with respect to SDOF and MDOF as described before. Secondly, the MDOF are blocked out by setting $\boldsymbol{q}_\mathrm{m} = 0$ resulting in an auxiliary system

$$\mathbf{M}_\mathrm{ss}\ddot{\boldsymbol{q}}_\mathrm{s}(t) + \mathbf{K}_\mathrm{ss}\boldsymbol{q}_\mathrm{s}(t) = 0, \tag{16}$$

of which the eigenmodes φ_i are calculated subsequently. Analogue to the modal condensation, the reduced modal matrix $\mathbf{\Phi}_\mathrm{r}$ is established by selecting certain eigenmodes as described by Siedl (2008). The SDOF is again linked to the MDOF via the static correlation and the transformation matrices are complemented to

$$\mathbf{W} = \mathbf{V} = \begin{bmatrix} \mathbf{\Phi}_\mathrm{r} & \mathbf{\Phi}_\mathrm{s} \\ 0 & \mathbf{I} \end{bmatrix} \tag{17}$$

with $\mathbf{\Phi}_\mathrm{s} = -\mathbf{K}_\mathrm{ss}^{-1}\mathbf{K}_\mathrm{sm}$ and $\mathbf{\Phi}_\mathrm{r} = \begin{bmatrix} \varphi_1, & \varphi_2, & \cdots, & \varphi_k \end{bmatrix}$ as mentioned in Sect. 3.1 and Sect. 3.2, respectively.

4 Krylov subspace methods

Originally, Krylov subspace methods were developed for iteratively solving large and sparse linear systems of equations. Multiple methods are available, for example Generalized-Minimal-Residual-Method (GMRES), Biconjugate-Gradient (BiCG,) Conjugate-Gradien-Square (CGS) and Transpose-Free-Quasi-Minimal-Residual (TFQMR), which can be found e.g. in Meister (2005), Kanzow (2005), Van der Vorst (2003) and Bai (2000) to name only a few.

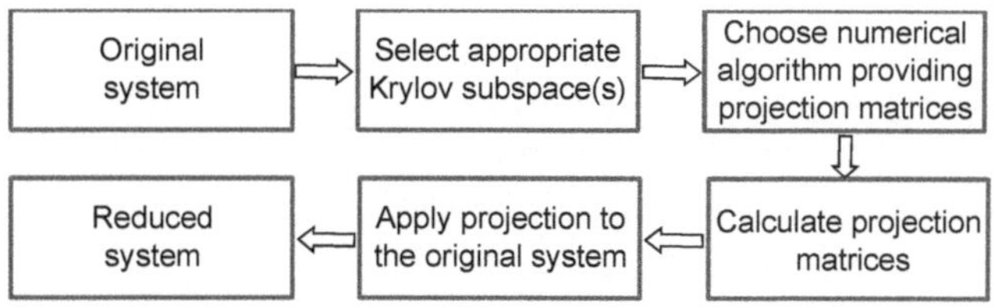

Figure 3. Key steps of dimensional reduction using Krylov subspaces.

In case of MOR, Krylov subspaces are used to form the projection matrices $\mathbf{V}$ and $\mathbf{W}$ in Eq. (8). The general approach is based on the approximation of the transfer behavior of the original system by means of the transfer function $\mathbf{H}(s)$ in the frequency domain given in Eq. (4) and Eq. (5), respectively. MOR using Krylov subspaces is also well-known as moment matching, meaning that a certain number of moments of the reduced and the original system is equal. Hence, the moments of the transfer function and their matching in the sense of an approximation play a decisive role and are described below. Figure 3 illustrates the key steps using Krylov subspaces.

In Table 2, the main advantages and disadvantages are listed for MOR using Krylov subspaces, given in Salimbahrami (2006) and Eid (2009).

4.1 Transfer function and its moments

Considering the transfer function in Eq. (5) for the MIMO case, rearranging results in

$$\mathbf{H}(s) = -\mathbf{C}(\mathbf{I} - s\mathbf{A}^{-1}\mathbf{E})^{-1}\mathbf{A}^{-1}\mathbf{B}. \tag{18}$$

Taking into account the Neumann series, as described by Werner (2009) and Eid (2009),

$$(\mathbf{I} - s\mathbf{A}^{-1}\mathbf{E})^{-1} = \sum_{j=0}^{\infty}(\mathbf{A}^{-1}\mathbf{E}s)^j, \tag{19}$$

one obtains the following Taylor series

$$\mathbf{H}(s) = \sum_{j=0}^{\infty} -\mathbf{C}(\mathbf{A}^{-1}\mathbf{E})^j\mathbf{A}^{-1}\mathbf{B}s^j = \sum_{j=0}^{\infty} -\mathbf{M}_j^0 s^j. \quad (20)$$

This power series is called MacLaurin series, with $\mathbf{M}_j^0 = \mathbf{C}(\mathbf{A}^{-1}\mathbf{E})^j\mathbf{A}^{-1}\mathbf{B}$ called moments of the transfer function and its expansion point being $s = 0$. Depending on the expansion point, the moments and its corresponding MOR schemes are defined as:

- Expansion point $s = 0$ results in a MacLaurin series with its moments defined by

  $$\mathbf{M}_j^{s=0} = \mathbf{C}(\mathbf{A}^{-1}\mathbf{E})^j\mathbf{A}^{-1}\mathbf{B}. \quad (21)$$

  The corresponding MOR procedure is called *Padé-Approximation*.

- Expansion point $s = s_k$ results in a Taylor series with its moments defined by

  $$\mathbf{M}_j^{s=s_k} = \mathbf{C}((\mathbf{A}-s_k\mathbf{E})^{-1}\mathbf{E})^j(\mathbf{A}-s_k\mathbf{E})^{-1}\mathbf{B}. \quad (22)$$

  The corresponding MOR procedure is called *Shifted Padé-Approximation*, *Rational Interpolation* or *Multi-point Padé-Approximation*.

- Expansion point $s \to \infty$ results in a Markov series with its moments defined by

  $$\mathbf{M}_j^{s\to\infty} = \mathbf{C}(\mathbf{E}^{-1}\mathbf{A})^j\mathbf{E}^{-1}\mathbf{B}. \quad (23)$$

  The corresponding MOR procedure is called *Partial Realisation*.

The users choice of the expansion point concerning its location and number in case of the Shifted Padé-Approximation, which is also suitable for multiple expansion points, affects the approximating characteristics. In the context of MOR, Padé-Approximation matches the behavior in the low frequency range whereas Partial Realisation fullfils in the high frequency range and Rational Interpolation in the user specified one. The usage of moment matching in sparse, large scale systems is due to the simplicity of the involved operations to compute these moments, namely matrix vector multiplication and matrix inversion. Details may be found in Eid (2009), Lehner and Eberhard (2006), Grimme (1997) and Salimbahrami (2006).

4.2 Krylov subspaces and moment matching

To provide the projection matrices $\mathbf{V}$ and $\mathbf{W}$, Krylov subspaces are employed to match a certain number of the previously defined moments. A Krylov subspace is spanned by a succession of vectors as

$$\mathbf{K}_q(\mathbf{A},\boldsymbol{b}) = span\{\boldsymbol{b},\mathbf{A}\boldsymbol{b},\mathbf{A}^2\boldsymbol{b},...,\mathbf{A}^{q-1}\boldsymbol{b}\}, \quad (24)$$

where $\mathbf{A} \in \mathbb{R}^{n\times n}$ is a matrix, $\boldsymbol{b} \in \mathbb{R}^n$ is called starting vector and $\boldsymbol{b},\mathbf{A}\boldsymbol{b},\mathbf{A}^2\boldsymbol{b},...,\mathbf{A}^{q-1}\boldsymbol{b}$ are the generated basic vectors spanning the q-th Krylov subspace. The first linearly independent basic vectors form the basis of this Krylov subspace. In case multiple starting vectors have to be considered, a q-th order block Krylov subspace can be rendered precisely by

$$\mathbf{K}_q(\mathbf{A},\mathbf{B}) = span\{\mathbf{B},\mathbf{A}\mathbf{B},\mathbf{A}^2\mathbf{B},...,\mathbf{A}^{q-1}\mathbf{B}\}, \quad (25)$$

where $\mathbf{A} \in \mathbb{R}^{n\times n}$, $\mathbf{B} = [\boldsymbol{b}_1,...,\boldsymbol{b}_p] \in \mathbb{R}^{n\times p}$ is the starting matrix containing the p linearly independent starting vectors. If $p = 1$, one obtains the standard Krylov subspace in Eq. (24) with the starting vector $\boldsymbol{b}$.

For the reduced order system in Eq. (8), the basic vectors of a suitable Krylov subspace can be utilized to find the projection matrices $\mathbf{V}$ and $\mathbf{W}$. An explicit calculation of the moments to match is not neccessary. As carried out in Koutsovasilis (2009) and Salimbahrami (2006) an input block Krylov subspace $\mathbf{K}_{q1}(\mathbf{A}^{-1}\mathbf{E},\mathbf{A}^{-1}\mathbf{B})$ and output block Krylov subspace $\mathbf{K}_{q2}(\mathbf{A}^{-T}\mathbf{E}^T,\mathbf{A}^{-T}\mathbf{C}^T)$ are to be used for an expansion point $s = 0$ to match some moments in Eq. (21).

For $s = s_k$, the input block Krylov subspace is $\mathbf{K}_{q3}((\mathbf{A}-s_k\mathbf{E})^{-1}\mathbf{E},(\mathbf{A}-s_k\mathbf{E})^{-1}\mathbf{B})$ and the output block Krylov subspace $\mathbf{K}_{q4}((\mathbf{A}-s_k\mathbf{E})^{-T}\mathbf{E}^T,(\mathbf{A}-s_k\mathbf{E})^{-T}\mathbf{C}^T)$. If the colmuns of the matrix $\mathbf{V}$ form a basis of $\mathbf{K}_{q1}$ and $\mathbf{K}_{q3}$, respectively, and the matrix $\mathbf{W}$ is chosen such that $\mathbf{A}_r = \mathbf{W}^T\mathbf{A}\mathbf{V}$ is non-singular, then the first $\frac{q1}{m}$ and accordingly $\frac{q3}{m}$ moments match. A typical choice is $\mathbf{W} = \mathbf{V}$, named Galerkin (orthogonal) projection, otherwise $\mathbf{W} \neq \mathbf{V}$ Petrov-Galerkin (oblique) projection. Using only the input Krylov subspace is known as one-sided Krylov subspace method. Two-sided methods applying input and output Krylov subspaces match $\frac{q1}{m} + \frac{q2}{m}$ and accordingly $\frac{q3}{m} + \frac{q4}{m}$ moments.

To calculate the desired projection matrices $\mathbf{V}$ and $\mathbf{W}$, a wide range of numerical, mostly iterative algorithms is avaiable. The most popular ones are Arnoldi algorithm, two-sided Arnoldi algorithm and Lanczos algorithm which are not further investigated in this paper. Details and further algorithms can be found in Bechtold et al. (2007), Meister (2005), Grimme (1997), Krohne (2007), Salimbahrami (2006), Antoulas et al. (2001), Eid (2009) and Bai (2002).

5 Control theory methods based on singular value decomposition

Methods using the singular value decomposition (SVD) aim for an approximation of matrices by lower ranked matrices, as carried out by Hackbusch (2009). The reduced order model (ROM) is found by deleting those system states that are both least controllable and observable. Figure 4 gives an overview of the key steps using singular value decomposition. In Table 3, the main advantages and disadvantages are listed for MOR using SVD methods.

Table 3. Characteristics of SVD methods.

Advantages	Disadvantages
– invariant to representation	– only for small systems
– applicable to nonlinear systems	– new methods may have convergence troubles
– comprising all relevant system parts	– computationally expensive (Lyapunov equations)
– existence of an error bound	
– preservation of stability	
– fully automatable	
– SVD supported in Matlab	

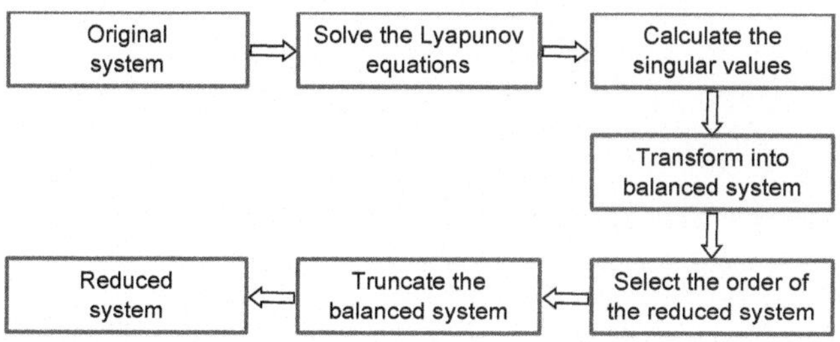

Figure 4. Key steps of dimensional reduction using singular value decomposition (SVD).

5.1 Controllability and observability of a system

Considering the system in Eq. (2) represented by

$$\dot{x}(t) = \hat{\mathbf{A}}x(t) + \hat{\mathbf{B}}u(t),$$
$$y(t) = \mathbf{C}x(t), \tag{26}$$

the two important control theory quantaties controllability $\mathcal{P}$ and observability $\mathcal{Q}$ can be introduced, as done by Lienemann (2006), via the Gramians

$$\mathcal{P} = \int_0^\infty e^{\hat{\mathbf{A}}t}\hat{\mathbf{B}}\hat{\mathbf{B}}^T e^{\hat{\mathbf{A}}^T t} dt \tag{27}$$

and

$$\mathcal{Q} = \int_0^\infty e^{\hat{\mathbf{A}}^T t}\mathbf{C}^T\mathbf{C}e^{\hat{\mathbf{A}}t} dt. \tag{28}$$

The concept of controllability examines how the states $x(t)$ of a system are linked to its inputs $u(t)$ whereas observability deals with the connection between the states $x(t)$ and its outputs $y(t)$, as delineated by Eid (2009).

Under certain conditions, described e.g. by Bechtold et al. (2007) and Koutsovasilis (2009), the Gramians can be found by solving the two system related Lyapunov equations given by

$$\hat{\mathbf{A}}\mathcal{P} + \mathcal{P}\hat{\mathbf{A}}^T + \hat{\mathbf{B}}\hat{\mathbf{B}}^T = 0,$$
$$\hat{\mathbf{A}}^T\mathcal{Q} + \mathcal{Q}\hat{\mathbf{A}} + \mathbf{C}^T\mathbf{C} = 0. \tag{29}$$

5.2 The singular value decomposition

As known from linear algebra and described by Gugel (2009) and Hackbusch (2009), any matrix $\mathbf{A} \in \mathbb{R}^{m\times n}$ can be decomposed into the product of three matrices, more precisely an orthogonal matrix $\mathbf{U}$, a diagonal matrix $\mathbf{\Sigma}$ and the transpose of an orthogonal matrix $\mathbf{V}$ such that

$$\mathbf{A} = \mathbf{U}\mathbf{\Sigma}\mathbf{V}^T = \sum_{i=1}^{k} \sigma_i u_i v_i^T \tag{30}$$

where the matrix

$$\hat{\mathbf{\Sigma}} = \begin{bmatrix} \sigma_1 & \cdots & 0 \\ \vdots & \ddots & \vdots \\ 0 & \cdots & \sigma_k \end{bmatrix}, \mathbf{\Sigma} = \begin{bmatrix} \hat{\mathbf{\Sigma}} & 0 \\ 0 & 0 \end{bmatrix} \in \mathbb{R}^{m\times n} \tag{31}$$

comprises the singular values $\sigma_i = \sqrt{\lambda_i(\mathbf{A}^T\mathbf{A})} \geq \sigma_{i+1} \geq 0$ on its diagonal in descending order with rank($\mathbf{A}$) $=k$ and $k = \min\{m,n\}$. The left singular vectors $\mathbf{U} = [u_1\ u_2 \ldots u_m] \in \mathbb{R}^{m\times m}$ are the orthonormal vectors of $\boldsymbol{AA}^T$, the right singular vectors $\mathbf{V} = [v_1\ v_2 \ldots v_n] \in \mathbb{R}^{n\times n}$ are the orthonormal vectors of $\mathbf{A}^T\boldsymbol{A}$. The term in Eq. (30) is called a singular value decomposition (SVD) of A and provides a basis for finding a lower ranked matrix $\mathbf{A}_r$ which best approximates $\mathbf{A}$ such that

$$\mathbf{A}_r = \sum_{i=1}^{r} \sigma_i u_i v_i^T. \tag{32}$$

The task is to find an optimal approximation in a certain system norm, eg. the spectral norm ($\mathcal{H}_2$-norm), as specified by Antoulas and Sorensen (2001), Hackbusch (2009) and Dahmen and Reusken (2008), via the Schmidt-Mirsky theorem which results in the minimisation of the approximation error such that

$$\min_{\text{rank}(\mathbf{A}_r)\leq\text{rank}(\mathbf{A})} \|\mathbf{A} - \mathbf{A}_r\|_{\mathcal{H}_2} = \sigma_{k+1}(\mathbf{A}). \tag{33}$$

In the following, the most commonly used SVD procedures are specified, namely balanced truncation approximation, singular perturbation approximation and Hankel norm approximation. They distinguish in the rating of the approximation error and the procedure of its minimization, see Benner et al. (2004).

5.3 Balanced truncation approximation

In case of a stable system, balanced truncation approximation (BTA) assumes that the system is balanced first and afterwards truncated. Based on the system representation given in Eq. (26), the balanced partition of the system matrices and state vector can be quoted as

$$\begin{bmatrix}\dot{x}_1(t)\\ \dot{x}_2(t)\end{bmatrix}=\begin{bmatrix}\hat{\mathbf{A}}_{11} & \hat{\mathbf{A}}_{12}\\ \hat{\mathbf{A}}_{21} & \hat{\mathbf{A}}_{22}\end{bmatrix}\begin{bmatrix}x_1(t)\\ x_2(t)\end{bmatrix}+\begin{bmatrix}\hat{\mathbf{B}}_1\\ \hat{\mathbf{B}}_2\end{bmatrix}\mathbf{u}(t)$$
$$y(t)=\begin{bmatrix}\hat{\mathbf{C}}_1 & \hat{\mathbf{C}}_2\end{bmatrix}\begin{bmatrix}x_1(t)\\ x_2(t)\end{bmatrix} \tag{34}$$

with $\hat{\mathbf{A}}_{11}\in\mathbb{R}^{r\times r}$, $\hat{\mathbf{B}}_1\in\mathbb{R}^{r\times m}$ and $\hat{\mathbf{C}}_1\in\mathbb{R}^{p\times r}$. At all times, a balancing transformation is feasible for minimal order systems. The McMillan degree of the system is specified as the order of any minimal state-space realization of the transfer-function matrix, as stated by Sasane (2002) and Baur and Benner (2008).

The singular values are the eigenvalues of the observability $\mathcal{P}$ and controllability $\mathcal{Q}$ Gramians: $\sigma_i=\sqrt{\lambda_i(\mathcal{PQ})}$ which are called Hankel singular values (HSV) of $\mathbf{\Sigma}$, e.g. described by Gugercin and Antoulas (2004) and Antoulas and Sorensen (2001). The Gramians are equal and diagonal. The balanced truncation approximation is predicted on transforming the state-space-system into a balanced realization. Those state variables that are least observable and controllable, and therefore related to the smallest Hankel singular values, are truncated with respect to the error bound between original and reduced transfer function

$$\|\mathbf{H}(s)-\mathbf{H}(s)_{\mathrm{r}}\|_{\mathcal{H}_\infty}\leq 2\sum_{r+1}^{m}\sigma_i. \tag{35}$$

The ROM obtained by BTA is related to

$$\dot{x}_1(t)=\hat{\mathbf{A}}_{11}x_1(t)+\hat{\mathbf{B}}_1\mathbf{u}(t),$$
$$y(t)=\hat{\mathbf{C}}_1x_1(t). \tag{36}$$

as mentioned by Obinata and Anderson (2000). BTA involves an approximation error in the low-frequency region as mentioned by Bechtold et al. (2007), Obinata and Anderson (2000) and Baur and Benner (2008). For details and aspects in numerical calculation see e.g. Benner and Quintana-ort'i (2004), Bechtold et al. (2007), Koutsovasilis (2009), Antoulas (2005) and Gugercin and Antoulas (2004).

5.4 Singular perturbation approximation

The singular perturbation approximation (SPA) is closely related to the balanced truncation approximation and ensures zero error at zero frequency and is applicable to nonlinear systems. Refering to the expression in Eq. (34) and as explained by Liu and Anderson (1989) and Antoulas et al. (2001) the reduced system is identified by

$$\dot{x}_{\mathrm{r}}(t)=\tilde{\mathbf{A}}x(t)+\tilde{\mathbf{B}}u(t),$$
$$y_{\mathrm{r}}(t)=\tilde{\mathbf{C}}x(t), \tag{37}$$

with the reduced system matrices $\tilde{\mathbf{A}}=\hat{\mathbf{A}}_{11}-\hat{\mathbf{A}}_{12}\hat{\mathbf{A}}_{22}^{-1}\hat{\mathbf{A}}_{21}$, $\tilde{\mathbf{B}}=\hat{\mathbf{B}}_1-\hat{\mathbf{A}}_{12}\hat{\mathbf{A}}_{22}^{-1}\hat{\mathbf{B}}_2$ and $\tilde{\mathbf{C}}=\hat{\mathbf{C}}_1-\hat{\mathbf{C}}_2\hat{\mathbf{A}}_{22}^{-1}\hat{\mathbf{A}}_{21}$. In contrast to BTA, SPA has an approximation error at high frequencies, but it matches at low frequencies, see Obinata and Anderson (2000).

5.5 Hankel norm approximation

The Hankel norm approximation (HNA) employs the so-called Hankel norm which is defined as the maximal Hankel singular value of the system given in Eq. (26):

$$\|\mathbf{H}(s)\|_H=\sqrt{\lambda_{\max}(\mathcal{PQ})}=\sigma_{\max}. \tag{38}$$

The aim is to minimize the approximation error between the original system transfer function $\mathbf{H}(\mathbf{s})$ and the reduced one $\mathbf{H}_{\mathrm{r}}(s)$ with order r via the Hankel norm

$$\|\mathbf{H}(s)-\mathbf{H}_{\mathrm{r}}(s)\|_{\mathcal{H}_2}=\sigma_{r+1}(\mathbf{H}(s))\leq\|\mathbf{H}(s)-\tilde{\mathbf{H}}_{\mathrm{r}}(s)\|_H, \tag{39}$$

as stated by Glover (1984), Green and Limebeer (1995), Bechtold et al. (2007), Benner et al. (2004), Antoulas (2005) and Antoulas and Sorensen (2001) where also the computation of the optimal Hankel norm approximation is performed. In Eq. (39) $\tilde{\mathbf{H}}_{\mathrm{r}}(s)$ denotes all transfer functions of McMillan degree less than or equal to r.

6 Conclusion: benefits and aims for model order reduction in compliant mechanisms

In this work, system theoretical basics are reviewed which constitute the origin for model order reduction (MOR). Three different procedures, namely modal reduction, Krylov subspaces and singular value decomposition (SVD) based methods, are described with their associated characteristics.

The application of MOR can be motivated by challenges arising in the numerical characterisation of compliant mechanisms (CM) which can be traced back to the numerousness of ODEs required to specify their behavior in an accurate way. In most cases, large finite element models arise to capture a detailed representation of the geometrical domain and the dynamical performance of the CM. This circumstance is unfeasible, particularly with regard to further investigations such as simulation, optimization and control. For this reason, MOR in CMs seems to be a purposeful approach to generate efficient models and would benefit the design and development as well as analysis process of CMs.

The specific feature of CMs exists in their mechanical topology as they consist of rigid parts connected by flexure hinges. This characteristic has effects on the potentiality of MOR and requires further investigations to provide a suitable algorithm. Especially, the algorithm has to notably incorporate the distinct differences between the compliant and rigid parts of the structure with regard to their mechanical properties and influence on the systems transfer behavior. Moreover, having a small error between original and reduced system implying a global error bound, the procedure should preserve the passivity as well as stability, has to be automatable, stable and computational efficient also for high order systems.

In future work, different MOR schemes applied to CM will be investigated and reasonable combinations among them will be analysed to benefit from their advantages. Furthermore, a specific adjustment and upgrading is desired. Out of this, the aim is to provide a reduction procedure meeting the special circumstances of CMs.

Acknowledgements. This work is supported by the German Research Foundation (DFG) as part of the research project "SPP 1476 – Small machine tools for small workpieces" which is gratefully acknowledged.

Edited by: J. A. Gallego Sánchez
Reviewed by: two anonymous referees

References

Antoulas, A.: Approximation of large-scale dynamical systems, Society for Industrial Mathematics, 2005.

Antoulas, A. and Sorensen, D.: Approximation of large-scale dynamical systems: An Overview, Int. J. Appl. Math. Comput. Sci, 11, 1093–1121, 2001.

Antoulas, A., Sorensen, D., and Gugercin, S.: A survey of model reduction methods for large-scale systems, in: Structured matrices in mathematics, computer science, and engineering: proceedings of an AMS-IMS-SIAM joint summer research conference, Vol. 280, p. 193, 2001.

Bai, Z.: Templates for the solution of algebraic eigenvalue problems, Society for Industrial Mathematics, 2000.

Bai, Z.: Krylov subspace techniques for reduced-order modeling of large-scale dynamical systems, Appl. Numer. Math., 43, 9–44, 2002.

Baur, U. and Benner, P.: Gramian-Based Model Reduction for Data-Sparse Systems, SIAM J. Sci. Comput., 31, 776–798, 2008.

Bechtold, T., Rudnyi, E., and Korvink, J.: Fast simulation of electro-thermal MEMS: efficient dynamic compact models, Springer Verlag, 2007.

Benner, P. and Quintana-ort'i, E. S.: G.: Parallel model reduction of large-scale linear descriptor systems via Balanced Truncation, in: LECT NOTES COMPUT SC, Proc. 6th Intl. Meeting VECPAR04, 28–30 June 2004, 65–78, 2004.

Benner, P., Quintana-Orti, E., and Quintana-Orti, G.: Computing optimal Hankel norm approximations of large-scale systems, in: Decision and Control, 2004, CDC. 43rd IEEE Conference on, Vol. 3, 3078–3083, 2004.

Bennini, F.: Ordnungsreduktion von elektrostatisch-mechanischen Finite Elemente Modellen für die Mikrosystemtechnik, Ph.D. thesis, Universitätsbibliothek der Technischen Universität, 2005.

Craig, R. and Bampton, M.: Coupling of substructures for dynamic analysis, AIAA journal, 6, 1313–1319, 1968.

Dahmen, W. and Reusken, A.: Numerik für Ingenieure und Naturwissenschaftler, Springer, 2008.

Eid, R.: Time Domain Model Reduction by Moment Matching, Ph.D. thesis, TU München, 2009.

Gasch, R. and Knothe, K.: Strukturdynamik, Bd. 2, Springer, New York, NY, USA, 1989.

Glover, K.: All optimal Hankel-norm approximations of linear multivariable systems and their L-error bounds, Int. J. Control., 39, 1115–1193, 1984.

Green, M. and Limebeer, D.: Linear robust control, Prentice Hall, 1995.

Grimme, E.: Krylov projection methods for model reduction, Ph.D. thesis, Citeseer, 1997.

Gugel, D.: Ordnungsreduktion in der Mikrosystemtechnik, Ph.D. thesis, Universitätsbibliothek der TU Chemnitz, 2009.

Gugercin, S. and Antoulas, A.: A survey of model reduction by balanced truncation and some new results, Int. J. Control., 77, 748–766, 2004.

Hackbusch, W.: Hierarchische Matrizen, Springer-Verlag Berlin Heidelberg, 2009.

Kanzow, C.: Numerik linearer Gleichungssysteme: direkte und iterative Verfahren, Springer, 2005.

Koutsovasilis, P.: Model Order Reduction in Structural Mechanics Coupling the Rigid and Elastic Multi Body Dynamics, Ph.D. thesis, TU Dresden, 2009.

Koutsovasilis, P. and Beitelschmidt, M.: Comparison of model reduction techniques for large mechanical systems, Multibody Syst. Dyn., 20, 111–128, 2008.

Krohne, K.: Order reduction of finite-volume models and its application to microwave device optimization, Ph.D. thesis, Swiss Federal Institue of Technology Zurich, 2007.

Lehner, M. and Eberhard, P.: Model Reduction in Flexible Multibody Systems, At.-Autom., 54, 170–177, 2006.

Lienemann, J.: Complexity reduction techniques for advanced MEMS actuators simulation, Ph.D. thesis, Universitätsbibliothek Freiburg, 2006.

Liu, Y. and Anderson, B.: Singular perturbation approximation of balanced systems, Int. J. Control., 50, 1379–1405, 1989.

Meister, A.: Numerik linearer Gleichungssysteme: eine Einführung in moderne Verfahren, Vieweg + Teubner Verlag, 2005.

Obinata, G. and Anderson, B.: Model reduction for control system design, Springer Verlag, 2000.

Rickelt-Rolf, C.: Modellreduktion und Substrukturtechnik zur effizienten Simulation dynamischer, teilgeschädigter Systeme, Ph.D. thesis, TU Braunschweig, 2009.

Rudnyi, E. and Korvink, J.: Model order reduction for large scale engineering models developed in ANSYS, in: Lect. Notes Comput. Sc., 349–356, Springer, 2006.

Salimbahrami, B.: Structure preserving order reduction of large scale second order models, Ph.D. thesis, TU München, 2006.

Sasane, A.: Hankel norm approximation for infinite-dimensional systems, Springer, 2002.

Siedl, D.: Simulation des dynamischen Verhaltens von Werkzeugmaschinen während Verfahrbewegungen, Herbert Utz Verlag, 2008.

Stelzmann, U., Groth, C., and Müller, G.: FEM für Praktiker-Band 2: Strukturdynamik, Expert Verlag, 2002.

Van der Vorst, H.: Iterative Krylov methods for large linear systems, Cambridge University Press, 2003.

Werner, D.: Funktionalanalysis, Springer, 2009.

Wulfsberg, J., Redlich, T., and Kohrs, P.: Square Foot Manufacturing: a new production concept for micro manufacturing, Prod. Engineer., 4, 75–83, 2010.

24

Kinematic design of a finger abduction mechanism for an anthropomorphic robotic hand

L.-A. A. Demers and C. Gosselin

Département de génie mécanique, Université Laval, Québec, QC, G1V 0A6, Canada

Abstract. This paper presents the kinematic design of an abduction mechanism for the fingers of an underactuated anthropomorphic robotic hand. This mechanism will enhance the range of feasible grasps of the underactuated hand without significantly increasing its complexity. The analysis of the link between the index finger and the third finger is first assessed, where the parameters are studied in order to follow the amplitude constraint and to minimize the coordination error. Then, the study of the mechanism joining the third finger and the little finger is summarized. Finally, a prototype of the finger's abduction system is presented.

This paper was presented at the IFToMM/ASME International Workshop on Underactuated Grasping (UG2010), 19 August 2010, Montréal, Canada.

1 Introduction

This paper presents work currently underway aiming at the design of an underactuated hand for an anthropomorphic robot. Underactuation has been chosen to simplify the control of this hand and to minimize its cost in order to facilitate its use by a humanoid robot. The 16-degree-of-freedom hand will therefore need four actuators to control its closing and opening motion as well as its abduction movement. The analyses of underactuation in the fingers and thumb were already presented (Demers and Gosselin, 2009, 2010). In order to maximize the range of feasible grasps, a mechanism for the abduction of the fingers is developed here. Abduction is the action of pulling a member away from a median axis. According to Boutan and Casoli (2005), a typical grasp, the spherical grip, is associated with maximum separation of fingers "made in relation to the anatomical axis of the hand through the 3rd ray", i.e., the middle finger. Therefore, the abduction mechanism of the fingers should leave the middle finger fixed. Thereby, the thumb could perform a pinch with it when they are in opposition.

2 Prescribed abduction movements

In regard to measuring the magnitude of the spacing between the fingers, no source using a universal goniometer was found by Norkin and White (2009). For their part, Smahel and Klímová (2004) traced the contour of the fingers apart to measure the maximum angle between adjacent fingers. Also, Gurbuz et al. used an electronic inclinometer to measure active abduction of metacarpophalangeal joints (Gurbuz et al., 2006). Since no trend is apparent from these articles, we used the method proposed by Smahel and compared our results with those of other studies. We then concluded that a maximal abduction of 25° between each finger would be sufficient, which represents a range of 50° for the little finger.

It was also noted in the literature (Tubiana and Thomine, 1990; Kapandji, 2005) that the rays, the extensions of the fingers, all converge to a point at the base of the palm when the fingers are fully abducted, as illustrated in Fig. 1. Added to the fact that their width limits the minimum distance between them, it is possible to calculate the position of the axes of rotation, $O_{0,i}$, for the abduction of the metacarpals. Assuming that the distance between the centres of the fingers must be 25.4 mm and that the maximum abduction be 25°, we can calculate the distance from the meeting point p_0 of the rays on the middle finger's axis, that is 54.47mm. These dimensions are shown in Fig. 1. Furthermore, the metacarpophalangeal joints, $O_{1,i}$, in outspread position must be equidistant from the point of intersection of rays, p_0. For fingers of same

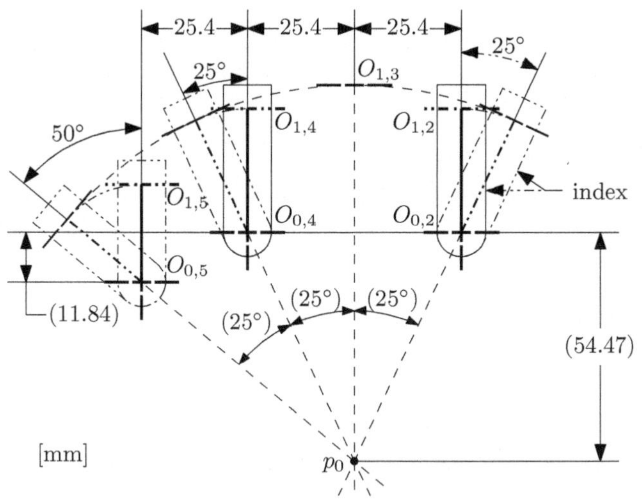

Figure 1. Abduction movement of the metacarpals.

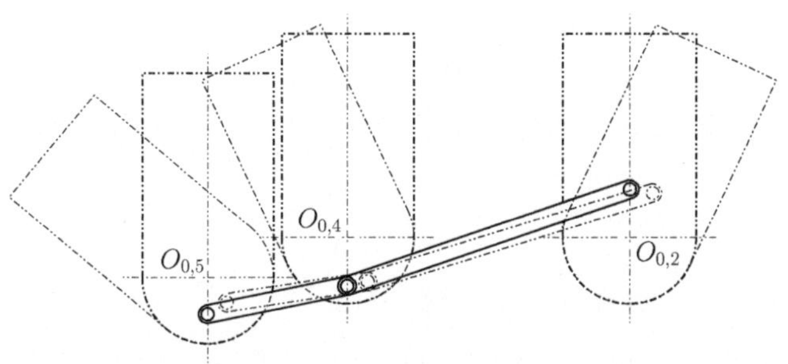

Figure 2. Linkage connecting the metacarpals.

dimension this specification allows the tips to be at the same distance during the closing, which is useful for a spherical grip. This feature is not necessary when making a cylindrical grip where the fingers are adjacent and wrapped around the object.

In order to simplify the architecture of the hand, abduction of the fingers and thumb should be performed with a single actuator. The mechanism of abduction of the thumb was presented in another article (Demers and Gosselin, 2010), but it is important to emphasize that its metacarpal is driven by a worm gear. This input can also be used to drive the abduction of the fingers. The design of the mechanism for the abduction of the fingers therefore consists in developing a link between the index and ring fingers, followed by a link between the ring and little fingers. Since simplicity and stiffness in the connection is required for a stable spherical grasp, two 4-bar mechanisms will be used, as shown in Fig. 2.

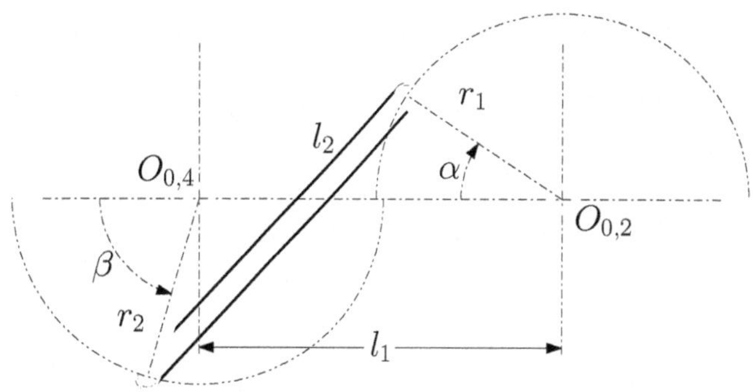

Figure 3. Linkage connecting the metacarpals of the index and ring fingers.

3 Linkage between the index and ring fingers

Since a 4-bar mechanism has been chosen to connect the index and ring fingers, it is now interesting to analyze the relationship between the input and output angles, that is to say between α and β according to Fig. 3. We must thus determine the parameters of the mechanism that will allow the ring finger to follow the abduction of the index, in other words the range of motion $\Delta\beta$ between the initial and final position of the ring finger is equal to the range of motion of the index, $\Delta\alpha$. Also, it is preferable that this abduction be performed with a somewhat synchronized motion of the fingers, so $\delta\beta \approx \delta\alpha$ during the trajectory.

With Fig. 3, it is possible to establish that

$$l_2^2 = (r_2\cos\beta + l_1 - r_1\cos\alpha)^2 + (r_1\sin\alpha + r_2\sin\beta)^2 \quad (1)$$

and using equal radii, $r_1 = r_2 = r$, one can obtain

$$A_\beta\cos\beta + B_\beta\sin\beta = C_\beta \quad (2)$$

where

$$A_\beta = \cos\alpha - \frac{l_1}{r} \quad (3)$$

$$B_\beta = -\sin\alpha \quad (4)$$

$$C_\beta = 1 + \frac{l_1^2 - l_2^2}{2r^2} - \frac{l_1}{r}\cos\alpha. \quad (5)$$

In a typical linkage synthesis application, Eq. (1) is written for a series of prescribed input-output pairs until a sufficient number of conditions is obtained. The equations are then solved using, for instance, nonlinear least squares techniques. However, in this work, only a small set of constraints will first be imposed, thereby leaving some design freedom in the problem. Then, a detailed analysis will be performed in order to explore the possibility of minimizing the coordination error throughout the motion range, without arbitrarily prescribing a large number of input-ouput points.

Equation (2) is readily solved and leads to two values of angle β associated with the input angle α for given link lengths. Since the abduction range of motion of the index and ring fingers should both be equal to 25°, thus $\Delta\beta = \Delta\alpha = 25°$. It is thus necessary to establish a relationship between these amplitudes and ratios l_1/r or l_2/r in order to synthesize a mechanism that will meet this condition.

3.1 Amplitudes of abduction

Using the initial, β_0, medial, β_1, and final, β_2, angles of the ring finger as well as those of the index, α_0, α_1 and α_2, relationships between these angles can be established.

$$\beta_2 - \beta_0 = \Delta\beta = \Delta\alpha = \alpha_2 - \alpha_0 \quad (6)$$

Consequently

$$\beta_2 = \beta_1 + \frac{\Delta\beta}{2} = \beta_1 + \frac{\Delta\alpha}{2} \quad (7)$$

$$\beta_0 = \beta_1 - \frac{\Delta\beta}{2} = \beta_1 - \frac{\Delta\alpha}{2} \quad (8)$$

$$\alpha_2 = \alpha_1 + \frac{\Delta\alpha}{2} \quad (9)$$

$$\alpha_0 = \alpha_1 - \frac{\Delta\alpha}{2}. \quad (10)$$

It is important to note that the median angles α_1 and β_1 do not necessarily occur simultaneously. Indeed, at this point, Eq. (2) is not required to be satisfied for α_1 and β_1. Equation (2) can be rewritten to highlight the constant term for a given mechanism, which leads to

$$\cos(\alpha + \beta) + \frac{l_1}{r}(\cos\alpha - \cos\beta) = 1 + \frac{l_1^2 - l_2^2}{2r^2}. \quad (11)$$

From the initial configuration (α_0,β_0) to the final one (α_2,β_2), the left hand side of Eq. (11) must always give the same result for the mechanism to be assembled. Therefore, one can write

$$\cos(\alpha_0 + \beta_0) + \frac{l_1}{r}(\cos\alpha_0 - \cos\beta_0) = \cos(\alpha_2 + \beta_2) + \frac{l_1}{r}(\cos\alpha_2 - \cos\beta_2). \quad (12)$$

Substituting Eq. (7) to Eq. (10) into Eq. (12), one can obtain

$$0 = -2\sin(\alpha_1 + \beta_1)\sin\Delta\alpha + 2\frac{l_1}{r}(\sin\beta_1 - \sin\alpha_1)\sin\frac{\Delta\alpha}{2}. \quad (13)$$

The latter equation is satisfied if either $\Delta\alpha = 2n\pi$ where $n \in \mathbb{N}$, so the mechanism does not move or makes n turns, or $\sin(\alpha_1 + \beta_1) = 0$ and $\sin\beta_1 - \sin\alpha_1 = 0$ for any l_1/r. The latter conditions are satisfied if and only if $\alpha_1 + \beta_1 = (2n-1)\pi$. Since the four-bar mechanism is assembled using the inverted branch and $\Delta\beta = \Delta\alpha$, then it follows from Eq. (7) to Eq. (10) that

$$\alpha_1 + \beta_1 = \pi \quad (14)$$

$$\alpha_0 + \beta_2 = \pi \quad (15)$$

$$\alpha_2 + \beta_0 = \pi. \quad (16)$$

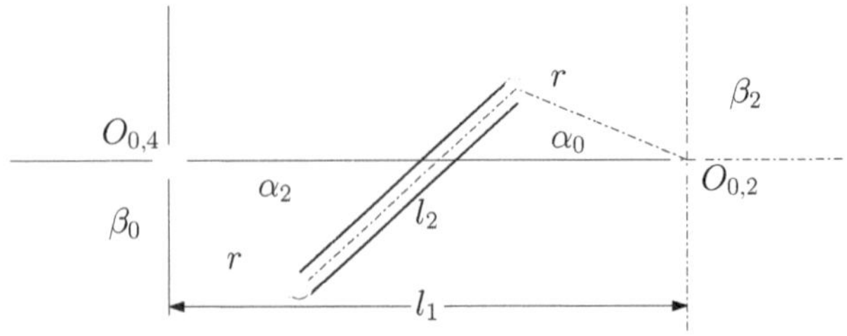

Figure 4. Linkage connecting the metacarpals of the index and ring fingers in its initial position (α_0,β_0).

Equations (15) and (16), described in Fig. 4, can then be substituted into Eq. (1), leading to

$$\left(\frac{l_2}{r}\right)^2 = \left(\frac{l_1}{r} - \cos\alpha_0 - \cos\alpha_2\right)^2 + \left(\sin\alpha_0 + \sin\alpha_2\right)^2 \quad (17)$$

which can be rewritten as

$$\left(\frac{l_2}{r}\right)^2 = \left(\frac{l_1}{r}\right)^2 - 4\frac{l_1}{r}\cos\alpha_1\cos\frac{\Delta\alpha}{2} + 4\cos^2\frac{\Delta\alpha}{2}. \quad (18)$$

For a given range of abduction $\Delta\alpha$, it is then possible to use the above equation to calculate the ratio l_2/r with l_1/r and α_1.

3.2 Minimizing the coordination error

In addition to imposing that the difference between the initial and final position of the ring finger is equal to that of the index, i.e. $\Delta\beta = \Delta\alpha$, it is required that the abduction motion of the fingers be somewhat synchronized, that is to say $\delta\beta \approx \delta\alpha$ over the motion range. It is therefore necessary to minimize the coordination error e_β between the orientation of the ring and index fingers. This can be described as the absolute difference between the angular displacements, i.e.

$$e_\beta = \left|(\beta - \beta_0) - (\alpha - \alpha_0)\right| = \left|\beta - \alpha - (\beta_1 - \alpha_1)\right| \quad (19)$$

and with Eq. (14):

$$e_\beta = \left|\beta - \alpha - \pi + 2\alpha_1\right|. \quad (20)$$

The error varies with the configuration of the mechanism, i.e. according to α and β. The parameters that can be modified in order to minimize the error e_β are l_1/r and α_1, since $\Delta\alpha$ is constrained and α varies according to $\alpha_1 - \Delta\alpha/2 \le \alpha \le \alpha_1 + \Delta\alpha/2$. The minimization of the error can be performed in several ways.

It is first interesting to trace the contours of the coordination error e_β for $\alpha_0 \le \alpha \le \alpha_2$ and various α_1 using a prescribed ratio l_1/r, as shown in Fig. 5. To minimize the error on the interval $\alpha_0 \le \alpha \le \alpha_2$, we can examine the evolution of the mean and the maximum error on this interval. Figure 6 is obtained using again $l_1/r = 5$. It can be observed in Fig. 5 that a minimum can be achieved for the error e_β. The value of α_1 for this minimum is denoted α_g in Fig. 6 and we note that

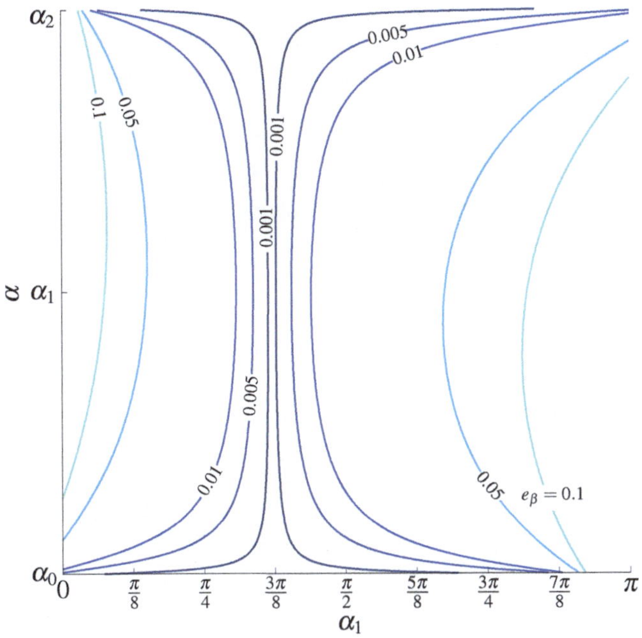

Figure 5. Coordination Error e_β as a function of the median angle α_1 for $\alpha_0 \leq \alpha \leq \alpha_2$ if $l_1/r = 5$.

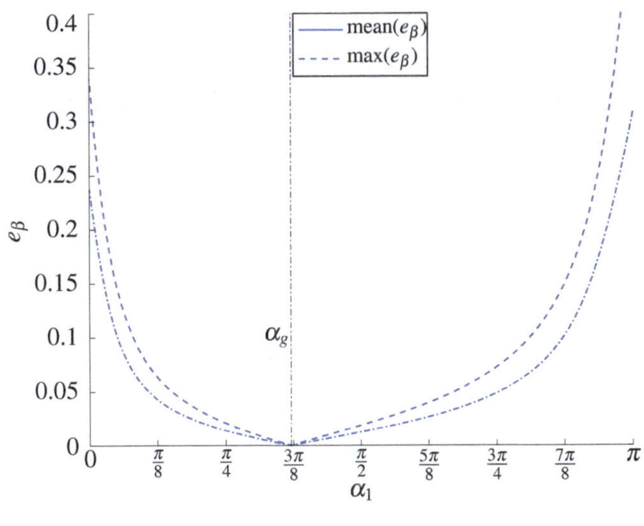

Figure 6. Mean and maximum error in the interval $\alpha_0 \leq \alpha \leq \alpha_2$ as a function of α_1 for $l_1/r = 5$.

the maximum as well as the average error over the interval is then minimized.

Since the trend of coordination error can be followed with its maximum value as well as its mean, it is now possible to analyze the effects of changing α_1 along with l_1/r, as shown in Fig. 7. Low levels contours $(0.01, 0.025, 0.04, 0.055, 0.07, 0.085)$ were added to higher levels $(0.1, 0.2, 0.3, 0.4, 0.5)$ to indicate the rapid growth of the error following the deviation from the ideal value α_g.

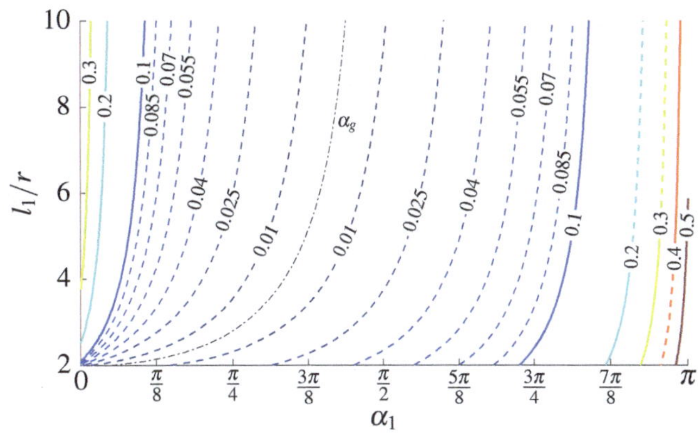

Figure 7. Maximum error in the interval $\alpha_0 \leq \alpha \leq \alpha_2$ as a function of α_1 and l_1/r.

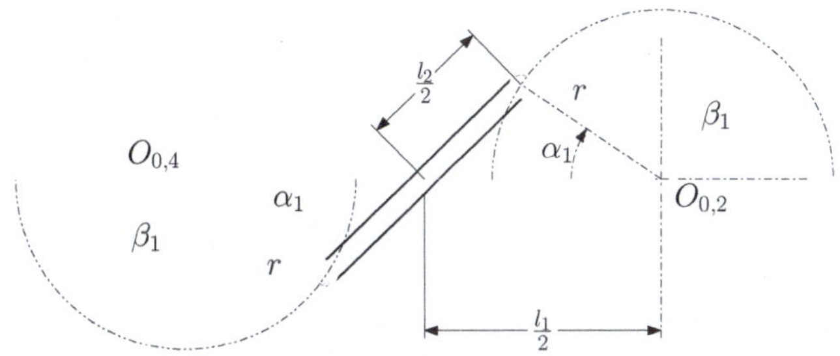

Figure 8. Linkage between the index finger and ring finger if the middle positions occur simultaneously.

Assuming that the median angles α_1 and β_1 occur simultaneously and using Eq. (14), we can conclude that the centre of the link of length l_2 is located at the centre point between the two axes of rotation as sketched in Fig. 8. It is then possible to deduce the angle α_1.

$$\cos\alpha_1 = \frac{1 + \left(\frac{l_1}{2r}\right)^2 - \left(\frac{l_2}{2r}\right)^2}{l_1/r} \tag{21}$$

Using Eq. (21), one can factor Eq. (18) into two terms to emphasize its roots

$$0 = \left[\left(\frac{l_1}{r}\right)^2 - \left(\frac{l_2}{r}\right)^2 - 4\cos\frac{\Delta\alpha}{2}\right]\left[1 - \cos\frac{\Delta\alpha}{2}\right] \tag{22}$$

where we find again the trivial case $\Delta\alpha/2 = 2n\pi$. Besides, it is obvious that an equation to calculate l_2/r can be obtained from the first term, which gives:

$$\frac{l_2}{r} = \sqrt{\left(\frac{l_1}{r}\right)^2 - 4\cos\frac{\Delta\alpha}{2}} \tag{23}$$

and we can finally change Eq. (21) to calculate the angle α_1 as a function of $\Delta\alpha/2$ and l_1/r.

$$\cos\alpha_1 = \frac{1 + \cos\frac{\Delta\alpha}{2}}{l_1/r} \tag{24}$$

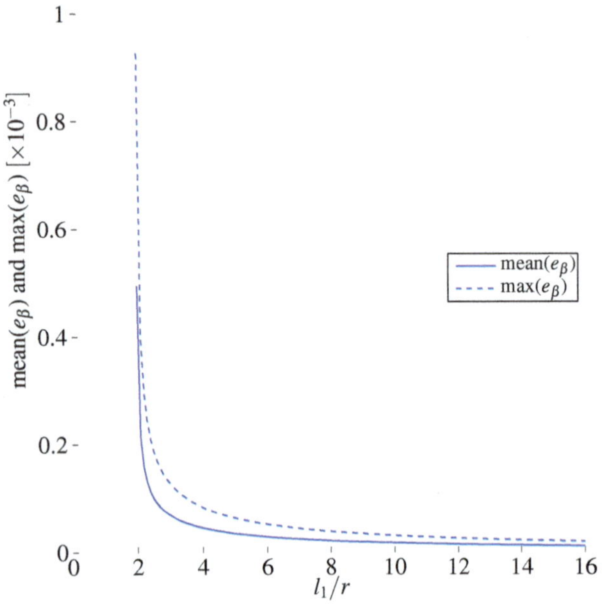

Figure 9. Mean and maximum error in the interval $\alpha_0 \le \alpha \le \alpha_2$ according to l_1/r.

The median angle α_1 obtained must be used if one wants to reduce the coordination error between the orientation of the ring finger and index finger. It is this angle, i.e. α_g, which is indicated on Fig. 6 and Fig. 7. Indeed, if we use $\Delta\alpha = 25°$ and $l_1/r = 5$, we obtain $\alpha_1 \approx 1.164 \approx 3\pi/8$, which is the value of α_g already found for these parameters.

Since Eq. (24) gives a relationship between the two parameters of the optimization, minimizing the error e_β on the interval $\alpha_0 \le \alpha \le \alpha_2$ depends now only on the ratio l_1/r. It is then possible to obtain Fig. 9 by tracing the evolution of the mean and the maximum error following this parameter. It is obvious that for values of $l_1/r > 3$ the maximum and average of the error e_β become negligible (max $(e_\beta) < 0.13 \times 10^{-3}$). Moreover, from $l_1/r = 5$, the slope of the curve max (e_β) remains greater than -0.015×10^{-3}. Since the distance l_1 between the axes of the index and ring fingers is equal to 50.8 mm, we can assume that a ratio $l_1/r = 5.08$ is feasible in practice, because the distance between two joints is then $r = 10$ mm. For guidance, we obtain $l_2/r \approx 4.68$, thus $l_2 \approx 46.8$ mm and $\alpha_1 \approx 1.171$.

The error obtained using this ratio is presented in Fig. 10 on a logarithmic scale. One can observe that the error is minimized over the interval $\alpha_0 \le \alpha \le \alpha_2$. In fact, the magnitude of the error generated is negligible in practice. Moreover, since we have imposed $\Delta\beta = \Delta\alpha$, we obtain $e_\beta = 0$ when $\alpha = \alpha_0$ and $\alpha = \alpha_2$. The error is also zero when $\alpha = \alpha_1$, because this angle is now calculated so that β_1 is achieved simultaneously with α_1. With Eq. (19), it is obvious that the error is null if one uses $\alpha = \alpha_1$ and $\beta = \beta_1$. Finally, the evolution of the angle β as a function of the angle α using $l_1/r = 5$ is plotted in Fig. 11.

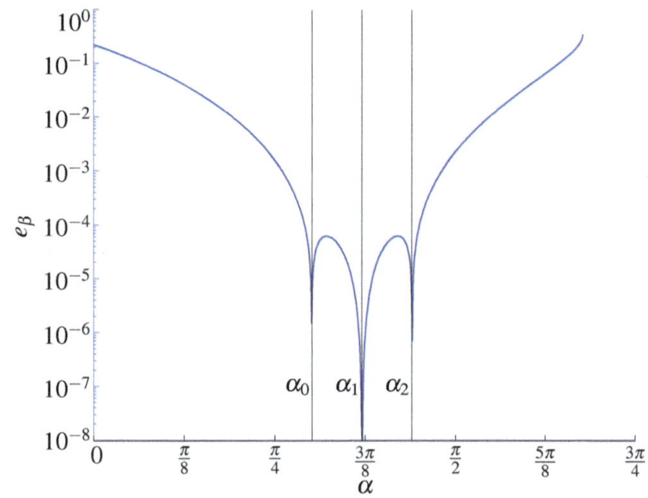

Figure 10. Error e_β as a function of α with $l_1/r = 5$.

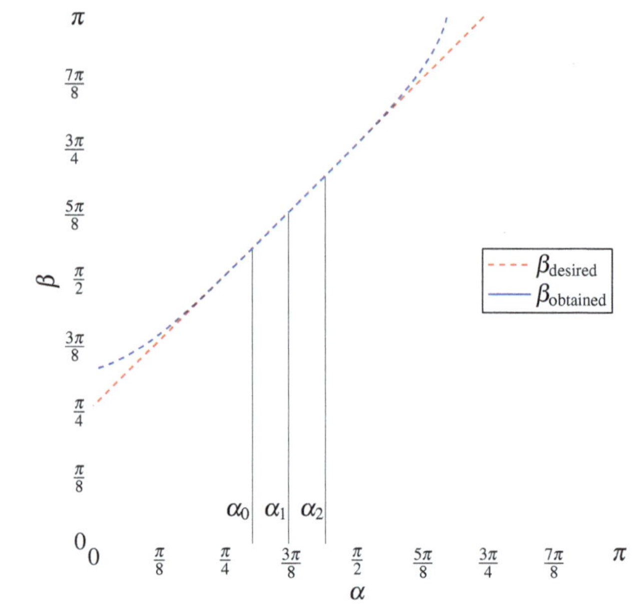

Figure 11. Angle β desired and obtained as a function of angle α with $l_1/r = 5$.

4 Linkage between the ring finger and the little finger

Since the parameters of the linkage between the index and ring fingers are fixed, it is now possible to analyze the linkage between the ring and little fingers, as shown in Fig. 12.

By following the same approach used to analyze the four-bar mechanism connecting the index and ring fingers, it is possible to determine from Fig. 12 that

$$l_4^2 = (r_3\cos\gamma + l_3 - r_2\cos\beta)^2 + (r_3\sin\gamma + l_5 - r_2\sin\beta)^2 \tag{25}$$

which yields

$$A_\gamma\cos\gamma + B_\gamma\sin\gamma = C_\gamma \tag{26}$$

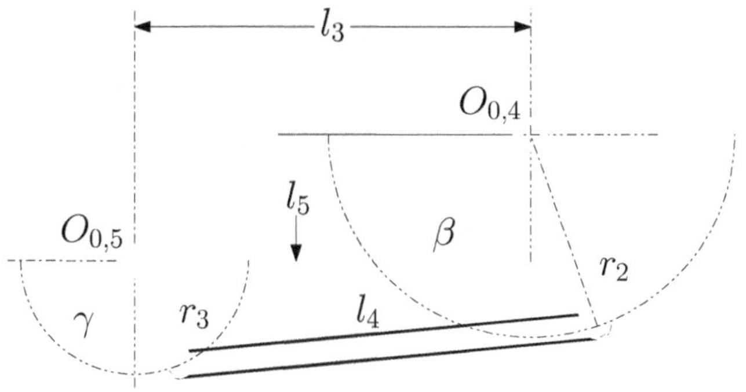

Figure 12. Linkage between the ring finger and the little finger metacarpals.

with

$$A_\gamma = 2r_3(r_2\cos\beta - l_3) \tag{27}$$
$$B_\gamma = 2r_3(r_2\sin\beta - l_5) \tag{28}$$
$$C_\gamma = r_2^2 + r_3^2 + l_3^2 + l_5^2 - l_4^2 - 2r_2(l_3\cos\beta + l_5\sin\beta). \tag{29}$$

Equation (26) is readily solved for γ if all other parameters are given. Since β_1 and r_2 of this linkage are equal to β_1 and r_2 of the previous linkage in order to simplify the mechanism, it is preferable to use links of different lengths in order to keep parameter r_3 for the optimization.

Again, the median angle γ_1 giving the lowest error is first established, which will subsequently be used to determine r_3 and l_4. Since the amplitude of the range of motion for the little finger is 50°, we have $\Delta\gamma = 2\Delta\beta = 50°$.

4.1 Amplitudes of abduction

Initial, γ_0, medial, γ_1, and final, γ_2, angles of the little finger are assumed to be similar to those of the ring finger, $\beta_0, \beta_1, \beta_2$. This allows to establish relations similar to those derived in the previous section:

$$\gamma_2 - \gamma_0 = \Delta\gamma = 2\Delta\beta = 2(\beta_2 - \beta_0) \tag{30}$$

thus

$$\gamma_2 = \gamma_1 + \frac{\Delta\gamma}{2} = \gamma_1 + \Delta\gamma \tag{31}$$
$$\gamma_0 = \gamma_1 - \frac{\Delta\gamma}{2} = \gamma_1 - \Delta\gamma \tag{32}$$

$$\beta_2 = \beta_1 + \frac{\Delta\beta}{2} \tag{33}$$
$$\beta_0 = \beta_1 - \frac{\Delta\beta}{2}. \tag{34}$$

Again, the median angles β_1 and γ_1 do not necessarily occur simultaneously. Following the same method as in the previous case, it is possible to rewrite Eq. (26) as $g(\beta,\gamma) = l_6$ where

$$\begin{aligned} g(\beta,\gamma) &= r_2 r_3\cos(\beta-\gamma) + l_3(r_2\cos\beta - r_3\cos\gamma) \\ &\quad + l_5(r_2\sin\beta - r_3\sin\gamma) \end{aligned} \tag{35}$$
$$l_6 = \frac{r_2^2 + r_3^2 + l_3^2 + l_5^2 - l_4^2}{2}. \tag{36}$$

Since l_6 is constant for a given mechanism, the function $g(\beta,\gamma)$ must also be constant. Consequently, it is possible to use the two positions where the angles are known, i.e., the initial (β_0,γ_0) and final (β_2,γ_2) configurations, to find a relationship between γ_1 and r_3 not involving l_4.

One has:

$$0 = g(\beta_0,\gamma_0) - g(\beta_2,\gamma_2) \tag{37}$$

which can be rewritten as

$$\begin{aligned} 0 &= r_2\big(r_3\sin(\gamma_1-\beta_1) + l_3\sin\beta_1 - l_5\cos\beta_1\big)\sin(\Delta\beta/2) \\ &\quad + r_3\big(l_5\cos\gamma_1 - l_3\sin\gamma_1\big)\sin\Delta\beta. \end{aligned} \tag{38}$$

If $\sin(\Delta\beta/2) \neq 0$, i.e. $\Delta\beta \neq 2n\pi$ where $n \in \mathbb{N}$, which implies that the mechanism does not move or makes n complete turns, then it is possible to obtain

$$\begin{aligned} 0 &= r_2\big(r_3\sin\gamma_1\cos\beta_1 - r_3\cos\gamma_1\sin\beta_1 + l_3\sin\beta_1 - l_5\cos\beta_1\big) \\ &\quad + r_3\big(l_5\cos\gamma_1 - l_3\sin\gamma_1\big)2\cos(\Delta\beta/2) \end{aligned} \tag{39}$$

and finally

$$\frac{r_2}{r_3} = \frac{r_2\sin\beta_1 - 2l_5\cos\frac{\Delta\beta}{2}}{l_3\sin\beta_1 - l_5\cos\beta_1}\cos\gamma_1 + \frac{-r_2\cos\beta_1 + 2l_3\cos\frac{\Delta\beta}{2}}{l_3\sin\beta_1 - l_5\cos\beta_1}\sin\gamma_1. \tag{40}$$

Knowing the values of $\Delta\beta, \beta_1, r_2, l_3$ and l_5, it is possible to use Eq. (40) to calculate r_3 depending of γ_1 and vice versa.

With r_3 and γ_1, it is then possible to calculate l_4 using the loop closure equation of the mechanism in the initial position (β_0,γ_0):

$$\begin{aligned} l_4^2 &= (r_3\cos(\gamma_1-\Delta\beta) + l_3 - r_2\cos(\beta_1-\Delta\beta/2))^2 \\ &\quad + (r_3\sin(\gamma_1-\Delta\beta) + l_5 - r_2\sin(\beta_1-\Delta\beta/2))^2. \end{aligned} \tag{41}$$

This allows to plot the lengths r_3 and l_4 in terms of γ_1 as presented in Fig. 13, where $\Delta\beta = 25°$, $\beta_1 = 1.97$, $r_2 = 10$ mm, $l_3 = 25.4$ mm and $l_5 = 11.8$ mm were used.

Knowing for γ_1 the value of r_3 and l_4, it is possible to calculate angle γ using Eq. (26) as well as the coordination error e_γ between the little finger and the ring finger, where

$$e_\gamma = \big|(\gamma-\gamma_0) - 2(\beta-\beta_0)\big| = \big|\gamma - 2\beta - (\gamma_1 - 2\beta_1)\big|. \tag{42}$$

It is finally possible to see the coordination error on the interval $\beta_0 \le \beta \le \beta_2$ according to the median angle γ_1 chosen, as shown in Fig. 14. Again, it is interesting to see the evolution of the mean and the maximum error on the useful range of β: this can be done with Fig. 15. The error is then minimized when $\gamma_1 = \gamma_{1g}$, which represents the median angle for which the error becomes zero when $\beta = \beta_1$, as seen in Fig. 14.

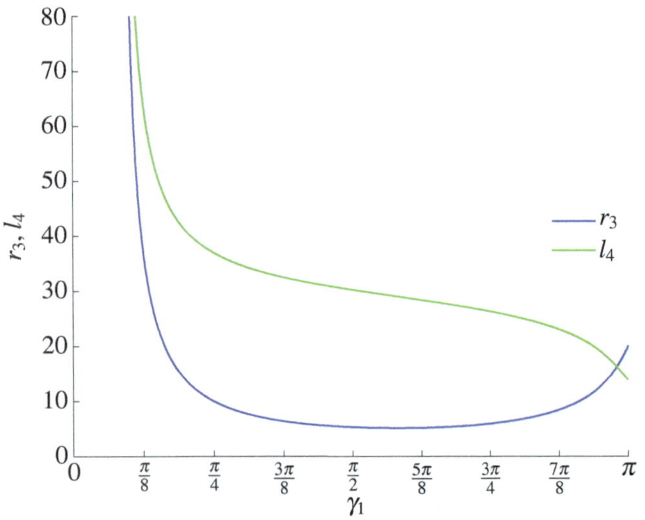

Figure 13. Lengths r_3 and l_4 as a function of angle γ_1.

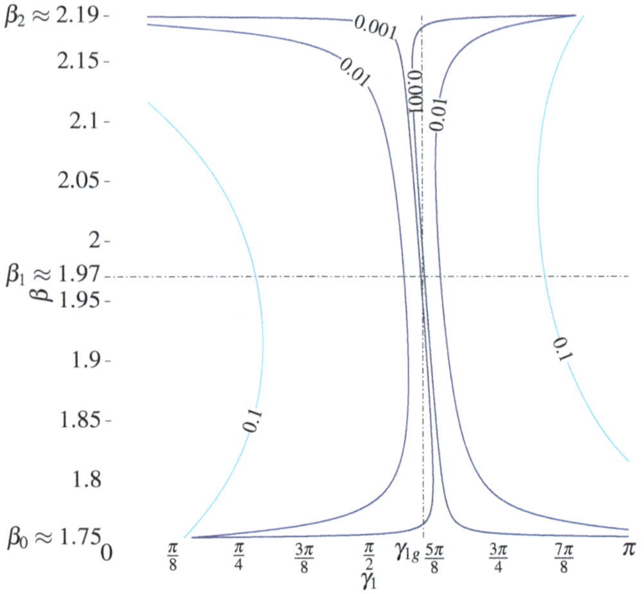

Figure 14. Error e_γ as a function of γ_1 and β.

One can obtain γ_{1g} with $0 = g(\beta_1, \gamma_{1g}) - g(\beta_0, \gamma_0)$, i.e.,

$$\begin{aligned} 0 &= r_2 r_3 \cos(\beta_1 - \gamma_{1g}) + l_3 (r_2 \cos\beta_1 - r_3 \cos\gamma_{1g}) \\ &\quad + l_5 (r_2 \sin\beta_1 - r_3 \sin\gamma_{1g}) - r_2 r_3 \cos(\beta_0 - \gamma_0) \\ &\quad - l_3 (r_2 \cos\beta_0 - r_3 \cos\gamma_0) - l_5 (r_2 \sin\beta_0 - r_3 \sin\gamma_0) \end{aligned} \tag{43}$$

which can be rearranged as

$$\begin{aligned} 0 &= \cos\gamma_{1g}\Big(r_2\big(\cos\beta_1 - \cos\beta_2\big) - l_3\big(1 - \cos\Delta\beta\big) - l_5 \sin\Delta\beta\Big) \\ &\quad + \sin\gamma_{1g}\Big(r_2\big(\sin\beta_1 - \sin\beta_2\big) + l_3 \sin\Delta\beta - l_5\big(1 - \cos\Delta\beta\big)\Big) \\ &\quad + \frac{r_2}{r_3}\Big(l_3 \cos\beta_1 - l_3 \cos\beta_0 + l_5 \sin\beta_1 - l_5 \sin\beta_0\Big) \end{aligned} \tag{44}$$

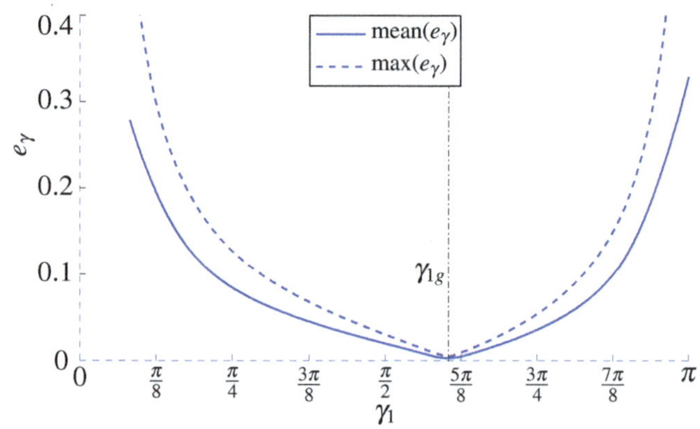

Figure 15. Mean and maximum error in the interval $\beta_0 \le \beta \le \beta_2$ as a function of γ_1.

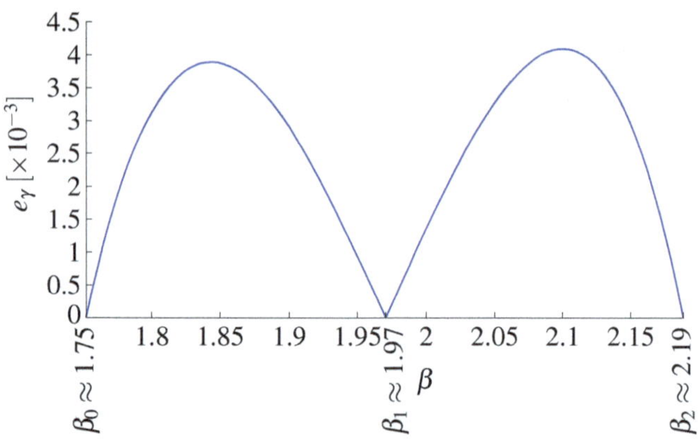

Figure 16. Error e_γ as a function of angle β.

then, substituting Eq. (40) into Eq. (44), one obtains:

$$A_{\gamma_{1g}} \cos\gamma_{1g} + B_{\gamma_{1g}} \sin\gamma_{1g} = 0. \tag{45}$$

where

$$\begin{aligned} A_{\gamma_{1g}} &= r_2\big(\cos\beta_1 - \cos\beta_2\big) - l_3\big(1 - \cos\Delta\beta\big) - l_5 \sin\Delta\beta \\ &\quad + d_{\gamma_{1g}}\big(r_2 \sin\beta_1 - 2 l_5 \cos\Delta\beta/2\big) \end{aligned} \tag{46}$$

$$\begin{aligned} B_{\gamma_{1g}} &= r_2\big(\sin\beta_1 - \sin\beta_2\big) + l_3 \sin\Delta\beta - l_5\big(1 - \cos\Delta\beta\big) \\ &\quad + d_{\gamma_{1g}}\big(-r_2 \cos\beta_1 + 2 l_3 \cos\Delta\beta/2\big) \end{aligned} \tag{47}$$

$$d_{\gamma_{1g}} = \frac{l_3(\cos\beta_1 - \cos\beta_0) + l_5(\sin\beta_1 - \sin\beta_0)}{l_3 \sin\beta_1 - l_5 \cos\beta_1}. \tag{48}$$

which is readily solved for γ_{1g}. Using the same parameters as used for Fig. 13, we obtain $\gamma_{1g} \approx 1.90$ and the associated values $r_3 \approx 5.1$ mm and $l_4 \approx 28.6$ mm. The error e_γ is presented in Fig. 16. One can observe that the error is zero at $\beta = \beta_0, \beta_1$ and β_2. Finally, the angle γ is plotted in Fig. 17 as a function of angle β.

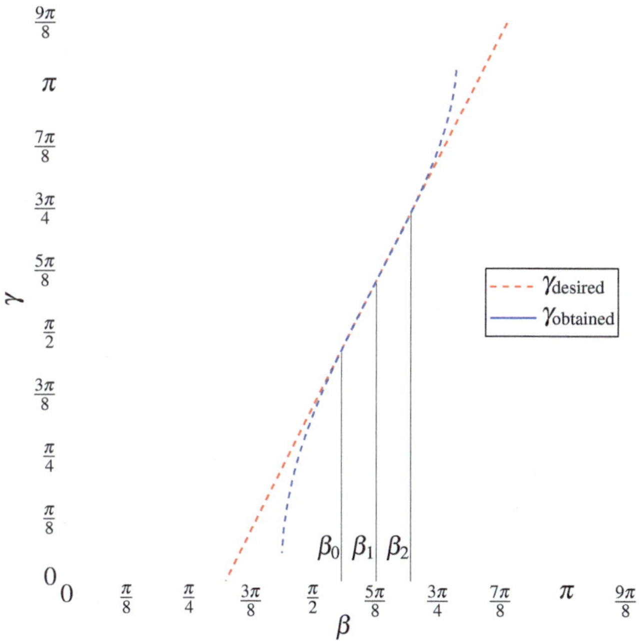

Figure 17. Angle γ desired and obtained according to β.

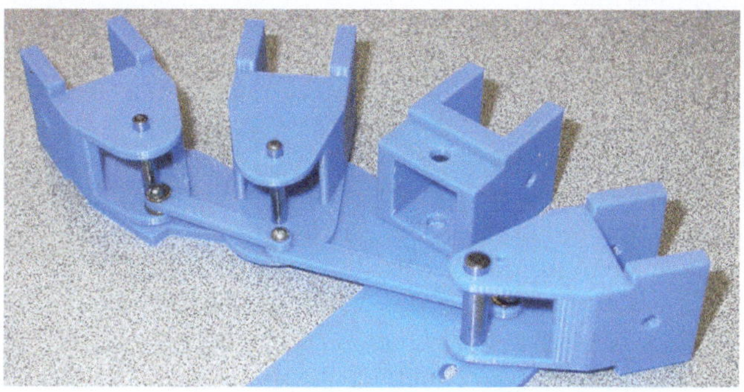

Figure 18. Prototype of the finger abduction system.

5 Prototype

Using the values found in this study, the prototype shown in Fig. 18 was built with rapid prototyping. One can note the similarity with the schematic presented in Fig. 2, on which the fixed middle finger metacarpal has been added. Again, the input consisting of a worm gear connecting with the index metacarpal is not shown. This mechanism will enhance the range of feasible grasps of the underactuated hand without significantly increasing its complexity.

6 Conclusions

The kinematic design of the finger abduction mechanism for an underactuated anthropomorphic robotic hand has been presented. The parameters of the linkage between the index finger and the third finger has been analyzed, followed by those of the linkage between the ring finger and the little finger. We therefore obtain a simple architecture that minimizes the coordination error among the abduction angles.

A robotic hand using this abduction mechanism is being built. It will use underactuation within its fingers and among them and will be tendon-driven (Demers and Gosselin, 2009, 2010). Testing of its grasping capabilities will then be conducted in order to validate this design.

Acknowledgements. This work was supported by the Fonds québécois de la recherche sur la nature et les technologies (FQRNT), the Natural Sciences and Engineering Research Council of Canada (NSERC) as well as by the Canada Research Chair Program.

Edited by: J. L. Herder
Reviewed by: two anonymous referees

References

Boutan, M. and Casoli, V.: Mains & Préhensions entre fonctions et anatomie, Sauramps médical, Montpellier, 2005.

Demers, L.-A. A. and Gosselin, C.: Kinematic design of an ejection-free underactuated anthropomorphic finger, in: Proceedings of the IEEE International Conference on Robotics and Automation (ICRA 2009), Kobe, Japan, 12–17 May 2009, 2086–2091, 2009.

Demers, L.-A. A. and Gosselin, C.: Kinematic design of a thumb metacarpal for an anthropomorphic robotic hand, in: Proceedings of the ASME International Design Engineering Technical Conferences (IDETC 2010), Montréal, Canada, 15–18 August 2010.

Gurbuz, H., Mesut, R., and Nesrin Turan, F.: Measurement of active abduction of metacarpophalangeal joints via electronic digital incinometric technique, Ital. J. Anat. Embryol., 111, 9–14, 2006.

Kapandji, A. I.: Physiologie articulaire : schémas commentés de mécanique humaine, Vol. 1, Maloine, 6 Edn., 2005.

Norkin, C. C. and White, D. J.: Measurement of Joint Motion: A Guide to Goniometry, Philadelphia, 4 Edn., 2009.

Smahel, Z. and Klímová, A.: The influence of age and exercise on the mobility of hand joints: 1: Metacarpophalangeal joints of the three-phalangeal fingers, Acta Chir. Plast., 46, 81–88, 2004.

Tubiana, R. and Thomine, J. M.: La main : Anatomie fonctionnelle et examen clinique, Masson, Paris, 1990.

A handy new design paradigm

B. Bergelin, B. Slaboch, J. Sun, and P. A. Voglewede

Department of Mechanical Engineering, Marquette University, Milwaukee, Wisconsin 53233, USA

Abstract. In light of technological advances, researchers have lost sight of robotic grippers/end effectors design intent. In a semi-structured environment the biomimetic approach is impractical due to the high complexity of the mechanism and control algorithms. Current industrial grippers are robust, but lack the flexibility that allows for in hand manipulation. The authors believe that underactuated grippers provide the best approach to allow for in hand manipulation along with being rugged enough for an industrial setting. Thinking of the robotic gripper and the robotic arm as one system (as opposed to two separate subsystems), one is capable of using the degrees of freedom of the robot in conjunction with that of the gripper to provide the desired motion profile without the complexity of running two subsystems. This paper will outline where recent grippers have failed and will introduce a new design paradigm for grippers along with several underactuated gripper ideas.

This paper was presented at the IFToMM/ASME International Workshop on Underactuated Grasping (UG2010), 19 August 2010, Montréal, Canada.

1 Introduction

Researchers have focused on replicating human dexterity for unstructured environments, like those of prosthetic hands and humanoids. Due to the immensity of this task, researchers typically have divided this field into robotic "arms" and "hands". The work on such grippers has been further divided into subcategories: grasping, manipulation, and the mechanical and algorithmic designs of both of these.

This discrete approach is inherently flawed. While much work has gone into optimizing and understanding these subfields, the overall goal of creating a system to pick up and manipulate objects has been somewhat lost. The subfields have been sub-optimized, and as such each subsystem must contain actuators and sensors that do not depend on the other subsystem. For example, when manipulating an object using finger gaiting, an excess of fingers must be used to ensure that the remaining fingers fully constrain the object while the degrees of freedom (DOF) of arm remains un-utilized. When combining the solutions from each subsystem, the result is a complicated, high DOF system with multiple sensors, algorithms, and control structures which is not practical for most applications.

While one may be tempted to say the best approach is to design the entire system holistically, the task is too large for any one research group to tackle. A better approach is to narrow the scope of the problem and fully integrate the capabilities of each subsystem together. Thus, for this paper, we consider structured and semi-structured environments (like those in industrial settings) which would allow the research to focus on several routine tasks. We also focus on the design of end effectors (admittedly a discrete approach), but one that inherently views the gripper as a mechanism that must work with the robot arm (instead of simply attached to it), with its environment (instead of simply in it), and with the object (instead of simply working on it). Inherent in this design paradigm are grippers which are simple, robust, and functional.

In Sect. 2, this paper will first review the state-of-the-art grippers and show how the current biomimetic paradigm aggravates this approach. Section 3 illustrates a further explanation of this new design paradigm and its basic tenets. We will then introduce some rudimentary grippers that are a step in the right direction in Sect. 4. In Sect. 5, it will be discussed how the prototypes implement the new paradigm. Section 6 concludes with an explanation of the potential research areas that arise from this approach and from the elementary designs.

2 Background

Many impressive robotic/prosthetic hands have been designed (for some historical reviews see Okamura et al. (2000); Bicchi (2000); Boubekri et al. (2002); Mason et al. (1985); Kato et al. (1987); Mason (2001)). The most notable include the Utah/MIT hand (Jacobsen et al., 1984), the Salisbury Hand (aka the Stanford/JPL hand) (Mason et al., 1985), and the IOWA hand (Yang et al., 2004). All of these hands have taken a very biomimetic approach to the problem. They consist of a fixed palm with movable fingers. Other research has focused on more rudimentary hands including the NYU hand (Demmel et al., 1988), the planar STYX (Murray et al., 1990), the Barrett Hand (Townsend, 2000), and a similar hand introduced by Biagiotti et al. (2003). These hands are suited for the grasping and manipulation of unknown objects using a variety of different grasp types (Cutkosky, 1989).

In these devices, manipulation of the object is typically done in one of two ways: finger gaiting (moving the fingers around the exterior of the grasped object while maintaining a stable grasp) or regrasping (placing the object on a surface and picking it up again). Each technique is inherently slow and requires extensive computational power to be implemented. While the engineer has been able to mimic the human hand in a kinematic sense, these designs fall short in developing a robust system capable of surpassing the capabilities of the human hand. Once again, the suboptimized hand does not work on solving the entire problem.

These hands can be contrasted to industrial grippers where simplicity and robustness are key. Some of these include those introduced by Tella et al. (1982) and Brown et al. (1999) (similarly by Balan et al., 2003). While these grippers are easy to use, they are purely for grasping with very limited manipulation. In the semi-structured industrial environment, these grippers work very well. While they do not allow for complete manipulation of the object, they do provide a fast and efficient method of grasping parts.

In order to circumvent the problem of high DOF systems, underactuated grippers have been developed. The recent work on underactuated grasping is a step in the right direction of creating a better system. Underactuated grasping recognizes the inherent problem with traditional approaches and creates grippers that decrease the number of actuators. Most of these approaches combine joints of a "finger" together with "tendons" and/or springs to give the "fingers" the capability to form to the grasped object shape. This reduces the DOF of the end effector, but does not allow it to be manipulated. Some rudimentary manipulation is gained by rolling on the fingers, but this is limited.

The future of grasping (and underactuated grasping) is a recombination of grasping and manipulation. Some initial work has involved using passive joints in conjunction to simultaneously grasp and manipulate. Several such devices are a fur picking gripper (Doulgeri et al., 2002), a pivoting gripper (Carlisle et al., 1994), a low friction gripper (Goldberg et al., 1992, 1993), and a postal bag gripper (Kazerooni et al., 2004). These grippers allow for some limited manipulation combined with the grasping task.

3 Proposed new philosophy of grasping

While the existing work in grippers is admirable, the future of grasping and manipulation lies in integrating the gripper into the system to which it is attached and its environment. The new philosophy of grasping and manipulation we are proposing consists of five major tenets:

1. *An effective robotic end effector does not necessarily need to look like a human hand.* This is consistent with a move away from biomimetics to bioinspiration. There needs to be a renewed focus on understanding the problem and moving away from prostheses and humanoids where form trumps function. This tenet needs to be prevalent throughout the research.

2. *In order to make an effective robotic end effector, the number of DOF must be minimal.* While technology has shown greater capability to handle larger number of DOF, this cannot be used as a crutch to lead one back to burdensome designs. Too many designs assume that multiple actuators and joints can be easily managed. Time has always shown the "keep it simple" method is most effective.

3. *The end effector should be assumed to be on a robot arm.* While this tenet may appear obvious, most end effector designs do not use the arm other than a positional device to locate the end effector. The DOF of the end effector can be decreased if one assumes that the robot arm can play a role in the grasping and manipulation.

4. *The grasped object and the environment are integral parts of the system.* Current thought immobilizes the object and then manipulates it. However, the object itself has properties which allow it to be more easily grasped and manipulated. For example, the center of gravity of an object can be utilized as a force for manipulation. Additionally, the environment offers fixed structures to which the hand can be used to move the object.

5. *The end effector structure must possess multiple functions.* Current end effectors typically have specialized the function of each structure (fingers provide a point contact). However, in underactuated grasping, the "fingers" can provide point contact as well as an enveloping grasp. This dual functionality is critical to the design of new end effectors to decrease the number of DOF.

In this new philosophy, we have created what we believe to be some steps in this direction.

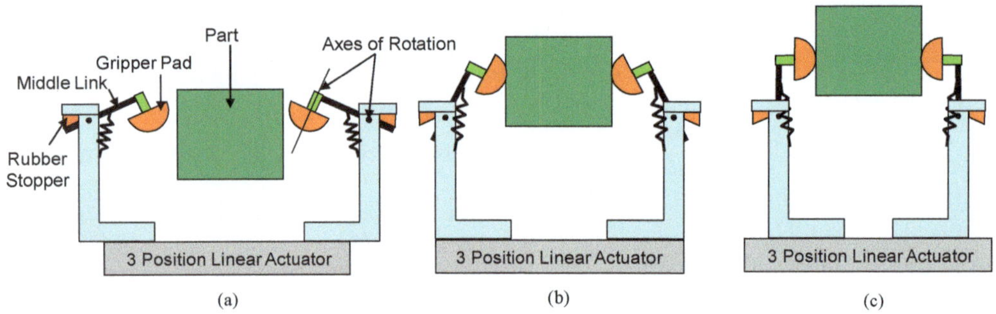

Figure 1. Parallel jaw gripper with singularities.

4 Preliminary work in this paradigm

This section will outline the work done at Marquette University in terms of this new paradigm.

Idea 1: Parallel Jaw Gripper with Singularities

In many applications, one needs two different motion profiles. For example, one may need translation followed by rotation. In this design (Fig. 1), a part is approached as in (a), and the part is gripped as in (b). Due to the rolling contact at the fingers (consider each gripper pad as a wheel), the part can move in or out of the page rolling on the gripper pads. When the linear actuator moves to configuration (c), in which the mechanism encounters a singularity, the part is no longer allowed to translate, but rotation is allowed around the axis between the point contacts. A prototype of this design is shown in Fig. 2.

The actuator only needs to have one DOF, and even further, only needs three positions to perform the operations. There are passive joints outfitted with springs that provide the necessary contact forces to ensure that the part does not slip. Thus, much like classical underactuated graspers, there is a coupling between the DOF and the object.

The issue with such an approach is how to provide the motion to the object since there is no actuation on it. This can be done using an external fixed finger and the remaining six DOF of the robot on which the end effector is attached. Thus, in total, there are seven DOFs that need to be controlled drastically decreasing the *total* DOF of the system. Therefore, the system DOF is decreased and the system performs the same operation. Additionally, the grasp and the manipulation are designed as one; they are indistinguishable in the design.

Idea 2: Parallel Jaw Gripper with Changing Joints

A possible variant of the design which allows for fixed positioning of the object is to modify the gripper pads such that the translational axis is replaced by planar contacts such that the block is fully constrained (Fig. 3). In the second stage (b), due to the friction between the block and the planar contacts, the block cannot move and is fully constrained. When the linear actuator is moved again, the ball bearings contact the block and due to the reduced friction as in (c), allow rotation similar to Idea 1. The force for this motion can again be provided by an external fixed finger and using the robotic arm itself. The prototype can be seen in Fig. 4.

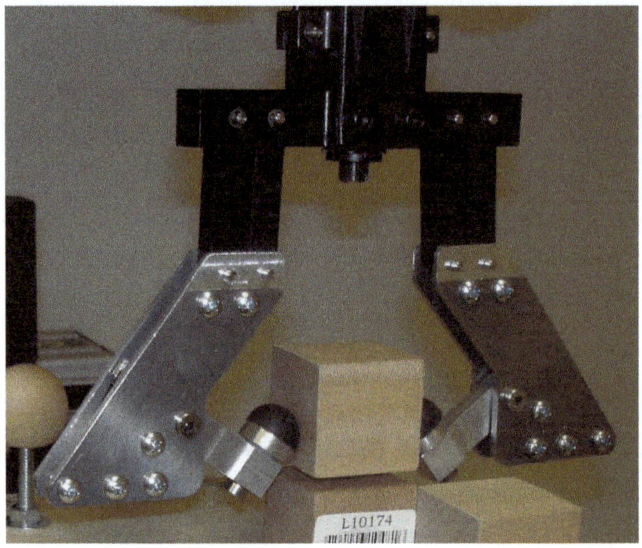

Figure 2. Prototype of parallel jaw gripper with singularities.

The design is modular in that it can allow different types of motion by a simple end effector change. While this is not ideal for unstructured environments, it is very feasible from an industrial standpoint. The end effector can be changed for different parts easily and allows for quick manipulation of the object without finger gaiting or regrasping. This type of device can be better considered as a metamorphic mechanism.

Idea 3: Parallel Jaw Gripper with Spring Actuated Joints

A third variation of the design utilizes spring plungers (Fig. 5) (typically used for positioning parts in fixtures). In position (b), the spring plungers are compressed which allows for planar contacts between the part and the friction

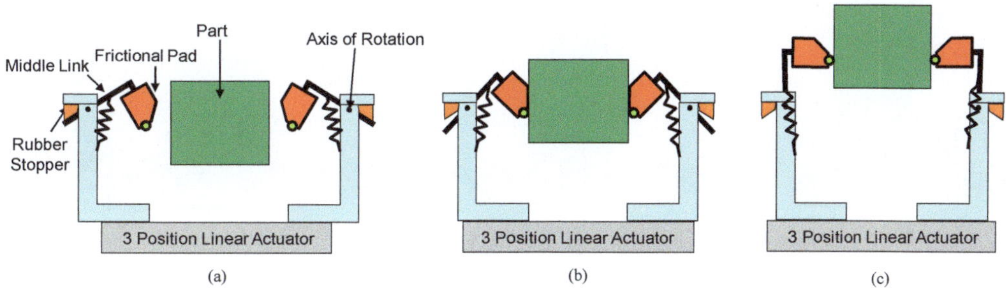

Figure 3. Parallel jaw gripper with changing joints.

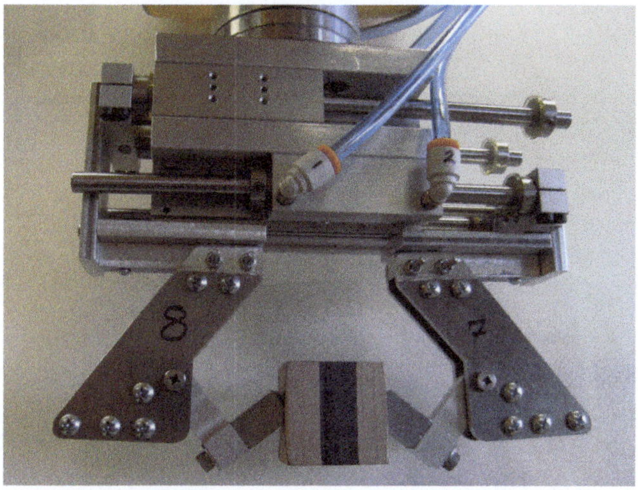

Figure 4. Prototype of parallel jaw gripper with changing joints.

pad. When the linear actuator is moved to position (c), a point contact is formed. The part can then be rotated using an external fixed finger or from gravitational force (depending on grasp location). The advantage of this design is that the system remains in a stable equilibrium position throughout its operation and is quite rigid. The prototype is shown in Fig. 6.

5 Discussion

Each of the proposed end effector designs follows the newly proposed philosophy of grasping. Rather than describing each gripper in detail, it will be shown how these novel end effectors follow this new paradigm.

1. *An effective robotic end effector does not necessarily need to look like a human hand.* While each of the grippers have hand-like components ("fingers" and a "palm"), they do not necessarily look or act like a human hand. In fact, each gripper uses linear motion of the base, as opposed to rotational actuation. This linear motion is much more typical of industrial grippers, which are designed to be simple and robust. There is also no coupled motion of the joints (typical of most underactuated grippers). The end effectors outlined in this paper, while not appearing hand-like, are robust but also retain some of the dexterity to which a biomimetic gripper may possess.

2. *In order to make an effective robotic end effector, the number of DOF must be minimal.* Complex anthropomorphic hands are inherently slow and difficult to implement due to the required computational power. Each of the proposed grippers can be controlled with a single actuator, and thus reduces the complexity of the system as a whole. The implementation of passive components allows the grippers to gain or lose DOF. The proposed designs are intended to provide the system with an added DOF which can be gained or lost without the complexity of a biomimetic hand (i.e., number of joints actuated and intricate control algorithms). The DOF of the gripper are changed solely by controlling a single linear actuator. This allows for the robustness, simplicity, and speed needed in an structured setting.

3. *The end effector should be assumed to be on a robot arm.* Attaching the end effector to a robot arm allows one to utilize the robot's DOF when performing manipulation tasks. All of the grippers, when incorporated into the system, use the DOF provided by the robotic arm, along with the end effector's single actuator, to achieve the desired in-hand manipulation. Also, rotation of the grasped object is provided; the axis to which the object rotates orients it from facing into the "palm" to out of the "palm". This is typically a shortfall of most robotic systems. This idea follows closely with Tenet 4.

4. *The grasped object and the environment are integral parts of the system.* In each of the proposed gripper designs, an external fixed "finger" may be used in conjunction with the DOF of the robot to achieve the desired amount of rotation for the part. This is related to Tenet 2, as it decreases the number of actuators and Tenet 3 in that it utilizes the robotic arm's DOF to provide this motion. In each design the grasped object and fixed finger (part of the "environment") are critical in achieving this in-hand manipulation.

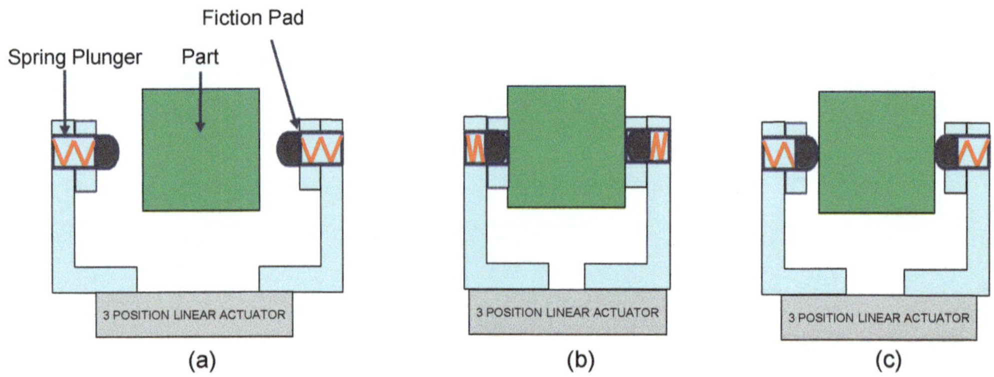

Figure 5. Parallel jaw gripper with spring actuated joints.

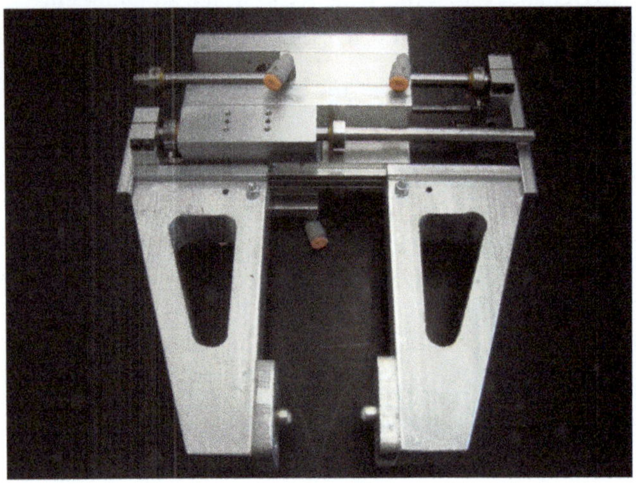

Figure 6. Prototype of parallel jaw gripper with spring actuated joints.

5. *The end effector structure must possess multiple functions*. Each novel parallel jaw gripper can perform multiple functions. The grippers can be used for pick-and-place operations that do not require any in-hand manipulation. In addition, the new grippers can also be used to provide rotation of a part about the gripper's contact points. Combining both of these functions into a single gripper can potentially reduce costs and improve efficiency.

While it is clear that these novel end effectors are not yet optimal, it is our belief that this method exhibits great potential for future research. The next section outlines future research directions and potential applications for the new design paradigm.

6 Future research directions

The future of grasping is in the development of the tools and techniques for the synthesis of these mechanisms. There exists a wealth of tools for the design of four bar mechanisms and some for other planar mechanisms. There even has been fundamental work on designing closed chain mechanisms in general to provide certain motion characteristics. However, these tools do not exist for grippers and mechanisms that change depending on their pose. Here we present specific research directions that can potentially lead to advances in robotic grasping.

1. *Synthesis of reconfigurable mechanisms*

   Reconfigurable mechanisms are those in which the DOF of the mechanism changes depending on its configuration. This paper has presented novel gripping mechanisms that utilize variable topology joints. Analytical tools such as graph theory and adjacency matrices may be used to analyze reconfigurable mechanisms but general synthesis techniques have yet to be developed.

2. *Expansion of analysis tools using screw theory*

   Screw theory has been a useful tool in the synthesis of non-reconfigurable mechanisms. Expanding these techniques to reconfigurable mechanisms could lead to significant advances in the design of underactuated robotic grippers.

3. *Motion planning algorithms*

   Current motion planning algorithms are written so that the object grasped avoids contact with obstacles as it is transferred from one location to another. However, external obstacles (i.e., a fixed finger) have the potential to aid in manipulation of an object without adding extra actuators. New motion planning techniques need to incorporate environmental factors in conjunction with the DOF of the robot.

7 Conclusions

We believe that there needs to be a fundamental shift in the way researchers approach end effector design for industrial applications. Unless used as a prosthesis, end effectors need not look or act as a human hand. It seems as though robotic end effectors continue to increase in complexity while their main goal remains unchanged. Through implementation of the proposed design paradigm, robotic grippers can achieve in-hand manipulation while decreasing the amount of actuators and remaining robust. By using the five tenets proposed in the new design paradigm, three end effector designs and corresponding prototypes were presented. They are simple, robust and capable of achieving their objectives through use of only a three position linear actuator. Underactuated grasping, while still underdeveloped, is a step in the right direction in designing grippers which possess the ability to manipulate objects in-hand and still maintain the durability and simplicity desired for industrial settings.

Edited by: J. L. Herder
Reviewed by: two anonymous referees

References

Balan, L. and Bone, G.: Automated gripper jaw design and grasp planning for sets of 3D objects, J. Robotic Syst., 20(3), 147-162, 2003.

Biagiotti, L., Melchiorri, C., and Vassura, G.: A dextrous robotic gripper for autonomous grasping, Ind. Robot, 30(5), 449-458, 2003.

Bicchi, A.: Hands for dextrous manipulation and robust grasping: A difficult road toward simplicity, IEEE T. Robotic. Autom., 16(6), 652-662, 2000.

Boubekri, N. and Chakraborty, P.: Robotic grasping: Gripper designs, control methods and grasp configurations – a review of research, Integrated Manufacturing Systems, 13(7), 520–531, 2002.

Brown, R. G. and Brost, R.: A 3-D modular gripper design tool, IEEE T. Robotic. Autom., 15(1), 174-186, 1999.

Carlisle, B., Goldberg, K., Rao, A., and Wiegley, J.: A pivoting gripper for feeding industrial parts, Proceedings of the 1994 IEEE International Conference on Robotics and Automation, Vol. 2, 1650-1655, 1994.

Cutkosky, M.: On grasp choice, grasp models, and the design of hands for manufacturing tasks, IEEE T. Robotic. Autom., 5(3), 269–279, 1989.

Demmel, J., Lafferriere, G., Schwartz, J., and Sharir, M.: Theoretical and experimental studies using a multifinger planar manipulator, Proceedings of the 1998 IEEE International Conference on Robotics and Automation, 390-395, 1988.

Doulgeri, Z. and Fahantidis, N.: Picking up flexible pieces out of a bundle, IEEE Robot. Autom. Mag., 9(2), 9-19, 2002.

Goldberg, K. and Furst, M.: Low friction gripper, March, United States Patent Number 5098145, 1992.

Goldberg, K. and Furst, M.: Low friction gripper, February, United States Patent Number 5186515, 1993.

Jacobsen, S., Wood, J., Bigger, K., and Iverson, E.: The Utah/MIT hand: Work in progress, Int. J. Robot. Res., 4(3), 21–50, 1984.

Kato, I. and Sadamoto, K.: Mechanical Hands Illustrated, Hemisphere Publishing Corp., New York, 1987.

Kazerooni, H. and Foley, C.: A practical robotic end-effector for grasping postal sacks, ASME Journal of Dynamic Systems, Measurement, and Control, 126, 154-161, 2004.

Mason, T.: Mechanics of RoboticManipulation, MIT Press, 2001.

Mason, M. and Salisbury, J.: Robot Hands and the Mechanics of Manipulation, MIT Press, Cambridge, MA, 1985.

Murray, R. and Sastry, S.: Control experiments in planar manipulation and grasping, Proceedings of the 1990 IEEE International Conference on Robotics and Automation, 624-629, 1990.

Okamura, A. M., Smaby, N., and Cutkosky, M.: An overview of dextrous manipulation, Proceedings of the IEEE International Conference on Robotics and Automation, 255-262, 2000.

Tella, R., Birk, J., and Kelley, R.: General purpose hands for bin-picking robots, IEEE T. Syst. Man Cyb., SMC-12(6), 828-837, 1982.

Townsend, W. T.: The Barrett Hand grasper programmably flexible part handling and assembly, Ind. Robot, 27(3), 181-188, 2000.

Yang, J., Pitarch, E., Abdel-Malek, K., Patrick, A., and Lindkvist, L.: A multi-fingered hand prosthesis, Mech. Mach. Theory, 39(6), 555-581, 2004.

Type synthesis of freedom and constraint elements for design of flexure mechanisms

H.-J. Su[1] and C. Yue[2]

[1]Department of Mechanical and Aerospace Engineering, Ohio State University, Columbus, Ohio, 43210, USA
[2]Department of Mechanical Engineering, University of Maryland Baltimore County, Baltimore, Maryland, 21250, USA

Correspondence to: H.-J. Su (su.298@osu.edu)

Abstract. In this paper, we present the type synthesis of freedom and constraint elements for design of general flexure mechanisms. As an important step in the conceptual design stage, the goal of type synthesis is to qualitatively determine the topology or connectivity of flexure elements and rigid bodies in a mechanism. The synthesis procedure presented here is based on a recently emerging screw theory based approach for flexure mechanisms. We first categorize a list of commonly used atomic flexure primitives including blades, wires, notches and bellow springs etc. We then derive their twist and wrench matrices that mathematically represent their freedom and constraint spaces. The synthesis procedure rigorously follows screw algebra. Freedom elements including R-joints and P-joints are defined as basic motion elements that allow a single rotation or a single translation. By using parallel structures of these flexure primitives, eleven designs of R-joints and eight designs of P-joints are systematically synthesized. As a duality, constraint elements including P-constraints and R-constraints remove a single translation or rotation. In contract to freedom elements, we synthesized serial chains of flexure primitives and obtained six designs of P-constraints and three designs of R-constraints. These freedom and constraint elements form a catalogue of basic building blocks for designing more complex flexure mechanisms. At last we utilize four design examples to demonstrate how to synthesize hybrid structures with serial and parallel combination of these elements.

1 Introduction

Compliant mechanisms (Howell, 2001) or flexure mechanisms (Smith, 2000; Smith and Chetwynd, 1992), formed by a set of rigid bodies connected with compliant elements, produce a defined motion through elastic deformation of their compliant elements. They are widely used in various precision instruments and machines such as nano-manipulators Culpepper and Anderson (2004), nano-positioners (Chen and Culpepper, 2006; Brouwer et al., 2010; Yao et al., 2008; Dong et al., 2008) and precision manufacturing machines (Varadarajan and Culpepper, 2007a,b).

The design of flexure mechanisms has been an ad hoc process that heavily relies on designers' experience and intuition that are typically built up over years of training. One approach often used in precision engineering community is called the constraint-based approach (Blanding, 1999; Hale, 1999; Awtar and Slocum, 2007). In a recent conference tutorial by Henein (2011), the author enumerated a list of flexure bearing designs based on their degree of freedom (DOF), Grubler mobility and degree of hyperstaticity (DOH). Many authors have attempted to systemize the constraint based approach. Recently this approach has been further formalized into the Freedom and Constraint Topology (FACT) framework (Hopkins, 2007a; Hopkins and Culpepper, 2010a,b). Essentially the constraint-based design approach and the FACT approach are mathematically equivalent to screw theory (Ball, 1998; Hunt, 1978; Phillips, 1984, 1990; Davidson and Hunt, 2004) that has been widely used in kinematics community for various problems such as type synthesis of parallel mechanisms (Kong and Gosselin, 2010) and mobility analysis of rigid body mechanisms (Huang et al., 2008).

In recognizing this intrinsic connection between the constraint based approach and screw theory, a series of work (Su et al., 2009; Su and Tari, 2010, 2011; Hopkins and Culpepper, 2010c; Yu et al., 2010; Su, 2011) on screw theory based approach for type synthesis and analysis of flexure mechanisms have been done. This approach is completely based on screw algebra (Dai and Jones, 2001, 2003), which can be easily implemented in computer programs for truly automating the design process of flexure mechanisms, especially the conceptual design stage. However currently these work have been focusing on mechanisms with relatively simple topologies or mobility analysis (rather than synthesis). The type synthesis of general flexure mechanisms for any specified mobility is still yet to be done.

In this paper, we present a systematic methodology based on screw theory for the type synthesis of general flexure mechanisms with serial, parallel or hybrid (combination of serial and parallel) topologies. First, a list of commonly used flexure primitives is studied. Then they are used to build basic freedom and constraint elements. At last these freedom and constraint elements are used for constructing more complex flexure mechanisms.

2 Screw theory overview

In this section, we first review basic concepts of screw theory as a background preparation for the following sections.

2.1 Twists and wrenches

A flexure mechanism subject to a general load represented by a wrench $\hat{W}$ undergoes an instantaneous motion represented by a twist $\hat{T}$. From the freedom and constraint point of view, twists represent allowable motions while wrenches represent forbidden motions. Both twist $\hat{T}$ and wrench $\hat{W}$ are 6 by 1 column vectors, written as

$$\hat{T}=\begin{Bmatrix}\mathbf{\Omega}\\ \mathbf{V}\end{Bmatrix}=\begin{Bmatrix}\mathbf{\Omega}\\ \mathbf{c}\times\mathbf{\Omega}+p\mathbf{\Omega}\end{Bmatrix}, \tag{1}$$

$$\hat{W}=\begin{Bmatrix}\mathbf{F}\\ \mathbf{M}\end{Bmatrix}=\begin{Bmatrix}\mathbf{F}\\ \mathbf{c}\times\mathbf{F}+q\mathbf{F}\end{Bmatrix}, \tag{2}$$

where p and q are called pitches of twist and wrenches. And $\hat{T}$ and $\hat{W}$ satisfy the so called reciprocal condition:

$$\hat{T}\circ\hat{W}=\mathbf{\Omega}\cdot\mathbf{M}+\mathbf{V}\cdot\mathbf{F}=0. \tag{3}$$

A general rotational or translational freedom respectively corresponds to a twist with zero or infinite pitch, written as

$$\hat{T}_R=\begin{Bmatrix}\mathbf{\Omega}\\ \mathbf{c}\times\mathbf{\Omega}\end{Bmatrix},\quad \hat{T}_P=\begin{Bmatrix}\mathbf{0}\\ \mathbf{V}\end{Bmatrix}. \tag{4}$$

Similarly a general rotational or translation constraint removes a rotation or translation along a particular direction. They respectively correspond to a wrench with infinite or zero pitch, written as

$$\hat{W}_R=\begin{Bmatrix}\mathbf{0}\\ \mathbf{M}\end{Bmatrix},\quad \hat{W}_P=\begin{Bmatrix}\mathbf{F}\\ \mathbf{c}\times\mathbf{F}\end{Bmatrix} \tag{5}$$

For convenience, we define six principal twists as the rotation and translations about the three coordinate axes,

$$\begin{aligned}
\hat{R}_x&=\begin{pmatrix}1&0&0&0&0&0\end{pmatrix}^T\\
\hat{R}_y&=\begin{pmatrix}0&1&0&0&0&0\end{pmatrix}^T\\
\hat{R}_z&=\begin{pmatrix}0&0&1&0&0&0\end{pmatrix}^T\\
\hat{P}_x&=\begin{pmatrix}0&0&0&1&0&0\end{pmatrix}^T\\
\hat{P}_y&=\begin{pmatrix}0&0&0&0&1&0\end{pmatrix}^T\\
\hat{P}_z&=\begin{pmatrix}0&0&0&0&0&1\end{pmatrix}^T
\end{aligned} \tag{6}$$

Similarly, we also define six principal wrenches as the rotational and translational constraint about the three coordinate axes

$$\begin{aligned}
\hat{F}_x&=\begin{pmatrix}1&0&0&0&0&0\end{pmatrix}^T\\
\hat{F}_y&=\begin{pmatrix}0&1&0&0&0&0\end{pmatrix}^T\\
\hat{F}_z&=\begin{pmatrix}0&0&1&0&0&0\end{pmatrix}^T\\
\hat{M}_x&=\begin{pmatrix}0&0&0&1&0&0\end{pmatrix}^T\\
\hat{M}_y&=\begin{pmatrix}0&0&0&0&1&0\end{pmatrix}^T\\
\hat{M}_z&=\begin{pmatrix}0&0&0&0&0&1\end{pmatrix}^T
\end{aligned} \tag{7}$$

2.2 Coordinate transformation of twists and wrenches

The coordinate transformation of a twist or wrench is calculated as

$$\hat{T}'=[\mathbf{Ad}]\hat{T},\quad \hat{W}'=[\mathbf{Ad}]\hat{W}, \tag{8}$$

where $\hat{T},\hat{W}$ and $\hat{T}',\hat{W}'$ correspond to the twist and the wrench before and after the transformation. And $[\mathbf{Ad}]$ is the so-called 6×6 adjoint matrix, written as

$$[\mathbf{Ad}]=\begin{bmatrix}R&0\\ DR&R\end{bmatrix} \tag{9}$$

where $[\mathbf{R}]$ is a 3 by 3 rotation matrix and $[\mathbf{D}]$ is the 3 by 3 skew-symmetric matrix defined by the translational vector $\boldsymbol{d}=(d_x,d_y,d_z)^T$. They have the form

$$[\mathbf{R}]=\begin{bmatrix}\mathbf{x}&\mathbf{y}&\mathbf{z}\end{bmatrix},\quad [\mathbf{D}]=\begin{bmatrix}0&-d_z&d_y\\ d_z&0&-d_x\\ -d_y&d_x&0\end{bmatrix}$$

When the coordinate transformation is applied to a principal screw (twist or wrench), the resultant screw will along the axes of the new coordinate system, e.g.

$$\hat{R}'_x = [\mathbf{Ad}]\hat{R}_x = \begin{Bmatrix} \mathbf{x} \\ d\times\mathbf{x} \end{Bmatrix} \quad (10)$$

$$\hat{P}'_x = [\mathbf{Ad}]\hat{P}_x = \begin{Bmatrix} \mathbf{0} \\ \mathbf{x} \end{Bmatrix} \quad (11)$$

3 Flexure primitives

A flexure primitive is defined as an "atomic" flexure mechanism that consists of only one flexure element and zero intermediate body. They cannot be further divided into substructures. Recently Hopkins (2012) presented three commonly used flexures and their freedom and constraint spaces using FACT approach. In this section, we first categorize a more comprehensive set of commonly used flexure primitives and derive their freedom and constraint spaces. Then we will discuss a general synthesis methodology for constructing serial and parallel kinematic chains of these flexure primitives.

3.1 Commonly used flexure primitives

According to the mobility or the rank of their twist system, we can categorize the most commonly used flexure primitives as shown in Table 1.

1. A notch hinge, denoted by symbol "R", allows a rotation about the centerline and constraints other motions. The shape of the cross section may be circular, elliptical, hyperbolic etc. Meanwhile short beams, living hinges that have one dimension significantly smaller than others can function as a notch hinge. A split tube which is a tube sliced along its longitudinal direction also allows a single rotation about its axis. If we define the axis of the R-joint to be the z-axis, its twist and wrench matrices can be written as

$$[T_r] = \begin{bmatrix} \hat{R}_z \end{bmatrix}, \quad [W_r] = \begin{bmatrix} \hat{F}_x & \hat{F}_y & \hat{F}_z & \hat{M}_x & \hat{M}_y \end{bmatrix} \quad (12)$$

2. A spherical notch or short wires/rod, denoted by "S", allows three rotations and constrains three translations. Kinematically it is equivalent to a serial chain of three R-joints, i.e. S=3R. Their twist and wrench matrices are

$$[T_s] = \begin{bmatrix} \hat{R}_x & \hat{R}_y & \hat{R}_z \end{bmatrix}, \quad [W_s] = \begin{bmatrix} \hat{F}_x & \hat{F}_y & \hat{F}_z \end{bmatrix} \quad (13)$$

3. Blade/sheet flexures also called leaf springs, denoted by "B", allow two rotations and one translation, i.e. $B = 2R - P$. The rotational symmetric cylinders and disc rings also have the same freedom and constraint spaces. As shown in Table 1, we define the normal of the blade as the y-axis and the longitudinal direction being the x-axis. The two rotations are about two in-plane axes (x, z) due to beam torsion and beam bending respectively. And the translation is along the normal direction y, caused by beam bending. From the constraint point of view, a blade removes one rotation and two translations. Their twist and wrench matrices are

$$[T_b] = \begin{bmatrix} \hat{R}_x & \hat{R}_z & \hat{P}_y \end{bmatrix}, \quad [W_b] = \begin{bmatrix} \hat{F}_x & \hat{F}_z & \hat{M}_y \end{bmatrix} \quad (14)$$

4. A long wire/rod flexure, denoted by "W", removes the translation along its axis and allows the other five motions. A corner blade (a folded sheet) also provides a single constraint along its fold line. Their freedom space is equivalent to three R-joints and two P-joints, i.e. W=3R-2P. Mathematically its corresponding twist and wrench matrices are

$$[T_w] = \begin{bmatrix} \hat{R}_x & \hat{R}_y & \hat{R}_z & \hat{P}_y & \hat{P}_z \end{bmatrix}, \quad [W_w] = \begin{bmatrix} \hat{F}_x \end{bmatrix} \quad (15)$$

5. And lastly, a bellow spring, denoted by "B_s", removes a single rotation along its axis with a freedom space denoted by B_s=2R-3P. Its twist and wrench matrices are

$$[T_{bs}] = \begin{bmatrix} \hat{R}_x & \hat{R}_y & \hat{P}_x & \hat{P}_y & \hat{P}_z \end{bmatrix}, \quad [W_{bs}] = \begin{bmatrix} \hat{M}_z \end{bmatrix} \quad (16)$$

Table 1 summarizes the aforementioned flexure primitives and their freedom space and twist and wrench matrices. These primitives are basic building blocks for constructing more complex flexure systems.

3.2 Serial chains of flexure primitives

The freedom space of a rigid body represents all of its allowable motion in space. For any given flexure element or building block, its freedom space is a f system that is represented by a twist matrix $[\mathbf{T}]$ formed by f independent twists, written as

$$[\mathbf{T}] = \begin{bmatrix} \hat{T}_1 & \hat{T}_2 & \cdots \hat{T}_f \end{bmatrix} \quad (17)$$

When a mechanism is formed by m flexure elements or building blocks that are connected in serial, its twist matrix (freedom space) can be obtained by combining the twist matrix of each building block column-wise, mathematically written as

$$[\mathbf{T}] = [Ad_1T_1 \quad Ad_2T_2 \quad \cdots \quad Ad_mT_m] \quad (18)$$

where $[Ad_j]$ represent the coordinate transformation from the j-th building block to the functional stage, defined in (9). This formulation has been previously presented by Su (2011). Also see similar work by Hopkins (2007b); Hopkins and Culpepper (2011).

By applying a column-wise reduction, we can easily obtain a set of independent $r = \text{rank}(T)$ twists that forms the basis of the freedom space. r is also called the mobility or degree-of-freedom of the mechanism. Any set of r independent twists called basis can span the entire freedom space.

Table 1. The motion and constraint spaces of commonly used flexure primitives.

Flexure	Freedom Symbol	[T]	[W]
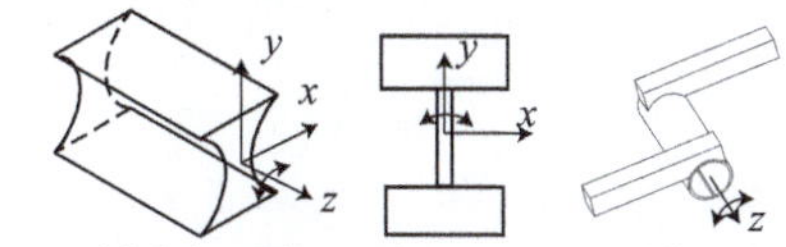 Notch/Living Hinge Short Beam Split Tube	R	$\left[\hat{R}_z\right]$	$\left[\hat{F}_x \quad \hat{F}_y \quad \hat{F}_z \quad \hat{M}_x \quad \hat{M}_y\right]$
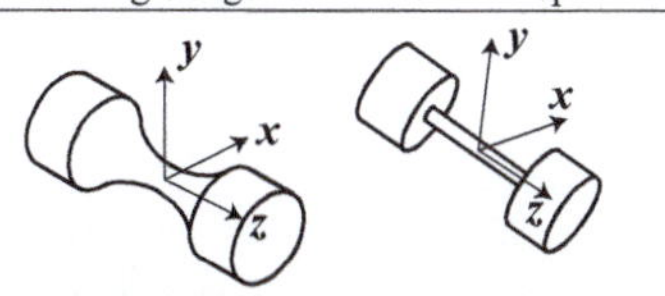 Spherical Notch Short Wire/Rod	S=3R	$\left[\hat{R}_x \quad \hat{R}_y \quad \hat{R}_z\right]$	$\left[\hat{F}_x \quad \hat{F}_y \quad \hat{F}_z\right]$
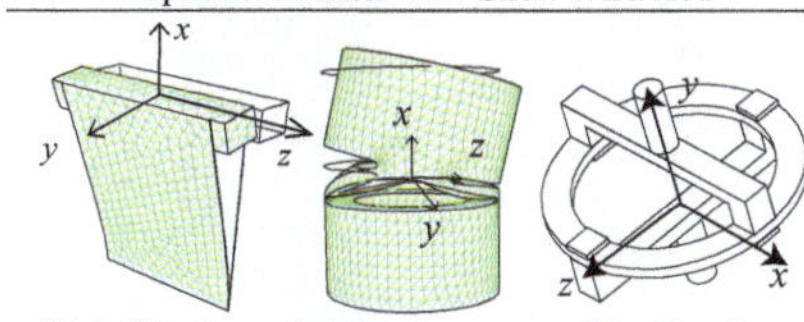 Blade/Sheet/ Leaf Spring; Rotational Symmetric Cylinder; Disc Coupling	B=2R-P	$\left[\hat{R}_x \quad \hat{R}_z \quad \hat{P}_y\right]$	$\left[\hat{F}_x \quad \hat{F}_z \quad \hat{M}_y\right]$
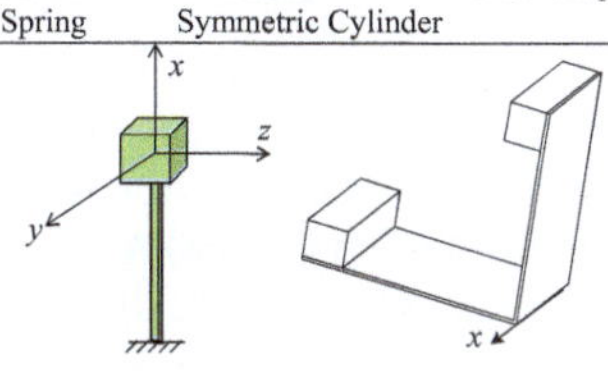 Long Wire/Rod Corner Blade	W=3R-2P	$\left[\hat{R}_x \quad \hat{R}_y \quad \hat{R}_z \quad \hat{P}_y \quad \hat{P}_z\right]$	$\left[\hat{F}_x\right]$
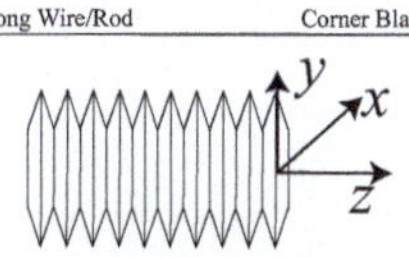 Bellow Spring	B_s=2R-3P	$\left[\hat{R}_x \quad \hat{R}_y \quad \hat{P}_x \quad \hat{P}_y \quad \hat{P}_z\right]$	$\left[\hat{M}_z\right]$

The transformation from one basis to another is done by a linear operation to the twist matrix.

A freedom space can be spanned by different sets of basis twists. This can be very useful in design practices. For instance, consider a freedom space formed by rotations about two parallel axes. The twist matrix of these two rotations is

$$[\mathbf{T}] = \left[\hat{T}_{R_1} \quad \hat{T}_{R_2}\right] = \begin{bmatrix} \mathbf{\Omega} & \mathbf{\Omega} \\ \mathbf{c}_1 \times \mathbf{\Omega} & \mathbf{c}_2 \times \mathbf{\Omega} \end{bmatrix} \tag{19}$$

Subtracting the second column from the first one yields

$$[\mathbf{T}] \doteq \begin{bmatrix} \mathbf{\Omega} & \mathbf{0} \\ \mathbf{c}_1 \times \mathbf{\Omega} & (\mathbf{c}_2 - \mathbf{c}_1) \times \mathbf{\Omega} \end{bmatrix} \tag{20}$$

where $\doteq$ represents a column-wise linear operation. Note the second column of the above matrix represents a translation along the direction normal to both the rotation axis and the line $\mathbf{c}_1\mathbf{c}_2$. Basically this means that a serial chain of two rotations is equivalent to a serial chain of a rotation and a translation. See the top part of Fig. 1.

Similarly, a serial chain of three (non-coplanar) parallel rotations is equivalent to a rotation plus two orthogonal translations. They are formulated in screws as

$$[\mathbf{T}] = \left[\hat{T}_{R_1}\hat{T}_{R_2} \quad \hat{T}_{R_3}\right] = \begin{bmatrix} \mathbf{\Omega} & \mathbf{\Omega} & \mathbf{\Omega} \\ \mathbf{c}_1 \times \mathbf{\Omega} & \mathbf{c}_2 \times \mathbf{\Omega} & \mathbf{c}_3 \times \mathbf{\Omega} \end{bmatrix} \tag{21}$$

$$\doteq \begin{bmatrix} \mathbf{\Omega} & \mathbf{0} & \mathbf{0} \\ \mathbf{c}_1 \times \mathbf{\Omega} & (\mathbf{c}_2 - \mathbf{c}_1) \times \mathbf{\Omega} & (\mathbf{c}_3 - \mathbf{c}_1) \times \mathbf{\Omega} \end{bmatrix}, \tag{22}$$

where the second and third columns represent translations. See the bottom part of Fig. 1 for illustration of this case.

3.3 Parallel chains of flexure primitives

The constraint space of a rigid body represents all the forbidden motions of the body subject to a constraint arrangement. In screw theory, a constraint space can be represented by a wrench matrix $[\mathbf{W}]$ combined by c independent wrenches, written as

$$[\mathbf{W}] = \left[\hat{W}_1 \quad \hat{W}_2 \quad \cdots \quad \hat{W}_c\right] \tag{23}$$

The constraint space of a flexure mechanism formed by m building blocks connected in parallel is given by a wrench matrix that is obtained by assembling the wrench matrix of each building block, written as

$$[\mathbf{W}] = [Ad_1 W_1 \quad Ad_2 W_2 \quad \cdots \quad Ad_m W_m] \tag{24}$$

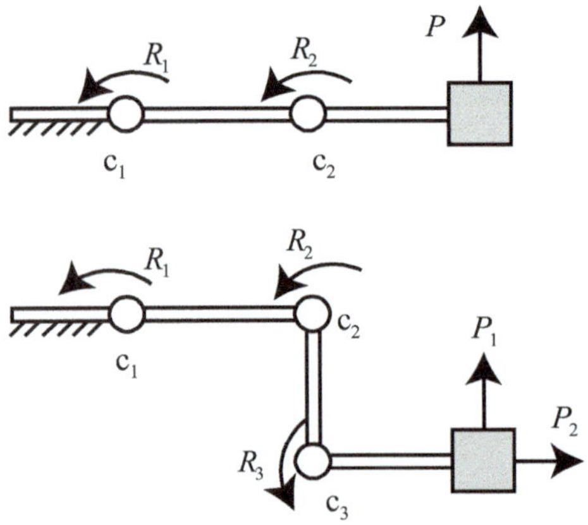

Figure 1. Equivalent freedom spaces. (Top) A serial chain of two parallel rotations produces a translation. (Bottom) A serial chain of three (non-coplanar) parallel rotations produce two translation.

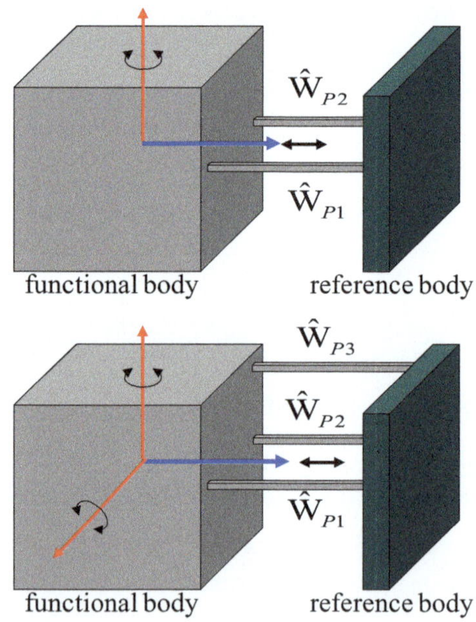

Figure 2. Equivalent constraint spaces. (Top) A parallel chain of two parallel translational constraints exerts a translational plus a rotational constraint to the functional body. (Bottom) A parallel chain of three (non-coplanar) parallel translational constraints exerts a translation plus two rotational constraints. The arrows represent the removed motions.

Again, matrices $[\mathbf{Ad}_j]$ represent the coordinate transformation from j-th building block to the functional stage.

Similar to serial chains of flexures, the constraint space of two parallel translational constraints assembled in parallel (Fig. 2) is represented by two force wrenches with identical direction **F**

$$[\mathbf{W}] = \left[\hat{W}_{P_1} \quad \hat{W}_{P_2}\right] = \begin{bmatrix} \mathbf{F} & \mathbf{F} \\ \mathbf{c}_1 \times \mathbf{F} & \mathbf{c}_2 \times \mathbf{F} \end{bmatrix} \tag{25}$$

Subtracting the second column from the first one yields

$$[\mathbf{W}] = \begin{bmatrix} \mathbf{F} & \mathbf{0} \\ \mathbf{c}_1 \times \mathbf{F} & (\mathbf{c}_2 - \mathbf{c}_1) \times \mathbf{F} \end{bmatrix} \tag{26}$$

Note the second column of the above matrix represents a rotational constraint along the direction perpendicular to both the constraint axis **F** and the line $\mathbf{c}_1\mathbf{c}_2$. Basically this says that a parallel chain of two translational constraints removes one rotation and one translation.

And a parallel chain of three (non-coplanar) parallel translational constraints removes two rotations and one translation. This is illustrated with the following mathematical formulation

$$[\mathbf{W}] = \left[\hat{W}_{P_1}\hat{W}_{P_2}\hat{W}_{P_3}\right] = \begin{bmatrix} \mathbf{FFF} \\ \mathbf{c}_1 \times \mathbf{F}\mathbf{c}_2 \times \mathbf{F}\mathbf{c}_3 \times \mathbf{F} \end{bmatrix} \tag{27}$$

$$\doteq \begin{bmatrix} \mathbf{F00} \\ \mathbf{c}_1 \times \mathbf{F}(\mathbf{c}_2 - \mathbf{c}_1) \times \mathbf{F}(\mathbf{c}_3 - \mathbf{c}_1) \times \mathbf{F} \end{bmatrix} \tag{28}$$

4 Synthesis of freedom elements

In this section, we use flexure primitives listed in Table 1 to design flexure joints that have only one DOF, rotational (R) or translational (P). To design freedom elements, we use parallel chains of primitives to remove unwanted freedoms.

4.1 Synthesis of R-joints $\hat{T}_R$

The freedom space of an R-joint is given by $\hat{T}_R$ shown in (4). Without loss of generality, we assume the axis of R-joint through the origin, i.e. $\mathbf{c} = (0,0,0)^T$. This gives us

$$\hat{T}_R = \begin{Bmatrix} \mathbf{\Omega} \\ \mathbf{0} \end{Bmatrix} \tag{29}$$

Our goal is to remove the other five unwanted motions: two rotations and three translations. Since we consider parallel structures of at least two limbs and each limb must apply 1–5 constraints to the functional body, primitive “R” (notch hinge) cannot be used in the design as it is indeed a R-joint. This leads us total 11 possible combinations with the other four primitives B, S, W and B_s. Among them, cases 1-5 use no bellow springs and cases 6–11 use at least one bellow spring.

Use flexure primitives B and W only

The first three cases concerns designs with primitives B and W only: 2B, B-2W, 5W. These are the most commonly used designs.

4.1.1 Case 1: 2B

The synthesis with two parallel blades (2B) is described as the following. As shown in Fig. 3, let us denote the

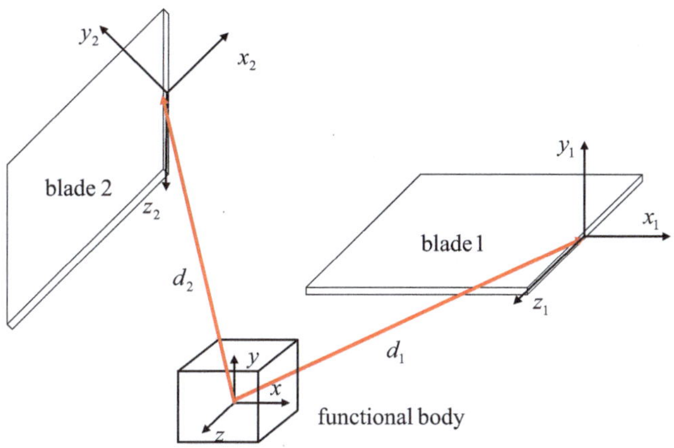

Figure 3. Synthesis of R-joints with a parallel connection of two blades. The two blades are rigidly connected to the functional body.

coordinate transformation of two blades relative to the functional body by $[\mathbf{Ad}_1]$ and $[\mathbf{Ad}_2]$ which are the six by six adjoint matrices with the form of (9). By applying the coordinate transformation to both blades, their wrench matrices are calculated by (10,11) as

$$[\mathbf{W}_i] = [\mathbf{Ad}_i][\mathbf{W}_b] = \begin{bmatrix} \mathbf{x}_i & \mathbf{z}_i & \mathbf{0} \\ d_i \times \mathbf{x}_i & d_i \times \mathbf{z}_i & \mathbf{y}_i \end{bmatrix}, \quad i = 1, 2, \tag{30}$$

where $[\mathbf{W}_b]$ is the wrench matrix of blade written in (14). And vectors $\mathbf{x}_i, \mathbf{y}_i, \mathbf{z}_i, d_i$ denote the orientation and position of i-th blade.

Requiring the reciprocity of the $[\mathbf{W}_i]$ with $\hat{T}_R$ yields the necessary condonations regarding to the orientation and position of the blades

$$\boldsymbol{\Omega} \cdot (d_i \times \mathbf{x}_i) = 0, \implies \mathbf{x}_i \cdot (d_i \times \boldsymbol{\Omega}) = 0 \tag{31}$$

$$\boldsymbol{\Omega} \cdot (d_i \times \mathbf{z}_i) = 0, \implies \mathbf{z}_i \cdot (d_i \times \boldsymbol{\Omega}) = 0 \tag{32}$$

$$\boldsymbol{\Omega} \cdot \mathbf{y}_i = 0. \tag{33}$$

Simplifying the first two conditions yields

$$d_i \cdot \mathbf{y}_i = 0, \quad \boldsymbol{\Omega} \cdot \mathbf{y}_i = 0. \tag{34}$$

The first equality is interpreted as that the translation d_i must in the blade plane $\mathbf{x}_i\mathbf{z}_i$. The second condition means that $\mathbf{y}_i$, the normal direction of the blade, must be perpendicular to the axis of the R-joint ($\boldsymbol{\Omega}$). If $\boldsymbol{\Omega} = (1,0,0)^T$, $\mathbf{y}_i$ must be in the plane yz of the functional body. Picking any two independent directions in the yz plane leads us a solution, e.g.

$$\mathbf{y}_1 = (0,1,0)^T, \quad \mathbf{y}_2 = (0,0,1)^T \tag{35}$$

Once $\mathbf{y}_i$ is determined. The axes of $\mathbf{x}_i$ and $\mathbf{z}_i$ are chosen arbitrarily as long as the two blades are not co-planar (redundant).

Depending on how $\mathbf{x}_i$ and $\mathbf{z}_i$ are chosen for each blade, we can obtain three possible designs. If $\mathbf{z}_1 = \mathbf{z}_2 = (1,0,0)^T$, we have a design with both blades subject to bending, shown in Fig. 4a. This is the most often used flexure hinge design. If $\mathbf{x}_1 = \mathbf{z}_2 = (1,0,0)^T$, we have a design with blade 1 subject to twisting and blade 2 subject to bending, Fig. 4b. If $\mathbf{x}_1 = \mathbf{x}_2 = (1,0,0)^T$, we have a design with both blades subject to twisting, Fig. 4c.

4.1.2 Case 2: B-2W

This design evolves from the case 2B. Since a blade allows two rotations and one translation, we just need to remove one rotation and one translation with two wires. We have shown in (27) that two parallel constraints remove one translation and one rotation. Therefore, replacing one blade of the 2B design with two parallel wires yields a B-2W design, shown in Fig. 4d.

4.1.3 Case 3: 5W

In order to obtain a hinge design with five wires, we simply replace the blade of the B-2W design with three co-planar wires. This is due to the fact that a blade is equivalent to three co-planar wires. See Fig. 4e. An alternative synthesis procedure using screw algebra for the case five wires can be found in Su and Tari (2010).

Use flexure primitives S, B, W

Now we consider designs using primitives S, B, W with at least one S-joint. There are three possible combinations: 2S, S-2W, B-S.

However the case B-S is not qualified as a R-joint for the following reasons. Without loss of generality, we let the coordinate system align with the local coordinate system of the blade shown in Table 1. And we denote the position of S-joint by $\boldsymbol{d} = (d_x, d_y, d_z)^T$. Therefore the wrench matrix of the blade and the S-joint are

$$[\mathbf{W}_b] = \begin{bmatrix} \hat{R}_x & \hat{R}_z & \hat{P}_y \end{bmatrix}, \; [\mathbf{W}'_s] = \begin{bmatrix} \boldsymbol{i} & \boldsymbol{j} & \boldsymbol{k} \\ \boldsymbol{d} \times \boldsymbol{i} & \boldsymbol{d} \times \boldsymbol{j} & \boldsymbol{d} \times \boldsymbol{k} \end{bmatrix} \tag{36}$$

where vectors $\boldsymbol{i}, \boldsymbol{j}, \boldsymbol{k}$ are unit vectors along three coordinate axes. And $[\mathbf{W}'_s]$ is the wrench matrix with an appropriate coordinate transformation, i.e. $[\mathbf{W}'_s] = [\mathbf{Ad}][\mathbf{W}_s]$. We write the wrench matrix of the parallel structure B-S and apply a linear operation to obtain

$$[\mathbf{W}_{bs}] = \begin{bmatrix} \mathbf{W}_b & \mathbf{W}'_s \end{bmatrix} = \begin{bmatrix} 1 & 0 & 0 & 1 & 0 & 0 \\ 0 & 0 & 0 & 0 & 1 & 0 \\ 0 & 1 & 0 & 0 & 0 & 1 \\ 0 & 0 & 0 & 0 & -d_z & d_y \\ 0 & 0 & 1 & d_z & 0 & -d_x \\ 0 & 0 & 0 & -d_y & d_x & 0 \end{bmatrix} \tag{37}$$

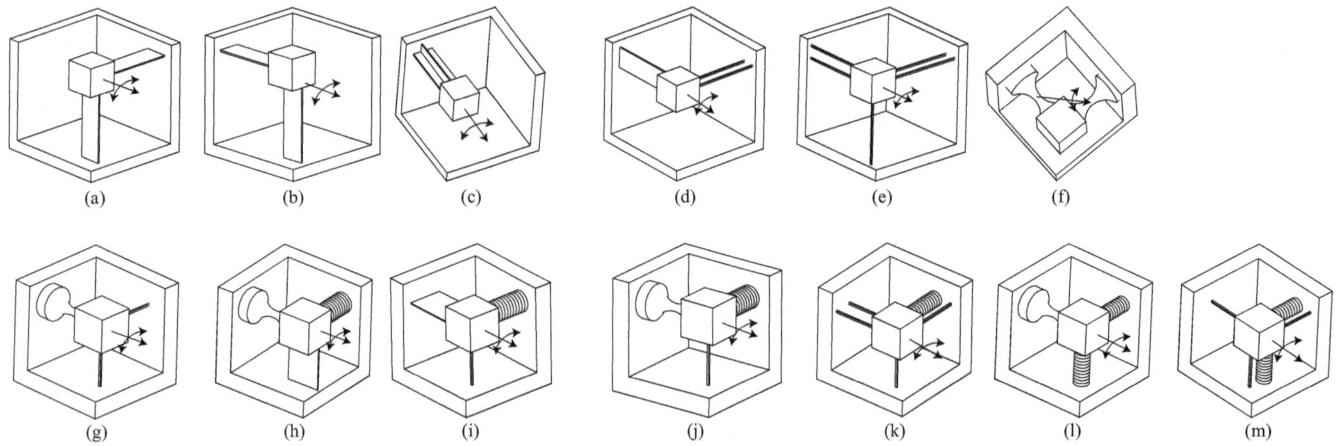

Figure 4. Various designs of R-joints with flexure primitives: B, W, S and B_s. **(a–c)** case 1 with two blades (BB). **(d)** case 2 with one blade and two wires (B-2W). **(e)** case 3 with five wires (5W). **(f)** case 4 with two spherical notches (2S). **(g)** case 5 with one spherical notch with two wires (S-2W). **(h)** case 6 with one bellow spring, one blade and one spherical notch (B_s-B-S). **(i)** case 7 with one bellow spring, one blade and one wire (B_s-B-W). **(j)** case 8 with one bellow spring, one spherical notch and one wire (B_s-S-W). **(k)** case 9 with one bellow spring and four wires (B_s-4W). **(l)** case 10 with two bellow springs and one spherical notch ($2B_s$-S). **(m)** case 11 with two bellow springs and three wires ($2B_s$-3W). The double arrow arcs represent the rotation allowed by flexure R-joints. The box represents the functional body.

$$\doteq \begin{bmatrix} 1 & 0 & 0 & 0 & 0 & 0 \\ 0 & 0 & 0 & 0 & 1 & 0 \\ 0 & 1 & 0 & 0 & 0 & 0 \\ 0 & 0 & 0 & 0 & -d_z & d_y \\ 0 & 0 & 1 & 0 & 0 & 0 \\ 0 & 0 & 0 & -d_y & d_x & 0 \end{bmatrix} \tag{38}$$

From Eq. (38), we can draw two conclusions. (1) If $d_y \neq 0$, $[\mathbf{W}_{bs}]$ has a full rank which means that the body is fully constrained by this structure. (2) If $d_y = 0$, i.e. the S-joint is on the blade plane, $[\mathbf{W}_{bs}]$ has a rank of 2 since the 4th and 6th column would be zeros. As a matter of fact, its reciprocal freedom space is formed by the two rotations allowed by the blade flexure. Therefore, parallel structures of B-S cannot form a R-joint design.

4.1.4 Case 4: 2S

The 2S case is trivial. When two S-joints are connected in parallel, this results in a single rotation about the line connecting to their joint centers, Fig. 4f.

4.1.5 Case 5: S-2W

The case S-2W is synthesized as the following. The S-joint itself removes three translations. We just need to remove two additional rotations with two wire flexures. By using a wire parallel to y-axis with a non-zero offset $\boldsymbol{d} = (d,0,0)^T$, the rotation $\hat{R}_z$ is removed. Similarly, a wire parallel to z-axis with a non-zero offset $\boldsymbol{d} = (d,0,0)^T$ removes the rotation $\hat{R}_y$. Mathematically, the wrench matrix of S-joint and the two wires are

$$[\mathbf{W}_s] = \begin{bmatrix} 1 & 0 & 0 \\ 0 & 1 & 0 \\ 0 & 0 & 1 \\ 0 & 0 & 0 \\ 0 & 0 & 0 \\ 0 & 0 & 0 \end{bmatrix}, \quad [\mathbf{W}_{2w}] = \begin{bmatrix} 0 & 0 \\ 1 & 0 \\ 0 & 1 \\ 0 & 0 \\ 0 & -d \\ d & 0 \end{bmatrix} \tag{39}$$

The constraint space of S-2W is obtained by combining $[\mathbf{W}_s]$ and $[\mathbf{W}_{2w}]$, to which we apply a column-wise reduction and obtain

$$[\mathbf{W}_{s\text{-}2w}] = \begin{bmatrix} \mathbf{W}_s & \mathbf{W}_{2w} \end{bmatrix} = \begin{bmatrix} 1 & 0 & 0 & 0 & 0 \\ 0 & 1 & 0 & 0 & 0 \\ 0 & 0 & 1 & 0 & 0 \\ 0 & 0 & 0 & 0 & 0 \\ 0 & 0 & 0 & 0 & -d \\ 0 & 0 & 0 & d & 0 \end{bmatrix} \tag{40}$$

Obviously its reciprocal twist matrix is $\hat{R}_x$ which corresponds to a single rotation about the x-axis. This design is shown in Fig. 4g.

Use flexure primitives B_s, S, B, W

At last, we synthesize the design cases with at least one bellow spring. If one B_s flexure is used, there are four possible design cases: B_s-B-S, B_s-B-W, B_s-S-W, B_s-4W. And if two bellow springs are used, there are two additional combinations: $2B_s$-S and $2B_s$-3W. Note $2B_s$-B is not possible as B removes one rotation and $2B_s$ remove another two rotation which results in a freedom space with no rotation. Let us discuss these six cases in the following.

4.1.6 Case 6: B_s-B-S:

From Eq. (38), we have concluded that a B-S parallel structure may result in a freedom space of two rotations if the center of S-joint is on the blade plane. Now we just need to use a bellow spring B_s to remove one rotation in order to yield a freedom space of only one rotation, hence a hinge design in Fig. 4h.

4.1.7 Case 7: B_s-B-W:

As we know one blade has a freedom space of 2R-P. To obtain a single rotation, we can use a bellow spring to remove one rotation and a wire to remove the translation. See this design in Fig. 4i.

4.1.8 Case 8: B_s-S-W:

The S-joint has a freedom space of 3R. To obtain a single rotation, we can use a bellow spring to remove one rotation then use a wire offset to the center of the S-joint to remove the second rotation. This results a design shown in Fig. 4j.

4.1.9 Case 9: B_s-4W:

To synthesize this case, we first use three wires aligning the three coordinate axes to remove three translations. This leaves us three rotations. We then remove two more rotations by two more wires. By Eq. (26), we know that two parallel constraints remove one additional rotation. Therefore we let a fourth wire be parallel to one of the three wires aforementioned. This fourth wire removes a rotation about the normal line to the plan formed by the two parallel wires. At last, a bellow spring is used to remove another rotation. This results a hinge design shown in Fig. 4k.

4.1.10 Cases 10: $2B_s$-S

The synthesis of $2B_s$-S case is simple. The S-joint removes three translations and two bellow springs remove two of the three rotations, Fig. 4l.

4.1.11 Cases 11: $2B_s$-3W

This case is also trivial. First the three wires aligning the three coordinate axes remove three translations. And we then use two bellow springs remove two rotations. This design is shown in Fig. 4m.

4.2 Synthesis of P-Joints $\hat{T}_P$

Now let us synthesize P-joints with flexure primitives listed in Table 1. The freedom space of an P-joint is given by a single twist

$$\hat{T}_P = \begin{Bmatrix} \mathbf{0} \\ \mathbf{V} \end{Bmatrix} \tag{41}$$

where vector $\mathbf{V}$ represent the translational direction of P-joint.

We are interested in simple parallel structures with at least two limbs that remove three rotations and two translations. And each limb must apply at least one constraint and allow at least one translational motion for the functional body. Therefore primitives "R" and "S" are not qualified as they allow no translation. This leaves three possible primitives: B, W and B_s from which we can have eight possible cases. Three of them use no bellow springs and the other five use at least one bellow spring.

Designs with flexures B, W only

Considering only flexure B and W, we have three cases: 2B, B-2W and 5W.

4.2.1 Case 1: 2B

For designs with two blades, the reciprocity conditions of $\hat{T}_P$ in (41) with the wrench matrices $[\mathbf{W}_i]$ in (30) yields

$$\mathbf{V} \cdot \mathbf{x}_i = 0, \quad \mathbf{V} \cdot \mathbf{z}_i = 0, \quad i = 1,2 \tag{42}$$

from which we conclude that $\mathbf{V}$ must be parallel to both $\mathbf{y}_i$ (normal of the blade plane). This means that two blades must be parallel with the normal of the blade plane along the axis of P-joint. This is the well known parallel sheet design of P-joints shown in Fig. 5a.

4.2.2 Case 2: B-2W

The design with one blade and two wires can be easily obtained by replacing one blade with two wires that are parallel to the blade plane. Figure 5b.

4.2.3 Case 3: 5W

This design is obtained from the B-2W design by replacing one blade with three co-planar wires and the other one with two wires. This case can also be computationally synthesized with screw theory. Assume the P-joint is along the x direction, i.e. $\mathbf{V} = (1,0,0)^T$. First of all, compute the reciprocal wrench matrix of $\hat{P}$ using linear algebra as

$$[\mathbf{W}] = \begin{bmatrix} 0 & 0 & 0 & 0 & 0 \\ 1 & 0 & 0 & 0 & 0 \\ 0 & 1 & 0 & 0 & 0 \\ 0 & 0 & 1 & 0 & 0 \\ 0 & 0 & 0 & 1 & 0 \\ 0 & 0 & 0 & 0 & 1 \end{bmatrix} \tag{43}$$

We then apply a column-wise linear operation to this wrench matrix to obtain

$$[\mathbf{W}'] = \begin{bmatrix} 0 & 0 & 0 & 0 & 0 \\ 1 & 1 & 0 & 0 & 1 \\ 0 & 0 & 1 & 1 & 0 \\ 0 & 0 & 0 & 0 & 1 \\ 0 & 0 & 0 & 1 & 0 \\ 0 & 1 & 0 & 0 & 0 \end{bmatrix} \tag{44}$$

Each column of $[\mathbf{W}']$ is a force wrench which can be realized with a long wire flexure (W). The 5W design is shown in Fig. 5b. See the work by Su and Tari (2010) for an alternative synthesis procedure for this case.

Designs with flexures B, W and B_s

There are five cases with at least one bellow spring: B_s-B-W, B_s-4W, 2B_s-B, 2B_s-3W, 3B_s-2W. We discuss each in the following.

4.2.4 Case 4: B_s-B-W

A blade B has a freedom space of 2R-P. To obtain a P-joint design, we must remove the two rotations. First we use a wire flexure with axis being parallel and has a non-zero distance to the blade plane to remove one rotation. The second rotation is removed by a bellow spring flexure. See Fig. 5d.

4.2.5 Case 5: B_s-4W

This design can be obtained from the case B_s-B-W by replacing the blade with three co-planar wires. A computational way to synthesize this case is applying an alternative linear operation to the wrench matrix $[\mathbf{W}]$ in (43). This time we would like to have four force wrenches and one couple wrench. The new wrench matrix is written as

$$[\mathbf{W}''] = \begin{bmatrix} 0 & 0 & 0 & 0 & 0 \\ 1 & 1 & 0 & 0 & 1 \\ 0 & 0 & 1 & 0 & 0 \\ 0 & 0 & 0 & 0 & 1 \\ 0 & 0 & 0 & 1 & 0 \\ 0 & 1 & 0 & 0 & 0 \end{bmatrix} \tag{45}$$

of which the fourth column is a couple wrench and the other four are force wrenches. Realizing each force wrench with a wire and the couple wrench with a bellow spring yields the design of B_s-4W. See Fig. 5e.

4.2.6 Case 6: 2B_s-B

This case is trivial since the two bellow springs remove the two rotations of the blade. This leaves a freedom space with a single translation, hence a P-joint design shown in Fig. 5f.

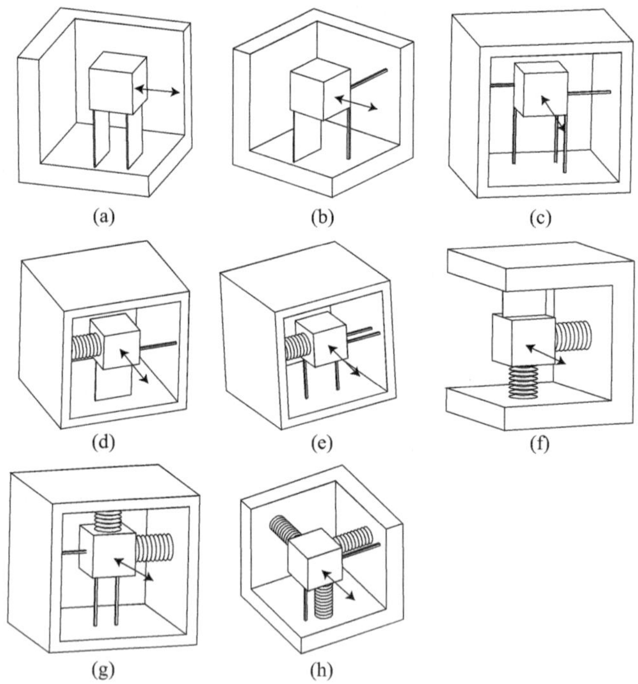

Figure 5. Various designs of P-joints with parallel structures of flexure primitives: B, W, B_s. **(a)** case 1 with two blades (2B). **(b)** case 2 with one blade and two wires (B-2W). **(c)** case 3 with five wires. **(d)** case 4 with one bellow spring, one blade and one wire (B_s-B-W). **(e)** case 5 with one bellow spring and four wires (B_s-4W). **(f)** case 6 with two bellow springs and one blade (2B_s-B). **(g)** case 7 with two bellow springs and three wires (2B_s-3W). **(h)** case 8 with three bellow springs and two wires (3B_s-2W). The arrowed lines indicate the direction of translation. The box represents the functional body.

4.2.7 Case 7: 2B_s-3W

This case evolves from case 2B_s-B by replacing the blade with three co-planar wires. See Fig. 5g.

4.2.8 Case 8: 3B_s-2W

For this case, the 3 bellow springs remove three translations and two wires remove two translations. This design is shown in Fig. 5h.

5 Synthesis of constraint elements

As a duality to the design of freedom elements, in this section, we design basic constraint elements $\hat{W}_R$ and $\hat{W}_P$ which remove one rotation and one translation respectively. These constraint elements are most often used in designing parallel structures for removing a particular set of motions. The basis approach is to use a serial chain of flexure primitives listed in Table 1 to construct the complementary freedom space of the constraint element to be synthesized.

5.1 Synthesis of P-constraints $\hat{W}_P$

A P-constraint element removes a translational freedom of the functional body while allows all other motions. The constraint space of the translational constraint is given by a pure force given in (5).Without loss of generality, we assume the axis of translational constraint through the origin of the coordinate system, i.e. $\mathbf{c} = (0,0,0)^T$. That is

$$\hat{W}_P = \begin{Bmatrix} \mathbf{F} \\ \mathbf{0} \end{Bmatrix} \tag{46}$$

whose complementary freedom space consists of three rotations plus two translations, i.e. 3R-2P.

To synthesize the translational constraint $\hat{W}_P$, we consider serial chains of at least two flexure primitives to make up the complementary freedom space of $\hat{W}_P$. We do not include the long wire (W) flexure in the design as itself is indeed a translational constraint. Also bellow springs (B_s) are not qualified either as they allow three translations. As a result, only three types of primitives S, B and R are considered in design. Based on the degree of freedom, there are six possible combinations: S-S, S-B, S-2R, B-B, B-2R and 5R. However these primitives must satisfy some restriction on their relative orientation and position. In what follows, we derive the geometric conditions for each case.

5.1.1 Case 1: S-S

By intuition, we know that a serial chain of two S-joints removes one translation along the line connecting the center of the joints. Here let us give a mathematical proof with screw theory. Without loss of generality, we assume one S-joint being at the origin of the coordinate system and the other one being at a distance of d along x-axis, i.e. $\boldsymbol{d} = (d,0,0)^T$. The twist matrices of these two blades are

$$[\mathbf{T}_1] = \begin{bmatrix} i & j & k \\ \mathbf{0} & \mathbf{0} & \mathbf{0} \end{bmatrix}, \quad [\mathbf{T}_2] = \begin{bmatrix} i & j & k \\ d\times i & d\times j & d\times k \end{bmatrix} \tag{47}$$

By the formulation for serial structures in Eq.(18), the twist matrix of the S-S chain is written as

$$[\mathbf{T}] = \begin{bmatrix} \mathbf{T}_1 & \mathbf{T}_2 \end{bmatrix} = \begin{bmatrix} i & j & k & i & j & k \\ \mathbf{0} & \mathbf{0} & \mathbf{0} & d\times i & d\times j & d\times k \end{bmatrix} \tag{48}$$

Substituting $\boldsymbol{d} = (d,0,0)^T$ and subtracting the last three columns from the first three columns yield

$$[\mathbf{T}'] = \begin{bmatrix} i & j & k & \mathbf{0} & \mathbf{0} & \mathbf{0} \\ \mathbf{0} & \mathbf{0} & \mathbf{0} & \mathbf{0} & dk & -dj \end{bmatrix} \tag{49}$$

which indicates three rotations and two translation along y- and z-axes, i.e. the translation along x-axis is removed. See Fig. 7a. The arrow line represents the removed translation.

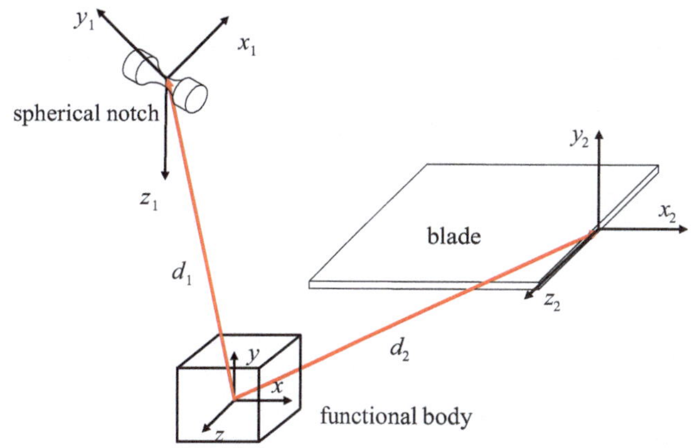

Figure 6. Synthesis of P-constraint with a serial chain of one blade (B) and one spherical notch (S). The blade is connected with the spherical notch via an intermediate body (not drawn)

5.1.2 Case 2: S-B

Figure 6 shows a serial chain of S and B flexures. Applying a general coordinate transformation to the twist matrices of the blade and the spherical notch in (13, 14) yields

$$[\mathbf{T}_s] = \begin{bmatrix} \mathbf{x}_1 & \mathbf{y}_1 & \mathbf{z}_1 \\ d_1\times\mathbf{x}_1 & d_1\times\mathbf{y}_1 & d_1\times\mathbf{z}_1 \end{bmatrix}, \tag{50}$$

$$[\mathbf{T}_b] = \begin{bmatrix} \mathbf{x}_2 & \mathbf{z}_2 & \mathbf{0} \\ d_2\times\mathbf{x}_2 & d_2\times\mathbf{z}_2 & \mathbf{y}_2 \end{bmatrix} \tag{51}$$

Since both freedom spaces are complementary to the prescribed translational constraint, $\hat{W}_P$ in (46) is reciprocal to both $[\mathbf{T}_s]$ and $[\mathbf{T}_b]$. The reciprocity condition of S-joint leads to

$$\mathbf{F}\cdot(d_1\times\mathbf{x}_1) = 0 \quad \Longrightarrow \quad \mathbf{x}_1\cdot(\mathbf{F}\times d_1) = 0 \tag{52}$$
$$\mathbf{F}\cdot(d_1\times\mathbf{y}_1) = 0 \quad \Longrightarrow \quad \mathbf{y}_1\cdot(\mathbf{F}\times d_1) = 0 \tag{53}$$
$$\mathbf{F}\cdot(d_1\times\mathbf{z}_1) = 0 \quad \Longrightarrow \quad \mathbf{z}_1\cdot(\mathbf{F}\times d_1) = 0 \tag{54}$$

which are reduced to

$$\mathbf{F}\times d_1 = \mathbf{0}, \quad \Longrightarrow \quad d_1 = d_1\mathbf{F} \tag{55}$$

This implies that the position of the S-joint must be along the direction of $\mathbf{F}$.

And the reciprocity requirement for the blade leads to

$$\mathbf{F}\cdot(d_2\times\mathbf{x}_2) = 0 \quad \Longrightarrow \quad \mathbf{x}_2\cdot(\mathbf{F}\times d_2) = 0 \tag{56}$$
$$\mathbf{F}\cdot(d_2\times\mathbf{z}_2) = 0 \quad \Longrightarrow \quad \mathbf{z}_2\cdot(\mathbf{F}\times d_2) = 0 \tag{57}$$
$$\mathbf{F}\cdot\mathbf{y}_2 = 0 \tag{58}$$

which are reduced to

$$\mathbf{F}\times d_2 = d_2\mathbf{y}_2, \quad \mathbf{F}\cdot\mathbf{y}_2 = 0 \tag{59}$$

Substituting (55) into (59) yields

$$d_1\times d_2 = d_1 d_2\mathbf{y}_2, \quad \mathbf{F}\cdot\mathbf{y}_2 = 0 \tag{60}$$

The first equality implies that the spherical notch must be on the blade plane. And the second equality means that the normal of the blade must be perpendicular to the direction of **F**. The design is shown in Fig. 7b.

5.1.3 Case 3: S-2R

For this design case, we first recognize that an S-joint provides three rotations (3R). We just need to make up two more translations with two notch hinges (R). We have shown in Sect. 3.2 that two parallel rotations produce a translation and three parallel rotations produce two translations. Therefore we can have two possible designs. One design utilizes two perpendicular R-joints. Each R-joint is parallel to one rotation axis of the S-joint. The direction of the translational constraint is along the direction that is perpendicular to both R-joint axes. See Fig. 7c.

The other design uses two parallel R-joints which together with one rotation of the S-joint produce two translations. The S-joint and two R-joints axes are co-planar. The direction of the translational constraint is parallel to the R-join axis. And the constraint line passes through the center of S-joint. See Fig. 7d. Note the functional body cannot translate along the direction indicated with its orientation fixed.

5.1.4 Case 4: B-B

The synthesis of B-B follows a similar procedure of the case S-B. For this case, both blades must satisfy the conditions in Eq. (59), i.e.

$$\boldsymbol{d}_1 = d_1\mathbf{y}_1 \times \mathbf{F}, \quad \mathbf{F}\cdot\mathbf{y}_1 = 0 \tag{61}$$

$$\boldsymbol{d}_2 = d_2\mathbf{y}_2 \times \mathbf{F}, \quad \mathbf{F}\cdot\mathbf{y}_2 = 0 \tag{62}$$

where we have applied a vector algebra operation. This means that the normal of both blades must be perpendicular to the direction **F**. However the two blades must not be parallel to each other as otherwise they would be redundant. The intersection line of the blades is the constraint line. Depending on how the blades are oriented, we can have two possible designs. In the design shown in Fig. 7e, the constraint line is along the longitude direction of the blades. And in the design shown in Fig. 7f, the constraint line is along the width direction of the blades.

5.1.5 Case 5: B-2R

Since a blade has two rotations and one translation, we can make up one rotation and one translation with two R-joints. Obviously one R-joint must be along the normal of the blade plane in order to make up the third rotation. The second R-joint is parallel to the first R-joint. These two parallel R-joints produce one additional translation along the local z-axis of the blade. See Fig. 7g.

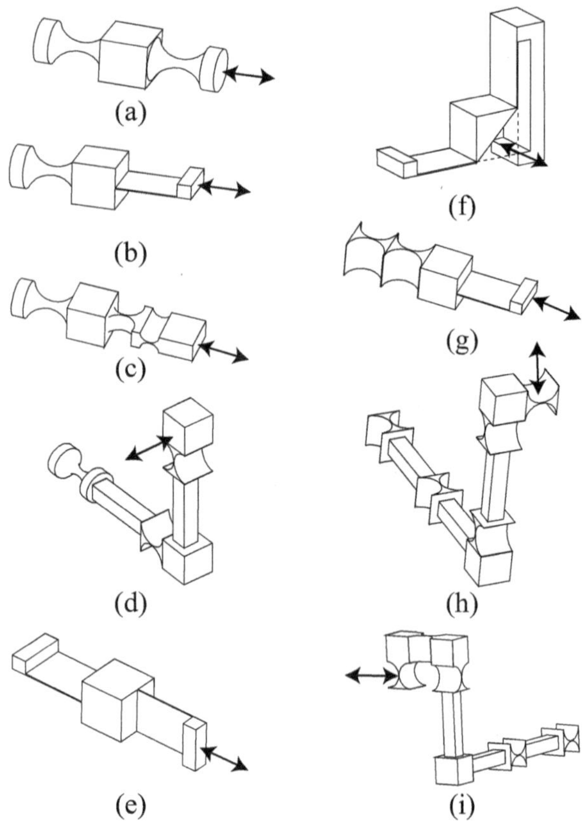

Figure 7. Various designs of P-constraint $\hat{W}_P$ with serial chains of flexure primitives S, B and R. The double head arrows represent the direction of the constrained translation. One end of the chains is fixed. **(a)** case 1 with two spherical notches (2S). **(b)** case 2 with one spherical notch and one blade (S-B). **(c–d)** case 3 with one spherical notch and two notch hinges (S-2R). **(e–f)** case 4 with two intersecting blades (2B). **(g)** case 5 with one blade and two notch hinges (B-2R). **(h–i)** case 6 with five notch hinges (5R).

5.1.6 Case 6: 5R

Synthesis of this case is simple. To produce two translations with a serial chain of 5R, we can have two possible designs by using parallel rotations. In one design, we have two R-joints parallel to one plane and another two R-joints parallel to a second plane. This design takes advantage of the fact that two parallel rotations produce a translation as proven previously. The direction of the translational constraint is along. See Fig. 7h.

And the other design uses three parallel R-joints which produce two translations in the plane normal to the axis of the R-joints. The direction of the translational constraint is along the the axis of the three parallel R-joints. See Fig. 7i.

5.2 Synthesis of R-Constraints $\hat{W}_R$

A R-constraint element removes a single rotation while allows all other motions. The constraint space of a rotational constraint is pure couple wrench $\hat{W}_R$ given in (5), which we copy here for convenience

$$\hat{W}_R = \begin{Bmatrix} \mathbf{0} \\ \mathbf{M} \end{Bmatrix} \tag{63}$$

Our goal is to use serial chains of at least two flexure primitives to design $\hat{W}_R$. That is to make up its complementary freedom space which consists of two rotations (R) and three translations (P). Since a bellow spring is itself a rotational constraint, we do not consider it in our design. And wires and spherical notches are also excluded in design as their freedom space already consists of three rotations. Therefore we only consider primitives B and R. There are three potential designs B-B, B-2R and 5R.

5.2.1 Case 1: B-B

To synthesize a serial chain of B-B for a rotational constraint, we first write the twist matrix of the blades as

$$[T_i] = \begin{bmatrix} \mathbf{x}_i & \mathbf{z}_i & \mathbf{0} \\ d_i \times \mathbf{x}_i & d_i \times \mathbf{z}_i & \mathbf{y}_i \end{bmatrix}, \quad i = 1,2 \tag{64}$$

They must be reciprocal to $\hat{W}_R$ in (63). This leads us the following necessary conditions

$$\mathbf{M} \cdot \mathbf{x}_i = 0, \quad \mathbf{M} \cdot \mathbf{z}_i = 0 \tag{65}$$

which implies that

$$\mathbf{y}_1 = \mathbf{y}_2 = \mathbf{M} \tag{66}$$

Therefore, we conclude that two blades must be parallel to each other with their y-(normal) axis along $\mathbf{M}$. The two parallel blades must be separated by a nonzero distance in order to produce sufficient translations in three directions. This design is shown in Fig. 8a.

5.2.2 Case 2: B-2R

Tee design of B-2R is synthesized as the following. The B flexure has a freedom space of 2R-P. We have to use two R-joints to make up two extra translations. And the two translations must be in the blade plane. The only solution is to have one R-joint parallel to the bending direction of the blade and the other R-joint parallel the torsion direction of the blade. And they should be separated by nonzero distance. This design is shown in Fig 8b.

5.2.3 Case 3: 5R

This case is trivial as there is only one possible design. That is, three R-joints are parallel to one plane and the other two R-joints parallel to a second plane. And these two planes are perpendicular to each other. See Fig. 8c.

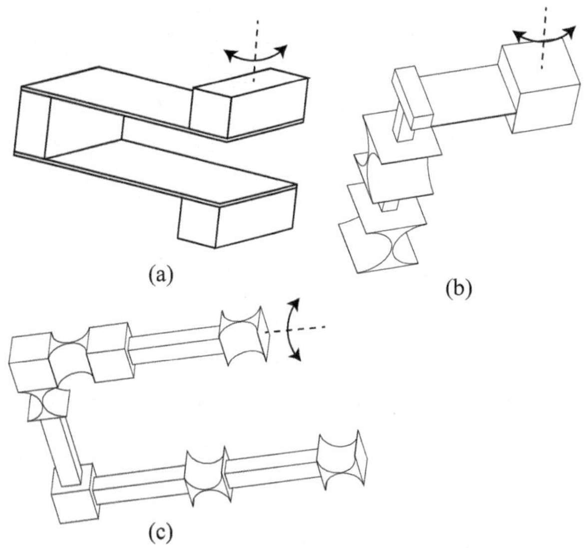

Figure 8. Various designs of the rotational constraint $\hat{W}_R$ with serial chains of flexure primitives B and R. The arrowed arcs represent the rotation constrained by the flexure design. One end of the chains is fixed.

6 Synthesis of hybrid structures

We can build more complex flexure mechanisms with hybrid structures of flexure primitives together with the freedom and constraint elements synthesized in the previous sections. Here a hybrid structure is a structure with both serial and parallel connections. As a matter of fact, many spatial flexure mechanisms in practice are in hybrid structures (Yao et al., 2008; Dong et al., 2008). In this section, we use four examples to demonstrate how to design hybrid structures.

6.1 Hybrid designs of joints and constraints

In the previous sections, we have presented synthesis of R-joints, P-joints, R-constraints, P-constraints with flexure primitives in Table 1. Although these procedures concern designs of notch hinges (R), blades (B), spherical notches (S), long wires (W) and bellow springs (B_s), designers should be aware that there may be multiple choices for these primitives. For instance, a blade flexure of any of these designs can be replaced by a rotational symmetric cylinder or a disc coupling since they have identical freedom and constraint space. Similarly, a wire flexure can be substituted with a corner blade and a spherical notch with a short wire, a notch hinge with a short mean or a split tube. Moreover, a notch hinge of any design can be replaced with the R-joint designs listed in Fig. 4.

These designs are by no means the complete list of the basic freedom and constraint elements as many flexure mechanisms are in hybrid structures which are combinations of serial and parallel chains. For instance, Fig. 9a shows the well known parallelogram 4-bar design with four notch hinges. This structure is considered as a hybrid structure that is

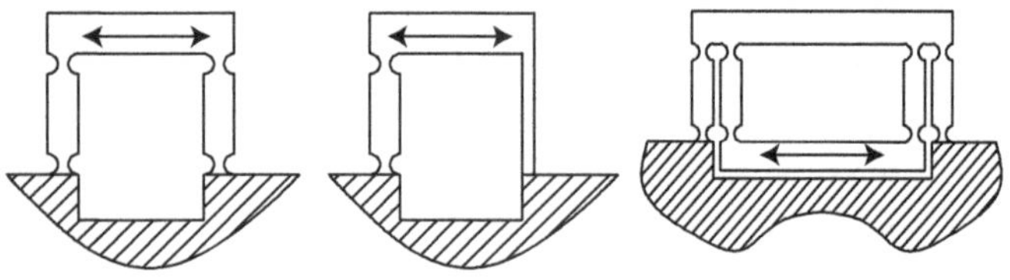

Figure 9. Designs of P-joints with hybrid structures.

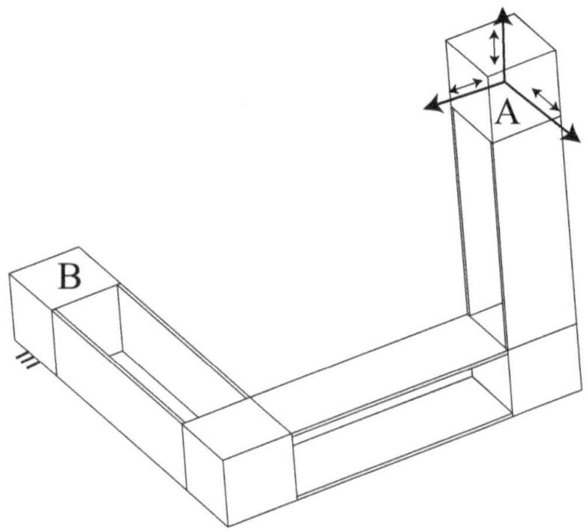

Figure 10. A serial chain of three P-joints connected with two intermediate bodies. Each P-joint is a parallel blade design shown in Fig. 5a. Body B is the fixed base. Body A is the functional body. The arrow lines represent three translations along three coordinate axes.

formed by two identical limbs assembled in parallel. Each limb is a serial chain of two R-joints. It is not hard to prove that this hybrid structure allows only one translation indicated in the figure. If we replace one of the RR chain with a blade flexure, we obtain another design shown in Fig. 9b. And Fig. 9c shows a double parallelogram design that is formed by a serial connection of two parallelogram 4-bar design. This design is widely used to reduce parasitic errors and increase the displacement stroke.

6.2 Synthesis of serial chains with freedom elements

As we discussed previously, we can use a serial chain of R-joints and P-joints to construct a freedom space with any combination of DOF. For instance, a serial chain of three P-joints can make up a freedom space of three translations. If we use the double parallel flexure structure for the P-joint design, we obtain a serial PPP design with three decoupled translations. See Fig. 10. Of course, one can further increase the structural stiffness with assembling multiple serial chains of PPP in parallel (Awtar et al., 2011).

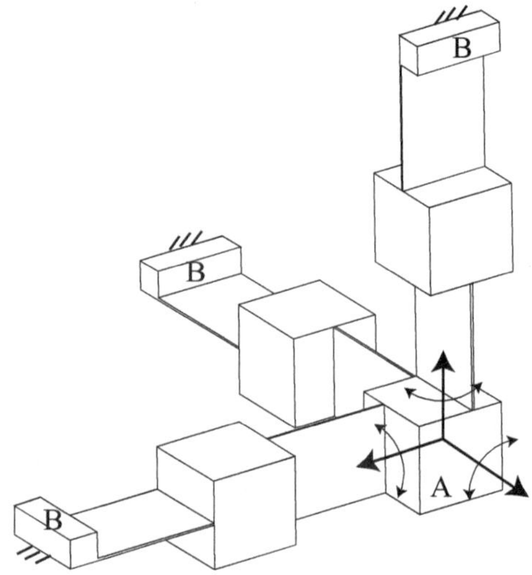

Figure 11. A parallel kinematic chain of three translational constraints results in a design with three rotations about the center of the functional body A. The body B are fixed.

6.3 Synthesis of parallel chains with constraint elements

In duality, one can also synthesize parallel chains with the constraint elements synthesized in the previous sections. For instance, if we would like to design a parallel structure with three rotations, we just need to use three translational constraint elements to remove all translations. If we choose the BB design in Fig. 7e for all three translational constraints, we obtain the design shown in Fig. 11. The functional body A can rotate about its center relative to the base body B while its translations are constrained.

As another example, we would like to design a parallel structure with three translations. We just need to use three rotational constraint elements to remove all rotations. If we choose the BB design in Fig. 8a for all three rotational constraints, we obtain the design shown in Fig. 12. The functional body A can translate in all directions while its rotations are constrained.

7 Conclusions

A list of commonly used flexure primitives is first categorized. A complete list of rotational and translational freedom elements is synthesized with parallel connections of these primitives. In duality, translational and rotational constraint elements are synthesized with serial chains of flexure primitives. These freedom and constraints elements form a catalogue of basic building blocks for constructing more complex flexure mechanisms. We have also demonstrate with examples the use of these building blocks with design of hybrid structures for any combinations of freedom. The importance of this work lies in the fact that it formalizes the

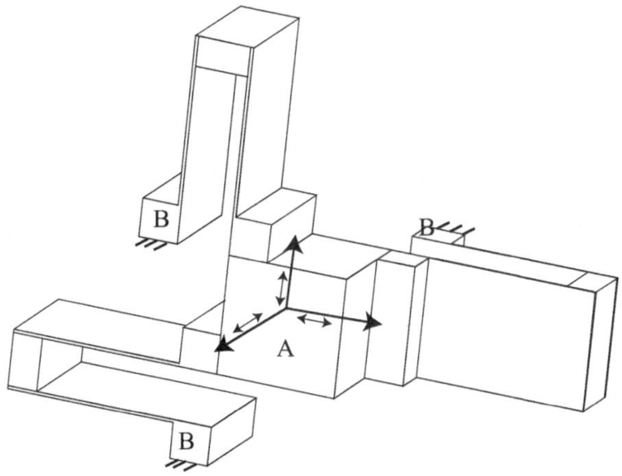

Figure 12. A parallel kinematic chain of three rotational constraints result in a design with three translations along the direction indicated in the figure. The body B are fixed. The body A is the functional body.

type synthesis (type/toplogy selection) of flexure mechanisms which has been an ad hoc process. It also paves the way towards to design automation and systematic invention of new flexure machinery.

Acknowledgements. This material is based upon work supported by the National Science Foundation under Grant No: CMMI-1161841. Any opinions, findings, and conclusions or recommendations expressed in this material are those of the author(s) and do not necessarily reflect the views of the National Science Foundation.

Edited by: G. Hao
Reviewed by: two anonymous referees

References

Awtar, S. and Slocum, A. H.: Constraint-based design of parallel kinematic xy flexure mechanisms, J. Mech. Des.-T. ASME, 129, 816–830, 2007.

Awtar, S., Ustick, J., and Sen, S.: An XYZ Parallel Kinematic Flexure MechanismWith Geometrically Decoupled Degrees of Freedom, ASME Journal of Mechanisms and Robotics, 5, 119–126, 2011.

Ball, R. S.: The Theory of Screws, Cambridge University Press, Cambridge, England (Originally published in 1876 and revised by the author in 1900, now reprinted with an introduction by H. Lipkin and J. Duffy), 1998.

Blanding, D. L.: Exact Constraint: Machine Design Using Kinematic Processing, ASME Press, New York, NY, 1999.

Brouwer, D., de Jong, B., and Soemers, H.: Design and modeling of a six DOFs MEMS-based precision manipulator, Precis. Eng., 34, 307–319, 2010.

Chen, S.-C. and Culpepper, M. L.: Design of a six-axis micro-scale nanopositioner-μHexFlex, Precis. Eng., 30, 314–324, 2006.

Culpepper, M. L. and Anderson, G.: Design of a low-cost nanomanipulator which utilizes a monolithic, spatial compliant mechanism, Precis. Eng., 28, 469–482, 2004.

Dai, J. S. and Jones, J. R.: Interrelationship between screw systems and corresponding reciprocal systems and applications, Mech. Mach. Theory, 36, 633–651, 2001.

Dai, J. S. and Jones, J. R.: A Linear Algebraic Procedure in Obtaining Reciprocal Screw Systems, J. Robot. Syst., 20, 401–412, 2003.

Davidson, J. K. and Hunt, K. H.: Robots and Screw Theory: Applications of Kinematics and Statics to Robotics, Oxford University Press, New York, NY, 2004.

Dong, J., Yao, Q., and Ferreira, P. M.: A novel parallel-kinematics mechanism for integrated, multi-axis nanopositioning: Part 2: Dynamics, control and performance analysis, Precis. Eng., 32, 20–33, 2008.

Hale, L. C.: Principles and Techniques for Designing Precision Machines, Ph.D. thesis, MIT, Cambridge, MA, 1999.

Henein, S.: Flexures: simply subtle, tutorial on the design of flexure-mechanisms, Second International Symposium on Compliant Mechanisms, IFToMM/ASME CoMe 2011, delft, the Netherlands, 2011.

Hopkins, J.: Modeling and Generating New Flexure Constraint Elements, in: Proc. of the 12th EUSPEN International Conference, 424–428, Stockholm, Sweden, 2012.

Hopkins, J. B.: Design of parallel flexure systems via freedom and constraint topologies (FACT), MS, Massachusetts Institute of Technology, Cambridge, MA, 2007a.

Hopkins, J. B.: Design of flexure-based motion stages for mechatronic systems via freedom, actuation and constraint topologies (FACT), Ph.D. thesis, Massachusetts Institute of Technology, Cambridge, MA, 2007b.

Hopkins, J. B. and Culpepper, M. L.: Synthesis of multi-degree of freedom, parallel flexure system concepts via Freedom and Constraint Topology (FACT) – Part I: Principles, Precis. Eng., 34, 259–270, 2010a.

Hopkins, J. B. and Culpepper, M. L.: Synthesis of multi-degree of freedom, parallel flexure system concepts via freedom and constraint topology (FACT). Part II: Practice, Precis. Eng., 34, 271–278, 2010b.

Hopkins, J. B. and Culpepper, M. L.: A screw theory basis for quantitative and graphical design tools that define layout of actuators to minimize parasitic errors in parallel flexure systems, Precis. Eng., 34, 767–776, 2010c.

Hopkins, J. B. and Culpepper, M. L.: Synthesis of precision serial flexure systems using freedom and constraint topologies (FACT), Precis. Eng., 35, 638–649, 2011.

Howell, L. L.: Compliant Mechanisms, Wiley-Interscience, New York, NY, 2001.

Huang, Z., Liu, J., and Zeng, D.: A general methodology for mobility analysis of mechanisms based on constraint screw theory, Science in China Series E: Technological Sciences, 52, 1337–1347, 2008.

Hunt, K. H.: Kinematic Geometry of Mechanisms, Oxford University Press, New York, NY, 1978.

Kong, X. and Gosselin, C. M.: Type Synthesis of Parallel Mechanisms, Springer, softcover reprint of hardcover 1st Edn. 2007 Edn., 2010.

Phillips, J.: Freedom in Machinery. Volume 1, Introducing Screw Theory, Cambridge University Press, Cambridge, UK, 1984.

Phillips, J.: Freedom in Machinery. Volume 2, Screw Theory Exemplified, Cambridge University Press, Cambridge, UK, 1990.

Smith, S. and Chetwynd, D.: Foundations of Ultra-Precision Mechanism Design, CRC Press LLC, 1992.

Smith, S. T.: Flexure: Element of Elastic Mechanisms, CRC Press LLC, London, UK, 2000.

Su, H.-J.: Mobility Analysis of Flexure Mechanisms via Screw Algebra, ASME Journal of Mechanisms and Robotics, 3, 041010, doi:10.1115/1.4004910, 2011.

Su, H.-J. and Tari, H.: Realizing Orthogonal Motions With Wire Flexures Connected in Parallel, J. Mech. Des.-T. ASME , 132, 121002, doi:10.1115/1.4002837, 2010.

Su, H.-J. and Tari, H.: On Line Screw Systems and Their Application to Flexure Synthesis, ASME Journal of Mechanisms and Robotics, 3, 011009, doi:10.1115/1.4003078, 2011.

Su, H.-J., Dorozhkin, D. V., and Vance, J. M.: A Screw Theory Approach for the Conceptual Design of Flexible Joints for Compliant Mechanisms, ASME Journal of Mechanisms and Robotics, 1, 041 009.1–041 009.8, 2009.

Varadarajan, K. M. and Culpepper, M. L.: A dual-purpose positioner-fixture for precision six-axis positioning and precision fixturing: Part I. Modeling and design, Precis. Eng., 31, 276–286, 2007a.

Varadarajan, K. M. and Culpepper, M. L.: A dual-purpose positioner-fixture for precision six-axis positioning and precision fixturing: Part II. Characterization and calibration, Precis. Eng., 31, 287–292, 2007b.

Yao, Q., Dong, J., and Ferreira, P. M.: A novel parallel-kinematics mechanisms for integrated, multi-axis nanopositioning: Part 1. Kinematics and design for fabrication, Precis. Eng., 32, 7–19, 2008.

Yu, J. J., Li, S. Z., Pei, X., Su, H.-j., Hopkins, J. B., and Culpepper, M. L.: Type Synthesis Principle and Practice of Flexure Systems in the Framework of Screw Theory: Part I – General Methodology, Proceedings of the ASME 2010 International Design Engineering Technical Conferences and Computers and Information in Engineering Conference, 543–552, doi:10.1115/DETC2010-28783, 2010.

A proposal for benchmark tests for underactuated or compliant hands

G. A. Kragten, C. Meijneke, and J. L. Herder

Faculty of Mechanical, Maritime and Materials Engineering, Dept. of BioMechanical Engineering, Delft University of Technology, Mekelweg 2, 2628 CD Delft, The Netherlands

Abstract. There is a lack of agreement in the literature as to what exactly quantifies the performance of underactuated hands. This paper proposes two benchmark tests to measure the ability of underactuated hands to grasp different objects and the ability to hold the objects when force disturbances apply. The first test determines the smallest and largest cylindrical object which can be successfully grasped in an enveloping grasp or in a pinch grasp. The second test provides the maximal allowable force which can be applied to a grasped object without loosing it. A setup was constructed consisting of standard components. Exemplary tests were applied to the Delft Hand 2. The proposed benchmark tests are representative to quantify the performance of pick and place operations with underactuated hands. The results of the tests can be applied to evaluate, compare, and improve the performance of robotic hands.

This paper was presented at the IFToMM/ASME International Workshop on Underactuated Grasping (UG2010), 19 August 2010, Montréal, Canada.

1 Introduction

The need for adaptive grasping, the reduction of actuators, weight, costs, and the reduction of control complexity are reasons to choose for underactuated robotic hands instead of fully actuated ones (e.g. Laliberté and Gosselin, 1998; Wu et al., 2009). Prototypes presented in the literature show that such hands are indeed able to grasp a large variety of objects, while the hand is driven even with a single motor (e.g. Gosselin et al., 2008). However, comparison of the performance of these prototypes based on the literature is almost impossible. Actually, there is hardly any agreement about the definition of performance. This means that the proposal of future research directions or design improvements substantiated by the results presented in literature seems impossible.

In previous research (Kragten and Herder, 2010), we made an overview of performance metrics which were applied to the design or evaluation of underactuated hands. Based on these metrics, two new, functional metrics were compiled to quantify the performance to pick and place different objects. They firstly quantify the range of object sizes that can be successfully grasped by an underactuated hand (*ability to grasp*), and secondly quantify the allowable force disturbances on the grasped objects (*ability to hold*). Cylindrically shaped objects were used in the definitions, because such objects are easy for parameterization, for contact modeling, and for reproduction in experiments. In addition, the shape of many objects in daily life can be approximated by a cylinder (van Lunteren and van Lunteren-Gerritsen, 1989). A static grasp model was presented in (Kragten and Herder, 2010) to calculate these performance metrics. However, a test to measure these performance metrics for prototypes of underactuated hands does not exist.

The objective of this paper is to propose benchmark tests to assess the performance of underactuated or compliant hands. It was aimed to define tests that are easily reproducible, that only need standard components, and quantify the performance regarding the ability to grasp and to hold objects. The results of such tests can be used to compare the performance of underactuated or compliant hands, and to measure the effect of design modifications. The tests proposed in this paper are limited to planar grasping but allow extension to spatial grasps.

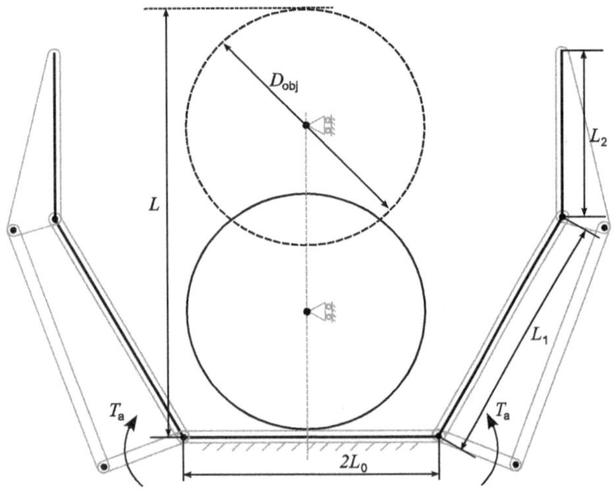

Figure 1. Initial positions of a freely moving cylindrical object with respect to the palm of a hand to determine the ability to successfully grasp this object. The palm position is represented by the solid object, and the pinch position by the dashed object. D_{obj} is the diameter of the object, $2L_0$ is the width of the palm, L_1 is the length of the proximal phalanx, L_2 is the length of the distal phalanx, L is the length of the finger. A torque T_a applies to the base of the fingers.

The structure of the paper is as follows. First a benchmark test to measure the range of objects which can be successfully grasped is proposed. In this test, grasping by enveloping or pinching the objects is distinguished. Then a benchmark test to measure the maximum allowed force disturbances applied to a grasped object is proposed. A setup to perform both tests is described, on which experiments with the Delft Hand 2 (DH-2) (Meijneke et al., 2010) are executed. The results of these tests are presented and discussed.

2 Methods

In the following sections, the benchmark tests and the experimental setup are defined.

2.1 Benchmark test 1: ability to grasp objects

It was proposed to quantify the range of object sizes that can be successfully grasped by the smallest and largest cylinder for which grasp equilibrium exists (Kragten and Herder, 2010). The performance metric was defined as follows:

$$Q_{\text{grasp}} = \frac{\frac{\pi}{2}\Delta D_{\text{obj}}}{2L+2L_0} \tag{1}$$

where ΔD_{obj} is the difference between the diameter of the largest and smallest graspable object, $L = L_1 + L_2$ is the length of one finger and $2L_0$ the palm width of the hand, see also Fig. 1. This metric can assume values between 0 and 1 for frictionless grasping, where 1 is regarded as the best performance. To measure this performance, freely moving, cylindrical objects of various sizes are grasped to assess the smallest and largest size for which grasp equilibrium exists. We propose to distinguish between an *enveloping grasp* where the object is enclosed by the fingers and near the palm, and a *pinch grasp* where the object is grasped by the distal phalanges only. This distinction was made, because objects which have to be grasped from the top normally have to be grasped by the distal part of the fingers. Also smaller objects are more likely to be grasped by the distal phalanges while large objects are more likely to be enveloped by the fingers.

The proposed benchmark test is as follows. Place a rigid cylinder with a radius D_{obj} on a surface or platform with low friction. Position the palm of the hand against the object – for enveloping grasping – or on a distance L from the object – for pinch grasping, while the fingers are in a fully open configuration, see Fig. 1. Apply a constant torque to actuate the fingers and to grasp the object. Determine if the object is successfully grasped (e.g. equilibrium of the fingers and the object). Repeat the test with different sizes of cylinders to determine the smallest and largest object for which a enveloping grasp or pinch grasp exist.

For this test, cylinders in the sizes of 40, 50, 55, 60, 63, 75, 90, 110, 115, 120 mm diameter were available. Currently, only experiments with symmetric hands were conducted. Therefore, freely moving of the objects along the line of symmetry was sufficient, which was realized by a linear guide (THK, RSR 9KM).

2.2 Benchmark test 2: ability to hold objects

It was proposed to determine the minimal force needed to pull a grasped object all the way out of the hand in order to quantify the maximal allowable external forces on the object (Kragten and Herder, 2010). The performance metric was defined as follows:

$$Q_{\text{hold}} = \frac{F}{T_a/L} \tag{2}$$

where F is the minimal force needed to pull a object out of the hand, T_a is the constant actuation torque applied at the base of the fingers as defined in Birglen et al. (2008, p. 35), and L is the total length of the finger. A high value of Q_{hold} is regarded as a good performance. The force needed to pull an object out of the hand depends on the pull direction. Therefore, different disturbance directions have to be tested. The worst-case direction determines the maximum allowed external force on the object.

The proposed benchmark test is as follows. Place a cylinder with a diameter D_{obj} against the palm of the hand. Slowly move the object along a straight line in the disturbance direction u, and allow the object to move in the perpendicular direction v, see Fig. 2. Measure the force needed to pull the object and determine the maximum force along this trajectory. Repeat this measurement for increasing Φ. Determine the minimum of the maxima of all the force-displacement characteristics.

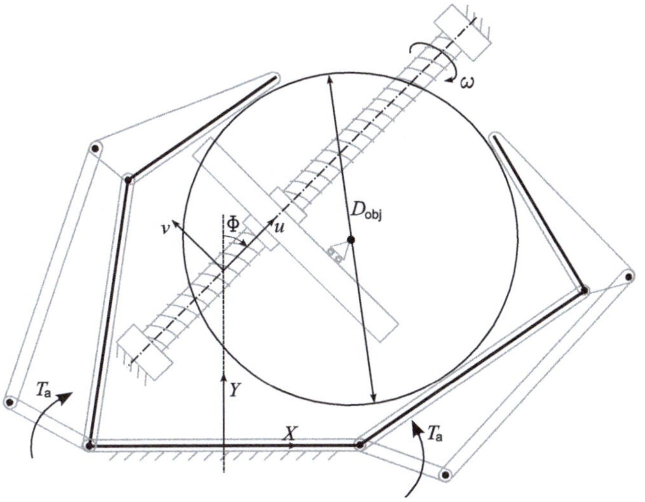

Figure 2. Schematic drawing of the setup where a cylindrical object with a diameter D_{obj} is pulled out of the hand at a constant slow speed ω in the direction u, while the fingers are actuated at a constant torque T_a. The object is free to move in the direction v, which is perpendicular to u. The resultant of the contact forces on the object in the pull direction u is measured.

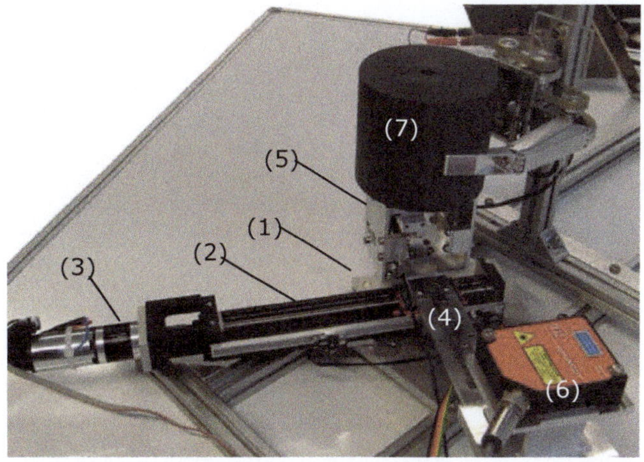

Figure 3. Photo of the setup for benchmark test 2 where (1) is a turning table, (2) is a ball screw spindle, (3) is the motor to drive the spindle; (4) is a linear guide perpendicular to the spindle; (5) is the force sensor supported by a construction of leaf springs; (6) is an optical displacement sensor; and (7) is a cylindrical object.

A setup was built for the benchmark test to determine the minimal force needed to pull the object out of the hand. A picture of this setup is shown in Fig. 3. The setup consists of a turning table (1), where Φ can be adjusted between −90 and 90 degrees in steps of 15 degrees. On top of this turning table, a linear ball screw spindle (2) is mounted (Hiwin, KK40, pitch 1 mm, stroke 200 mm), which is driven at a constant speed of 5 mm/s by a brushed motor (3) (Maxon, A-Max26). The speed was controlled by a universal motion interface (National Instruments, UMI-7772) and LabView 8.2 on a standard PC. A linear guide (4) (THK, RSR 9KM) was mounted perpendicular to the spindle. The force sensor (5) (FUTEK, LSB200, capacity 110 N) was placed on top of this guide to eliminate measuring the friction in the linear drive. This sensor measured only the resultant of the contact forces that were applied to the object in the direction of the linear drive. All bending moments and forces in other directions on the object were supported by a construction of leaf springs. On top of this force sensor, objects of different size can be mounted. The position of the object in the pull direction was measured by the encoder of the motor. The perpendicular position was optically measured by a laser sensor (6) (optoNCDT, ILD1300-100) that moved along with the linear guide. The force and position data were collected at a sample rate of 20 Hz.

2.3 Experiments with the Delft Hand 2

By way of example, the benchmark tests were used to quantify the performance of the Delft Hand 2 (DH-2). This hand consists of three underactuated fingers each having 2 phalanges. The driving mechanism of the finger is inspired by the four bar linkage of the SARAH (Birglen et al., 2008, Ch. 7). More details about the design and construction of this hand can be found in a companion paper (Meijneke et al., 2010).

First the range of objects that could be grasped by this hand (benchmark test 1) was measured. Performance calculations comparable to those presented in (Kragten and Herder, 2010) predicted that objects with a diameter between 58 and 130 mm could be grasped in an enveloping grasp type. It was predicted that the fingers would even pull the object towards the palm and have equilibrium in an enveloping grasp when the objects were initially located at a finger length distance from the palm.

The ability to hold (benchmark test 2) was measured for an object with ⌀115 mm. The object was pulled a distance of 30 mm in the directions $\Phi = 0°, 45°, 90°$, while the force and the object displacement were measured. All experiments were repeated three times, and the maximum measured force was determined for each test. The waiting time between the experiments was at least 30 s to minimize the effect of temperature differences of the motor on the performance.

3 Experimental results

The smallest graspable object in the Delft Hand 2 (DH-2) was $D_{obj,min} = 55$ mm. The largest available object of $D_{obj,max} = 120$ mm could also be grasped by the hand. When the objects were located at the pinch grasp position (see Fig. 1), the fingers still pulled the objects towards the palm, and an enveloping grasp resulted. The Q_{grasp} of the DH-2 is at least 0.39 according to Eq. (1) and the results in Table 1.

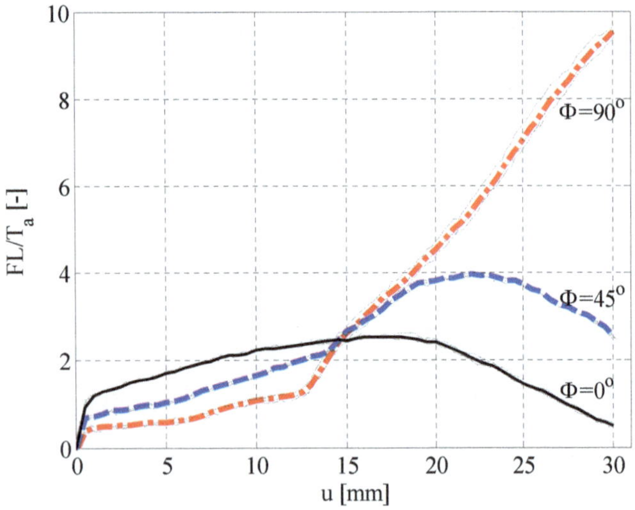

Figure 4a. Mean pull force F as a function of the displacement u of an object ($D_{obj} = 115$ mm) in three different directions Φ (see also Fig. 2). F is normalized to the actuation torque T_a and the finger length L. The gray, thin lines represent the raw data.

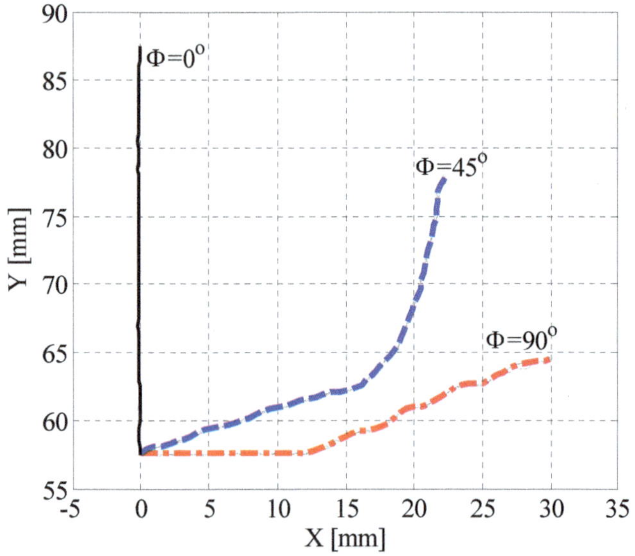

Figure 4b. Mean displacement trajectories in the X-Y-frame of an object ($D_{obj} = 115$ mm) while it was pulled in three different directions Φ (see also Fig. 2). F is normalized to the actuation torque T_a and the finger length L. The gray, thin lines represent the raw data.

Table 1. Results of benchmark test 1, showing that in both object positions the smallest and largest available objects could be successfully grasped.

	Palm Position	Pinch Position
D_{obj} min	55	55
D_{obj} max	120	120

Table 2. Results of benchmark test 2, showing the minimal force needed to extract an object of $D_{obj} = 115$ mm all the way out of the hand in three different directions (standard deviation between brackets). The second line shows the force normalized by the actuation torque ($T_a = 557$ Nmm) and finger length ($L = 100$ mm).

	$\Phi = 0°$	$\Phi = 45°$	$\Phi = 90°$
F	14.3 N ($\sigma = 0.08$)	22.1 N ($\sigma = 0.15$)	53.4 N ($\sigma = 0.27$)
Q_{hold}	2.56	3.94	9.46

A better performance could have been measured if larger objects were available. It was observed that objects smaller than $D_{obj,min}$ were also pulled towards the palm of the hand, but due to the mechanical limits in the joints, the fingers could not maintain contact with the object.

The force-displacement graph of the object of $D_{obj} =$ 115 mm pulled in the directions of $\Phi = 0°, 45°, 90°$ are shown in Fig. 4a, where the force is normalized by the actuation torque of the finger ($T_a \approx 557$ Nmm) and the length of the finger ($L = 100$ mm). In addition, the trajectory of the object during the disturbance is shown in Fig. 4b. The mean of the maximum force of the three repeated experiments was calculated for each disturbance direction and is shown in Table 2. So, the maximum allowed force disturbance on this object is 14.3 N at the current actuation torque and finger length. This means that, according to Eq. (2), the Q_{hold} of this hand and object is 2.56.

4 Discussion

Two benchmark tests were proposed in this paper. The first one determines the range of object sizes which can be grasped by an underactuated or compliant hand. The second test measures the minimal force needed to pull the grasped objects out of the hand. The first benchmark test quantifies the adaptability of the hand to grasp different items. The larger this range, the better the hand is capable of grasping various objects, which is useful in unstructured environments. The second benchmark test quantifies the maximal allowable external force that can be applied to the grasped objects in different directions. This force can be interpreted as the maximum allowed weight of the object (payload), or as the maximum acceleration of the robot arm – multiplied by the mass of the object – as a function of the orientation of the hand. Since the measured force was normalized by the actuation torque, the result of the test also shows the effectiveness at which the actuation torque is distributed to the phalanges. Finally, the test showed the worst-case disturbance direction. This result might be input for the design to improve the performance of the hand at this scenario, or it might be input for the application of the robot hand to prevent this scenario.

The benchmark tests were applied to the DH-2. The tests could be done with standard components and were simple to execute. The force-displacement characteristics showed a small variation between the experiments. For this hand, pulling the object straight out of the hand is the worst-case scenario. 14.3 N was required to extract the object, while the fingers were actuated at $T_a = 557$ Nmm. This means, for example, that the weight of an object with ⌀115 mm must be less than 1.5 kg when the orientation of the hand is with the fingers vertically downward.

During the design phase of the DH-2, the force-displacement characteristics were calculated based on a static, frictionless model (Meijneke et al., 2010). However, the calculated forces were much smaller than the measured results. We observed that friction in the joints strongly limited the extension of the distal phalanges with respect to the proximal phalanges once the fingers have enveloped the object. When the object was pulled out of the hand, the fingers only rotated about the proximal joint. The calculations, however, showed that both joints would rotate if there were no friction in these joints. In addition, contact friction was also neglected in the grasp model. Hence, we assume that the large deviation between the calculated and measured results for the second benchmark test with the DH-2 is mainly caused by neglecting friction in the model. Clearly, friction should not be neglected by default for adequate simulation results.

Rigid, cylindrical objects were used to simplify the reproduction of the tests. The assumption of cylindrical objects is also common in the literature. However, to verify if the results obtained with cylindrical objects are representative, preliminary benchmark tests were executed with square, non-rigid objects of the sizes 52×52, 58×58, and 75×75 mm. For non-cylindrical objects, the orientation has to be specified. The initial chosen orientations of the object were (i) with one side parallel to the palm and (ii) with one corner pointing to the palm. The DH-2 was able to grasp all these objects in both orientations from the palm position and the pinch position. When grasping the smallest square in the orientation parallel to the palm, the fingers were almost at their mechanical limits. This means that the order of magnitude of the size of graspable squares equals the size of cylindrical objects. The force needed to pull the squares out of the hand strongly depended on their initial orientation. This means that the maximum allowed external force on cylindrical objects cannot directly be applied to objects with comparable sizes but other shapes. For future work, the limitations of the use of cylindrical objects to calculate or measure the performance have to be investigated in more detail.

The proposed benchmark tests are limited to planar grasp cases, which is common in the grasping literature. The necessity to extend to spatial tests must be investigated in future work. Furthermore, the proposed benchmark test do not capture other issues like, for instance, the prevention of object damaging, the safety of the hand or the sensitivity of the successful grasp range to initial position deviations of the object (Dollar et al., 2010). Future research is needed to translate the important functional requirements of robotic hands to performance metrics that can be calculated by relatively simple grasp models and measured by tests that can be easily reproduced.

5 Conclusions

Two benchmark tests were defined to quantify the performance of underactuated or compliant hands to pick and place different objects. The adaptability of robotic hands to grasp various objects is quantified and tested by measuring the smallest and largest cylindrical object for which a grasp equilibrium exists (benchmark test 1). The maximum allowed weight or external disturbing force on a grasped object is quantified and tested by the minimal force needed to pull the object all the way out of the hand (benchmark test 2).

The benchmark tests are successfully applied to the Delft Hand 2 (DH-2). The tests showed that this hand can grasp objects between at least ⌀55 and ⌀120 mm without any feedback control or active adjustment of the hand. The maximal allowed disturbing force on an object of ⌀115 mm is 14 N when applied perpendicular to the palm up to 52 N when applied parallel to the palm, while the fingers were actuated by a constant torque of 0.56 Nm. The proposed benchmark tests can be applied to evaluate, compare, and improve the performance of robotic hands.

Acknowledgements. This work was carried out as part of the FALCON project under responsibility of the Embedded System Institute with Vanderlande Industries as the industrial partner. This project is partially supported by the Dutch Ministry of Economic Affairs within the framework of the Embedded System Institute (BSIK03021) program.

We gratefully acknowledge the help of Rik van den Hoeven and Jos van Driel to construct the setup for the experiments.

Edited by: J. L. Herder
Reviewed by: two anonymous referees

References

Birglen, L., Laliberté, T., and Gosselin, C. M.: Underactuated Robotic Hands, Springer Tracts in Advanced Robotics, Vol. 40, Springer-Verlag, Berlin, Heidelberg, 2008.

Dollar, A. M., Jentoft, L. P., Gao, J. H., and Howe, R. D.: Contact sensing and grasping performance of compliant hands, Auton. Rob., 28(1), 65–75, 2010.

Gosselin, C. M., Pelletier, F., and Laliberté, T.: An anthropomorphic underactuated robotic hand with 15 dofs and a single actuator, IEEE International Conference on Robotics and Automation ICRA 2008, 749–754, 2008.

Kragten, G. A. and Herder, J. L.: The Ability of Underactuated Hands to Grasp and Hold Objects, Mech. Mach. Theory, 45(3), 408–425, 2010.

Laliberté, T. and Gosselin, C. M.: Simulation and Design of Underactuated Mechanical Hands, Mech. Mach. Theory, 33(1–2), 39–57, 1998.

van Lunteren, A. and van Lunteren-Gerritsen, G. H. M.: On the use of prostheses by children with a unilateral congenital forearm defect, Journal of Rehabilitation Sciences, 2(1), 10–12, 1989.

Meijneke, C., Kragten, G. A., and Wisse, M.: Design and Performance Assessment of an Underactuated Hand for Industrial Applications, Mech. Sci., in press, 2010.

Wu, L., Carbone, G., and Ceccarelli, M.: Designing an underactuated mechanism for a 1 DOF finger operation, Mech. Mach. Theory, 44(2), 336–348, 2009.

Permissions

All chapters in this book were first published in MS, by Copernicus Publications; hereby published with permission under the Creative Commons Attribution License or equivalent. Every chapter published in this book has been scrutinized by our experts. Their significance has been extensively debated. The topics covered herein carry significant findings which will fuel the growth of the discipline. They may even be implemented as practical applications or may be referred to as a beginning point for another development.

The contributors of this book come from diverse backgrounds, making this book a truly international effort. This book will bring forth new frontiers with its revolutionizing research information and detailed analysis of the nascent developments around the world.

We would like to thank all the contributing authors for lending their expertise to make the book truly unique. They have played a crucial role in the development of this book. Without their invaluable contributions this book wouldn't have been possible. They have made vital efforts to compile up to date information on the varied aspects of this subject to make this book a valuable addition to the collection of many professionals and students.

This book was conceptualized with the vision of imparting up-to-date information and advanced data in this field. To ensure the same, a matchless editorial board was set up. Every individual on the board went through rigorous rounds of assessment to prove their worth. After which they invested a large part of their time researching and compiling the most relevant data for our readers.

The editorial board has been involved in producing this book since its inception. They have spent rigorous hours researching and exploring the diverse topics which have resulted in the successful publishing of this book. They have passed on their knowledge of decades through this book. To expedite this challenging task, the publisher supported the team at every step. A small team of assistant editors was also appointed to further simplify the editing procedure and attain best results for the readers.

Apart from the editorial board, the designing team has also invested a significant amount of their time in understanding the subject and creating the most relevant covers. They scrutinized every image to scout for the most suitable representation of the subject and create an appropriate cover for the book.

The publishing team has been an ardent support to the editorial, designing and production team. Their endless efforts to recruit the best for this project, has resulted in the accomplishment of this book. They are a veteran in the field of academics and their pool of knowledge is as vast as their experience in printing. Their expertise and guidance has proved useful at every step. Their uncompromising quality standards have made this book an exceptional effort. Their encouragement from time to time has been an inspiration for everyone.

The publisher and the editorial board hope that this book will prove to be a valuable piece of knowledge for researchers, students, practitioners and scholars across the globe.

List of Contributors

S. Krut
LIRMM (CNRS & Univ. Montpellier 2), Montpellier, France

V. Bégoc
LIRMM (CNRS & Univ. Montpellier 2), Montpellier, France
B+ Equipment, Gemenos, France

D. N. Nenchev
Graduate School of Engineering, Tokyo City University, Tamazutsumi 1-28-1, Setagaya-ku, Tokyo 158-8557, Japan

N. Giri
Department of Electrical and Computer Engineering, Clemson University, Clemson, SC, USA

I. Walker
Department of Electrical and Computer Engineering, Clemson University, Clemson, SC, USA

T. Laliberté
Département de Génie Mécanique, Université Laval, Québec, QC, G1V 0A6, Canada

M. Baril
Département de Génie Mécanique, Université Laval, Québec, QC, G1V 0A6, Canada

F. Guay
Département de Génie Mécanique, Université Laval, Québec, QC, G1V 0A6, Canada

C. Gosselin
Département de Génie Mécanique, Université Laval, Québec, QC, G1V 0A6, Canada

R. Seifried
Institute of Engineering and Computational Mechanics, University of Stuttgart, Pfaffenwaldring 9, 70569 Stuttgart, Germany

W. Blajer
Faculty of Mechanical Engineering, Technical University of Radom, ul. Krasickiego 54, 26-600 Radom, Poland

M. Richard
Ecole Polytechnique Fédérale de Lausanne, Laboratoire de Systémes Robotiques (LSRO), Station 9, 1015 Lausanne, Switzerland

R. Clavel
Ecole Polytechnique Fédérale de Lausanne, Laboratoire de Systémes Robotiques (LSRO), Station 9, 1015 Lausanne, Switzerland

L. Bruzzone
University of Genoa, DIME, Via Opera Pia 15A, 16145 Genoa, Italy

G. Quaglia
Politecnico di Torino, DIMEAS, Corso Duca degli Abruzzi, 24, 10129 Turin, Italy

R. M. Fowler
Department of Mechanical Engineering, Brigham Young University, Provo, UT 84602, USA

L. L. Howell
Department of Mechanical Engineering, Brigham Young University, Provo, UT 84602, USA

S. P. Magleby
Department of Mechanical Engineering, Brigham Young University, Provo, UT 84602, USA

P. J. Kyberd
Institute of Biomedical Engineering, University of New Brunswick, Fredericton, New Brunswick, E3B 4A3, Canada

A. Clawson
Institute of Biomedical Engineering, University of New Brunswick, Fredericton, New Brunswick, E3B 4A3, Canada

B. Jones
Institute of Biomedical Engineering, University of New Brunswick, Fredericton, New Brunswick, E3B 4A3, Canada

G. Hao
Department of Electrical and Electronic Engineering, School of Engineering, University College Cork, Cork, Ireland

L. Birglen
Ecole Polytechnique of Montreal, Department of Mechanical Engineering, Montreal, QC, H3T 1J4, Canada

C. Quennouelle
Laboratoire de robotique, Département de génie mécanique, Université Laval, Québec, QC, Canada

M. E. Abdallah
General Motors R&D, Manufacturing Systems Research Lab, Warren, MI 48090, USA

C. W. Wampler
General Motors R&D, Manufacturing Systems Research Lab, Warren, MI 48090, USA

H. C. Greenberg
Department of Mechanical Engineering, Brigham Young University, Provo, UT 84602, USA

M. L. Gong
Department of Mechanical Engineering, Brigham Young University, Provo, UT 84602, USA

S. P. Magleby
Department of Mechanical Engineering, Brigham Young University, Provo, UT 84602, USA

L. L. Howell
Department of Mechanical Engineering, Brigham Young University, Provo, UT 84602, USA

B. Herman
Université catholique de Louvain, Institute of Mechanics, Materials, and Civil Engineering, Center for Research in Energy and Mechatronics, Louvain-la-Neuve, Belgium
Université catholique de Louvain, Louvain Bionics, Louvain-la-Neuve, Belgium

A. Devreker
Université catholique de Louvain, Institute of Mechanics, Materials, and Civil Engineering, Center for Research in Energy and Mechatronics, Louvain-la-Neuve, Belgium
K. U. Leuven, Department of Mechanical Engineering, Division Production engineering, Machine design and Automation, Leuven, Belgium

F. Richer
Université Pierre et Marie Curie - CNRS, Institute for Intelligent Systems and Robotics, Paris, France

A. Hassan Zahraee
Université Pierre et Marie Curie - CNRS, Institute for Intelligent Systems and Robotics, Paris, France

J. Szewczyk
Université Pierre et Marie Curie - CNRS, Institute for Intelligent Systems and Robotics, Paris, France

M. Safar Beiranvand
Institute of Petroleum Engineering, College of Engineering, University of Tehran, Tehran, Iran

P. Mohammadmoradi
Department of Chemical and Petroleum Engineering, Sharif University of Technology, Tehran, Iran

B. Aminshahidy
Institute of Petroleum Engineering, College of Engineering, University of Tehran, Tehran, Iran

B. Fazelabdolabadi
Institute of Petroleum Engineering, College of Engineering, University of Tehran, Tehran, Iran

S. Aghahoseini
Department of Petroleum Engineering, Azad University of Kharg Island, Kharg Island, Iran

M. Riedel
Department of Mechanism Theory and Dynamics of Machines (IGM), RWTH Aachen University, Aachen, Germany

M. Nefzi
Department of Mechanism Theory and Dynamics of Machines (IGM), RWTH Aachen University, Aachen, Germany

B. Corves
Department of Mechanism Theory and Dynamics of Machines (IGM), RWTH Aachen University, Aachen, Germany

J. Lassooij
Delft University of Technology, Faculty of Mechanical, Maritime and Materials Engineering, Department of Biomechanical Engineering, Mekelweg 2, 2628 CD Delft, The Netherlands

N. Tolou
Delft University of Technology, Faculty of Mechanical, Maritime and Materials Engineering, Department of Biomechanical Engineering, Mekelweg 2, 2628 CD Delft, The Netherlands

G. Tortora
The BioRobotics Institute, Scuola Superiore Sant'Anna, Piazza Martiri della Libertà 33, 56127 Pisa, Italy

S. Caccavaro
The BioRobotics Institute, Scuola Superiore Sant'Anna, Piazza Martiri della Libertà 33, 56127 Pisa, Italy

A. Menciassi
The BioRobotics Institute, Scuola Superiore Sant'Anna, Piazza Martiri della Libertà 33, 56127 Pisa, Italy

J. L. Herder
Delft University of Technology, Faculty of Mechanical, Maritime and Materials Engineering, Department of Biomechanical Engineering, Mekelweg 2, 2628 CD Delft, The Netherlands

C. Meijneke
Faculty of Mechanical, Maritime and Materials Engineering, Dept. of BioMechanical Engineering, Delft University of Technology, Mekelweg 2, 2628 CD Delft, The Netherlands

G. A. Kragten
Faculty of Mechanical, Maritime and Materials Engineering, Dept. of BioMechanical Engineering, Delft University of Technology, Mekelweg 2, 2628 CD Delft, The Netherlands

M. Wisse
Faculty of Mechanical, Maritime and Materials Engineering, Dept. of BioMechanical Engineering, Delft University of Technology, Mekelweg 2, 2628 CD Delft, The Netherlands

N. Docquier
Université catholique de Louvain, Center for Research in Mechatronics, B 1348 Louvain-la-Neuve, Belgium
Postdoctoral Researcher of the Fonds de la Recherche Scientifique - FNRS

A. Poncelet
Université catholique de Louvain, Center for Research in Mechatronics, B 1348 Louvain-la-Neuve, Belgium

P. Fisette
Université catholique de Louvain, Center for Research in Mechatronics, B 1348 Louvain-la-Neuve, Belgium

D. Che
Key Laboratory for Advanced Materials Processing Technology, Ministry of Education, Department of Mechanical Engineering, Tsinghua University, Beijing 100084, China

W. Zhang
Key Laboratory for Advanced Materials Processing Technology, Ministry of Education, Department of Mechanical Engineering, Tsinghua University, Beijing 100084, China

M. Rösner
Institute for Mechanics, Helmut-Schmidt-University/ University of the Federal Armed Forces Hamburg, Holstenhofweg 85, 22043 Hamburg, Germany

R. Lammering
Institute for Mechanics, Helmut-Schmidt-University/ University of the Federal Armed Forces Hamburg, Holstenhofweg 85, 22043 Hamburg, Germany

L.-A. A. Demers
Département de génie mécanique, Université Laval, Québec, QC, G1V 0A6, Canada

B. Bergelin
Department of Mechanical Engineering, Marquette University, Milwaukee, Wisconsin 53233, USA

B. Slaboch
Department of Mechanical Engineering, Marquette University, Milwaukee, Wisconsin 53233, USA

J. Sun
Department of Mechanical Engineering, Marquette University, Milwaukee, Wisconsin 53233, USA

P. A. Voglewede
Department of Mechanical Engineering, Marquette University, Milwaukee, Wisconsin 53233, USA

H.-J. Su
Department of Mechanical and Aerospace Engineering, Ohio State University, Columbus, Ohio, 43210, USA

C. Yue
Department of Mechanical Engineering, University of Maryland Baltimore County, Baltimore, Maryland, 21250, USA